# The Neural Crest, Second Edition

T0291498

The development of an organism from a single cell to a fertilized egg is one of the most marvelous and beautiful of all nature's creations. In the context of a rapidly advancing field of study, this fully revised edition of *The Neural Crest* contains the most current information about a structure unique to the vertebrate embryo and which has only a transient existence in early embryonic life. The neural crest's component cells migrate throughout the embryo and home to specific sites, where they differentiate into a large variety of cell types, including the peripheral nervous system, pigment cells, endocrine cells, and mesenchymal derivatives. The latter provide a large contribution to the vertebrate head as it gives rise to nearly all the skull and facial skeleton. The ontogeny of the neural crest embodies the most important issues in developmental biology, as the neural crest is considered to have played a crucial role in evolution of the vertebrate phylum.

Fifteen years ago, research on the neural crest centered on avian and amphibian embryos. Included are new data that analyze neural crest ontogeny in murine and zebrafish embryos, two species that enable both genetic studies and single-cell analysis. Coverage of recent advances in our understanding of markers of neural crest cell subpopulations is also included, and a full chapter is devoted to cell lineage analysis, a field that has also significantly contributed to an understanding of the phenotypic segregation of neural crest cell subpopulations.

Since the first edition, perhaps the major research breakthrough has been the introduction of molecular biology to neural crest research, enabling the elucidation of many molecular mechanisms of neural crest development. These include effects of growth factors and receptors of the neurotrophin family, the endothelin family, bone morphogenetic proteins, etc. Likewise, elucidation of the molecular mechanisms leading to neural crest migration in the embryo characterizes the type of frontline research covered in this revised edition.

This book is essential reading for students and researchers in developmental biology, cell biology, and neuroscience.

Nicole M. Le Douarin is Professor at the Collège de France in Paris and Director of the Institute of Cellular and Molecular Embryology of the Collège de France and CNRS.

Chaya Kalcheim is Professor in the Department of Anatomy and Cell Biology at the Hebrew University of Jerusalem-Hadassah Medical School.

# Developmental and cell biology series

SERIES EDITORS

Jonathan B. L. Bard, *Department of Anatomy, Edinburgh University*
Peter W. Barlow, *Long Ashton Research Station, University of Bristol*
David L. Kirk, *Department of Biology, Washington University*

The aim of the series is to present relatively short critical accounts of areas of developmental and cell biology where sufficient information has accumulated to allow a considered distillation of the subject. The fine structure of cells, embryology, morphology, physiology, genetics, biochemistry and biophysics are subjects within the scope of the series. The books are intended to interest and instruct advanced undergraduates and graduate students and to make an important contribution to teaching cell and developmental biology. At the same time, they should be of value to biologists who, while not working directly in the area of a particular volume's subject matter, wish to keep abreast of developments relevant to their particular interests.

# The Neural Crest
# Second Edition

Nicole M. Le Douarin
*Institute of Cellular and Molecular Embryology, CNRS and Collège de France*

Chaya Kalcheim
*University of Jerusalem-Hadassah Medical School*

CAMBRIDGE
UNIVERSITY PRESS

CAMBRIDGE UNIVERSITY PRESS
Cambridge, New York, Melbourne, Madrid, Cape Town, Singapore,
São Paulo, Delhi, Dubai, Tokyo

Cambridge University Press
The Edinburgh Building, Cambridge CB2 8RU, UK

Published in the United States of America by Cambridge University Press, New York

www.cambridge.org
Information on this title: www.cambridge.org/9780521122252

First published 1999
This digitally printed version 2009

A catalogue record for this publication is available from the British Library

Library of Congress Cataloguing in Publication data

Le Douarin, N.
    The neural crest/Nicole M. Le Douarin and Chaya Kalcheim. – 2nd
ed.
        p.      cm. – (Developmental and cell biology series : 36)
    Includes bibliographical references and index.
    1. Neural crest.   I. Kalcheim, Chaya. 1955–   . II. Title.
III. Series
QL938.N484L4   1999                                    98-53256
573.8´6387 – dc21                                         CIP

ISBN 978-0-521-62010-9 Hardback
ISBN 978-0-521-12225-2 Paperback

Additional resources for this publication at www.cambridge.org/9780521122252

*A mes petits-enfants*
*Adrien, Héléna, Antoine et Alexandre*

Nicole M. Le Douarin

*To my mother*
*To Avi, Einat and Yoav*

Chaya Kalcheim

# Contents

# Foreword

The development of an organism from a single cell, the fertilized egg, is one of the most marvelous and beautiful of all nature's creations. The last decades have seen an explosion in our understanding of the mechanisms involved in development, largely due to the application of the techniques of molecular biology. The language of cells is molecules.

The neural crest presents us with a system in which almost all the basic features of development are present – pattern formation, determination, directed migration, and differentiation. It is truly remarkable in the diversity of cell types to which it gives rise – neural, mesenchymal, endocrine, and pigment – and all these end up away from their site of origin. Its development requires to know how genes control the behavior of the cells. We need to know to what extent the crest's diversity is laid down early on and to what extent it is multi-potential. We also need to know the role of intrinsic developmental programs and their relation to external signals. A particular problem with the crest is how its migration to specific sites is directed. The creation of the crest during vertebrate evolution was a major event and has profound implications for human development.

The early understanding of the development of the crest relied on the techniques of classical embryology, such as removal of the crest from the early embryo and noting the effects, but a crucial step was the ability to mark the neural crest population reliably and so follow their development. The discovery by Nicole Le Douarin of the quail nuclear marker transformed the field, for it opened up the way to finding out what the crest gives rise to and about the development of its cells. All this now "classical" work is dealt with in the 1982 edition of *The Neural Crest*.

This present volume provides a detailed and most impressive account of the current state of the field. It is not just that numerous genes and signals controlling neural crest development have been identified, but it is the combination of the older techniques of experimental embryology with molecular approaches. Cloning experiments have illuminated both the multipotential capacity of the early neural crest as well as some restriction. It provides one of the best systems for understanding the directed migration of cells to distant sites, and it is now clear that both positive and negative signals are involved. Understanding the nature of crest migration has provided new insights into the origin of the segmental nature of the peripheral nervous system.

Many of the advances have been due to Le Douarin and her coworkers, and it is not surprising that the sight of crest cells down the microscope still provides Le Douarin with an aesthetic pleasure. It remains an exciting area for study.

Lewis Wolpert is Professor of Biology as Applied to Medicine in the Department of Anatomy and Developmental Biology, University College London, London, UK.

# Preface

The first edition of *The Neural Crest* appeared in 1982. As I wrote in the preface at that time, it "marked a step in a venture which started twelve years before when I noticed that the nucleus of quail cells has a characteristic structure and can be distinguished at first glance from its chick counterpart." Quail–chick chimeras used to follow the behavior and fate of neural crest cells in the avian embryo had proved an efficient tool to decipher the participation of this structure in forming key structures that constitute the body of higher vertebrates.

During the period of time separating these two editions, developmental biology has considerably benefited from the recent advances in cell and molecular biology. Many new investigators have become interested in the ontogeny of the neural crest; certain questions raised in the early work have been answered and new avenues of research are now open. The present book is an illustration of such an evolution, as two friends and colleagues, former mentor and student, have joined their expertise and deep interest in the subject to reevaluate our knowledge on the ontogeny of this fascinating embryonic structure. Our ambition is not only to provide the reader with the most up-to-date information but hopefully also to raise the interest of newcomers to the field.

Both Chaya Kalcheim and myself wish to express our deep gratitude to the colleagues and friends who helped us during the preparation of this book. The contribution of our students and collaborators who have read chapters, provided insightful suggestions, and furnished original photographs has been invaluable. Our thanks go to Gérard Couly, Catherine Ziller, Anne Grapin, Anne Eichmann, Alan Burns, Elisabeth Dupin, Laure Lecoin, Ron Goldstein, Nitza Kahane, Gilat Brill, and Anat Braffman.

I am particularly grateful to Dr Françoise Dieterlen for her constant and friendly support in this venture and to Professor Joseph Bagnara who took time to carefully read two chapters of this book. Professor Mike Gershon will find here an expression of our deep appreciation for reading the chapter on the autonomic nervous system, his domain of outstanding expertise.

Many authors of published and unpublished work have generously provided us with permission to reproduce their data and, in certain cases, with original photographs. We both thank them deeply.

We are indebted to the very dedicated collaborators who, in the Nogent laboratory, have efficiently helped both of us in preparing the book: Chrystèle Guilloteau for the typescript, Marcelle Gendreau and Charmaine Herberts who have devoted all their skill to preparing the bibliography and reference chasing. Françoise Viala, Francis Beaujean, Sophie Gournet, and Hélène San Clémente have prepared photographs and artwork. Chaya Kalcheim and I are happy to thank them all for their very professional and friendly contributions.

Nicole M. Le Douarin
*Nogent-sur-Marne, March 15, 1998*

# General Introduction

Although the neural crest is a discrete structure which comprises only a few cells and exists transiently in the early embryo, it embodies most of the crucial issues of developmental biology. Its highly pluripotent component cells yield an astonishing variety of cell types: from bones, tendons, connective and adipose tissues, and dermis to melanocytes, neurons of many kinds, glial, and endocrine cells. The neural crest forms according to a rostrocaudal gradient along the body axis and releases free-moving mesenchymal-like cells that follow definite migration routes at precise times of development, finally reaching target embryonic sites where they settle and differentiate. What triggers this migratory behavior? Which signals do these cells require to move along these definite pathways and what causes them to stop and accumulate? These are among the most puzzling problems raised by morphogenesis, problems that go beyond embryology and concern the stability of the histiotypic state and its disruption in metastasis.

The neural crest was first described in the chick embryo by His (1868) as a "*Zwischenstrang*," a strip of cells lying between the dorsal ectoderm and the neural tube. This observation prompted a number of studies, essentially performed in the amphibian embryo, which provided the grounds for the first recognition of the neural crest as a "remarkable embryonic structure" designated as such by Newth (1951). The classical contributions resulting from this pioneering work were reviewed in a comprehensive monograph by Hörstadius in 1950. After these early studies, the ontogeny of the neural crest in higher vertebrates remained largely unresolved for many years. However, the introduction of cell labeling with tritiated thymidine in the 1960s by Chibon and Weston marked the beginning of a new era. Visualization of neural crest cells became possible in amphibian and avian embryos. This cell-labeling method, however, allowed only a transient follow-up of the cells as sequential cell divisions progressively diluted the radioactive isotope. This limitation was completely overcome upon introduction of the quail–chick marker, which provided a stable means of tracing the migration and fate of neural crest progenitors exiting from the neural primordium along the entire axis. This technique was subsequently utilized to elucidate the contribution of the neural crest to a large number of tissues and structures in the avian embryo. Moreover, it revealed a high level of plasticity in the development of neural crest derivatives, the fates of which are largely influenced by the environment to which the crest cells are subjected during migration and in the embryonic sites to which they home. The wealth of data provided by this technique throughout the 1970s was reviewed by Nicole Le Douarin in *The Neural Crest*. Since the publication of this monograph in 1982, other methods have been developed whereby individual cells or groups of crest cells can be labeled in a non-invasive manner in all vertebrate classes from fish to mammals. With the advent of molecular biology and developmental genetics, considerable progress has occurred in our understanding of the mechanisms controlling neural crest ontogeny, and a second edition of *The Neural Crest* seemed timely.

In the early 1980s, data concerning the migration processes of neural crest cells had been gathered on two classes of vertebrates only, amphibians and avians. Investigations now include fish and mammals. It has become apparent that the major crest migratory paths have been conserved between fish, amphibia, birds, and mammals in spite of the significant differences in the number of migrating cells. A better understanding has been acquired concerning the nature of the extracellular matrix through which the crest cells move, as well as about the relationships between cells and matrix. Notwithstanding this progress, little is still known about the mechanisms of epithelial–mesenchymal conversion, a key process that triggers the onset of crest cell migration, and of the mechanisms responsible for the choice of the homing sites by the crest-derived cells.

Concerning the mechanisms of cell diversification into the various derivatives of the neural crest, significant advances have been made, although it is worth stressing that our understanding of lineage segregation is still in its initial stages. Some answers have been found regarding both the fate and state of commitment of neural crest cells at different times of development. Owing to the development of elegant *in vivo* lineage tracing techniques of single progenitors and of *in vitro* cloning procedures, it is now clear that neural crest stem cells exist and that diversification of neural crest derivatives present many similarities to the processes through which hemopoietic lineages are generated. The existence of a totipotent neural crest progenitor has also been demonstrated in the avian embryo although its autoreplicating capabilities have to be proven. On the other hand, neural crest progenitors endowed with self-renewal capacities have been found in mammals, but they have not been shown to be totipotent.

The last decade has witnessed the identification and cloning of a vast number of molecules recognized to play important roles in the survival of postmitotic neurons during the critical period of programmed cell death. Some of these factors have been shown to be active earlier in development in important aspects of the ontogeny of the peripheral nervous system, namely proliferation, differentiation, and survival of distinct neuronal and glial cell types.

The formation of color patterns of the vertebrate skin and its appendages and the regulation of color changes in vertebrates are topics that pertain to neural crest ontogeny, constituting a vast field which has benefited from recent spectacular advances. This progress has relied heavily on the collection over the years of a number of mouse mutants remarkable for their pigment patterns. With the advent of molecular genetics, the underlying mechanisms and interactions leading to pigmentation are currently being unraveled.

In mice, some of the genes mutated in several models for neurocristopathies, such as piebaldism and Hirschprung's disease, have been cloned. Moreover, null mutants are currently being created that lack specific genes. These approaches, in association with refined cellular techniques, provide insight into the etiology of the diseases as well as the mechanisms of normal development.

Another expanding field in neural crest research concerns its capacity to yield mesenchymal derivatives similar to those arising from the mesoderm, with the exception of blood and blood vessel endothelia. It has been established that the head is essentially of ectodermal origin via the neural plate, which gives rise to brain and to most of the skull and connective tissues. We now know much more about the genetic basis of head morphogenesis due to the discovery of genes that encode transcription factors and growth factors involved in head and face morphogenesis. Nearly every month a new regulatory pathway is disclosed which affects cephalic and facial derivatives of the neural crest. Nevertheless, the complexity of head morphogenesis makes it difficult to predict when an integrated picture of these pathways will emerge.

The advances summarized above are only part of the enormous progress achieved since the first edition of *The Neural Crest*. This progress has been possible thanks to the advent of new markers, which can follow subsets of neural crest cells throughout ontogeny at the molecular and cellular levels in increasingly refined ways. Going hand in hand with modern developments, the quail–chick marker, widely used to trace the migration and fate of crest cells, is still serving the field, whether applied in isolation or in combination with the detection of specific proteins or genes.

In every way, it has been both a challenge and a pleasure to write this book. Our hope is to convey to the readers our enthusiasm and curiosity for what the next decade will bring to the field.

# 1

# Methods for Identifying Neural Crest Cells and Their Derivatives

## 1.1 Introduction

A prerequisite for studying the individualization, migration, and fate of neural crest cells is the availability of specific markers that allow this cell type to be identified throughout ontogeny. This is due to the fact that in the majority of vertebrate species the movements of embryonic cells cannot be directly followed *in situ* under the microscope because the embryo contains too many cells. Moreover, the migratory neural crest cells do not exhibit generally recognizable features which would allow them to be distinguished from the other embryonic cells. Teleost embryos, however, constitute a model system particularly amenable to study neural crest ontogeny since, at least in certain species, crest cells can be distinguished in living samples by direct observation. Thus, several species of fishes have been used to study neural crest development (see, e.g., Newth, 1956; Lamers *et al.*, 1981; Langille and Hall, 1988a,b; Sadaghiani and Vielkind, 1989, 1990). In zebrafish (*Brachydanio rerio*), the blastomeres and their progeny are nearly transparent and the embryo has relatively few cells, making it possible to recognize them individually and to follow their migration. Moreover, in this species, crest cells have a large size and can be easily labeled, as described below, and even ablated or transplanted heterotopically (see Eisen and Weston, 1993, for a review).

As mentioned above, experimental procedures had to be devised in order to disclose the dynamic behavior of neural crest cells in higher vertebrates. These methods relied on two principles: either the source of neural crest cells, the neural folds, were removed prior to crest cell emigration and their fate was inferred from the deficiencies consecutive to the operation, or the neural fold-derived cells were recognizable by natural or artificial markers which allow the

1

crest cells to be traced up to their definitive target sites in the embryo. In recent years, some distinctive molecular markers have been identified in neural crest cells. This was useful for the validation of pre-existing evidence based on embryonic grafting experiments of the neural primordium.

## 1.2   Extirpation or *in situ* destruction of the neural crest

In this technique, the neural fold or the entire neural primordium is either removed or destroyed *in situ* by electrocauterization at a defined level of the neural axis. The deficiencies subsequent to the operation provide information on the presumptive fate of the excised territory. This experimental approach has often been used for studying the migration of neural crest cells in lower and higher vertebrates (see reviews of Hörstadius, 1950; Weston, 1970) and has yielded significant information. A serious drawback of this technique, however, is the fact that extirpation of embryonic areas triggers regulatory mechanisms which may restore the deletions. This is particularly true for the neural crest which is endowed with a high power of regeneration after partial removal of the neural fold. This will be discussed in detail in Chapter 3. In addition, tissue interactions commonly occur between embryonic structures and play a decisive role in many histogenetic processes. Removal of embryonic tissues may disturb such developmental mechanisms. Thus, lack of certain structures or their abnormal morphogenesis may result primarily from such disturbances rather than from the actual removal of the primordia. A striking example is provided by the role of axial organs, neural tube and notochord in ensuring the survival of somitic cells (Teillet and Le Douarin, 1983).

These remarks show that, although some fundamental knowledge concerning the role of the neural crest in embryogenesis has been acquired through the use of this technique, the results obtained are relatively crude. Much more precise and detailed information on the migration process and on the fate of neural crest cells has been obtained by means of cell marking experiments.

## 1.3   Differential identification of neural crest cells throughout ontogeny

Identifying premigratory and migrating neural crest is of interest from several viewpoints. Firstly, it provides knowledge of the exact source of the future migratory cells and what migration routes they take. It further allows the molecular characteristics which make these routes favorable for crest cell translocation to be investigated. Secondly, following the crest-derived cells throughout migration until they stop and differentiate in their target sites leads to knowledge of their derivatives and allows a fate map along the neural axis to be drawn. In fact, constructing fate maps is a prerequisite to studying the mechanisms underlying morphogenesis and organogenesis in most cases. Two approaches permit visualization of neural crest cells throughout ontogeny: one takes advantage of intrinsic molecular markers they may have, the other applies extrinsic markers to the cells prior to their departure from their site of origin. Since the beginning of research on the neural crest, enormous progress has been made along these two lines.

### 1.3.1   Intrinsic markers of neural crest cells

*1.3.1.1   Pioneer work based on cytological criteria.* Pioneer studies during the first half of this century, essentially carried out in the amphibian embryo, have taken advantage of a variety of naturally occurring and easily observable cell characteristics to follow neural crest cell movements. Various intrinsic markers such as cytoplasmic inclusions (e.g., yolk granules, pigments), nuclear staining properties, and differences in cell size (Raven, 1937; Triplett, 1958) have been used. For example, Raven (1936, 1937), taking advantage of differences in cell size between *Triturus* and *Amblystoma*, carried out xenoplastic reciprocal grafting experiments by means of which he could gain some insight into the origin of sympathetic ganglia from the neural crest. These markers may be recognized in regions where neural crest cells accumulate, but cannot be used to identify isolated cells and therefore are of no help in following them during migration.

Before modern techniques became available to distinguish molecular markers expressed by neural crest cells (as reported below), several authors took advantage of certain histologically distinguishable features of neural crest cells to study their behavior. Thus, Stone (1932) noted that crest cells in *Amblystoma* contained vesicles which stained blue when exposed to Nile blue sulfate. Milaire (1959) found that, in the mouse, crest cells could be distinguished from other cells by their high RNA content when stained with methyl green-pyronin. Nichols (1981) devised a histological technique which, after appropriate fixation and further toluidine blue staining, reveals in a striking manner the neural crest cells emigrating from the neural fold and undergoing the characteristic epitheliomesenchymal transition in the mouse embryo (Nichols, 1981).

*1.3.1.2   Molecular markers of neural crest precursor cells.* The presumptive neural crest cells have several characteristics which differentiate them from their neighbors, the future cells of the neural tube and of the superficial ectoderm. First, they are endowed with migratory properties; second, they have a different fate. This is reflected by a number of molecular features or intrinsic markers that are more or less stable, depending on their nature, but which allow crest progenitors to be identified at various stages of their ontogeny.

A. AVIAN NEURAL CREST CELLS EXPRESS THE ENZYME ACETYLCHOLINESTER-ASE. Although not a strictly specific marker for neural crest cells, because it is also present in other embryonic cell types (e.g., in the ectoblast during gastrulation; Drews, 1975; Rama-Sastry and Sadavongvivad, 1979), acetylcholinesterase becomes evident in the avian embryo in the neural folds and later in the neural crest while it becomes clearly individualized after neural tube closure (Fig. 1.1). Individually migrating crest cells also contain this protein at least during the early stages of their migration (Cochard and Coltey, 1983). The function of this enzyme in neural crest cells is not clear but may be related either to the fact that, at the stage of migration, neural crest cells synthesize acetylcholine through the enzyme choline acetyltransferase (Smith *et al.*, 1977, 1979), or to a variety of non-catalytic functions known to be undertaken by acetylcholinesterase (Layer and Willbold, 1995; Sternfeld *et al.*, 1998).

B. THE NEURAL CREST AND THE HNK-1/NC1 EPITOPE. With the advent of monoclonal antibody (Mab) technology and in an attempt to further characterize the cells composing the neural crest, several authors developed antibodies against either neural crest cells themselves or their derivatives. In the latter case, the aim was to see to what extent, and from what stage in ontogeny, phenotypic molecular markers were expressed in the precursor cells contained in the crest itself.

Several interesting reagents were obtained in an attempt to identify neural crest cells in the chick or quail embryo. The most useful of them in the avian embryo was the Mab which recognizes the HNK-1/NC1 epitope because it is carried by a large majority of the neural crest cells soon after they have left the neural primordium. The NC1 Mab was obtained by immunizing a mouse with crude extracts of 8-day-old (E8) quail ciliary ganglion (Vincent and Thiery, 1984). One of the clones resulting from this immunization produced an antibody showing a strong affinity for a membrane antigenic determinant carried by migrating neural crest cells but absent on their precursors contained in the neural fold. The epitope recognized by the NC1 antibody remains expressed in Schwann cells along the peripheral nerves and in ganglia where virtually all glial cells and some neurons are brightly labeled.

The neural crest origin of the cells which exhibited NC1 immunoreactivity was checked on quail–chick chimeras (see below and Fig. 1.1) in which a fragment of quail neural tube had been isotopically implanted into a chick embryo before the onset of neural crest cell emigration at this level (Le Douarin *et al.*, 1984a). It could be seen that the large majority of neural crest cells carry the NC1 epitope when they are migrating in both quail and chick. The flux of neural crest cells exiting from the encephalic vesicles is NC1-positive when they start migrating and are in a subectodermal position. However, the cephalic crest cells which are endowed with mesectodermal differentiation capacities and reach a ventral position in the branchial arches and facial structures (see Chapter 3) rapidly lose their immunoreactivity which thus remains restricted to the neural derivatives of the neural crest (Vincent and Thiery, 1984).

The melanoblastic precursors which take the dorsolateral pathway of migration and colonize the skin express differently the NC1 antigen according to the species considered. In the chick they are NC1-positive (Erickson *et al.*, 1992) and much less so in the quail (Nataf *et al.*, 1993). Subsequently, the NC1

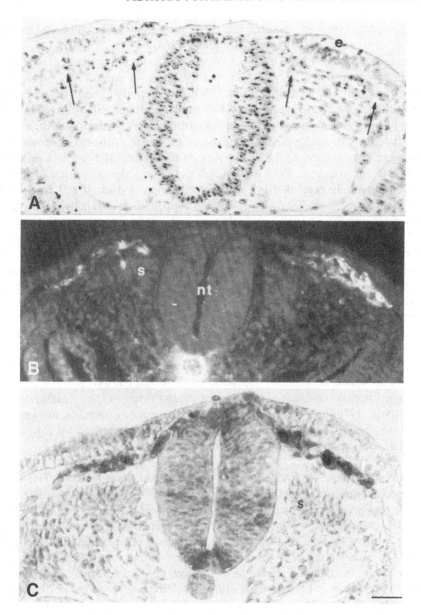

**Figure 1.1** Crest cell migration at the posterior rhombocephalic (vagal) level. (A) Postotic level, Feulgen–Rossenbeck DNA staining of 15-somite chimeric embryo. (B) First somite level, HNK-1/NC1 labeling, 16-somite quail embryo. (C) Third somite level, acetylcholinesterase activity, 14-somite chick embryo. Neural crest cells identified by the nuclear, membrane, and cytoplasmic markers migrate in the acellular space located underneath the ectoderm (e). At the level of the first 3–4 somites, this migration takes place essentially between the ectoderm and the top of the somites (s) with little if any ventral progression between neural tube (nt) and somites. Bar = 25 μm. (From Le Douarin *et al.*, 1984a.)

epitope appears on various other cell types such as in neural tube and in the perichondrium (Tucker, 1984; Tucker *et al.*, 1984).

Soon after the NC1 Mab had been obtained, it was shown that its specificity was the same as that of a Mab prepared by Abo and Balch (1981) against a human lymphoid cell line with natural killer (NK) activity (Tucker *et al.*, 1984). Then the epitope recognized by HNK-1 (and very likely also by NC1) was found to be a glucuronic acid-containing carbohydrate carried by several surface glycoproteins and glycolipids (Chou *et al.*, 1986). Among those are the membrane glycoprotein of the Ig-like superfamily called BEN or SC1 or DM-GRASP (Pourquié *et al.*, 1992, and references therein), the expression of which is developmentally regulated in the nervous system and other cell types and SMP (Schwann cell myelin protein) which is expressed on glial cells in the avian embryo (Dulac *et al.*, 1988, 1992; Dupin *et al.*, 1990; Cameron-Curry *et al.*, 1993). Moreover, certain human neuropathies are characterized by the presence of IgM recognizing the HNK1 epitope (Chou *et al.*, 1985, 1986). Most of the studies relying on the recognition of the HNK1/NC1 epitope were done using the antibody produced by the HNK-1 hybridoma cell line. However, it should be noted that no reactivity with mouse neural crest cells could ever be detected with either the HNK-1 or the NC1 Mabs while they were used in several studies in the rat embryo (Erickson *et al.*, 1989; Pomeranz *et al.*, 1993; Chalazonitis *et al.*, 1994).

In conclusion, the HNK-1/NC1 Mab has been instrumental in studying the early steps of neural crest cell migration in the avian embryo both in normal and experimental conditions (see, e.g., Bronner-Fraser, 1986a; Kalcheim and Le Douarin, 1986) but, since it is neither a permanent nor a specific marker for these cells, the HNK-1/NC1 Mab is not appropriate to determine the long term fate of neural crest cells and to establish fate maps.

C. TRANSCRIPTION FACTORS EXPRESSED IN NEURAL CREST CELLS. The *Slug* gene encoding a zinc finger transcription factor is strongly expressed in the neural crest precursor cells prior to the onset of migration (see Fig. 2.1). The *Slug* gene remains active in the crest cells particularly in the head, at least during the early stages of their migration. Thus, the presence of *Slug* transcripts in neural crest cells precedes the onset of immunoreactivity for the HNK-1 Mab. No detailed study has so far been done to find out when, during their differentiation, the various neural crest derivatives stop transcribing the *Slug* gene.

The *Slug* gene was first isolated from a screen of an E2 chick embryo cDNA library with a probe from *Xsna*, the *Xenopus* homolog of the *Drosophila Snail* gene known as a zinc finger-containing nuclear protein (Nieto *et al.*, 1994). A high level of sequence similarity is observed in the finger region between chicken *Slug* and two genes of the *Drosophila* Snail family, *Escargot* and *Snail* (69% and 80%, respectively), but no significant sequence conservation was observed in the other domains of the protein.

The *Slug* gene is interesting because it is expressed in the vertebrate embryo in two systems at the time when the cells initiate a migration phase by going through an epitheliomesenchymal transition: the primitive streak and the neural crest. *Slug* expression has also been recently detected in the growing limb bud (Ros *et al.*, 1997).

D. GROWTH FACTOR RECEPTORS EXPRESSED IN THE NEURAL CREST. Inactivation of the genes encoding EDN3 (endothelin-3) and its heptahelical G protein-coupled receptor EDNRB (endothelin receptor B) in the mouse resulted in phenocopies of the long-described mouse mutants *lethal spotted* (*ls*) and *piebald lethal* (*s^l*), respectively, in which skin pigmentation and gut innervation are severely impaired (Greenstein-Baynash *et al.*, 1994; Hosoda *et al.*, 1994; see also Gariepy *et al.*, 1996, in the rat). This led to the demonstration that *EDN3* and *EDNRB* genes were allelic to the genes mutated in *ls* and *s^l*, respectively (Greenstein-Baynash *et al.*, 1994; Hosoda *et al.*, 1994). Cloning of the quail homolog of EDNRB (QEDNRB) allowed Nataf *et al.* (1996) to explore the sites where this gene is expressed in the embryo. A prominent level of QEDNRB transcripts is present in the neural crest precursors of the neural fold prior to the outset of migration as well as in the outflow of crest cells exiting the neural primordium at both the head and trunk levels (Fig. 1.2). This marker remains present in most derivatives of the neural crest except the mesectoderm and the melanocytic lineages. Its absence in the latter cell type was paradoxical since, in the mouse, pigmentation is affected in EDNRB mutants. It was subsequently shown that, in the avian embryo, the melanocytic cells taking the dorsolateral pathway express another closely related receptor designated EDNRB2 (Lecoin *et al.*, 1998).

Other growth factor receptors of the protein tyrosine kinase type are also present in the neural fold and maintained at least for a while in the neural crest cells as they exit the neural primordium and start migrating. Many of them have a regionalized distribution. Such is the case for the platelet-derived growth factor receptor α (PDGFRα) which is present only in the cephalic neural fold from the mid-diencephalic level down to the level of somite 5 in quail and chick embryos. PDGFRα transcripts are later on present only in a subpopulation of cephalic neural crest cells which are likely to correspond to the precursors of mesectodermal cells since, in the mouse, a mutation in the PDGFRα gene, designated *Patch*, results in cephalic defects of the mesenchymal neural crest derivatives.

The c-ret receptor, the targeted mutation of which results in the absence of neural crest-derived gut innervation, is also expressed in the neural fold prior to the onset of the crest cell migration at the level of somites 1–7 in the chick embryo, which corresponds to the source from which the enteric ganglion cells and plexuses emerge (Le Douarin and Teillet, 1973; Robertson and Mason, 1995).

Several other molecular markers have an even more restricted distribution in transverse segments of the neural fold along the anteroposterior axis, especially at the cephalic level of the neural crest. They will be described in forthcoming chapters (e.g., Chapter 3).

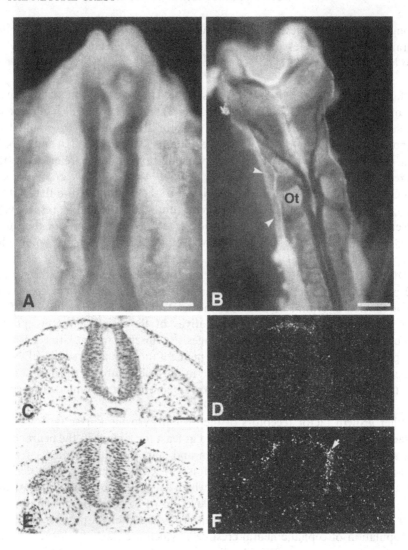

**Figure 1.2** QEDNRB expression in quail embryos. (A, B) Whole-mount embryos after hybridization with digoxygenin-labeled probes. At 4ss (A), cephalic neural folds express EDNRB. The anteriormost region of the neural folds, which does not yield neural crest cells, does not express EDNRB. Bar = 125 µm. At 14ss (B), neural crest cells migrating rostral and caudal to the otic vesicle (Ot) (arrowheads) express EDNRB. Bar = 266 µm. (C, D) QEDNRB expression by trunk neural crest cells after radioactive *in situ* hybridization on transverse sections of quail embryos at 27ss. Section in C (bright field) and D (dark field) is more caudal than in E (bright-field) and F (dark-field). Trunk neural crest cells express EDNRB when they are still in the neural folds (D) and when they migrate in the dorsoventral pathway (arrow). Bar = 60 µm. (Reproduced, with permission, from Nataf *et al.*, 1996. Copyright (1996) National Academy of Sciences, U.S.A.)

Another receptor that is expressed in migrating crest cells is the low-affinity neurotrophin receptor p75. Expression of p75 has been used to study migration of murine and rat neural crest cells as HNK1 was for avian and rat embryos (Stemple and Anderson, 1992; Bannerman and Pleasure, 1993; Fariñas et al., 1996; Chalazonitis et al., 1997). p75 is a general marker for murine migrating crest cells in cranial, vagal, and trunk areas, and highlights the migratory pathways previously defined with independent techniques such as DiI labeling (Brill, Melkman, and Kalcheim, unpublished observations). Finally, trkC MRNA that encodes a receptor for the neurotrophin NT3 is expressed in avian neural crest cells located in the premigratory domain and thereafter constitutes a selective marker for the ventrally migrating crest cells and some of their neural derivatives. It is, however, not present on melanocytes (Kahane and Kalcheim, 1994; Henion et al., 1995; Lefcort et al., 1996; see also Fig. 4.1).

More recently, some markers were found that recognize neural crest cells in the mouse. These include the AP-2 transcription factor which is expressed in neural crest cells migrating from the cranial neural folds. Null mutation of the gene encoding this factor revealed its significance in development of craniofacial structures including some cranial ganglia (Schorle et al., 1996; Zhang et al., 1996).

### 1.3.2  Extrinsic markers applied to neural crest precursor cells

Vital dye staining of the premigratory crest cells is a long-used technique to study neural crest cell migration. In the pioneer studies carried out on amphibians, the donor embryo was treated with a dilute solution of dyes such as Nile blue sulfate, neutral red or Bismark brown, and stained areas of the neural fold were grafted into unlabeled recipients (Detwiler, 1937a,b; Hilber, 1943). This procedure suffered from a lack of specificity due to dye diffusion to neighboring cells and loss of label with time, together with a certain level of cytotoxicity.

The modern versions of this approach are much more efficient. First, they involve direct labeling of cells *in situ* without requiring transplantation, and call upon the lipophilic dyes DiI (1,1'-dioctadecyl-3,3,3',3'-tetramethylindocarbocyanine perchlorate) or DiO (3,3'-dioctadecyloxacarbocyanine perchlorate; Molecular Probes) (Stern, 1990; Selleck and Stern, 1991, 1992; Hatada and Stern, 1994) which are deposited on the surface of the cells, to be labeled. Since the dye is incorporated into the cell membrane, it is transferred to the daughter cells through a certain number of generations and provides a means of marking them for 2–4 days. Because it is passively transported through membranes, diffusion to neighboring cells also occurs, yet to a limited extent (C. Kalcheim, personal observation). This method was extensively used in chick and also in mouse embryos where the neural crest cells became labeled after the dye, diluted in isotonic sucrose, was injected into the lumen of the neural tube (e.g., Serbedzija et al., 1989, 1990) or deposited directly on the cells to be labeled (Lumsden et al., 1991).

Labeling of neural crest cells by this method can be combined with the explantation of segments of the trunk of E2 chick embryo in an organotypic culture system. This allows the direct observation for 2 days of the fluorescently labeled neural crest cells when they are migrating by using low-level light videomicroscopy (Krull *et al.*, 1995).

More refined to follow both migration and clonal descendance of single embryonic cells is the technique which consists in labeling single cells by iontophoretic injection of a recognizable and unharmful tracer. This has been achieved by injecting the enzyme horseradish peroxidase into embryonic cells of the *Xenopus* embryo (Jacobson, 1984).

Another cell lineage marker is lysinated rhodamine dextran (LRD) (Gimlich and Braun, 1985). Both molecules are large and membrane impermeant that are passed to progeny by cell division over a few cell cycles. They have been used successfully in avian (Bronner-Fraser and Fraser, 1988, 1989; Artinger *et al.*, 1995) and zebrafish embryos (Kimmel *et al.*, 1989; Warga and Kimmel, 1990).

Before modern vital dyes came into common use, radioisotopic labeling of the nucleus using tritiated thymidine ([$^3$H]TdR) represented real progress in neural crest studies. Introduced by Weston (1963) to study the development of the neural crest in the chick embryo and by Chibon (1964) for the same purpose in amphibians, it gave rise to a series of elegant studies (see also Weston and Butler, 1966; Weston, 1970; Chibon, 1966, 1970; Johnston, 1966; Noden, 1975, 1976, 1978a,b). The technique of marking nuclei with [$^3$H]TdR is particularly suitable for embryonic tissues in which growth by cell division is very active. However, this property is also one of its main limitations since the label rapidly becomes diluted as a result of cell proliferation. For example, in the study of the neural crest in bird and amphibian embryos, [$^3$H]TdR labeling can provide information only on events occurring within hours or at best within the first few days following incorporation of the radioisotope.

Since [$^3$H]TdR is concentrated by the cell's most radiosensitive structure, nuclear DNA, the question of toxicity had to be considered. According to Weston (1967), when [$^3$H]TdR was used at a specific activity of about 3.0 Ci/mmol to label the neural crest cells in the chick embryo, no abnormalities were detected either in the neural tube or in the labeled derivatives of the neural crest. One can therefore conclude that a dose of the isotope sufficient for marking the cells can be found which does not hinder the normal morphogenetic processes.

### 1.3.3 The quail–chick marker system

*1.3.3.1 The structure of the quail nucleus.* After hematoxylin staining, a large nucleolus is seen in the nucleus of all quail cell types, including mesenchymal cells. The unusual size of the quail nucleolus prompted one of us (Le Douarin, 1969, 1971a,b, 1973, 1974) to compare the structure of chick and quail nuclei after several nuclear staining procedures under light and electron microscopy.

The Feulgen–Rossenbeck reaction (Feulgen and Rossenbeck, 1924) applied to quail cells revealed that the large size of the nucleolus is related to the presence of a mass of heterochromatin (Fig. 1.3A). With the Unna–Pappenheim reaction the heterochromatin appears associated with the nucleolar RNA. The same disposition of chromatin material is observed in all embryonic and adult quail cells.

In contrast, the chromatin of chick cells is evenly distributed in the nucleoplasm, as a fine network with some dispersed chromocentres (Fig. 1.3B) The amount of nucleolar DNA is small and the nucleolus is not detectably stained by the Feulgen–Rossenbeck reaction. In certain cell types, however (e.g., hepatocytes or differentiating neuroblasts), the nucleolus contains a rather conspicuous shell of perinucleolar DNA, revealed by the Feulgen–Rossenbeck reaction as a ring around the unstained area of nucleolar RNA.

At the electron microscopic level, differences between quail and chick nuclei are as striking as those observed by light microscopy (Fig. 1.3C,D). Although certain variations of the quail nucleolar substructure occur during differentiation (see Le Douarin, 1971a,b, 1973, 1974), quail and chick cells can be distinguished at any stage of development in the neural crest and its derivatives.

A technique described by Bernhard (1968) using EDTA (ethylenediaminetetraacetic acid) treatment associated with uranyl acetate staining was also applied for analyzing further the structure of the quail nucleolus. With this technique RNA stains preferentially, while DNA and most of the proteins remain uncontrasted. The nucleolar RNA of the quail appeared located laterally in relation to the heterochromatic DNA in one to three masses in which strands of intranucleolar chromatin could be detected. Therefore it is possible to distinguish quail and chick cells at the electron microscopic level by the size and structure of their nucleoli in all tissues examined. Examples are those of adrenomedullary and calcitonin-secreting cells (Le Douarin, 1970; Le Douarin and Teillet, 1971). In the chick, these cells contain one or two small nucleoli 0.5–1.9 μm in diameter, the most frequent size being 1 μm. In the corresponding cells of the quail, the nucleolus is usually single and measures 2–4 μm when cut in its largest diameter. Of course, this criterion can be applied only to sections that include the central nucleolar regions, allowing species identification of only a certain percentage (about 50–60%) of cells in a tissue (Rothman et al., 1986).

Since the size of the nucleolus is significantly increased in cell types with active protein synthesis, nucleolar size alone cannot always be used to distinguish quail from chick cells. Feulgen–Rossenbeck staining and electron microscopic observation, with or without the help of the EDTA differential staining for nucleic acids, give indisputable results because they allow the unique feature of the quail cells (i.e. the heterochromatic condensation) to be visualized.

When quail cells are transplanted into a chick embryo, or associated with chick tissues in vitro in organotypic cultures, the cells from each species retain their nuclear characteristics and can be identified in the chimera after Feulgen–Rossenbeck staining. The natural nuclear labeling of quail cells is conspicuous

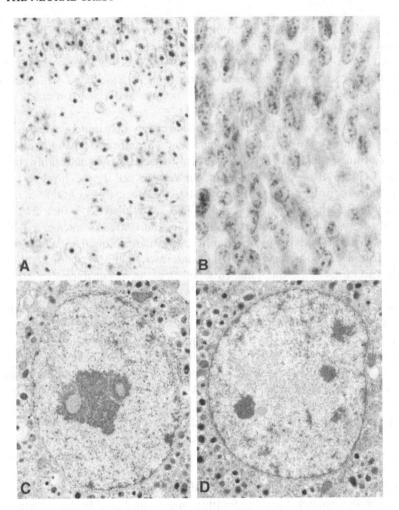

**Figure 1.3** Neuroblasts of 8-day quail (A) and chick (B) embryos stained with the Feulgen–Rossenbeck reaction. Note the large centronuclear condensation of heterochromatin in quail nuclei. In chick nuclei, the heterochromatin is dispersed in small chromocentres (×1100). Electron microscopy of quail (C) and chick (D) nuclei of the adreomedullary cells at 11 days of incubation. In quail cells the nucleoli are associated with the mass of heterochromatin, while in the chick they remain apart (×22 500).

enough to enable the identification of a single quail cell located in chick tissues, provided that the section includes the nucleolus.

Since the quail–chick marker system was devised, several species-specific antibodies have been produced which have greatly enhanced the power of this technique. The polyclonal chicken anti-quail serum prepared by Lance-Jones and Lagenaur (1987) enables virtually all quail cell types to be identified. In addition, the QCPN Mab (standing for quail non-chick perinuclear antigen) raised by Carlson and Carlson at the University of Michigan, is easier to use

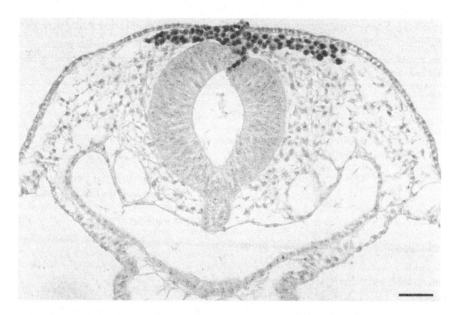

**Figure 1.4** Identification of quail cells in quail–chick chimeras by the QCPN antibody. Transverse section at the rhombencephalic level of a 8ss chimeric embryo. The host chick embryo at 5ss received isotopic graft of quail neural crest unilaterally following bilateral resection of the neural folds. Grafted neural crest cells migrate bilaterally, as revealed by immunoperoxidase staining with QCPN Mab. Bar = 45 μm (Reproduced, with permission of Company of Biologists Ltd., from Couly *et al.*, 1996.)

because it is produced in unlimited amounts (available at the Developmental Studies Hybridoma Bank (DSHB), Department of Biology, University of Iowa, Iowa City, IA 52242, USA).

The QCPN antigen appears as early as the blastula stage in all quail cells, and the Mab can be used for *in toto* staining of whole embryos at least up to E4 and later in embryonic regions or organs, thus providing a three-dimensional picture of the structures and tissues yielded by the graft. Visualization of QCPN Mab-stained cells can also be carried out on sections where it provides a very clear distinction of quail cells on a chick background even at the single cell level, as shown in Fig. 1.4. Other Mabs and nucleic probes which are both species- and cell type-specific are also available and useful to analyze the cell types yielded by either host or donor in quail–chick chimeras (see Table 1.1).

*1.3.3.2  Construction of quail–chick chimeras.* Several techniques have been used:

1. Grafting of fragments of the neural primordium (including the neural tube and the associated neural folds) between quail and chick embryos.

Table 1.1. *Species-specific antibodies and nucleic acid probes used to analyze quail–chick chimeras*

| Cell type | Quail | Chick |
|---|---|---|
| All | Chick anti-quail serum (Lance-Jones and Langenaur, 1987) QPCN (Carlson and Carlson, Hybridoma Bank) | |
| Neurons | QN (neurites) (Tanaka *et al.*, 1990) | 37F5 (neuronal cell bodies) 39B11 (neurites) (Takagi *et al.*, 1989) CN (neurites) (Tanaka *et al.*, 1990) |
| Glial cells | SMP: nucleic acid probe (Dulac *et al.*, 1992) | |
| Hemangioblastic lineage | MB1/QH1 (Péault *et al.*, 1983; Pardanaud *et al.*, 1987) | |
| MHC | TAC1 (CI II) (Le Douarin *et al.*, 1983) | TAP1 (CI II) (Le Douarin *et al.*, 1983) |
| T-cell markers | | $\alpha$TCR1 ($\gamma\delta$) (Chen *et al.*, 1988) $\alpha$TCR2 ($\alpha\beta$) (Cihak *et al.*, 1988) $\alpha$CT3 (Chen *et al.*, 1986) $\alpha$CT4 (Chen *et al.*, 1988) $\alpha$CT8 (Chen *et al.*, 1988) |

*Source*: Le Douarin *et al.* (1996).

2. Grafting of the quail neural fold from the cephalic region into chick embryos.
3. Injections of dissociated quail cells into chick embryos.

The operations are generally performed at stages preceding the onset of vascularization of the neural epithelium – i.e., during the second and third days of incubation in the chick and quail.

A. ISOTOPIC AND ISOCHRONIC GRAFTING OF FRAGMENTS OF THE NEURAL PRIMORDIUM. The technique of isotopic and isochronic grafting of fragments of the neural primordium is derived from that used by Weston (1963), who transplanted isotopically neural tube segments from chick embryos, in which cells in S-phase of the cell cycle had been labeled by [$^3$H]TdR, into normal unlabeled recipients of the same species.

Most often, in the interspecific quail–chick chimeras, the elected donor is a quail since it is easier to recognize quail cells, especially if they are isolated in a chick context, than vice versa. However, grafts in both directions are necessary to assess the validity of the experimental results.

The substitution can be done at any level of the neural axis at different stages of development. Closure of the neural tube occurs first at the presumptive level of mesencephalon at about the 5-somite stage (5ss) and then proceeds in both rostral and caudal directions to the anterior and posterior neuropores, respectively. Thus, the more posterior the level of the neuraxis to be studied, the older the embryo must be.

The fragment of the neural tube of the recipient is surgically removed from a definite level of the neuraxis using very fine microscapels obtained either by sharpening steel sewing needles on an Arkansas oiled stone or made from tungsten needles sharpened by electrolysis (Conrad et al., 1993).

The notochord and paraxial mesoderm are left in situ. Visualization of the embryo is obtained by injecting a suspension of Indian ink in Tyrode into the subblastodermal space rather than by using vital dyes. By density, the particles of Indian ink eventually fall down in the yolk and do not perturb normal development of the embryo. For isotopic and isochronic grafts the donor neural tube is removed from stage-matched embryos at the axial level chosen for excision in the recipient. In certain experiments heterotopic transpositions of the neural tube were performed. In those cases, donor and recipients were at different developmental stages (see, e.g., Le Douarin and Teillet, 1974; see also Le Douarin et al., 1996, for details).

Transpositions of rhombomeres along the anteroposterior axis are currently performed and involve only one half of the neural tube except the floor plate (Grapin-Botton et al., 1995, 1997). In experiments aimed at identifying the neural crest derivatives the donor neural tube must be dissociated from its surrounding tissues by trypsinization in order that it is freed from any contamination by mesodermal cells.

In the grafts performed at the level of the spinal cord, it appeared readily that well-known neural crest derivatives, such as sensory ganglia and sympathetic ganglionic chains, developed normally at the level of the graft and were derived from the implant. This suggested that in the chimeras the migration pattern of the grafted crest cells was indistinguishable from that in normal embryos. Though the technique involves interspecies combinations, no developmental abnormalities were observed in the host embryos when the graft had been done properly. This was confirmed by the fact that the graft-derived derivatives of the neural crest were normally positioned and differentiated in the host. The gross anatomy of the recipient embryo was normal and the only evident sign that the embryo was chimeric was the appearance of a transverse stripe of quail-like pigmented feathers on a White Leghorn chick background. This stripe corresponds to the skin area which was colonized by the melanocytic precursors emanating from the grafted segment of quail neural crest. Moreover, the immunolabeling of neural crest cells during the early steps of their migration by the HNK-1 Mab revealed a pattern similar to that conspicuously provided by the quail-into-chick chimeras (Le Douarin et al., 1984a).

An additional proof that development is not perturbed in the chimeras was given by transplantations of single rhombomeres (Couly *et al.*, 1996; Köntges and Lumsden, 1996). The neural crest cells released by the graft were evidenced in embryos immunostained by QCPN Mab at stage 14–19 of Hamburger and Hamilton (1951) (HH14–19) (Fig. 1.5). The pattern of neural crest cell migration was found to be similar to the pictures provided after DiI labeling of non-operated embryos as performed by Lumsden *et al.* (1991). Thus, "any possible delays due to embryonic healing thus appear negligible and do not systematically affect neural crest migratory pathways" (in Köntges and Lumsden, 1996). This was also evident if crest cell migration was compared in quail–chick chimeras and in intact embryos where the crest-derived cells were labeled by HNK-1 Mab or for acetylcholinesterase (Fig. 1.1).

The quail–chick neural chimeras were able to hatch and, after birth, exhibited a sensory-motor behavior apparently similar to that of normal birds of the same age (Fig. 1.6). However, after several weeks (the number of which varied according to the individuals) quail into chick neural chimeras developed a neurological syndrome due to the immune reaction of the host against the grafted quail cells (Kinutani *et al.*, 1986, 1989). This graft rejection process was significantly delayed with respect to the immune maturity of the chick which takes place during the first 2 weeks of postnatal life. The delayed immune reaction following neural grafts contrasted with that triggered by embryonic grafts of other tissue types. Thus, grafts of the anterior limb bud between E4 quail and chick embryos resulted in the normal development of the transplanted wing and in its prompt and acute rejection initiated sometime during the 2 postnatal weeks (Ohki *et al.*, 1987, 1989).

In the case of neural quail-into-chick chimeras in which progression of the immune disease has been the more thoroughly studied, it appeared that the first sites of inflammation involved the derivatives of the neural crest (i.e., the peripheral nerves and ganglia including the adrenal medulla). The part of the central nervous system (CNS) derived from the graft was invaded by T and B lymphocytes and other leukocytes only in a second step, after sensitization of the host immune system against both the quail-specific and the neural-specific antigens had occurred at the periphery. The immune attack extended in some cases to the host's CNS itself, thus consisting of a true autoimmune reaction similar to that observed in experimental allergic encephalomyelitis. The reason for both the delay and the pattern of the immunological syndrome of the chimeras must be related to the fact that the nervous system has a privileged immunological status due to the presence of the blood–tissue barrier in the CNS and to the low antigenicity of the cells derived from the neuroepithelium (neurons and glia) which express no or low levels of MHC antigens.

Grafts of quail neural primordium into chick embryos have been performed extensively at the level of the brain vesicles to study either their neural crest derivatives (Le Lièvre and Le Douarin, 1975; Le Lièvre, 1978) or problems of brain development (see Le Douarin, 1993, and Le Douarin *et al.*, 1997, for reviews). Chick or quail embryos carrying more or less extended regions of the brain from the other species are able to hatch as are birds operated at the spinal

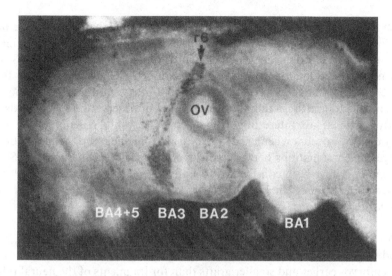

**Figure 1.5** Rhombencephalic region of a chimeric quail–chick embryo after whole-mount immunocytochemistry with QCPN Mab. The rhombomere 6 (r6) of chick host at 5ss was replaced isotopically and unilaterally by its quail counterpart. Twenty-four hours after the graft, quail cells immunoreactive for QCPN can be seen migrating from r6 into the third branchial arch. BA, branchial arch; OV, otic vesicle.

**Figure 1.6** Quail–chick chimera at 2 months of age, showing the quail-like pigmented feathers on the wings and chest. The feathers have been pigmented by melanocytes originating from the grafted neural primordium implanted at the brachial level (somite 15–20) in 22ss host embryo.

cord level (Balaban *et al.*, 1988; Hallonet *et al.*, 1990; Balaban, 1997). This has opened new avenues to study certain aspects of brain development, namely those related to the origin of some behavioral traits (Balaban *et al.*, 1988; Le Douarin, 1993; Le Douarin *et al.*, 1997; Balaban, 1997). However, the fact that the heterospecific chimeras eventually develop a most often fatal neurological syndrome due to graft rejection limits the period of observation of the birds to the early postnatal life. Interestingly, no or only a very mild reaction of the host toward the graft occurs when exchange of embryonic neuroepithelium is done in allogeneic combinations between chicks of distinct MHC haplotypes. Such brain chimeras have been constructed to study a genetic form of reflex epilepsy exhibited by a strain of chicken carrying an autosomic recessive mutation (see Batini *et al.*, 1996, for review).

B. GRAFTS OF THE QUAIL CEPHALIC NEURAL FOLD INTO CHICK EMBRYOS. The method of grafting cephalic neural fold is interesting since it allows operation on the embryos earlier and smaller grafts than for fragments of the neural tube. This method was first developed in birds by Johnston (1966) to study cephalic crest cell migration using radioisotopically labeled crest cells grafted into unlabeled embryos. Only relatively small fragments of the neural fold were involved in the operation generally performed at stage HH9.

Later, similar transplantations were carried out between quail and chick embryos in a number of studies (Johnston *et al.*, 1974; see Noden, 1978b; Narayanan and Narayanan, 1978a,b). Couly and coworkers have systematically investigated the developmental fate of the cephalic neural fold at the early stages of neurulation ranging from the time preceding the appearance of the first somite to that of the fifth somite (0–5ss). The neural fold segments involved in the operation were measured and positioned in the embryo using an ocular micrometer. Their size ranged from 50 to 300 μm according to the type of experiment performed (Fig. 1.7) (Couly and Le Douarin, 1985). These experiments were critical to establish the fate of the cephalic neural fold with precision (see Chapter 3).

C. SUPERNUMERARY GRAFTS OF QUAIL TISSUES INTO CHICK EMBRYOS. Supernumerary grafts of quail tissues into chick embryos involve fragments of quail neural crest and ganglia being inserted into one of the pathways of migration followed by crest cells (i.e., between the neural tube and the somites) in 2-day-old chick embryos.

This method was first devised to study the evolution of differentiating ciliary ganglion cells in the dorsal mesenchyme of the trunk (Le Douarin *et al.*, 1978) (see Fig. 1.8) and was found useful for other purposes (Le Douarin *et al.*, 1979; Le Lièvre *et al.*, 1980; Erickson *et al.*, 1980; Ayer-Le Lièvre and Le Douarin, 1982; Schweizer *et al.*, 1983; Fontaine-Pérus *et al.*, 1988; Rothman *et al.*, 1990, 1993b). A variation of this method consists in the injection of quail cells in the lumen of a somite of young chick embryos (Bronner and Cohen, 1979; Bronner-Fraser and Cohen, 1980; Bronner-Fraser *et al.*, 1980) or between

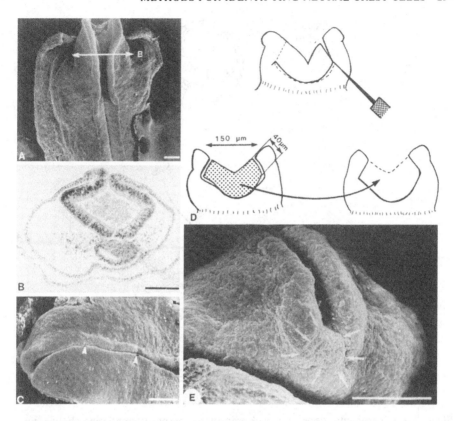

**Figure 1.7** (A) Dorsal view in scanning electron microscopy (SEM) of the head of a 3ss chick embryo. (B) At the same stage, transverse section at the presumptive level of the mesencephalon. No cephalic mesodermal cells are present dorsally within the neural fold. (C) 2.5 hours after the graft, SEM view of a chick embryo in which the right prosencephalomesencephalic neural fold (arrows) has been replaced at 3ss by its counterpart taken from a stage-matched quail embryo (Le Douarin *et al.*, 1993). Note the perfect incorporation of the transplant into the host neural fold. (D) Diagrammatic representation of the removal of the anterior neural ridge by microsurgery in 3ss embryos (Couly and Le Douarin, 1985). The size of the removed neural ridge fragment was about 150 × 40 μm. (E) SEM view of the anteriormost region of a chick embryo 2 hours after grafting the anterior neural fold. Arrows indicate the limits of the graft. Bar = 100 μm. (From Le Douarin *et al.*, 1996.)

the neural tube and the somites (Korade and Frank, 1996) to look thereafter at their dispersion.

D. CONSTRUCTION OF CHIMERIC TISSUES IN CULTURE. Direct association of the neural crest or of its derivatives (e.g., ganglia of the peripheral nervous system, PNS) with various types of tissues has been used to study a variety of problems. The two primordia are associated *in vitro* in organotypic culture and if the duration has to be prolonged for several days the explant may be thereafter

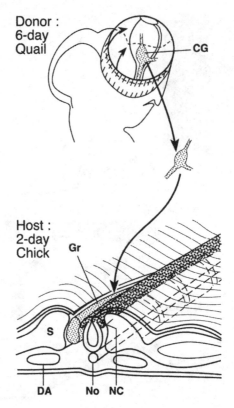

Donor :
6-day
Quail

CG

Host :
2-day
Chick    Gr

S

DA       No  NC

**Figure 1.8** Experimental design for back-transplanation of differentiating cholinergic ganglia from 4–6-day quail donor embryo into 2-day chick hosts. The ciliary ganglion (CG) is dissected and inserted, either whole or in pieces, into a slit made between the neural primordium and the somites (S). DA, dorsal aorta; Gr, graft; NC, neural crest; No, notochord. (From Le Douarin, 1982.)

transferred *in vivo* onto the chorioallantoic membrane of 6-day-old chick embryos. The *in vitro* culture period allows adhesion to occur between the two tissues whilst the *in vivo* phase of the experiment provides suitable conditions for growth and differentiation in an aneural embryonic environment. By this method, neural crest cells can be associated either with their normal target tissue or with embryonic rudiments that they normally do not meet. This can provide interesting information on the interactions of crest cells with various kinds of environment. Furthermore, this method may provide a more defined model of analysis than do neural crest derivatives developed *in situ*. For example, direct association of neural crest cells with the aneural hindgut in grafts onto the chorioallantoic membrane allows the differentiation of enteric ganglia to be studied in the absence of preganglionic innervation (Le Douarin and Teillet, 1974; Smith *et al.*, 1977; Teillet *et al.*, 1978). Thus, the glial phenotype characterized by the expression of SMP was shown by this method to be controlled by environmental cues (Dulac and Le Douarin, 1991).

E. PHENOTYPIC ANALYSIS OF THE CHIMERIC TISSUES. Since neural crest cells become widely distributed over the embryo and participate in numerous structures, it was of interest to know what cell types they give rise to in each of their target sites. In numerous cases, this can be achieved by combining methods which allow detection of quail or chick cells and others which identify a specific cell type.

A number of cytological techniques have been used for this purpose. For example, the silver impregnation technique has been applied to detect neurons in various kinds of ganglia. Two consecutive sections of the digestive tract are respectively stained, e.g., by Tinel's reaction to recognize neurons and Feulgen–Rossenbeck's reaction to identify quail or chick cells by the structure of the nucleus (Le Douarin and Teillet, 1974; Le Douarin et al., 1975; Fontaine et al., 1977).

Formaldehyde-induced fluorescence (FIF) for the detection of biogenic amines (Falck, 1962) has been applied in many experiments. On the other hand, the capacity to take up specific amino acids, (such as 3,4-dihydroxy-phenylalanine (DOPA) and 5-hydroxytryptophan (5-HTP), and decarboxylate them to produce the corresponding amine, dopamine and 5-hydroxytrypt-amine (5-HT), respectively, has been tested by intravascular injection of DOPA and 5-HTP 1 hour before tissue fixation for FIF (Fontaine, 1973; Le Douarin et al., 1974). In these experiments, sections are first observed in ultra-violet light, then postfixed in Zenker's fluid and stained by the Feulgen–Rossenbeck reaction for DNA. The sequence of techniques makes it possible to determine whether a given fluorogenic cell belongs to the chick or quail species.

Detection of [$^3$H]5-HT in enteric serotonergic neurons was also carried out in quail cells (Rothman et al., 1986).

A number of experiments have involved the detection of quail cells by their content of heterochromatin together with the application of immunocytochemical techniques to evidence the cell content in various polypeptides such as calcitonin, vasoactive intestinal peptide (VIP) and substance P (Fontaine-Pérus et al., 1981). In this case acridine orange rather than the Feulgen–Rossenbeck technique is used to evidence DNA since it does not alter the immunoreactivity of the polypeptides to be detected. Recently immunocytochemistry using two antibodies, one of which was QCPN Mab and the other specific for neurons, was used to identify the fate of sacral crest cells colonizing the hindgut (Burns and Le Douarin, 1998; see Fig. 5.4).

F. CODETECTION OF QUAIL CELLS AND mRNA IN CHIMERAS. Several techniques have been used to combine the advantages of the quail–chick marker and the search for specific gene activities by detection of their transcripts in grafted embryonic territories.

Grapin-Botton et al. (1995) transposed hemisections of the rhombencephalon corresponding to the length of two rhombomeres along the neuraxis by taking chick embryos as hosts and quails as donors. The genes under scrutiny were *Hox* genes that were revealed by radioactive *in situ* hybridization. Two

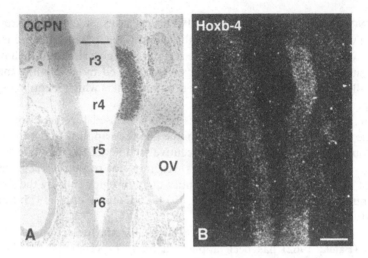

**Figure 1.9** Frontal sections of the rhombencephalic region of a chimeric embryo treated with immunocytochemistry for QCPN Mab (A, bright field) and radioactive *in situ* hybridization with *Hoxb-4* riboprobe (B, dark field) 2 days after unilateral graft of a quail rhombomere (r)8 to the position of r3/4 in a 5ss chick embryo. QCPN immunostaining shows the limits of the graft (A), which expresses *Hoxb-4* (B). OV, otic vesicle. Bar = 120 μm.

methods were successfully applied to recognize the transplanted neuroepithelium and the neural crest it produced together with *Hox* gene transcripts. A simple hematoxylin–eosin staining was applied to the sections after completion of the *in situ* hybridization procedure. Then, the large nucleolus could be distinguished in quail cells. It is worth mentioning that the standard Lison's nuclear stain (Couly *et al.*, 1993; Köntges and Lumsden, 1996) clearly shows quail nucleoli in black, cartilage in blue, and bone in red, and is usefully applied in a number of experiments with or without codetection of mRNA. However, the visualization and particularly the quality of the photomicrographs were not totally satisfactory. A better resolution was obtained by treating alternate sections with the QCPN Mab and *in situ* hybridization (Fig. 1.9).

# 2

# The Migration of Neural Crest Cells

## 2.1. Introduction

The neural crest segregates from the tips of the neural folds just before or shortly after they fuse to give rise to the neural tube. Prior to their migration, neural crest cells are an integral part of the neuroepithelium. Thanks to a morphogenetic conversion that causes these epithelial progenitors to become mesenchymal, the neural crest cells become motile and engage in migration. This process, which is highly complex both in space and time, leads the cells through defined pathways until reaching their final homing sites.

Research during the past years has made use of sensitive techniques (described in Chapter 1) to track the movements of these progenitors in relation to the equally dynamic neighboring structures. These studies have clarified the timing and pathways of migration of neural crest cells and related these parameters with their ultimate fate. At the same time, they raised fundamental questions as to the molecular mechanisms that regulate migration. Some examples are the as yet uncovered molecular cascade leading to the detachment of neural crest cells from the neuroepithelium and the factors that control the onset and arrest of migration, respectively.

A general problem about the mechanisms that control the migration of neural crest cells concerns the relationship that may exist between the directionality of migration and the degree of cell specification. More specifically, do neural crest cells migrate in a stochastic manner along the different paths to become specified by local cues present at their homing sites, or, alternatively, are they predetermined progenitors responding differentially to signals expressed along the pathways leading them to migrate in a directional fashion? Obviously, the above possibilities represent extreme cases, and additional testable combinations could also be envisaged. Furthermore, the biochemical nature of the migratory substrata and of the signals affecting cell migration also remains to be clarified as well as their mode of action. Recent studies have emphasized the notion that neural crest cell migration, much like neurite growth, is channeled both by permissive contact-mediated guidance cues and by specific chemorepellent molecules that exert their effects either via diffusion or cell contact.

## 2.2  Segregation of the neural crest from the neuroepithelium

### 2.2.1  Induction of the neural crest

During the process of neurulation, ectodermal cells thicken to form the neural plate. The lateral margins of the plate adjacent to the ectoderm undergo changes in cell shape and give rise to neural folds. Forces proposed to emanate from embryonic growth in the longitudinal axis and motile activity in the basal surfaces of the epithelial cells were proposed to cause the elevation of these folds and their ultimate fusion in the dorsal midline leading to the formation of the neural tube and the subsequent regeneration of the superficial ectoderm dorsally (see Jacobson *et al.*, 1986, and references therein).

Neural crest cells are classically considered to originate from the lateral ridges (neural folds) of the closing neuroepithelium. This is the case for the cephalic and truncal regions of avian, amphibian and murine embryos. In contrast, the neural tube of teleosts forms from a ventral thickening of the ectoderm that gives rise to a neural keel initially devoid of a central canal (Lamers *et al.*, 1981). In these embryos, the neural crest cells are thought to derive directly from the dorsal region of the neural keel that lies in apposition to the ectoderm. This situation observed in fish resembles the process of neurulation observed in the caudalmost region of the avian axis (see Schoenwolf and Smith, 1990; Schoenwolf, 1991; Catala *et al.*, 1995).

The initial studies dealing with the generation of neural crest cells were performed in amphibian embryos. Several models were proposed to explain the formation of the neural crest from the ectoderm. Exposure of ectoderm to different concentrations of an inducer was postulated to give rise either to epidermis, to neural plate, or to neural crest (Raven and Kloos, 1945). The validity of this model, however, was not further challenged. In a second model, it was postulated that the various ectodermal derivatives cited above arise as a result of a temporal change in competence of the ectoderm to respond to a neural inducer from the mesoderm (see Albers, 1987; Servetnick and Grainger, 1991). *In vivo* and *in vitro* experiments have now partially confirmed this view and concluded that, in *Xenopus*, the ectoderm loses competence to respond at least to neural fold induction by the end of gastrulation (Mancilla and Mayor, 1996). In a third line of experiments the neural plate and epidermis of different pigmented species of Urodeles were juxtaposed, giving rise to the formation of neural crest at the junction between the two tissues. These results suggested a third model which postulates that formation of the neural crest derives from an interaction between cells of the neural plate and the non-neural ectoderm (Rollhauser ter Horst, 1979; Moury and Jacobson, 1989, 1990).

Approaching the central question of the mechanisms of neural crest generation entirely depends upon the ability to distinguish these cells from the adjacent ectoderm and the CNS primordium, respectively. In early studies, this was attempted by direct observation of unstained specimens at the electron microscope level (see, e.g., Schroeder, 1970; Sadaghiani and Thiebaud, 1987, for *Xenopus* embryos). In avians, the activity of the enzyme acetylcholinesterase was utilized as a marker. In the head region, the neural fold and the adjacent band (four or five cells thick) of the lateral thin ectodermal layer are made up of strongly acetylcholinesterase-positive cells. This enzymatic activity is maintained later when the cells leave the neural epithelium, and during the early phases of migration. In the trunk, acetylcholinesterase activity is particularly conspicuous in the dorsal region of the closing neural tube and can be followed for a while in the cells that progressively leave the neural epithelium to form the neural crest (Cochard and Coltey, 1983).

The expression of acetylcholinesterase, however, is not restricted to the neural crest. Therefore, it was necessary to develop more specific markers that help define the earliest time of segregation of the neural crest lineage. More recently, two markers expressed in premigratory neural crest cells have become available: the Slug transcription factor (see Chapter 1) and the cadherin c-Cad-6B. Transcripts encoding this cadherin are specifically expressed in premigratory crest cells and are downregulated when the cells leave the neural primordium (Nakagawa and Takeichi, 1995). Slug appears in the extreme lateral region of the neural plate only after it has begun to fold and its expression is maintained to different extents during crest cell migration in both avian and amphibian embryos (see Fig. 2.1; see also Nieto *et al.*, 1994; Liem *et al.*, 1995; Mayor *et al.*, 1995; Mancilla and Mayor, 1996).

Using *Slug* expression as the earliest marker for neural crest emergence, the third theory discussed above was reevaluated. To this end, experiments were performed in both avian and amphibian embryos in which pieces of epidermal ectoderm and neural plate were brought into contact either *in vitro* or *in vivo*. Consistent with the proposed notion, this juxtaposition resulted in the generation of Slug-positive neural crest cells which were not observed in explants of each tissue separately (Dickinson *et al.*, 1995; Liem *et al.*, 1995; Mancilla and Mayor, 1996). Moreover, contact between the epidermal ectoderm and the neural plate was found to induce the expression of a variety of additional markers that localize in the dorsal regions of the neural tube such as *Wnt-1*, *Wnt-3a*, *Pax-3*, and *Dorsalin 1* (Dickinson *et al.*, 1995; Liem *et al.*, 1995). This competence of neural tissue to express Wnt genes in response to ectodermal signaling was found to be restricted to stage 8–10 chick embryos, whereas the upregulation of *Slug* occurred both at stages 4 and 8–10, regardless of the age of the ectoderm. These results suggest that the competence of the neural tissue to respond to ectodermal signals changes with time.

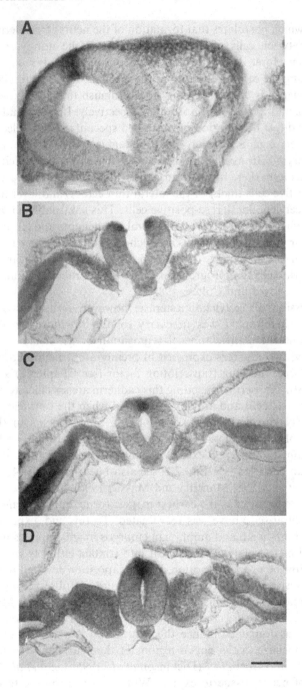

Furthermore, the inductive activity of the ectoderm could be mimicked *in vitro* by the addition of two distinct members of the TGFβ (transforming growth factor) gene family, BMP4 and BMP7, which are first expressed in the epidermal ectoderm of stage 10 chick embryos and then in the dorsal neural tube (Liem *et al.*, 1995). Thus, dorsalization of the neuroepithelium, including the generation of the neural crest lineage, requires active induction by factors derived from the non-neural ectoderm. This mechanism is analogous to the ventralization of neural tube properties, including the generation of motor neurons, which is induced by floor plate-derived *Sonic Hedgehog* (Roelink *et al.*, 1994; Ericson *et al.*, 1995).

Is the neural plate the only competent tissue to generate neural crest cells? Grafting LRD-labeled neural plate underneath the non-neural ectoderm of unlabeled *Xenopus* hosts yielded *XSlug*-positive cells of both labeled and unlabeled type (Mancilla and Mayor, 1996). Likewise, analogous recombinations between chick and quail tissues have shown the formation of neural crest cells bearing the two markers (Selleck and Bronner-Fraser, 1995). Thus, both the prospective neural plate and the epidermis can contribute to the formation of neural crest cells. Ectodermal cells might thus generate neural crest proceeding through an intermediate neural plate stage, at least until closure of the neural tube when the neural lineages (crest and tube) definitively segregate from the ectoderm (Selleck and Bronner-Fraser, 1995). This evidence suggests that, at least during the initial phases of neural crest induction before closure of the neural tube, some type of reciprocity in the transmission of inductive signals between the two tissues must occur rather than a unidirectional flow of information from the ectoderm to the neural plate. It is not known whether BMP family members are instrumental in these inductive processes prior to stage 10. The identity of possible additional inducers is at present unclear.

### 2.2.2 *Epithelial–mesenchymal transition and the onset of neural crest migration*

The next apparent stage in neural crest ontogeny is the onset of cell migration, which entails dramatic changes in cell–cell and cell–matrix interactions related to the generation of cell motility. Although the identity of the factors that

**Figure 2.1** The expression of Slug mRNA in neural crest cells. Transverse sections through a 15ss quail embryo after *in situ* hybridization with a digoxygenin-labeled avian specific Slug probe. (A) Mesencephalic region showing the presence of Slug transcripts in premigratory neural crest cells confined to the dorsal neural tube and in crest cells migrating underneath the ectoderm. (B–D) Trunk region showing three successive segmental levels. (B) Neural groove stage opposite the segmental plate mesoderm. Slug mRNA is confined to the tips of the neural folds. (C) Recently closed neural tube facing the segmental plate. After fusion of the neural folds, Slug is restricted to the dorsal part of the neuroepithelium. (D) Neural tube facing an epithelial somite showing strong Slug expression in premigratory crest cells. Bar = 60 μm.

trigger initial emigration of neural crest cells from the neuroepithelium is unknown, several prerequisites for the initiation of this process have been recognized (reviewed in Erickson, 1993a). Some of these are discussed below.

*2.2.2.1 The breakdown of the basal lamina over the dorsal portion of the neural tube.* According to Erickson (1987), a fully established basal lamina appears to be impenetrable to neural crest cells. The question is then whether neural crest cells actively digest the lamina overlying them and thereby are set free to migrate from the neural primordium. Neural crest cells have this digestive capacity as they produce a variety of proteolytic enzymes and particularly the plasminogen activator, as shown by Valinsky and Le Douarin (1985; see also Erickson *et al.*, 1992). Moreover, electron microscopic examination of mouse, chick, salamander, and zebrafish embryos has clearly shown that the basal lamina around the neural tube in the region that overlies the neural crest cells is actually incomplete and reconstitution of a continuous basal lamina does not take place until emigration of neural crest cells is completed (Martins-Green and Erickson, 1987; Raible *et al.*, 1992, and references therein). These results, which are observed in all species so far tested, strongly suggest that the presence of an incomplete basal lamina is important for emigration of neural crest cells. They, however, do not document the mechanism through which this process takes place.

*2.2.2.2 Cell–matrix interactions.* In axolotl embryos, grafting of extracellular matrix (ECM)-coated filters underneath the ectoderm in contact with the premigratory neural crest led to precocious crest cell emigration (Löfberg *et al.*, 1985). These results suggested that local changes in the ECM trigger the onset of neural crest dispersal (see also Section 2.2.2.4). In contrast to the situation in axolotl embryos, explanting avian neural tubes containing neural crest cells onto ECM substrates was unable to accelerate the timing of neural crest emigration (Newgreen and Gibbins, 1982).

Although avian neural crest cells can contribute to the production of specific ECM components (Kalcheim and Leviel, 1988), many of the ECM macromolecules required for later migration are already present in the right place prior to crest emigration. This is the case for fibronectin, laminin, and collagen types I and IV (Newgreen and Thiery, 1980; Duband and Thiery, 1982, 1987; Sternberg and Kimber, 1986a,b; Krotoski *et al.*, 1986; Duband *et al.*, 1986a,b; Martins-Green and Erickson, 1987; Krotoski and Bronner-Fraser, 1990). Consistent with a general requirement for ECM molecules in initiation of cell migration, injection of neutralizing antibodies to fibronectin, laminin–heparan sulfate proteoglycan complex or tenascin, into a lateral position with respect to the cranial neural tube, led to defects in neural crest emigration. A similar treatment at thoracic levels was, however, without effect (Bronner-Fraser, 1986b; Poole and Thiery, 1986; Bronner-Fraser and Lallier, 1988). The reasons for this regional discrepancy remain unclear. These may include differences in the responsiveness of cranial versus truncal crest cells to the same

ECM components, or technical limitations of the microinjection technique into the restricted space available at thoracic levels of the axis.

Many cell membrane receptors which bind specific motifs of the ECM molecules have been identified. They constitute a family of transmembrane molecules called integrins. In Table 2.1, the integrins expressed by neural crest cells are recorded. The expression of the $\beta 1$ subunit of the integrin hetero-dimeric receptor, which binds all the glycoproteins mentioned above, is wide-spread at all axial levels tested throughout embryonic development (Duband *et al.*, 1986a,b; Krotoski *et al.*, 1986; Sutherland *et al.*, 1993; Gawantka *et al.*, 1994). This expression pattern suggests that the differences reported in the responsiveness of cranial versus truncal crest cells are not likely to be mediated by the $\beta 1$ subunit alone. In contrast, the $\alpha 1$ integrin subunit, contained at least in the receptor complex that binds laminin and collagen type I, was not found to be prominent in the neural tube during neurulation and was still weak on migrating neural crest cells prior to overt neuronal differentiation (Duband *et al.*, 1992). In fact, little is known about the exact subunit composition of integrin receptor complexes in different regions, a limitation that makes it more difficult to interpret the results of loss-of-function studies. For instance, injection of neutralizing antibodies to the $\beta 1$ subunit or to various $\alpha$ subunits of the integrin receptors, led to an abnormal onset of crest cell migration when injected into the midbrain (Kil *et al.*, 1996), but results with knockout mice showed no effect of specific integrin deletions on neural crest migration (see review by Perris, 1997). Taken together, available data are consistent with the idea that the presence of an elaborated ECM may be a prerequisite for emi-grating cells to begin advancing outside the confines of the neural primordium, yet the mechanism of action of specific ECM components (identity of receptor complexes) and their possible role in neural crest emigration remain to be clarified.

*2.2.2.3 Changes in cell–cell adhesion.* Observation of avian embryos with transmission electron microscopy revealed that the apical zones of presumptive neural crest cells are joined by intercellular junctions at the stage of neural fold fusion, but such junctions disappear progressively about 5 hours before the onset of migration (Newgreen and Gibbins, 1982). To further investigate the molecular nature of these intercellular contacts, neural primordia were cultured in the presence of calcium blockers. This treatment reduced the cohesion between premigratory crest cells and caused their precocious emigration from the explanted neural tubes (Newgreen and Gooday, 1985). These results suggested that calcium-dependent cell–cell adhesions play a role in the main-tenance of the epithelial premigratory state of crest progenitors. The onset of migration can thus be provoked either by downregulation of calcium-dependent cell adhesion molecules of the cadherin family, or by biochemical changes in their properties, such as phosphorylation (see Newgreen and Minichiello, 1995). In fact, the expression of the cadherin c-Cad-6B is high at premigratory stages and is down-regulated during migration. Likewise, the expression of N-cadherin/A-CAM is high in the neural folds, and then avoids

Table 2.1. *Generalized distribution and proposed function of various ECM components during neural crest (NC) cell migration*

| Component | Timing of expression (along NC migratory routes) | | Role during NC cell migration[a] | Recognized domains (site) | Receptors |
|---|---|---|---|---|---|
| | NC cell migration | Postgangliogenesis | | | |
| Fibronectins[b] | No | Yes | Permissive[c] | CBD (RGD), IIICS (CS1) | $\alpha_v\beta_1$, $\alpha_4\beta_1$, $\alpha_x\beta_1$, $\alpha_1\beta_1$, $\alpha_7\beta_1$?, Galtase |
| Laminin-1 | Yes | Yes | Permissive | T8 | $\alpha_1\beta_1$ |
| Laminin-2 | No | Yes | Permissive | C-terminal rod[d] | $\alpha_1\beta_1$ |
| Laminin-4 | No | Yes | Permissive | C-terminal rod | $\alpha_1\beta_1$ |
| Laminin-8 | Yes | Yes[e] | Permissive | C-terminal rod | $\alpha_1\beta_1$ |
| Laminin-10 | No | Yes | ? | — | — |
| Tenascin-C | Yes | Yes | Non-permissive[f] | — | $\alpha_8\beta_1$? |
| Collagen type I | Yes | Yes | Permissive | CBr3, CBr7 $\alpha$1(I) chain | $\alpha_1\beta_1$ |
| Collagen type II | Yes | Yes | Non-permissive | — | — |
| Collagen type III | Yes | Yes | Non-permissive | — | — |
| Collagen type IV | Yes | Yes | Permissive | Triple helix, NC1 $\alpha$1–2 (IV) chains | $\alpha_x\beta_1$ |
| Collagen type V | No | Yes | Non-permissive | — | — |
| Collagen type VI | Yes | Yes | Permissive | Triple helix, GD $\alpha$1–3 (VI) chains | $\alpha_x\beta_1$ |
| Collagen type VII | No | No | Non-permissive | — | — |
| Collagen type VIII | No | Yes | Non-permissive | — | — |
| Collagen type IX | Yes | No | Non-permissive | — | — |
| Collagen type XI | Yes | Yes | Non-permissive | — | — |
| Collagen type XII-L/S | No | No | n.d. | — | — |
| Collagen type XIV | No | No | Non-permissive | — | — |
| Nidogen | Yes | Yes | Non-permissive | — | — |
| Fibulin-1 | Yes | Yes | Non-permissive | — | — |
| Fibulin-2 | No | Yes (?) | n.d. | — | — |
| Fibrilin-1 | No | Yes (?) | n.d. | — | — |
| Vitronectin | No (?) | Yes | n.d. | — | RGD ? |
| Link protein | No | Yes | n.d. | — | — |

Table 2.1 (*cont.*)

| Component | Timing of expression (along NC migratory routes) | | Role during NC cell migration[a] | Recognized domains (site) | Receptors |
|---|---|---|---|---|---|
| | NC cell migration | Postgangliogeneses | | | |
| Thrombospondin-2 | No | Yes | n.d. | — | — |
| Aggrecans | No | Yes | Inhibitory[g] | G1, G3 domains | Hyaluronan[h] |
| PG-M/versican-V0 | Yes | Yes | Non-permissive | — | C-terminus ? |
| PG-M/versican-V1 | Yes | Yes | (Non-permissive) | — | C-terminus ? |
| PG-M/versican-V2 | No | No | (Non-permissive) | — | C-terminus ? |
| PG-M/versican-V3 | Yes[e] | Yes[c] | (Non-permissive) | — | C-terminus ? |
| PG-Mersicans-like PGs[i] | Yes | Yes | (Non-permissive) | — | C-terminus ? |
| Perlecan | Yes | Yes | Non-permissive | — | — |
| Decorin | No | Yes | Non-permissive | — | — |
| Fibromodulin | No | Yes | Non-permissive | — | — |
| Biglycan | No | No | (Non-permissive) | — | — |
| Hyaluronan | Yes | Yes | Non-permissive | — | — |

*Notes:* [a]Including migration of NC derivatives. [b] Refers to fibronectins with and without the EDa, EDb and IIICS alternatively spliced modules/domains.
[c]Presumably the region corresponding to the "E8 domain" of EHS laminin-1. [d]Indicates that NC cells bind to the molecule and are capable of utilizing it as a migratory substrate to different extents.
[e]Thus far only determined at the mRNA level. [f]Indicates that NC cells are unable to interact with the molecule.
[g]Indicates that NC cells are capable of interacting with the molecule and this interaction causes a subsequent impedement of their movement. [h] Refers to cell surface-bound hyaluronan.
[i]May well correspond to novel alternative spliced variants of the PG-M/versican.

*Abbreviations:* PGs = proteoglycans; CBD = cell-binding domain comprising both the RGD and 1–2 synergistic cell recognition sites; IIICS = alternative spliced region of fibronectin embodying the primary binding site for $\alpha_4\beta_1$ and $\alpha_4\beta_7$ integrins denoted connecting segment 1 (CS1); T8 = smallest proteolytic fragment of the E8 domain of EHS laminin-1 retaining full cell attachment- and motility-promoting activity; CBr3 and CBr7 = cyanogen bromide-derived fragments from the $\alpha 1(I)$ chain; NC1 = C-terminal non-collagenous domain 1; GD = globular domains of all three constituent chains, $\alpha 1$–3 (VI); G1 domain = N-terminal hyaluronan-binding domain; Galtase = 1,4$\beta$-galactosyltransferase; n.d. = not determined.
*Source:* Kindly provided by R. Perris and reprinted in modified form from *Trends Neurosci.* **20**, Perris, R., The extracellular matrix in neural crest-cell migration, 23–31, Copyright (1997) with permission from Elsevier Science.

31

the premigratory crest and the migrating cells until being reexpressed at gang-liogenesis (Duband *et al.*, 1988; Bronner-Fraser *et al.*, 1992). The expression of N-CAM, a calcium-independent adhesion molecule, shuts down after initial emigration of neural crest cells (Thiery *et al.*, 1982; Akitaya and Bronner-Fraser, 1992). Consistent with these findings, ensuing studies have shown that loss of cell–cell adhesion mediated by calcium-dependent N-cadherin or by N-CAM accompanies detachment of the neural crest cells from the epithelial layer (Takeichi, 1988; Bronner-Fraser *et al.*, 1992; Takeichi *et al.*, 1997).

The notion that loss of intercellular adhesion correlates with the onset of migration does not exclude, however, the possibility that migratory cells may also be interconnected via their processes by specific cadherins. In fact, c-Cad-7 (Nakagawa and Takeichi, 1995) and U-cadherin (Winklbauer *et al.*, 1992) mediate the interaction among subsets of migrating neural crest cells and remain restricted after differentiation into specific crest derivatives. Thus, multiple cadherins are expressed by neural crest cells in a stage- and subpopulation-specific manner.

*2.2.2.4 Changes in the plane of cell division.* The cleavage plane formed during cytokinesis could separate daughter cells from apical junctions if the mitotic spindle were to form parallel to the surface of the neural tube epithelium. In support of such a notion, it was observed that 70% of all mitotic spindles in the dorsal neural tube are oriented such that one of the daughter cells would detach from the epithelium, whereas elsewhere along the neural tube the clea-vage planes are oriented so that the progeny retain a connection with the luminal surface. If this assumption is correct, then a factor stimulating the proliferation of premigratory crest cells could be part of the process leading to their emigration from the epithelium (Erickson and Perris, 1993).

*2.2.2.5 Generation of cell motility*
A. TGFβ FAMILY MEMBERS. Delannet and Duband (1992) have shown that the progression of migratory properties of neural crest cells could be accelerated by exposing premigratory crest cells to TGFβ1 and 2. More recently, it has been suggested that TGFβ family members may trigger the onset of crest cell migra-tion by affecting an initial stage of cell determination (Basler *et al.*, 1993; Liem *et al.*, 1995). The related protein Dorsalin-1 was also found to stimulate neural crest cell migration, although this factor is not expressed in the ectoderm and appears instead in the dorsal neural tube following initial neural crest specifi-cation (Basler *et al.*, 1993). The idea is that TGFβ-related proteins could induce the activation of a cascade of genes (cadherins, cytoskeletal components, tran-scription factors) required for initiation of neural crest cell migration. It is conceivable that the Slug transcription factor is one of the genes in this signal-ing pathway, firstly because its expression is induced by these molecules, and secondly because treatment of chick embryos with antisense oligonucleotides to Slug, resulted in the failure of epithelial–mesenchymal conversion of premigra-tory crest cells (Nieto *et al.*, 1994).

B. HEPATOCYTE GROWTH FACTOR/SCATTER FACTOR. The so-called "motility-inducing factors," such as hepatocyte growth factor/scatter factor (SF/HGF), are known to induce epithelial–mesenchymal transitions of several cell types. The SF/HGF exerts its activity through the c-met receptor, a tyrosine kinase receptor whose oncogenic variant has transforming activity (reviewed in Birchmeier and Birchmeier, 1994). Both HGF/SF and c-met show an overlapping expression pattern in murine neural crest cells (Andermarcher et al., 1996), and overexpression of the ligand in transgenic mice alters the normal differentiation of neural crest subpopulations, leading to skin hyperpigmentation and melanosis in the CNS (Takayama et al., 1996). This phenotype could be accounted for by a direct effect on pigment progenitor cells. Yet, the known scattering activities of this factor leave open the possibility of an early action on epithelial–mesenchymal conversion or on cell migration that could lead in turn to secondary defects in specific crest derivatives. Such possibilities are not supported, however, by results of deletion of the HGF/SF gene by homologous recombination which leads to defects in the liver and placenta, but neural induction and subsequent nervous system development appear to be normal at least at the level of analysis attained so far (Schmidt et al., 1995; Uehara et al., 1995).

C. PAX-3. The transcription factor PAX-3 is expressed in both the dorsal neural tube and the adjacent somites (Goulding et al., 1991). The mouse mutation Splotch, initially characterized by the presence of a white patch in the abdomen of heterozygotes (Russell, 1947), represents a deletion in the gene coding for PAX-3 (Kessel and Gruss, 1990; Epstein et al., 1991). Splotch mutants are characterized by defects in neural tube closure, and severe reduction or even absence of certain neural crest derivatives including pigment cells, sympathetic and spinal ganglia, enteric neurons, and cardiac structures. An in vitro study has suggested that these defects are due to a delay in the onset of neural crest emigration from the neural tube (Moase and Trasler, 1990). Another study performed in vivo found that crest cell emigration (or formation) was severely affected in the vagal and rostral thoracic areas, while virtually no cells emigrated from the tube more caudally (Serbedzija and McMahon, 1997). In the latter study neural crest cells were recognized after crossing heterozygote animals for the Splotch mutation with a stable line of transgenic mice which expresses a LacZ reporter gene under the control of the Wnt-1 enhancer. The Wnt-1/LacZ reporter gene, like the Wnt-1 gene, is expressed in much of the dorsal CNS both at cranial and spinal cord levels. β-Galactosidase activity in the neural crest cells derives from sustained expression of either the mRNA or the protein originally present in the dorsal neural tube (Echelard et al., 1994). To investigate whether the lack of migration of LacZ-positive neural crest cells was due to an intrinsic defect or to the absence of somitic or neural tube-derived PAX-3, the authors performed interspecies chimeras between mice and chick embryos. When neural tubes of both wild-type and Splotch mutants were grafted in the equivalent place of chick hosts, migration of crest cells occurred in the two cases, suggesting that crest cells in the Splotch mutant have the intrinsic capacity to form and migrate. However, when grafted into the lateral plate mesoderm of chick hosts, migration was observed from the wild-type mouse neural tube exclusively. This result indicates that successful emigration of crest cells requires an interaction between crest cells and the somites, and that the molecules mediating these interactions are defective in the *Splotch* mutation (Serbedzija and McMahon, 1997).

## 2.3    The cellular basis of neural crest migration

In this section we shall briefly overview the current knowledge on the pathways of migration of neural crest progenitors in different species, as revealed by the use of high-resolution techniques, intrinsic markers visualized by specific anti-bodies, or extrinsic marking techniques (DiI labeling, quail–chick grafts). For a more detailed account of specific migratory pathways, the reader is referred to Chapters 3–6, which discuss the development of cranial neural crest deriva-tives, sensory ganglia, autonomic ganglia, and melanocytes, respectively.

### 2.3.1    Pathways of migration related to fate

The easy accessibility of avian embryos to manipulation throughout embryo-genesis, as well as the availability of the quail–chick chimera technique, and of the HNK-1/NC1 antibody to detect migrating crest cells in intact embryos, has made the avian system the best studied so far in terms of the pathways and mechanisms that regulate the migration of neural crest cells.

*2.3.1.1    Migration in the head.* The main derivatives of the cranial crest are the entire facial and hypobranchial skeleton as well as distinct cell types in the cranial ganglia. In the head, notably in the forebrain, midbrain and hindbrain areas, where a loose mesoderm is present at the time of migration, the path of neural crest cell dispersion is essentially subectodermal (Fig. 2.1A). For instance, mesencephalic crest cells migrate rostrally over the eye primordia towards the nose and then laterally to give rise to the maxillary and mandibular arches. Migration at rhombencephalic levels follows several streams that are defined by the presence of physical barriers and leads essentially to the forma-tion of branchial arch components and contribution to cranial ganglia (see details in Chapter 3).

In the post-otic vagal area corresponding to the first five somites, several spatially and temporally distinct migratory pathways are apparent: a lateral path into branchial arches 4–5, which is an extension of the cranial ecto-mesenchyme, a lateral path along the aortic arches to give rise to the mesen-chyme of the heart outflow tract (see Chapter 3 for a detailed account of the above derivatives), a ventromedial migration pathway into the pharynx and foregut that gives rise to the enteric innervation and a further movement in the ventral direction towards the aorta to form the superior cervical ganglion (see

Chapter 4), a short ventral path that results in the colonization of cranial sensory ganglia, and finally a dorsolateral pathway followed by melanocyte precursors.

*2.3.1.2 Migration in the trunk.* In the trunk, two main directions of movement were recognized: first, a ventral pathway between neural tube and somites that gives rise to the sensory and sympathetic ganglia, to Schwann cells, and to chromaffin cells at somitic levels 18–24 (Le Douarin and Teillet, 1974). Second, a lateral path that is traversed by melanocyte progenitors. Several studies have clearly shown that migration in the trunk subserves the dynamic development of the somites. Thus, sequential stages of migration can be traced in the caudorostral direction of the axis of individual embryos (Loring and Erickson, 1987; Teillet *et al.*, 1987). This behavior is illustrated in Figs 2.1 and 2.2. At areas corresponding to the segmental plate, prior to somite individualization, neural crest cells are still confined to the dorsal part of the neuroepithelium, which may be already fused into a neural tube or still open, depending on embryonic age (Fig. 2.1B,C). At levels corresponding to the epithelial somites, some neural crest cells are still within the confines of the neuroepithelium (Fig. 2.1D), while others are apparent in a dorsolateral position between the tube and the somite (Fig. 2.2A). This ECM-rich, cell-free area gained a special designation as the "migration staging area," as it is considered to be the place where neural crest cells first "choose" between alternative pathways (reviewed by Weston, 1991, and more recently in Wehrle-Haller and Weston, 1997, with respect to its possible role as a primary site of cell specification).

A. THE VENTRAL PATHWAYS. Further rostral in the axis, the somites begin a process of epithelial–mesenchymal conversion. It is from this stage onwards that neural crest cells subdivide into different streams to interact with the mesoderm and give rise to different derivatives. Figure 2.2 summarizes the various pathways taken by crest cells with respect to somite development. In a first stage of somite dissociation, the formation of a sclerotomal mesenchyme has started in the ventrolateral aspect of the somite, whereas its mediodorsal portion still keeps an epithelial arrangement. At these levels, neural crest cells begin moving ventrally between the neural tube and somites until reaching the exit points of motor axons (Fig. 2.2C).

Neural crest cells first enter the somites upon formation of a dermomyotome and a sclerotome, between five and nine segments rostral to the most recently formed somite (Guillory and Bronner-Fraser, 1986). They follow two main ventral pathways, a slightly earlier ventral migration between adjacent somites (Fig. 2.2B) through which cells rapidly reach the dorsal aorta and give rise mainly to the primary sympathetic ganglia, and later to the aortic plexus. A second general route, which can be further subdivided into various pathways, is through the rostral half of the somites but not through their caudal moieties. Restriction of migration through the rostral domain was first recognized by Rickmann *et al.* (1985), and subsequently refined in several studies (Bronner-

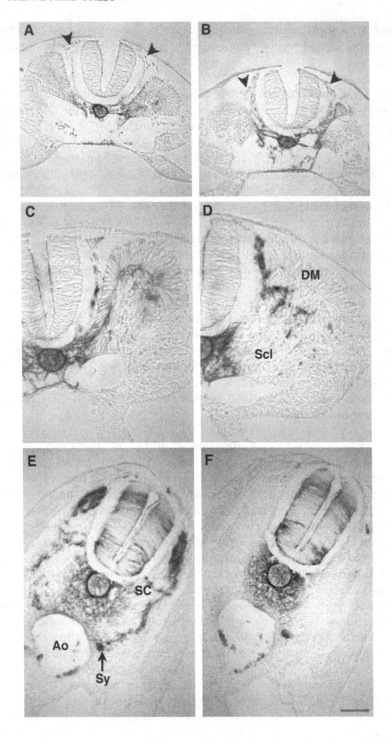

Fraser, 1986a; Teillet *et al.*, 1987). The above works have clearly shown that this initially segmented pathway of neural crest migration determines the metameric arrangement of the dorsal root ganglia (DRG), the sympathetic ganglia, and peripheral nerves with their accompanying Schwann cells (see Chapter 4 and Figs 4.1 and 4.6).

Within the rostral halves of each somite, several migratory routes are in fact distinguished. First, cells that initially arrived to a ventrolateral position with respect to the neural tube (see above) turn now laterally into the sclerotome, presumably accompanying the elongating axons and give rise to Schwann cells (Fig. 2.2E). Second, migration along the basement membrane that separates the dermomyotome from the sclerotome. Third, cell migration through the bulk of the sclerotomal mesenchyme (Fig. 2.2D). Lallier and Bronner-Fraser (1988) reported that in stage 15HH chick embryos approximately 65% of the migrating cells traverse the rostral sclerotome and intermingle with its cells. Tosney *et al.* (1994) found that neural crest cells initially prefer the myotome's basal lamina over the sclerotome as a migratory substrate. These authors have observed that crest cells located in the staging area "wait" until the formation of the dermomyotome and progress along its developing basal lamina on their way laterally. When they reach lateral regions, that are still devoid of basal lamina, they depart from the myotome's undersurface and enter the medial sclerotome. According to these authors, slightly later when a higher number of neural crest cells become engaged in migration, these cells penetrate the sclerotome directly because they are unable to establish direct contacts with the

**Figure 2.2** The migration of neural crest cells in the trunk of avian embryos revealed by HNK-1 immunolabeling. Transverse sections through a 30ss quail embryo immunostained with the HNK-1 antibody. In all sections, the notochord and paranotochordal area are HNK-1- positive. In these regions, the HNK-1 epitope is associated to a matrix-specific proteoglycan (see text). (A) Level of epithelial somites showing neural crest cells (arrowheads) located in a dorsolateral position with respect to the neural tube (staging area). At this stage and segmental level, neural crest cells do not yet express HNK-1. (B) Intersomitic region. HNK-1-negative neural crest cells migrate between two adjacent epithelial somites (arrowheads). (C) Level of an intermediate somite following its initial dissociation into sclerotome. The medial aspect of the somite is still epithelial. HNK-1-immunoreactive crest cells migrate ventrally between neural tube and somite. This pathway is restricted to the rostral half of each segment. (D) Following somite dissociation into dermomyotome (DM) and sclerotome (Scl), the first neural crest cells migrate in the rostral half of each segment underneath the dermomyotome. A few cells also penetrate the bulk of the sclerotome. (E) Section through the rostral somitic domain showing an advanced migratory stage. Neural crest cells coalesce into DRG in a dorsolateral position to the neural tube. Schwann cell (SC) progenitors are localized in ventral areas adjacent to the tube, lining along growing motor axons, other cells have migrated ventrally to the dorsal aorta (Ao) where they coalesce into the primary sympathetic ganglia (Sy). (F) Section through the caudal half of a somite showing the virtual absence of HNK-1 positive neural crest cells. Only a few cells can be seen adjacent to the dorsal neural tube and to the aorta. In both E and F, the HNK-1 epitope is also expressed by distinct cell types in the developing CNS. Bar = 60 μm.

basal lamina. Our own observations that the lateral migration underneath the dermomyotome precedes massive entry into the sclerotome are consistent with these findings.

B. THE LATERAL PATHWAY. Neural crest cells also migrate in the trunk along a lateral pathway between the ectoderm and the somites and, in avian embryos, they differentiate into melanocytes after invading the ectoderm. Although cells move initially in a mediolateral direction, further spreading of melanoblasts originating in a given segment occurs along several somites rostral and caudal of their site of origin. Moreover, unilateral grafts of half neural tubes of quail into chick hosts revealed that the pigmented stripes extended on both sides of the embryo, demonstrating a contralateral migration of the melanoblasts (Teillet, 1971a; Yip, 1986). Using the HNK-1 antibody (Erickson et al., 1992) and the MEBL-1 antibody (Kitamura et al., 1992) to immunostain intact avian embryos, it was shown that the onset of migration through the lateral pathway is delayed by 1 day approximately with respect to the ventral path. It has been proposed that the extracellular space dorsal to the epithelial dermo-myotome represents a barrier for neural crest entry. Peanut agglutinin lectin-binding activity and chondroitin 6-sulfate proteoglycan transiently localized in this region were suggested to mediate the inhibition to lateral migration. When the dermomyotome dissociates into dermis, neural crest cells proceed lateral-wards. This dissociation is accompanied by a reduction in the expression of the above molecules (Oakley and Tosney, 1991).

Another feature that characterizes migration of melanocyte progenitors is the lack of intrasegmental polarity. Nevertheless, some segmental variations are observed as migration over the body of the somites proceeds faster than over the intersomitic clefts. A day after the onset of migration through this pathway, melanoblasts invade the ectoderm (Erickson et al., 1992). At this stage, they become immunoreactive to the MelEM antibody, a marker for melanoblasts that recognizes a glutathione S-transferase subunit (Fig. 2.3; Nataf et al., 1993, 1995; see also Chapter 6).

### 2.3.2 Similarities and differences in the pathways of migration between different species

*2.3.2.1 Mammalian embryos.* The migration of neural crest cells in rat embryos was studied with the HNK-1 antibody (Erickson et al., 1989). This staining revealed a remarkable similarity in migration pathways between avians and rats. Cells were identified in the staging area, underneath the dermomyotome, and in the rostral sclerotome. Some limitations of the use of HNK-1 as a marker in this species were, however, also noticed: (1) melanocyte progenitors do not stain with the HNK-1 antibody in the rat so they could not be directly visualized; (2) in the rat, immunostaining in rostral areas of the axis becomes fibrillar and stops marking the crest and its progeny during advanced migration and gangliogenesis.

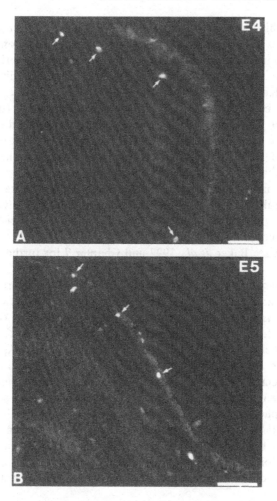

**Figure 2.3** Expression of the MelEM antigen in melanoblasts. Sections of E4 (A) and E5 (B) quails at the brachial level of the axis stained by the MelEM Mab. At E4, MelEM-positive cells (arrows) are seen underneath the ectoderm. Note that, at E5, some fluorescent cells (arrows) are already in the ectoderm. Bar = 50 µm. (Reproduced, with permission, from Nataf *et al.*, 1993.)

Analysis of crest cell migration in the mouse has been more difficult because of the lack of appropriate markers. In the absence of specific stains, Serbedzija *et al.* (1990) have injected mouse embryos from E8.5 to E10 with DiI into the lumen of the neural tube in order to label, among other cells, also the pre-migratory neural crest. The authors report that emigration from the tube starts at E8.5 and the cells migrate simultaneously along both the ventral path and the subectodermal space, in contrast to avians where a 1-day delay exists between the two routes (but see Weston, 1991; Wehrle-Haller and Weston, 1995).

Aside from this temporal discrepancy, the segmental pattern of migration that characterizes the behavior of trunk neural crest moving along the ventral pathways was consistently found in mice using DiI labeling (Serbedzija *et al.*, 1990). This was further confirmed by the use of p75 antibody staining (Melkman and Kalcheim, unpublished data), a more recently identified marker for murine crest cells. In the ventral path, early migrating cells (those injected with DiI between E8.5 and E9.5) attained the dorsal aortae and differentiated into sympathetic ganglia. A second wave of migratory cells (cells marked between E9 and E10) reached the DRG anlage. Based on these results, it was suggested that crest cells contribute to their derivatives in a ventral to dorsal order. Similar findings were reported for avian and zebrafish embryos. Moreover, several authors have invoked that the distinctive fates adopted by successive waves of emigrating cells are due to progressive restrictions in their developmental potential (Artinger and Bronner-Fraser, 1992; Raible and Eisen, 1994; but see Baker *et al.*, 1997 and Chapter 7 for further discussion).

*2.3.2.2 Amphibians.* The early knowledge of the pathways of crest cell migration in amphibians stems from grafting neural primordia between different species of *Xenopus* embryos, such as from *Xenopus borealis* into *Xenopus laevis* (Thiebaud, 1983), or grafting of fluorescent dextran-labeled *X. laevis* neural tubes into unlabeled hosts of the same species (Krotoski *et al.*, 1988). More recently, Collazo *et al.* (1993) have injected DiI into the neural tube and followed the movement of fluorescent cells in *X. laevis*. This was particularly useful for pigment cell progenitors as, in *X. laevis*, migration starts at stage 30 and melanophores first appear between stages 35 and 41. In contrast, in *Triturus* and *Amblystoma*, it was possible to follow the migration of a subset of chromatophores because they express intrinsic pigmentation already at the time of emigration from the neural tube (see Epperlein and Löfberg, 1990). Altogether, five distinct migratory pathways were identified in the above studies:

1. A dorsal pathway that gives rise to the dorsal pigment stripe and to ectomesenchyme of the fin. Later during development, both melanophores and xanthophores were shown to appear in the dorsal fin of the California newt *Tarica torosa*. Because each type of pigment cell behaves differently vis-à-vis the ECM, it was suggested that the xanthophores can actively invade the fin, whereas the melanophores most likely derive from local crest cells that invaded this region early during migration (Tucker and Erickson, 1986b).
2. A ventral pathway along the neural tube and the notochord. Cells migrating through this route give rise to sensory and sympathetic ganglia. DiI labeling experiments enabled recognition of their additional contribution to adrenomedullary cells. In addition, unlike the situation in avians and mammals, this path is also followed by presumptive pigment progenitors that later turn superficially. Another major difference between *Xenopus* and higher species is that ventral

migration occurs in streams between the neural tube and the caudal portion of each somite rather than through the rostral halves.

3. A minor pathway underneath the ectoderm that is followed by pigment cell progenitors. Whereas in avians and mammals this is the major pathway leading to pigment differentiation, in amphibians this is only a minor route.
4. A circumferential migration of crest cells into the ventral fin.
5. A ventral migration to the anus that subsequently populates the ventral fin. The last two pathways are restricted to the caudal portion of the trunk and were not observed in other species.

*2.3.2.3 Fish.* Because of the relative transparency of zebrafish embryos and the small number of crest cells, it is possible to mark them with a fluorescent tracer and to follow their movements in the living specimen using high-resolution fluorescence microscopy. Alternatively, neural crest cells can be readily distinguished at the electron microscopic level (see Fig. 2.4). No distinctive neural crest markers exist as yet for this species.

Similar to other vertebrates, zebrafish neural crest cells accumulate in a staging area dorsolateral to the neural tube along which they can move both rostrally and caudally to a limited extent of one segment (Raible *et al.*, 1992), comparable to the situation in avian embryos (Teillet *et al.*, 1987). From this region, the trunk cells disperse along two distinct pathways: dorsoventrally between the neural tube and the somites (ventral or medial pathway), and beneath the dorsal ectoderm (lateral pathway) (see review by Eisen and Weston, 1993). The main similarities and distinctive features of zebrafish neural crest migration in comparison to other vertebrates are discussed below.

A. THE MEDIAL PATHWAY. In avians, the patterns of neural crest migration and that of neurite growth are essentially segmental and depend upon differences in penetrability of rostral and caudal sclerotomal compartments (see above and also Section 2.4). Therefore, variations in the development of the sclerotome between species could result in differences in the behavior of migrating cells as well as of elongating motor axons. In avians, the sclerotome constitutes a significant portion of the somite and is adjacent to both the neural tube and the notochord. In contrast, in zebrafish, as in other teleost fishes and in amphibians, the sclerotome constitutes a much smaller part of the somite and is localized initially at a ventromedial position adjacent to the notochord. The myotomes instead occupy most of the volume of the segment and are adjacent to the neural tube.

The sclerotome in zebrafish arises from a cluster of cells located in the ventromedial part of each somite. Labeling individual cells within these clusters revealed that cells of the caudal part of the cluster develop into both sclerotome and muscle whereas cells of the rostral part of the cluster give rise to sclerotome only. The rostral sclerotomal precursors begin migrating 3–4 hours prior to migration of the caudal ones in a dorsal direction until they reach the level of

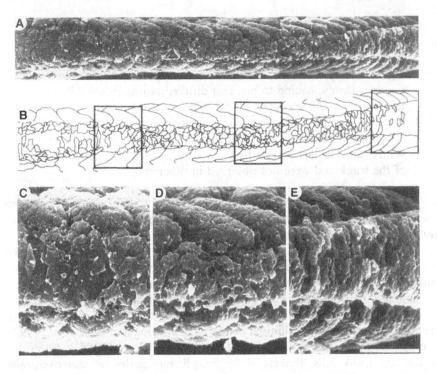

**Figure 2.4** Migration of neural crest cells in the zebrafish embryo. (A) Scanning electron micrograph showing sequential stages of neural crest migration in the zebrafish embryo. (B) Camera lucida drawing of A. (C–E) Higher magnification of the inserts in B. Zebrafish trunk neural crest cells segregate from the dorsal neural keel in a rostrocaudal sequence. Caudal is to the right. Bar = A,B: 125 μm; C–E: 62 μm. (Reproduced, with permission, from Raible *et al.*, 1992.)

the neural tube. Migration takes place along a path located in the middle of each somite and between the notochord and the myotome (Morin-Kensicki and Eisen, 1997). Migration of neural crest cells and motor axons was found to coincide with the localization of these rostral sclerotomal cells and all converge toward the middle of the somite (Fig. 2.5; see Raible *et al.*, 1992; Morin-Kensicki and Eisen, 1997). Thus, like in birds and mammals, some rostro-caudal differences can be detected during zebrafish sclerotome development that coincide with the behavior of crest cells and motor axons.

To directly test whether this colocalization has a functional implication to segmentation of DRG and motor axons, Morin-Kensicki and Eisen (1997) performed early ablations of the ventromedial cell cluster, and observed no difference in the segmental positioning of PNS derivatives. Based on these results, the authors proposed that, in zebrafish, short-range signals that influence PNS segmentation do not derive from sclerotome but rather from the myotomes which are appropriately positioned to provide such cues. Future manipulations of the myotome either by microsurgery or through the use of

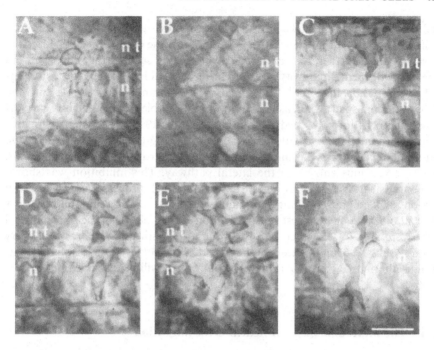

**Figure 2.5** In zebrafish, anterior sclerotome cells share a migration pathway with motor axons and neural crest cells. Individual sclerotome cells were labeled green by injection of fluorescein dextran and individual motor neurons and neural crest cells were labeled magenta by injection of rhodamine dextran; yellow indicates regions of overlap within a single focal plane. (A) Overlap between the ventrally extending axon of an identified motor neuron, CaP, and a dorsally migrating anterior sclerotome cell. (B–F) Time series of a single, ventrally migrating neural crest cell and a single dorsally migrating sclerotome cell followed over 6 hours in a living embryo. The sclerotome cell divided once during this time. n, notochord; nt, neural tube. Bar = 21 μm. (Reproduced, with permission of Company of Biologists Ltd, from Morin-Kensicki and Eisen, 1997.)

*For a colored version of this figure, see www.cambridge.org/9780521122252.*

embryos with mutations in somite derivatives will provide useful mechanistic and evolutionary insights into the processes of segmental migration and organogenesis.

B. THE LATERAL PATHWAY. Entry into the lateral path begins about 4 hours later than the onset of migration into the medial pathway (Raible *et al.*, 1992). This delay in migration through the lateral pathway appears to be consistent among several but not all species (Erickson *et al.*, 1992; see also previous sections). These observations raised the question as to whether differences in the choice of a pathway result from the molecular composition of the pathway itself or are due to intrinsic properties of the neural crest cells. LRD (lysinated rhodamine dextran) labeling of premigratory crest cells and follow-up of crest cell movement in living zebrafish embryos revealed that this delay is due to a collapsing effect of the lateral surface underneath the ectoderm that causes the paralysis, and further retraction of cell processes sent to that direction. In contrast, the medial pathway does not exert such an inhibition. These findings led to the notion that inhibitory signals delay entrance of crest cells into the subectodermal space whereas lack of inhibition enables the precocious entry of migrating cells into the medial path (Jesuthasan, 1996). Consistent with this interpretation, partial deletion of the somites, represented by the *spadetail* mutation that affects the movement of some gastrulating mesodermal cells, allows precocious entry into the lateral pathway. This inhibition was shown to be transient as a gradual loss in retraction of cell processes correlates with the onset of migration along the lateral route. Moreover, the inhibition was shown to be specific for crest cells, as Rohon-Béard cells from the dorsal neural tube were able to send their axons through this path prior to crest cell entry (Jesuthasan, 1996, and references therein). The molecular basis for this delay as well as alternative views that might account for it will be discussed in the next section.

## 2.4 The molecular basis of neural crest migration

As discussed in the previous sections, neural crest cells become confronted with a diversity of microenvironments during migration. Any attempt to understand the mechanisms that control the dispersion of these progenitors along the different pathways should take into account the fact that the molecular composition of the various pathways is highly heterogeneous as well as temporally dynamic, notions which are now being given a firm molecular basis. Molecules affecting differential migration belong to two main categories: specific ECM glycoproteins and proteoglycans as well as a particular three-dimensional conformation of the above molecules, and locally expressed ligands and their receptors.

How do these factors exert their effects upon the migrating cells? Classical studies of axon guidance have been instrumental in defining both short-range (cell–cell) and long-range (diffusion gradient) influences on the behavior of growth cones, and showed that each type of signaling mechanism may be either attractive or repulsive. In the case of growth cones, these influences may determine the ability of an axon to elongate and/or modulate its pattern of growth and branching (reviewed in Goodman and Shatz, 1993; Keynes and Cook, 1995).

Mechanisms analogous to those found to affect growth cone development are likely to direct as well the primary "choice" of neural crest cells between alternative pathways, and their subsequent movement along them. Moreover, neural crest cells could also actively participate in shaping the composition of their own migratory tracks. It is worth stressing that much less is known about the molecular basis of neural crest migration compared to the current knowledge on axon growth and guidance. The mechanisms unraveled so far point to a major role for short-range inhibitory signaling (ECM and receptor–ligand-mediated processes) in the guidance and arrest of neural crest cells, yet recent results challenge the existence of highly localized receptor–ligand interactions that lead to directional migration.

Another crucial issue that might shape the direction of migration of neural crest cells concerns their intrinsic heterogeneity. Cell lineage analysis of the neural crest shows that many of the migrating progenitors are multipotent whereas others are more restricted in their developmental potential (discussed in Chapter 7). Various scenarios could be postulated to account for differential migration based on this early segregation. First, progenitors with diverse degrees of commitment (heterogeneity among a given population or between cells that exit the neural tube in consecutive waves) might respond differently to a given migratory cue. A contrasting view would infer that, irrespective of the state of specification of a cell, migration is a stochastic event totally separable from a cell's history and from its later differentiation program.

Due to the highly dynamic nature of the crest cells and their environment, it is conceivable that more than one of the above mechanisms holds for shaping the properties of the migratory routes, and that migration along a given route is dictated by a delicate equilibrium between cell intrinsic and environmental signals.

### 2.4.1  *Delayed migration along the subectodermal pathway: transient inhibition or the need for cell specification*

The problem of the mechanisms that control migration of neural crest precursors along the lateral pathway has been extensively studied in the past few years. As discussed already, neural crest cell dispersion begins on the dorsolateral path about 24 hours after the onset of dispersal on the ventral track, and is concomitant with the dissociation of the epithelial dermomyotome into a mesenchymal dermis (Serbedzija *et al.*, 1989, 1990; Erickson *et al.*, 1992; Jesuthasan, 1996). These observations led to the notion that the environment of the epithelial dermomyotome is transiently inhibitory for migration. This hypothesis is further substantiated by the finding that high levels of versican-like glycoconjugates correlate with absence of migration while a decrease in the immunoreactivity for these compounds coincides with the onset of cell dispersion (Oakley *et al.*, 1994).

Additional correlations sustaining an inhibitory nature for these macromolecules stem from their expression in various tissues to which both neural crest cells and motor axons exhibit a repulsive behavior. These include the caudal half of the sclerotome (Keynes and Stern, 1984; Rickmann *et al.*, 1985; Tosney, 1987, 1988), the perinotochordal mesenchyme (Newgreen *et al.*, 1986; Tosney and Oakley, 1990; Pettway *et al.*, 1990, 1996) and the pelvic girdle primordium (Tosney and Landmesser, 1984).

Furthermore, mice carrying a deletion in the gene coding for the PDGFA (platelet-derived growth factor A) receptor (*Patch* mutants) which affects, among other cells, also the fate of melanoblasts, exhibit a profuse deposition of distinct basement membrane components as well as of interstitial proteoglycans along neural crest migratory routes (Morrison-Graham *et al.*, 1992; also see review by Perris, 1997).

Another example is provided by the white *Amblystoma mexicanum* mutant embryo in which neuronal precursors migrate normally but melanoblasts fail to disperse. When nucleopore membranes are covered with ECM material derived from the subectodermal space of a normal embryo and grafted into the equivalent site of a mutant embryo, neural crest migration is rescued along the lateral pathway. In contrast, grafting of membranes precoated with ECM from the mutants into wild-type specimens results in a failure to support migration in an otherwise normal environment. Using the same experimental approach, it was shown that the composition of the ECM of the ventral paths is similar in both wild-types and mutants (Löfberg *et al.*, 1989).

Altogether, these regional alterations in the ECM, whether they are part of a normal program of development or secondary to the deletion of a given gene in a natural or induced mutation, appear to be related to the ability or inability of neural crest cells to migrate.

Is the mere loss of an inhibitory ECM, which occurs concomitantly with the formation of the mesenchymal dermis, enough to account for the onset of migration of neural crest progenitors along the subectodermal path? Whereas permissiveness remains a possibility, recent results challenge the existence of more specific, attractive cues for crest progenitors along the path. Such a specificity requires two partners, one in the pathway itself and its complement on the concerned progenitors. This model requires as well that the responsive progenitors mature to a certain extent, perhaps even get fully specified to become melanocytes. Whether committed to a melanocyte fate or just partially segregated from the rest of the population (expressing distinctive receptors that do not necessarily reflect a restriction in developmental potential), such a specialization should include the ability to discriminate between alternative pathways, a precondition for directional migration.

It was recently suggested by Wehrle-Haller and Weston (1997) that the c-kit receptor and its cognate ligand Steel factor, which promotes the survival of melanocytes *in vitro* (Morrison-Graham and Weston, 1993; Lahav *et al.*, 1994; Reid *et al.*, 1995) play a role in directional migration of melanocyte progenitors. This proposal is based on the expression of c-*kit* by melanocyte precursors in the migration staging area and during migration in the skin, and the local synthesis of Steel mRNA in the dermatome prior to neural crest migration followed by localization to the ectoderm when melanoblasts have already entered the lateral path. Interestingly, in the *Steel* mutant, characterized by absence of Steel factor, melanocytes first appear in the migration staging area, as revealed by c-*kit* expression, but fail to migrate and then die. In the *Steel dickie* mutant that lacks membrane-bound factor but produces a soluble form, melanocytes initially disperse along the lateral pathway, but then die (Wehrle-Haller and Weston, 1995). Complementary ectopic expression of c-*kit* in the dermal mesenchyme, which is observed both in the *Patch* and *W-sash* mutations (Duttlinger *et al.*, 1993; Werle-Haller *et al.*, 1996), is compatible with initiation of dispersion of crest cells followed by their later disappearance. In both cases, ectopic c-kit was proposed to act in a dominant manner to compete with the endogenous c-kit present on crest cells for available ligand. As also stressed above, the *Patch* mutation entangles the production of an abnormal ECM, which may in combination with ectopic c-*kit* expression cause the resulting phenotype. These results are compatible with the view that the Steel factor / c-*kit* system is mostly required for the survival of premigratory or migrating melanocyte progenitors (see Chapter 6). To prove that Steel factor has a role in attraction it should be shown that it is distributed in a gradient-like fashion along the path, or that it has chemotactic properties *in vivo* as it has been found to display *in vitro* on other cell types (Blume-Jensen *et al.*, 1991; Sekido *et al.*, 1993). Interestingly, the expression of both factors in birds is delayed when compared to mouse embryos. In avian embryos, c-*kit* expression occurs 2 days after the crest cells have left the neural primordium, while Steel factor is expressed in the epidermis from E4 onward but not in the dermis (Lecoin *et al.*, 1995). This expression pattern is consistent with a role for c-kit/Steel in melanogenesis (Lahav *et al.*, 1994), but argues against the suggestion that these molecules affect, at least in the avian system, the directional migration of crest cells when they are still present in the staging area (see Chapter 6 for further discussion).

The state of specification of c-*kit*-expressing cells in the migration staging area is not known. For instance, these progenitors could coexpress transiently markers of ventrally migrating neural crest cells like trkC that characterizes a neurogenic population of crest cells (Kahane and Kalcheim, 1994; Henion *et al.*, 1995). Alternatively, the expression of these two receptors could already be segregated to distinct subpopulations. So, the above studies do not directly address the question as to whether full specification to a melanocytic fate is required for directional migration. Several studies have reported that multipotent progenitors and cells with neurogenic potential are encountered in the lateral migratory route, since, if isolated from the skin at E5 in the quail and cultured under appropriate conditions, some cells differentiate also along the glial and neuronal pathways (Sieber-Blum *et al.*, 1993; and see Wehrle-Haller and Weston, 1997).

A recent study by Erickson and Goins (1995) has shown that, when fully differentiated melanocytes are back-grafted into the pathways of young hosts, these cells penetrate the lateral pathway a day before the host cells, even facing an inhibitory environment such as the epithelial dermomyotome. If depleted of melanoblasts, cells only migrate ventrally. In addition, when young cells are grafted prior to their segregation into distinct phenotypes, they fail to enter the lateral track. These data suggest that neural crest cells can migrate along the subectodermal path at these early stages only if they are specified as melanocytes. That melanocytes can home preferentially to the skin was shown in a classical experiment by Weiss and Andres (1952) who injected heterogeneous cell populations from a colored strain of chicken intravascularly into White Leghorn hosts. This resulted in the appearance of colored patches in the plumage of the hosts, indicating that the injected cells traversed the blood vessels and homed in a specific manner to the skin. It is also worth mentioning that overexpression of N-myc by neural crest cells leads these cells to migrate preferentially along the ventral pathway and to differentiate into neuronal derivatives (Wakamatsu *et al.*, 1997). These studies provide support for a causal relationship between intrinsic cell specification to a given fate and directional migration but do not prove that this mechanism prevails in the naturally occuring crest cell dispersion process. Note that many results agree with the view that, when they migrate, neural crest cells are mostly pluripotent in nature and disperse within pathways that happen to be temporally compatible with their progression (see Chapter 7).

### 2.4.2   Molecular signals controlling segmental migration in the trunk

*2.4.2.1   Local changes in the extracellular matrix (ECM) of the migratory pathways.* Neural crest cells migrate along pathways composed of complex extracellular components. Therefore, their migration is likely to be regulated by the spatiotemporal distribution of these molecules, their supramolecular arrangement into complex three-dimensional matrices, and by the ability of the migrating cells to respond to these cues appropriately. Based on an extensive body of literature, ECM components can be generally divided into permissive and inhibitory with respect to their ability to support attachment and spreading of neural crest progenitors.

A. PERMISSIVE ECM-ASSOCIATED SIGNALS. Research along these lines has revealed that neural crest cells migrate through matrices composed of permissive sets of glycoproteins. In general, the directionality of migration of neural crest cells has been found to follow the orientation of ECM fibrils *in vivo* and also *in vitro* where modification of fiber orientation caused a consecutive change in the pattern of crest cell movement (Newgreen, 1989). A mechanism of contact guidance that requires specific recognition of the substrate's properties was proposed to account for the observed coalignment of crest cells with ECM fibrils. This mechanism seems more likely to account for crest cell migration than the alternative view whereby control of cell migration is exerted by available space between tissues or cells, because crest cells are able to pass through pores as small as 1 μm in diameter and even less, whereas the minimal size of intercellular spaces measured in embryos was always greater than 2 μm (Ebendal, 1976, 1977; see also Newgreen, 1989, and additional references therein).

Pioneer studies by Thiery and colleagues, led to the recognition that the glycoprotein fibronectin (FN) plays an important role in the migration of neural crest cells. The presence of FN is correlated *in vivo* with the migratory pathways of neural crest cells (reviewed in Duband *et al.*, 1986b), and in some cases the cessation of their migration is correlated with the disappearance of this molecule (Thiery *et al.*, 1982). Moreover, FN promotes cell attachment and dispersion *in vitro* (Rovasio *et al.*, 1983; Rogers *et al.*, 1990; Table 2.1). Furthermore, neural crest cells express integrin receptors that mediate binding to FN and are required for migration at least *in vitro* (see Section 2.2.2.2 and below). The interaction of FN with its receptor is mediated by a short sequence located in the central region of the molecule identified as the tetrapeptide RGDS. This peptide effectively competes with FN for attachment of neural crest cells and promotes itself cell attachment and spreading *in vitro*. Moreover, another sequence located in the N-terminus of the molecule, the connecting segment (CS1) site, was found to stimulate neural crest cell attachment but not to initiate spreading. The combined activity of both peptides was shown to give optimal locomotion (see Dufour *et al.*, 1988; Perris *et al.*, 1989).

In contrast to FN, the extent of neural crest migration on purified laminin substrates *in vitro* was maximal at relatively low concentrations of the molecule and was reduced at higher concentrations. Moreover, coupling of laminin to nidogen and to collagen type IV yielded maximal cell dispersion. The predominant motility-promoting activity of laminin was mapped to the E8 domain that possesses heparin-binding activity. Consistent with this result, migration on laminin was reduced after treatment of the substrates with heparin. As summarized in Table 2.1, diverse types of collagen, including collagen I, IV, and VI, are also able to mediate neural crest motility *in vitro* through $\beta 1$ integrins on the surface of the crest cells (Perris *et al.*, 1993a,b).

In fact, the attachment of both cranial and trunk neural crest cells to FN, laminin, and collagens can be efficiently blocked by treatment with anti-$\beta 1$ integrin antibodies, suggesting that this is a functional isoform expressed by crest cells in both regions. In contrast, neutralization of $\alpha 1$ integrins inhibited the attachment of trunk but not of cranial crest cells on laminin and collagen type 1 exclusively. These data, together with the results of antibody neutralization described in Section 2.2.2.2, suggest that distinct integrin heterodimers are present on trunk and cranial crest progenitors and that this heterogeneity in receptor composition mediates the interaction with distinct ECM molecules (Lallier and Bronner-Fraser, 1992; Lallier *et al.*, 1992, 1994).

What is the relevance of these findings to the migration of neural crest cells *in vivo* on ECM? Discrepancies appear when comparing the postulated roles of integrins on the migration of neural crest cells based on *in vitro* data and the observations stemming from neutralization studies of $\beta$ and $\alpha$ integrin subunits *in vivo* and from null mutations in murine embryos. The latter approaches have revealed no apparent abnormalities in cell migration or ganglion morphogenesis (reviewed by Perris, 1997), thereby suggesting either that integrins are not essential for crest migration in the embryos, that a high degree of plasticity or redundancy exists in the activity of integrin heterodimers, or that the integrin subunits critical in mediating interactions between ECM and neural crest were not identified as yet. Species differences may also exist in their function. Alternatively, multiple mechanisms for cell migration, some of them unrelated to cell–ECM attachment through integrin molecules, may regulate the migratory process.

B. INHIBITORY ECM-ASSOCIATED SIGNALS. Despite their permissive properties for neural crest migration, none of the above-mentioned glycoproteins was found to be differentially distributed in the somites, or to exert an active chemoattractant effect on neural crest cells. In contrast, several ECM components, most notably proteoglycans of different types, were found to localize specifically to the caudal half of the sclerotome, consistent with a possible inhibitory function for these molecules on neural crest entry into this domain. Tosney and coworkers (Oakley and Tosney, 1991; Oakley et al., 1994) have reported that tissues representing transient or permanent barriers for migration of crest cells and axonal elongation are rich in chondroitin 6-sulfate chains (see previous sections). In the ECM, chondroitin 6-sulfate chains are covalently linked to proteoglycan core proteins (Kjellen and Lindahl, 1991). The best described proteoglycan with restricted expression in the caudal sclerotomal domain is the cytotactin-binding protein (CTB-proteoglycan), whose core protein carries chondroitin sulfate chains and oligosaccharides recognized by PNA (see above and Tan et al., 1987). CTB-proteoglycan binds to cytotactin-tenascin, a glycoprotein expressed at epithelial somitic stages in the caudal somitic halves. Following somite dissociation this ligand shifts its distribution pattern to become restricted to the rostral halves of the sclerotome (Crossin et al., 1986; Tan et al., 1987, 1991; Mackie et al., 1989; Stern et al., 1989). This dynamic expression pattern was found to be independent from the presence of neural crest cells, as their ablation did not alter the programmed expression of cytotactin (Tan et al., 1987, 1991; Stern et al., 1989). Most interestingly, both cytotactin and its binding proteoglycan were found to inhibit the attachment and migration of crest cells on FN substrates. These observations led to the notion that cytotactin in the rostral sclerotome acts as a stop signal for some neural crest cells migrating in the rostral sclerotome, thereby facilitating their aggregation into ganglia. Another issue of relevance is the observation that neural crest cells themselves synthesize cytotactin. Thus, neural crest cells may interact with CTB-proteoglycan in a manner that restricts their migration into the rostral sclerotomal domains. The validity of the in vitro results remains to be challenged in the embryo, in particular in light of the observation that mice with null mutations in the cytotactin gene were shown to develop normally (Saga et al., 1992).

Chondroitin 6-sulfate-rich proteoglycans and certain keratan sulfate proteoglycans are initially distributed homogeneously along the entire somite yet localize to the caudal sclerotomal moiety concomitant with neural crest migration (Perris et al., 1991a). Another family of large aggregating proteoglycans includes, among other members, aggrecan and versican. Both aggrecan and versican failed to support the attachment and spreading of neural crest cells when immobilized on planar substrates and, moreover, counteracted the migration-promoting activities of other ECM components (Landolt et al., 1995; Perris et al., 1996; see also Table 2.1). Collagen type IX is another chondroitin sulfate-rich proteoglycan which exerts avoidance behavior of neural crest cells, and motor neurites in vitro (Ring et al., 1996). The activities of the above molecules were proposed to be mediated through their glycosaminoglycan side chains. Consistent with a putative effect on inhibition of migration, the V0 and the V1 splice variants of versican and collagen IX proteins were found to be present in the caudal sclerotome and in additional tissues that inhibit the entrance of neural crest cells (Landolt *et al.*, 1995; Ring *et al.*, 1996).

*2.4.2.2   Eph ligands and receptors.* Receptor tyrosine kinases (RTKs) have been shown to mediate cellular signals important for growth and differentiation. They form a large family of proteins that share several structural features such as a glycosylated extracellular ligand-binding domain, a transmembrane domain, and a conserved cytoplasmic catalytic domain (Yarden and Ullrich, 1988). Binding of ligands to the extracellular domain activates the cytoplasmic tyrosine kinase catalytic domain, and triggers tyrosine phosphorylation of several substrates in the cytoplasm (Ullrich and Schlessinger, 1990).

A number of transmembrane tyrosine kinases have been found to play key roles during development. Examples include the mouse c-*kit* (or *W*) proto-oncogene (Geissler *et al.*, 1988; see also previous section) and the *Drosophila* genes *sevenless* (Kramer *et al.*, 1991) and *torso* (Sprenger *et al.*, 1989), which are involved in pattern formation. Many RTKs have been shown to be developmentally regulated and predominantly expressed in embryonic tissues. These include receptors for the fibroblast growth factors (FGFRs) and for PDGF (PDGFRs), for neurotrophins (trk receptors), and for epidermal growth factor (EGFR).

A large screening effort carried out in several laboratories has led to the identification of several families of RTKs based on structural considerations. The Eph family (named for its first-described member by Hirai *et al.*, 1987) is the largest one with 13 distinct members (Tuzi and Gullick, 1994). The Eph family of RTKs is characterized by the presence of an immunoglobulin-like domain, 19 conserved cysteine residues and two fibronectin type III motifs (Skorstengaard *et al.*, 1986; Brümmendorf *et al.*, 1989). Several of these receptors have a segmental distribution during embryogenesis. Such is the case for Sek1 (Hek8 in human, and Cek8 in the chick) and for Sek2 expressed in definite rhombomeres and in early somites (Fig. 2.6; see also Chapter 3).

The members of the Eph receptor family were initially identified as "orphan receptors." Recently several proteins that bind these receptors have been molecularly cloned (Fig. 2.6; and Gale *et al.*, 1996). The striking feature common to the Eph family ligands is that they are all membrane-bound, either as transmembrane proteins or because they are attached to the surface via a glycosylphosphatidylinositol (GPI) linkage. Moreover, the Eph-family ligands are inactive (or even may act as antagonists) in soluble forms.

The fact that soluble forms of these ligands can be activated by dimerization led to the proposal that membrane attachment normally serves to facilitate their oligomerization (Davis *et al.*, 1994). Therefore, the strict requirement for membrane attachment ensures that receptor activation takes place by direct cell–cell contact. This is consistent with the fact that Eph RTKs usually exhibit a highly localized pattern of expression in the embryo. When experimentally

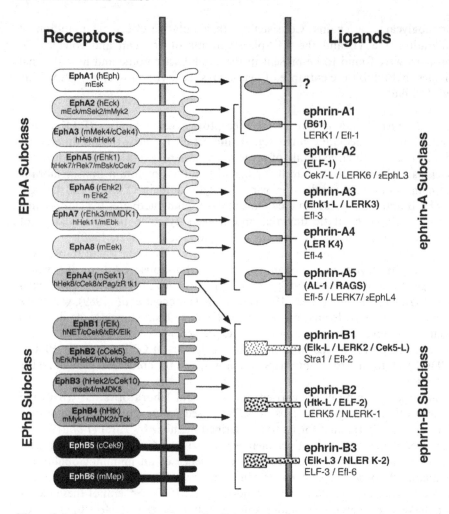

**Figure 2.6** Sequence homology trees for Eph receptors and ephrins. The ephrin-A ligands are GPI-anchored proteins, whereas the ephrin-B ligands are transmembrane proteins. Dendrograms were produced with the Clustal program, using the extracellular domains of the receptors, or the conserved core sequences of the ligands. (Unified nomenclature approved by the Eph Nomenclature Committee, 1997, kindly provided by Nick Gale.)

expressed ectopically Eph family receptors do not elicit conventional growth responses as do other RTKs such as the trk receptors used by the neurotrophins. They may instead be involved in axonal bundling or guidance by providing positional cues for the establishment of neuronal connections (see Tessier-Lavigne, 1995, and references therein).

The Eph-related RTKs and their so far identified ligands can each be grouped into two major specificity subclasses. Receptors in a given subclass bind most members of a corresponding ligand subclass but may also bind

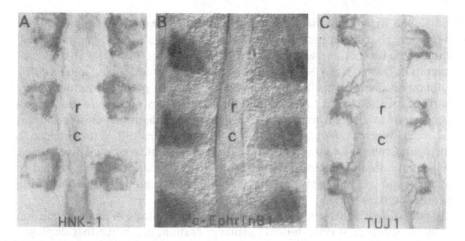

**Figure 2.7** Caudal half sclerotome expression of chick Ephrin-B1/Lerk2. At stage 18, crest cells (A) and motor axons (C) decorated by HNK-1 and TUJ1 antibodies, respectively, are seen in rostral somite halves (r), compared with caudal half restricted (C) c-Ephrin-B1 expression (B) in longitudinal sections. In rat/mouse embryos, Ephrin-B2 HTKL, but not Ephrin B1, is expressed in caudal somite halves. (Kindly provided by D. Anderson; reproduced, with permission, in modified form, from Wang and Anderson, 1997. Copyright Cell Press.)

ligands of the other subclass. The embryo is subdivided into domains defined by the exclusive expression of a given receptor subclass and its corresponding ligands. The interface between these domains corresponds to the sites where the receptors and their cognate ligands are expressed. This suggests that the Eph family is implicated in the formation of boundaries that may be of importance in defining the body plan (Fig. 2.6; Gale *et al.*, 1996).

*2.4.2.3   The role of Eph receptors and ligands in segmentation of neural crest cells in the trunk.* Using degenerate primers against conserved kinase domains and cDNA from neural crest cells, Wang and Anderson (1997) have detected the presence of the Nuk receptor whose expression was found to be relatively high on neural crest cells when compared to Sek-4, Sek-1, and Myk-1 mRNAs. Complementary expression of the transmembrane ligands HtkL and Lerk2 (ephrin-B2 and ephrin-B1, respectively) was found in the somites. In rat embryos, HtkL mRNA is expressed in the caudal half of the sclerotome and in the dermomyotome throughout the somite. In contrast, Lerk2 is present in the dermomyotome but shows no expression in caudal sclerotome. Interestingly, the chick homologs of these ligands have a somewhat different distribution; c-Lerk (c-ephrin-B1) displays a pattern similar to that of HtkL in the rat (caudal sclerotome and dermatome) (Fig. 2.7) whereas c-HtkL is present in endothelial cells of the aorta and between adjacent somites.

Expression of HtkL and Lerk2 to regions known to be inhibitory for neural crest cells, and the complementary expression of Nuk on the crest progenitors, led to the hypothesis that the somitic molecules might restrict neural crest migration via contact-mediated repulsion, due to their transmembrane presentation. This notion was confirmed using a stripe assay for testing cell migration. This assay consisted of treating nitrocellulose and anti-Fc antibody-coated coverslips with ligand–Fc fusion proteins in alternating narrow stripes. Neural tubes containing premigratory crest were placed perpendicular to the stripes and the behavior of emigrating crest cells was recorded. Whereas the high-density trail of neural crest cells attained close to the neural tube showed no preference for ligand-positive or -negative stripes, cells at the leading edges of the explants that are more dispersed tended to avoid ligand-containing stripes (Fig. 2.8), but showed no preference vis-à-vis a control Fc substrate.

Interestingly, when confronted with a uniform substrate of ligand–Fc fusion protein, neural crest cells showed normal migration, much like that displayed on fibronectin or control Fc proteins, suggesting that restriction to migration of neural crest cells on Eph ligand members, similar to axonal growth (Nakamoto et al., 1996) is not an absolute behavioral pattern but reflects a sensitivity to the presence of a boundary between relatively repulsive and permissive substrates. It is apparent from quail–chick chimeric studies that in vivo individual neural crest cells emigrating from the neural tube opposite caudal somitic halves remigrate longitudinally into the rostral areas but seem not to be confronted with a choice between rostral versus caudal somitic properties (Teillet et al., 1987). Thus, the behavior of trunk neural crest cells in vivo is likely to require a combination of cues that elicit both a repulsive behavior upon contact and act as inhibitory substrates for migration. If Eph ligands are not absolutely inhibitory for crest cell migration, as shown by the in vitro data, it could be envisaged that additional molecules contribute to the observed inhibition of migration into caudal somitic domains.

*2.4.2.4 F-spondin.* F-spondin was cloned in a screen for floor plate-specific genes using a subtractive hybridization technique (Klar et al., 1992). Analysis of the predicted amino acid sequence of this molecule (Klar et al., 1992) has revealed the presence of a signal sequence enabling its secretion from the cell. In addition, at its C-terminus, it bears six thrombospondin-like repeats that were implicated in mediating cell adhesion processes. Furthermore, a charged region is interposed between the thrombospondin repeats 5 and 6 containing a sequence characteristic of S-laminin that may function in neurite attachment and growth. The first, third, fifth and sixth repeats also contain clusters of basic residues that have been implicated in the binding of proteins to heparin and glycosaminoglycans. These features suggest that F-spondin is a molecule that associates to the ECM.

*In situ* hybridization for F-spondin has revealed that in avian, but not rodent embryos, F-spondin transcripts are localized in the paraxial mesoderm, in addition to its expression in the ventral midline observed along the entire axis. In 10ss embryos, F-spondin mRNA is homogeneously distributed in the somites, but from this stage onward its expression becomes enriched in the

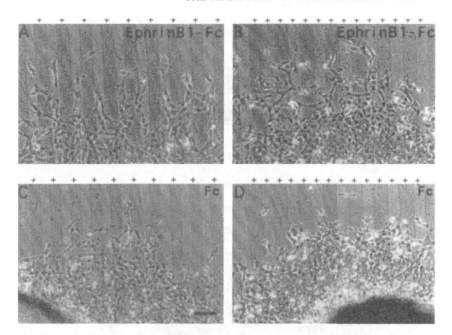

**Figure 2.8** Ephrin B1/Lerk2 fusion proteins inhibit rat neural crest cell migration in a stripe assay. Rat crest explants were cultured on coverslips coated with Ephrin-B1–Fc or Fc stripes. Each stripe is about 50 μm wide. (A, C) Ephrin-B1–Fc or Fc, respectively, was coated in alternating stripes (marked +), and subsequently visualized with alkaline phosphatase-conjugated anti-Fc antibodies after fixation of the explants. (B, D) Ephrin-B1–Fc or Fc respectively, was coated in all stripes marked +. The darker lanes were coated first and visualized by anti-BSA (bovine serum albumin) antibodies. Crest cell migration was biased toward ligand-free zones by coating alternating stripes (+) with Ephrin-B1–Fc (A), but not with Fc (C). This guidance was not observed when either ligand was coated in all stripes (B). Similar results were obtained for Ephrin-B2–Fc when tested in the above assays. Bar = 100 μm. (Kindly provided by D. Anderson; reproduced, with permission, in modified form, from Wang and Anderson, 1997. Copyright Cell Press.)

caudal halves of epithelial somites (Fig. 2.9A). Upon somite dissociation, F-spondin mRNA and protein remain enriched throughout the entire caudal sclerotome and dermomyotome and are also expressed within rostral somitic regions, at sites that are avoided by migrating neural crest cells, such as the paranotochordal mesenchyme and the dermomyotome. The structural features of F-spondin, together with its alternating expression pattern, raised the possibility that somite-derived F-spondin mediates some inhibitory properties of the somites on neural crest cell migration.

To test this idea, Debby-Brafman et al. (1999) have developed a sensitive and rapid in vitro bioassay that closely mimicks the in vivo behavior of neural crest progenitors with respect to the somites. Epithelial somites devoid of migrating neural progenitors were divided microsurgically into rostral somitic

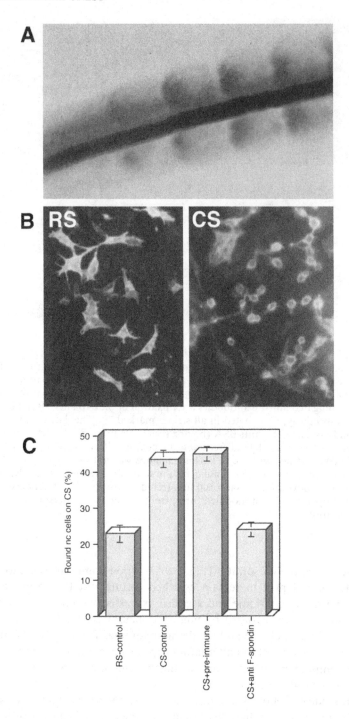

(RS) and caudal somitic (CS) halves. Like halves were pooled and cultured separately for 1 day. Addition of dissociated homogeneous neural crest cells on top of substrates of RS cells resulted in their rapid adhesion and acquisition of a fibroblastic morphology, characteristic of migrating cells. In contrast, most of the cells that had attached to the CS substrates failed to flatten, adopting instead a round morphology with few focal sites of attachment. Treatment of RS cultures with recombinant F-spondin caused the rounding up of neural crest cells which adopted a morphology similar to cells on the CS cultures (Fig. 2.9B). More significantly, pretreatment of CS cultures with neutralizing antibodies to endogenous F-spondin resulted in the flattening of the added crest cells like the picture observed on permissive RS cultures (Fig. 2.9C). In addition, overexpression of F-spondin in trunk explants resulted in lack of migration of neural crest cells into the sclerotome whereas neutralization of the protein *in vivo*, led to crest entry into otherwise inhibitory sites both in the rostral as well as caudal somitic halves. These results suggest that F-spondin mediates the inhibitory qualities of caudal and rostral somitic cells thereby contributing to the segmental patterning of crest migration and to the topographical segregation of crest cells within the rostral domains of the somite (Debby-Brafman *et al.*, 1999).

*2.4.2.5.    Other cell surface-bound molecules whose expression is associated with pathways inhibitory for neural crest migration.* Truncated cadherin (T-cadherin) is a calcium-dependent adhesion molecule that is anchored to the membrane through GPI (Ranscht and Dours-Zimmerman, 1991). In the mesoderm adjacent to the neural tube, T-cadherin is expressed in the caudal halves of the sclerotomes, where it is first apparent three segments rostral to the last formed somite concomitant with the invasion of the first neural crest cells into the rostral somite halves (Ranscht and Bronner-Fraser, 1991). This observation

**Figure 2.9** F-Spondin is expressed in the caudal domain of the somite and mediates the caudal somite inhibition of neural crest cell spreading. (A) Differential expression of F-spondin mRNA to the caudal third of each somite. *In situ* hybridization of a 23ss chick embryo showing the last four epithelial somites formed, found in continuity with the segmental plate. Note the expression of F-Spondin transcripts confined to the caudalmost third of each segment. The strong color along the axis represents the floor plate of the neural tube. Caudal is to the left. (B) Differential adhesion of neural crest cells to substrates composed of rostral versus caudal half-somites. Two-day-old neural crest clusters were dissociated into single cells and cultured on top of 1-day-old cultures of rostral (RS) or caudal (CS) somite halves. Cocultures were further incubated for an addititonal 75 min, washed and stained with the HNK-1 antibody. Note that, on RS substrate, neural crest cells are flat and fibroblastic, whereas on CS most cells adopt a round morphology and fail to flatten. (C) The percentage of neural crest (nc) cells with round morphology decreases on CS upon neutralization of endogenous F-Spondin. Pretreatment of CS cells with F-spondin antibodies, but not with preimmune serum, reduces the proportion of round crest cells to levels measured on a permissive RS substrate. (Provided by Debby-Brafman, Cohen, Klar and Kalcheim.)

readily suggested a possible implication of T-cadherin in avoidance of nerve growth and neural crest migration.

Though evidence for a possible role of this adhesion molecule in affecting neural crest behavior is still lacking, Fredette *et al.* (1996) have provided *in vitro* evidence for a negative effect of both soluble and substrate-bound T-cadherin on the extension of neurites emanating from spinal motor neurons and sympathetic neurons. Furthermore, this inhibition was overcome by treatment with neutralizing antibodies to the molecule.

Another set of cell surface-associated glycoconjugates (either membrane-bound, transmembrane proteins or ECM proteoglycans) that bind the lectin peanut agglutinin (PNA) has been identified in the caudal half of the sclerotome. Davies *et al.* (1990) have isolated, initially from chick somites and later from brain, a glycoprotein fraction that binds to PNA, thus bearing the Gal-$\beta$(1-3)-GalNAc disaccharide that was resolved into two major components (48 kDa and 55 kDa) exclusively found in caudal but not rostral somitic halves. Most importantly, the PNA-binding fraction was shown *in vitro* to cause the collapse of DRG growth cones. Although the relevant proteins are not yet cloned, other researchers have further confirmed the existence of a correlation between the binding of PNA to specific tissues and their inhibitory properties to crest cell migration and/or axonal growth. These include the caudal sclerotome, the dermomyotome, perinotochordal mesenchyme, pelvic girdle, and limb core (Hotary and Tosney, 1996, and references therein).

The first report to test whether PNA-binding molecules act on segmental migration of neural crest progenitors used an explant system of trunk regions which developed *ex ovo* for 2 days, and in which the migration of crest cells kept a segmental arrangement. Treatment of these explants with PNA caused a loss in the metameric migration of the crest cells which, under experimental conditions, entered both somite halves and slowed the rate of migration compared with control cultures (Krull *et al.*, 1995). Taken together, the available data suggest that the yet unknown PNA-binding molecules play a role both in segmental patterning of neurites and crest cells in the trunk. Moreover, they indicate that masking of these molecules may preclude the recognition by neural crest cells of additional signals expressed differentially in the somites.

## 2.5  Conclusions

In summary, segmental patterning of neural crest cells and their derivatives in the trunk of vertebrate embryos (at the exclusion of fish and amphibia) is likely to be regulated by inhibitory cues present at the right time in the caudal somitic halves which act to restrict the migration of neural crest progenitors to the complementary rostral domains. Also interesting is the exquisite channeling of neural crest cells through the rostral somitic domains, where migration also occurs through a permissive mesenchyme between inhibitory areas (the dermomyotome and the paranotochordal sclerotome). These signals include specific ECM components as well as cell surface-associated and transmembrane informative molecules. Future research should address the possibility of the existence of attractive signals in the rostral somite halves, and that of additional signals in the caudal domains. In addition, experimental paradigms should be devised to test which of these biochemically different types of molecules act as barriers for migration *in vivo* and how they integrate to control segmentation. Not less important, though, is the attempt to understand the possible role(s) of these macromolecules differentially distributed either in the rostral or caudal somitic domains in the differentiation of the somites and morphogenesis of the vertebrae.

# 3

---

# The Neural Crest: A Source of Mesenchymal Cells

## 3.1 Historical overview

### 3.1.1 The notion of mesectoderm

Kastschenko (1888) first reported that the neural crest is the source of mesenchymal cells in selacians. Soon after, Goronowitsch (1892, 1893) extended this notion to teleosts and birds. At the same time, Platt (1893, 1897), found that the mesenchymal cells forming the cartilage of the visceral arches and the dentine of the teeth were derived from the dorsal ectoderm of the head. She created the term "mesectoderm" for the mesenchyme of ectodermal origin, in contrast to the mesodermal mesenchyme which she called "mesentoderm." Nowadays, the term "ectomesenchyme" is often used to designate the mesenchymal cells derived from the neural crest. These findings were not immediately accepted and gave rise to a controversy reflecting how vigourous von Baer's germ layer theory was. Von Baer had pointed out in 1828 that homologous structures in different animals are derived from material belonging to the same germ layers; according to this view, the mesenchyme could arise only from mesoderm. It was not until the 1920s that thorough morphological observations and experimental work, carried out primarily in amphibians (Landacre, 1921; Stone, 1922, 1926, 1929), led to the unambiguous demonstration that a significant proportion of the mesenchymal cells of the body were actually derived from the ectodermal germ layer.

For many years the extent to which the neural crest generates mesenchymal derivatives was studied mostly in lower vertebrates. The role of the cephalic crest in the morphogenesis of the head and the hypobranchial region was recognized first, and the contribution of trunk neural crest to the dermis and fins was shown later by several authors, who used a variety of experimental techniques such as xenoplastic transplantation (Raven, 1931, 1936) or removal of the trunk crest (Du Shane, 1935).

The contribution of the neural crest to mesenchyme during normal development, as established in various classes of vertebrates, will be described here first. The considerable recent progress about the molecular mechanisms which control the development of structures of mesectodermal origin will be reviewed in the majority of this chapter.

### 3.1.2 *Pioneering work on the amphibian embryo and lower vertebrates*

The contribution of cephalic neural crest to the cranial skeleton was determined in amphibians through a number of experimental approaches such as cell-marking experiments (Hörstadius and Sellman, 1941, 1946; Chibon, 1966, 1967), xenoplastic transplantation (Raven, 1931, 1933, 1936; Harrison, 1935, 1938), or early extirpation of the neural crest (Stone, 1922, 1926, 1929). It is interesting to point out that Landacre (1921), working on *Amblystoma jeffersonianum*, was able to follow the first steps of neural crest cell migration simply by careful histological examination, since he could distinguish neural crest cells from the surrounding tissues through differences in cell size, staining affinities, pigment granules, and yolk globules. Further insights awaited experimental procedures used soon after.

The primitive site of origin of the various head ectomesenchymal derivatives was identified through two kinds of cell-marking experiments. One consisted of staining the neural ridge with vital dyes at the neurula stage, and following the migration of the stained cells directly beneath the ectoderm during the following days. This was performed by Hörstadius and Sellman (1941, 1946) in *Amblystoma jeffersonianum* and gave rise to fundamental results, which the authors confirmed by extirpation experiments.

In a thorough series of studies in *Pleurodeles* using [$^3$H]TdR-labeled crest cells implanted into a non-labeled host (Fig. 3.1), Chibon (1966, 1967) completed these data, and concluded that the neurocranium (i.e., the two parachordal cartilages, the basal plate, the trabeculae cranii (partly), and the auditory capsules) has a mesodermal origin, except for the trabeculae cranii which are mostly derived from the anterior head neural crest (sector 30°–50° for the anterior trabeculae and 50°–70° for the posterior trabeculae according to Chibon, 1966). In addition, some cells from the sector 50°–70° contribute to the parachordal cartilages and the basal plate (Fig. 3.1).

The visceral skeleton, in contrast, appeared to be derived essentially from the mesectoderm. It is formed of six visceral arches, and two basibranchials (copulae). The first arch is the mandibular. Its dorsal part, the palatoquadrate, is fused with the trabeculae by means of the processus ascendens. Its ventral part, Meckel's cartilage, is the cartilage of the lower jaw. The next posterior arches are the hyoid and the four gill arches. The five last arches fuse in the

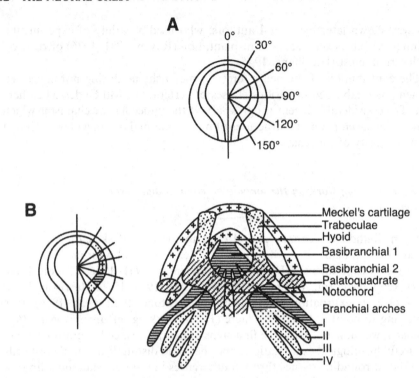

**Figure 3.1** (A) Subdivision of the cranial folds in *Pleurodeles* that serve as references to indicate the developmental fates of neural crest cells in normal and experimental conditions. (B) The contribution of the anteroposterior sectors of the neural fold to the different pieces of the cranial and visceral skeleton in *Pleurodeles* as deduced from transplantation experiments using [³H]TdR-labeled neural crest. (From Chibon, 1966.)

median line with the basibranchials 1 and 2 (Fig. 3.1). Except for the second basibranchial cartilage, the visceral skeleton is derived entirely from the neural folds.

The areas from which the visceral skeleton is derived are distributed anteroposteriorly in the sectors 70°–150° as defined by Chibon and indicated in Table 3.1. The forehead ridge (corresponding to sections 1 and 2 as defined by Hörstadius and Sellman, and to sector 0°–30° of Chibon) does not contribute at all to cartilage but gives rise to mesenchymal cells, the final fate of which has not been established. Bilateral excision of the neural folds at this level results in the absence of telencephalon and of the anterior part of the diencephalon, suggesting that most of the anterior neurectodermal ridge becomes incorporated into the brain.

Behind this region the neural folds are devoid of potentialities to form brain tissue and the areas from which the visceral skeleton is derived are distributed anteroposteriorly in the sectors 70°–150° as defined by Chibon (Table 3.1).

Although the normal fate of neural crest cells was not nearly as well documented in lower classes of vertebrates as in amphibians, some investigations

Table 3.1. *Presumptive territories of the visceral skeleton on the cranial neural crest in* Pleurodeles *embryo*

| Sectors of cranial neural crest | Skeletal derivatives of cranial crest |
| --- | --- |
| 0–30° | None |
| 30–50° | Anterior trabeculae |
| 50–70° | Palatoquadrate; some contribution to posterior trabeculae and basal plate |
| 70–100° | Meckel's cartilage; hyoid arches |
| 100–120° | Basibranchial 1; hypobranchial cartilages; anterior branchial arches |
| 120–150° | Posterior branchial arches |

*Source*: Chibon (1966, 1967).

have been performed in the lamprey (Newth, 1956) and in teleosts by Lopashov (1944) and more recently by Sadaghiani and Vielkind (1990) who used the HNK-1 Mab to follow neural crest cell migration in teleost embryos. Newth carried out xenoplastic transplantations of lamprey neural crest into the branchial region of newt neurulae. Differences in cell morphology in these two species allowed the testing of the capacity of differents parts of the neural crest to form cartilage. According to Newth, the cartilaginous ventral basket of the (young) 7-mm ammocoete, which is the latest stage analyzed, originates from the neural crest. On the other hand, Newth noticed that extirpation experiments resulted in the abnormal dilatation of the ventral arteries of the head and suggested that the connective wall of these vessels was, at least in part, of crest origin. This view was fully confirmed later by the experiments of Le Lièvre and Le Douarin (1975) and of Couly *et al.* (1993) on the avian embryo. The recent studies devoted to the zebrafish (*Brachydanio rerio*) neural crest will be described below.

The pioneering work carried out on the amphibian embryo also disclosed the contribution of the neural crest to the tooth papillae. Teeth develop from rudiments composed of oral epithelium and mesenchymal cells belonging to the first branchial arch. The epithelial cells of the tooth bud differentiate into enamel-secreting ameloblasts whereas mesenchymal cells form the dentine-secreting odontoblasts. As early as 1897, Platt proposed that both the odontoblasts and the pulp of the tooth papilla were derived from ectomesenchyme. Later, Adams (1924) and De Beer (1947) supported this idea. Stone (1926) and Raven (1931) observed that removal of the neural crest resulted in a considerable reduction in the number of teeth on the operated side. Furthermore, xenoplastic transplantation of *Amblystoma* neural crest into *Triturus* embryos (Raven, 1935) resulted in the presence of donor cells in the tooth papillae of the host. These findings were confirmed by others (Hörstadius and Sellman, 1946; Wagner, 1949).

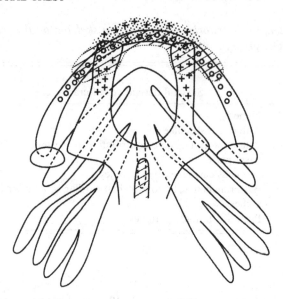

**Figure 3.2** Disposition of teeth in stage 38 (Gallien and Durocher, 1957) *Pleurodeles* larva. Superior teeth (area with crosses) are located anteriorly and under the trabeculae. Inferior teeth (area with circles) are on the anterior part of Meckel's cartilage. The stippled and hatched areas correspond to the parts of buccal epithelium originating from superior and inferior buccal ectoderm, respectively; non-shaded regions are endodermal. (From Chibon, 1970.)

A thorough study of the contribution of the neural crest to tooth formation was later made by Chibon (1966). After orthotopic grafts of [³H]TdR-labeled crest between 45° and 100° (Fig. 3.1), labeled crest cells were found in the tooth papillae. Palatini teeth were labeled when the operation involved the 30°–70° segment of the crest, whilst mandibular teeth papillae originated from a more posterior level (70°–100°) (Fig. 3.2). Crest cells from the graft could be seen not only in the operated sides but also contralaterally; this explains why a total absence of teeth can only be obtained after bilateral extirpation of the crest. In addition, with rare exceptions, heterotopic grafts of truncal crest at the head level did not yield teeth.

The respective roles played by the mesenchymal and epithelial components of the tooth rudiment were first studied in the amphibian embryo by Wagner (1949, 1955). He took advantage of the fact that urodele larvae of *Triturus* do not possess true teeth while those of the anuran *Bombina* do. He thus transplanted unilateral orthotopic grafts of cranial neural crest of *Bombina* onto *Triturus* at the neurula stage. The chimeric larva possessed teeth in which the dental papilla originated from *Bombina*. Therefore, the anuran neural crest cells were capable of inducing functional enamel organ in the urodele ectoderm which normally does not express this capacity at the larval stage. Reciprocally, the urodele neural crest grafted onto the anuran neurula yielded neural crest cells which normally migrated to the first branchial arch and differentiated into urodele-type visceral skeleton but failed to induce teeth formation. It thus appears that the primary tooth-generating information resides in the neural crest-derived mesenchyme.

A special mention should be made about the origin of mesenchymal cells that differentiate into scleroblasts of the postcranial dermal skeleton in fish and amphibians. Xenoplastic transplantations of trunk neural crest in the amphibian embryo (Raven, 1931, 1936) demonstrated the crest origin of connective cells that participate in the formation of the dorsal fin. This view was shared by Holtfreter (1935) and Detwiler (1937b). Du Shane (1935) showed that bilateral removal of trunk crest prevented the formation of the dorsal fin.

Morphogenesis of the fin is the result of tissue interactions between ectomesenchyme and dorsal ectoderm. The competence of the ectoderm to participate in fin formation is not strictly limited to its mediodorsal area, but extends to the flank ectoderm of the tail bud stage (Twitty and Bodenstein, 1941; Bodenstein, 1952).

Subsequent investigations in *Barbus conchorius* (Cyprinidae) by Lamers *et al.* (1981) showed that trunk crest-derived cells accumulate beneath the epithelium dorsally to the neural keel in association with neuromasts. However, whether these early aggregating cells are fated to become neural or skeletal tissue has not been determined. Smith *et al.* (1994) succeeded in labeling the dorsal aspect of the neural keel by focal DiI injection before trunk crest cell emigration in zebrafish embryos. Labeled cells were found in fin mesenchyme as well as in other well-known neural crest derivatives (i.e., peripheral neurons and melanocytes). These data have provided undisputable evidence according to which the early fin bud mesenchyme contains a contribution from the neural crest.

### 3.1.3  The modern era

Two periods have to be distinguished. The first is characterized by the advent of methods to label neural crest cells reliably *in vivo* in the avian embryo. This allowed the recognition of mesectodermal derivatives and their contribution to head morphogenesis in higher vertebrates. The second period started with the recognition of hindbrain segmentation into rhombomeres. This was at the origin of major conceptual advances in the field of craniofacial development. Thus, the behavior and patterning of neural crest-derived cells could be related to the expression of developmental genes at critical ontogenetic stages. The mode of action of these gene products is now being revealed so that our understanding of the molecular mechanisms underlying head morphogenesis is progressing rapidly.

## 3.2  The fate of the mesectodermal cells in the head

The development of the cephalic neural crest will first be reported, as revealed essentially in the avian embryo through various labeling techniques.

Experimental studies were also carried out in mammalian embryos cultured *in vitro* and on which the migratory crest cells were followed after DiI labeling.

It will become clear in the following report that the investigators aimed to discover first the nature and extent of the various mesectodermal derivatives of the neural crest and the relationships existing between them and the mesenchyme of mesodermal origin. Later, a more refined analysis of the origin of these derivatives on the neural axis was carried out and morphogenesis of the head structures was studied in relation to gene expression in the brain and neural crest derivatives.

### 3.2.1 Labeling the cephalic neural crest cells in the avian embryo

The early steps of cephalic crest cell migration were investigated in the chick embryo by implanting neural folds labeled with [³H]TdR into unlabeled embryos (Johnston, 1966; Noden, 1975), and also by interspecific grafts of the neural primordium between quail and chick embryos (Johnston *et al.*, 1974; Le Lièvre and Le Douarin, 1974, 1975; Le Lièvre, 1974, 1976, 1978; Noden, 1978a).

Due to the dilution of the radioisotope nuclear marker, the experiments using [³H]TdR did not permit a full identification of neural crest cell contribution to the definitive structures of the head and neck. The quail–chick chimeras were useful in this respect and showed that the role of ectomesenchyme in the ontogeny of the facial and hypobranchial structures is quantitatively more significant than previously suspected. In addition, the tissues arising from the cephalic neural crest were found to be very diversified.

The immunocytochemical analysis of neural crest cell migration at the cephalic level with the HNK-1/NC1 Mab clearly revealed that, at the prosencephalic, mesencephalic, and rhombencephalic levels, the neural crest migration is essentially subectodermal. The crest cells cover the cephalic paraxial mesoderm, with only a few of them entering the mesodermal sheet of cells (Fig. 1.1). At the hindbrain level, neural crest cell migration proceeds not as a continuous sheet of cells but rather in distinct strains (Fig. 1.2) corresponding to the branchial arches as previously described in amphibians (Hörstadius and Sellman, 1946).

At the forebrain and midbrain levels, the neural crest cells start migrating a short time before the complete closure of the neural tube. Closure of the tube begins at the junction of the prospective prosencephalon and mesencephalon, and proceeds both rostrally and caudally. Before spreading, the cephalic crest cells lose their epithelial arrangement and form a clearly distinguishable tightly packed mass (Fig. 3.3A). Their total dispersion at the brain level covers a period of about 8–9 hours (from 6ss to 13ss).

They first move in a relatively cell-free space beneath the dorsolateral head ectoderm. Later, they meet cells of the paraxial mesoderm and of the first somites. When midbrain crest cells reach their destination on the ventral side of the head and pharyngeal regions, they are in contact with the

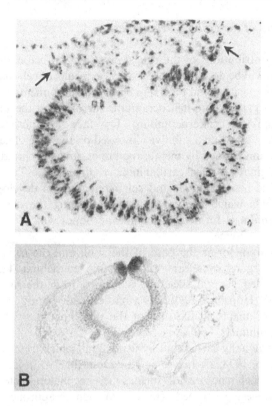

**Figure 3.3** (A) Transverse section in the midbrain of a 7-somite quail embryo. Feulgen–Rossenbeck staining; arrows indicate the neural crest (×425). (B) Transverse section of a 5ss quail embryo after whole-mount *in situ* hybridization with digoxygenin-labeled PDGFRA antisense riboprobe. Premigratory neural crest cells in the dorsal mesencephalic neuroepithelium express PDGFRA transcripts (×110). (From Eichmann *et al.*, unpublished.)

pharyngeal endoderm and the branchial superficial ectoderm. However, instead of proceeding ventrolaterally, some of the cephalic crest cells reaggregate quickly to form the primordia of cranial sensory ganglia (Johnston, 1966; Noden, 1975).

The cells destined to differentiate into ectomesenchyme lose their HNK-1/NC1 immunoreactivity as they reach their target sites in the embryo. This takes place during incubation days 3 and 4 (E3–4) in both chick and quail embryos (Tucker *et al.*, 1984).

The cephalic neural folds strongly expresses the PDGFRα gene even before they join in the midline, as do emigrating neural crest cells (Fig. 3.3B). Whether the PDGFRα-positive cells constitute a subpopulation of the migrating cephalic neural crest has not yet been established.

It is interesting to note that this receptor is not expressed in the truncal neural crest, suggesting that it may be restricted to mesectodermal progenitors.

### 3.2.2 Derivatives of the cephalic neural crest

The respective contributions of the neural crest-derived ectomesenchyme and of the mesoderm will be described here, based on results of cell-marking experiments carried out essentially in the avian embryo. Components of the skull are divided into two groups: the neurocranium, surrounding the brain, and the splanchnocranium (or viscerocranium) for face and branchial arches. Moreover, one can also find a subdivision based on the type of calcified tissue: the dermatocranium formed by membrane bones, and the chondrocranium in which bones originate from a cartilaginous rudiment (see Table 3.2). Thus, bones of both the neurocranium and splanchnocranium develop from cartilaginous rudiments which are considered as forming the chondrocranium. The dermatocranium is formed by membrane bones which ossify in a non-cartilaginous mesenchyme.

The neurocranium forms the base of the skull and the capsules that surround the sense organs: eyes, ears, and olfactory epithelium. The splanchnocranium, also called visceral skeleton, is formed from the branchial arches (Barghusen and Hopson, 1979). Dermatocranial bones enclose neurocranial structures, and form the roof of the skull and most of the bones of the splanchnocranium.

In the work originally carried out by using quail–chick transplantations (Le Lièvre, 1974, 1976, 1978; Le Lièvre and Le Douarin, 1975), large-sized grafts involving entire encephalic vesicles (such as mesencephalon, metencephalon, or myelencephalon) were performed. The goal of such an approach was to determine whether the various skeletal and connective structures of the head and neck were produced by populations of cells of either pure neural crest or mixed neural crest and mesodermal origin. The answer to this question was that the entire facial and hypobranchial skeleton was in fact entirely made up by neural crest cells.

Although the fate map established about 20 years ago from the work of Johnston et al. (1974), Le Lièvre (1974, 1976, 1978), Le Lièvre and Le Douarin (1975) and Noden (1978a) has been significantly refined since, it is still mostly valid (see below the results obtained by Couly et al., 1993, concerning the origin of the frontal and parietal bones which are entirely of neural crest origin, a fact that had not been perceived in the previous studies) and its main features are reproduced here in Table 3.3, where derivatives of the mesencephalic and rhombencephalic neural crest are recorded.

*3.2.2.1 Skeleton.* According to this early work, the mesencephalic and anterior rhombencephalic (also called metencephalic) neural crest cells form the skeleton of the upper and lower jaws, the palate, the tongue (i.e., the entoglossum and basihyal) and, with the posterior rhombencephalic crest, participate in the make-up of the preotic region. The hyoid bone originates from the mesencephalon, metencephalon, and myelencephalon down to the level of somite 4. The neural crest located caudally to the level of somites 4–5 does not yield mesectodermal cells in higher vertebrates (Table 3.3).

Table 3.2. *Constitution of the head skeleton*

| Neurocranium (skeleton of CNS) | | | Splanchnocranium (viscerocranium) (skeleton of the face and branchial arches) | |
|---|---|---|---|---|
| Chondrocranium | | Dermatocranium | Dermatocranium | Chondrocranium |
| *Paraxial mesodermal origin* | | *Neural crest origin* | *Neural crest origin* | *Neural crest origin* |
| *Somitic* | *Cephalic* | | | |
| • Occipital (basi-, exo-)<br>• Otic capsule (pars ampullaris) | • Supraoccipital<br>• Sphenoid (basipost- and pleuro-)<br>• Postorbital (orbital capsule)<br>• Otic capsule (pars ampullaris) | • Frontal<br>• Parietal<br>• Squamosal<br>• Sphenoid (basipre-)<br>• Otic capsule (pars cochlearis, parotic process) | • Nasal, vomer<br>• Maxilla, jugal, quadratojugal<br>• Palatine<br>• Pterygoid<br>• Dentary<br>• Opercular<br>• Angular<br>• Surangular | • Nasal capsule (ect-, mesethmoid, interorbital septum)<br>• Scleral ossicles<br>• Meckel's cartilage<br>• Quadrate<br>• Articular<br>• Hyoid (basihyal, entoglossum, basi-, epi-, ceratobranchial)<br>• Columella |

Table 3.3. *Fate of the mesectodermal cells deriving from various levels of mesencephalic and rhombencephalic neural crest*

| Origin of mesectoderm \ Differentiation of mesectodermal cells | Skeleton | | Loose connective tissue | Dermis | Arterial walls | | |
|---|---|---|---|---|---|---|---|
| | Lower jaw | Hyoid | | | Common carotid | Systemic artery | Pulmonary artery |
| Mesencephalon (mesencephalo-rhombencephalic constriction) | | | | | | | |
| Anterior rhombencephalon | | | | | | | |
| Posterior rhombencephalon (1–9 somite) | | | | | | | |

*3.2.2.2 Dermis, smooth muscles and connective components of striated muscles.* It was also found that the neural crest cells give rise to large areas of dermis. Although Raven (1931, 1936) had already recognized, from xenoplastic grafting experiments, that the neural crest contributes to the dermis in lower vertebrates, nothing was known in this respect in higher vertebrates before cell-marking experiments were performed to follow the long-term fate of neural crest cells. Not only is the connective dermis made up of ectomesenchymal cells but also the smooth arrector muscles associated with feathers, the subcutaneous adipose tissue, and the connective component of the striated muscles of the head (Le Lièvre and Le Douarin, 1975). Neural crest cells also generate smooth muscle cells which participate in the musculoconnective wall of the head blood vessels including that of the aortic arches.

The extent of the dermis of neural crest origin is represented in Fig. 3.4. It extends more dorsally than previously believed since the dermis overlying the frontal and parietal bones is derived from the prosencephalic and anterior mesencephalic neural crest (Le Lièvre and Le Douarin, 1975; Couly et al., 1993).

More recent studies have precisely documented the participation of neural crest cells to the connective tissue of the head. Mesectodermal cells and the presumptive myocytes of mesodermal origin develop in close contact with each other during development of the head musculature. All the connective compo-

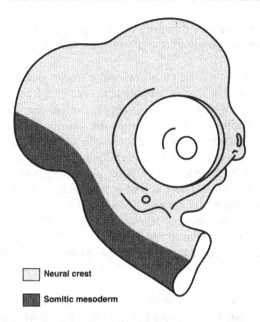

Neural crest

Somitic mesoderm

**Figure 3.4** Schematic representation of a chick head at E6 showing in light grey the extent of neural crest-derived dermis and in dark grey the region where the dermis is of somitic origin. This distribution is based on quail–chick chimera experiments.

nents of the ocular and facial muscles are neural crest-derived (Noden, 1986a; Couly *et al.*, 1992; Köntges and Lumsden, 1996).

Moreover, transplantation of trunk somites of quail in the place of cephalic paraxial mesoderm resulted in the formation of extrinsic ocular and visceral arch muscles with normal fiber aligments and attachments, thus showing that the information responsible for morphological patterning of striated muscles is provided by neural crest-derived connective cells (Noden, 1986a).

A similar analysis using the quail–chick chimera system was carried out for limb muscles. It was also shown that it is the connective tissue-forming mesenchyme and not the myogenic precursors which carry the patterning information for limb muscle morphogenesis – i.e. the segregation of individual muscles from common masses of myoblasts, the alignment of myotubes, and the establishment of definitive attachments (Chevallier *et al.*, 1977; Christ *et al.*, 1977; Chevallier and Kieny, 1982).

*3.2.2.3   Contribution of the cephalic neural crest to the glandular and lymphoid structures of the head and neck.* The mesenchymal components of the glands arising from the pharyngeal and buccal epithelium (pituitary and salivary glands, thyroid and parathyroids) have been found to be of crest origin.

Special attention was directed to the histogenesis of the thymus in the chimeric quail–chick embryos. For a long time the origin of the different cell components of this organ was uncertain and, due to the important role of the thymus in immune function, this problem was investigated within the general framework of primary lymphoid organ ontogeny (see review by Le Douarin *et al.*, 1984b).

When chick embryos receive an isotopic–isochronic graft of quail rhombencephalic primordium, the thymic rudiment is chimeric from its first developmental stages. The host endodermal cords, which are derived from the epithelial thymic rudiments arising from the third and fourth pharyngeal pouches, appear surrounded by a thin capsule of quail mesenchymal cells. Thereafter, the invasion of the thymic epithelium by connective tissue cells of quail origin can be followed. The latter are brought inside the epithelial cords together with the blood vessels whose endothelial cells are always derived from the host mesoderm. In none of the chimeric thymuses observed at the electron microscope level were endothelial cells found with the nuclear marker, which, in contrast, was present in the pericytes and the connective tissue lining the blood vessels. The reticular epithelial cells as well as lymphocytes of both cortex and medulla were found to be of the chick host type (Le Douarin and Jotereau, 1975).

The participation of mesenchymal cells of neural crest origin to the histogenesis of the thymus, thyroid, and parathyroids was confirmed by Bockman and Kirby (1984), who extirpated the neural folds over somites 1–5 bilaterally in stage HH9 and HH10 embryos. This resulted in strong reduction of thymus size. Moreover, heart defects (e.g., persistent truncus arteriosus and transposition of the great vessels), absence or hypoplasia of parathyroids and thyroid glands were associated with thymic defects.

The role of the mesenchyme in the development of mixed epitheliomesenchymal organ rudiments has been amply documented in the past. Concerning the thymus, Auerbach (1960) showed that the thymic epithelium fails to develop if it is separated from the thymic mesenchyme. Extirpation of the neural fold prevents the normal mesenchymal component of the glandular structures of pharyngeal origin reaching their destination. Thus, in the complete absence of mesenchyme, the branchial pouch endoderm which gives rise to the parathyroid glandular cords and to the thymic reticular epithelium does not differentiate. Similarly, the thyroid epithelium does not evolve if the neural crest-derived mesenchymal cells are missing.

The abnormalities resulting from neural crest ablation reproduce the array of defects observed in human Di George syndrome, which is characterized in its extreme form by the absence of thymus, parathyroids, reduced thyroid tissue, craniofacial abnormalities, and cardiac and arterial defects (Di George, 1968).

*3.2.2.4 Contribution of the neural crest to the cardiovascular system.* The capacity of the neural crest to yield mesenchymal cell types in the head (e.g., cartilage, bone, connective and adipose tissues, smooth muscles) which in other parts of the body are of mesodermal origin has raised the question as to whether endothelial cells of the head blood vessels are also of mesectodermal origin.

In fact, grafting of the chick cephalic neural crest into quail embryos generated chick facial mesenchyme which turned out to be vascularized exclusively by blood vessels whose endothelial cells were always of the quail host type. The reverse situation was found when quail neural crest was grafted into chick.

These results show that blood vessel endothelial cells and hemopoietic cells are exclusively derived from mesodermal precursors. Neural crest-derived mesenchymal cells give rise to the pericytes which associate with the capillaries that originate from the vascular buds of mesodermal origin. Moreover, the musculoconnective wall of the larger vessels is mesectodermal in nature in most cephalic structures (except those which are mesodermally derived) (Couly *et al.*, 1995).

As shown by Le Lièvre and Le Douarin (1975), if the whole rhomb-encephalic primordium of a quail is grafted into a chick, the walls of the brachiocephalic trunks and of the common carotid arteries (arising from the third branchial arch), as well as of the systemic aorta (derived from the fourth branchial arch) and the pulmonary arteries (originating in the sixth arch), are derived from the implanted quail cells (Fig. 3.5).

Another contribution of the neural crest to the cardiovascular system is the association of parasympathetic ganglia and nerves with persisting coronary arteries the walls of which are never neural crest-derived (Waldo *et al.*, 1994). At early developmental stages, many coronary arteries form, while most of them later on regress. This suggests that the parasympathetic innervation could be essential for the maintenance of the definitive coronary arteries.

In 1983, Kirby *et al.* showed that substitution of chick by quail neural crest at the level of somites 1–3 (at stage HH9) resulted in migration of quail cells into the aorticopulmonary septum. Removal of the same neural fold segment not followed by quail graft yielded abnormal hearts with common arterial outflow channels or transposition of the great vessels. The origin of the post-ganglionic parasympathetic innervation (i.e., the cardiac ganglia) from the neural crest facing the two first somites was also determined by the same method.

On the basis of ablation studies, the term "cardiac neural crest" was coined by Kirby *et al.* (1985) to designate the region of the neural fold located at the level of somites 1–3 and contributing massively to the cardiovascular system.

In a subsequent study, Kirby and colleagues (Phillips *et al.*, 1987) using quail–chick transplantations showed that, in fact, the neural crest from arch 4 contributes the largest number of cells to the aorticopulmonary and cono-truncal septa of the heart while the contribution from arches 3 and 6 to these structures is less extensive. The conclusion is that the term "cardiac neural crest" should include the level of the otic placode down to the anterior limit of somite 4. It was noticed that a significant mixing of cells originating from the various transverse levels of the neuraxis takes place during the migration process which leads the crest cells to arches 4–6 (Miyagawa-Tomita *et al.*, 1991). No reference to rhombomeres is made in this work where arches 2, 3, 4, and 6

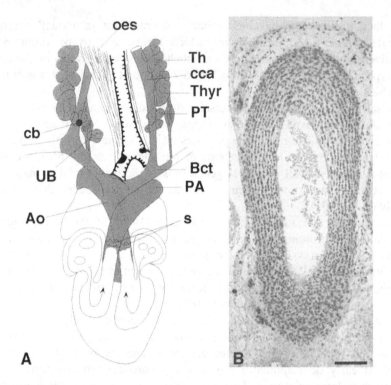

**Figure 3.5** (A) The walls of the large blood vessels derived from the aortic arches. The aortic (Ao) and pulmonary (PA) trunks, the brachiocephalic trunks (Bct), and the common carotid arteries (cca) are of neural crest origin, except for their endothelial layer which is mesodermal. Moreover, the connective components of the thymus (Th), thyroid (Thyr), and parathyroid (PT) glands are also neural crest-derived. The interventricular septum in its more cranial part and the sigmoid valves (s) contain neural crest-derived cells. (B) Transverse section of the aortic trunk stained with the QCPN Mab, showing that its musculoconnective wall is derived from the grafted neural crest corresponding to r8. UB, ultimobranchial body; cb, carotid body; Oes, oesophagus. Bar = 25 μm.

are considered to be seeded by crest from rostral myelencephalon to somite 3 included.

It was recently shown by Couly *et al.* (1998) that in fact neural crest cells from rhombomeres r6 to r8 inclusively (i.e., down to somite 4 included) contribute to the wall of the large arteries deriving from the aortic arches, and to the aorticopulmonary septum of the heart. The use of the QCPN Mab to identify quail cells revealed that the sigmoid valves of the aorta and pulmonary artery contain significant numbers of neural crest-derived cells (Fig. 3.5), which also penetrate into the interventricular septum but are totally absent in the interauricular septum and the mitral and tricuspid valves.

Although neural crest cells are essential for the persistence of the aortic arch arteries once blood flow is established and, when internal blood pressure requires the development of the tunica media, initiation of vessel formation by mesodermally derived endothelial cells takes place in "cardiac neural crest"-deficient embryos (Waldo *et al.*, 1996). Further data obtained by Kirby *et al.* (1993) supported the paramount role of the neural crest to normal cardiovascular development. Destruction of the "cardiac neural crest" by laser beam irradiation was followed by *in situ* injection of neural crest cells removed from the same region and subjected to *in vitro* culture in the presence of LIF (leukemia inhibitory factor). These cells, which in certain experiments were DiI labeled, are able to migrate to the heart and blood vessels, thus increasing the survival rate of the recipient embryos which otherwise frequently die due to the ablation of neural crest cells and subsequent cardiovascular defects. Culture of crest cells in the presence of LIF maintained the cells in an undifferentiated state and apparently increased their rescuing capacity in this experimental design.

*3.2.2.5 Contribution of the neural crest to the ocular and periocular tissues.* Although the embryonic origins of the neural retina, pigmented epithelium, and lens of the vertebrate eye have been known for a long time (see review by Coulombre, 1965), the origins of the other ocular components and of periocular tissues remained controversial until the work of Johnston, Le Douarin, and their colleagues.

Neural crest cells form all the skeletal and connective tissues adjacent to the medial, nasal, and lateral parts of the eye, including the endothelial and stromal cells of the cornea and of the orbit (Le Lièvre, 1976, 1978; Johnston *et al.*, 1979; Couly *et al.*, 1993). The first demonstration of the unique origin of the sclera from the mesencephalic crest is due to Le Lièvre (1976, 1978), who grafted the whole mesencephalic primordium of quail into chick and vice versa, and obtained sclera that was entirely of donor origin. While the myofibers of extrinsic ocular muscles are mesodermal (Couly *et al.*, 1992), the connective cells associated with these muscles are neural crest-derived. The ciliary muscles in contrast are formed by neural crest cells. As for the periocular blood vessels, their endothelium is made up of mesodermal cells, and the perivascular muscles and connective cells are ectomesenchymal in origin (Le Lièvre, 1976, 1978; Johnston *et al.*, 1979).

Ontogeny of the avian cornea, which has been extensively studied by Hay and her coworkers, provides an interesting model for following the migration of the neural crest cells, since ectomesenchyme plays a significant part in cornea formation (see Fig. 3.6).

The early chick cornea is composed of an acellular collagenous stroma lined with an anterior epithelium and a posterior "endothelium." Following lens formation, the superficial ectoderm starts to produce the primary stroma (Fig. 3.6) (Hay and Revel, 1969; Dodson and Hay, 1971). The only cells which enter the cornea during this period are macrophages.

From stage HH22, cells originating from the perioptic vesicle mesenchyme follow the macrophages and migrate along the posterior surface of the stroma using both the lens basement lamina and the stroma itself as substrata for their

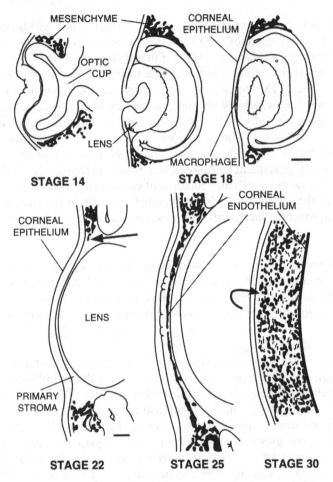

**Figure 3.6** Camera lucida drawings showing the early stages in development of the avian cornea. At the time that the lens placode begins to invaginate (stage HH14), the presumptive corneal epithelium lies over the lip of the optic cup. At 3 days of incubation, the lens vesicle pinches off and the overlying ectoderm becomes the corneal epithelium (HH18). Macrophages clean up debris associated with lens vesicle formation. By stage HH22 (4 days) the corneal epithelium has secreted the primary corneal stroma and the mesenchymal cells destined to become the corneal endothelium have started to invade the area (straight arrow). Endothelial cell migration is almost complete at stage HH25 ($4\frac{1}{2}$–5 days). During stage HH27 ($5\frac{1}{2}$–6 days) junctions between the endothelial cells are established and the primary stroma swells (not shown). It is then immediately invaded by the corneal fibroblasts. By stage HH30 ($6\frac{1}{2}$–7 days) the fibroblasts occupy all layers of the stroma except for a narrow juxtaepithelial zone (curved arrow). Bar = 50 μm (From Hay, 1980; courtesy of Dr E. Hay.)

migration (Bard *et al.*, 1975). They give rise to the posterior "endothelium" and stop entering the cornea when a continuous endothelium is formed. The endothelium so formed delimits the space that will become the anterior eye chamber while the more superficial part of the primary stroma remains acellular and is later converted to Bowman's capsule. Between stages HH27 and 28 the migration of perioptic mesenchymal cells into the cornea starts again and gives rise to the stromal fibroblasts.

Transplantation of quail mesencephalic primordia or midbrain neural folds has shown that the corneal endothelium (Johnston, 1974; Le Lièvre, 1976; Noden, 1976) as well as the stromal fibroblasts are derived from the neurectoderm.

*3.2.2.6  Formation of the meninges.* It was thought by the early investigators that all the structures of the nervous system including its developing membranes arose from the neural primordium. However, Kölliker (1880) and His (1903) noticed that the anlage of the meninges was made up primarily of a mesenchymal tissue and therefore concluded that, similar to the other mesenchymes of the body, the primary meninges are derived from the mesoderm.

In 1922, Oberling advanced the hypothesis, based on histopathological studies, that the pia–arachnoid membrane is a structure developmentally analogous to the Schwann cells of the peripheral nerves and therefore is derived from the neural crest. This hypothesis received experimental confirmation by the work of Harvey and Burr (1926) who carried out transplantations of pieces of neural primordium in *Amblystoma* with or without the neural crest. They demonstrated that certain ectodermal elements, derived mostly from the neural crest, take part in the formation of the leptomeninx. Later, Harvey *et al.* (1933) confirmed this view on both *Rana sphenocephala* and the chick embryo.

In fact, the work of Le Lièvre (1976) and of Couly and Le Douarin (1987) using the quail–chick transplantation technique showed that at the level of the prosencephalon the pachymeninx and the leptomeninx originate from the mesencephalic neural crest. Curiously, the mesencephalic crest cells, which migrate anteriorly and cover the forebrain, do not yield the midbrain meninges. The blood vessels irrigating the meninges were always lined by endothelial cells of host origin which invade the central nervous system (CNS) and the meninges.

The meninges of the midbrain itself and of the rest of the CNS were always found to be of host origin at whatever level the quail neural crest was implanted. They can therefore be considered as derived from the mesodermal germ layer.

*3.2.3  Analysis of the cephalic crest cell migration and fate in mammalian embryos*

*3.2.3.1  Cephalic cell migration studied in mouse and rat embryos.* Significant progress has been made in the last 10 years in this field due to the improvement of culture methods, enabling the survival of mouse or rat embryos in culture, and to the use of vital dyes such as DiI. Although the long-term fate of the labeled crest could not be studied in this type of experiment, it could be inferred from the results obtained in birds by using the quail–chick chimera system.

In the rat and mouse embryo, the cranial neural crest cells emigrate from the ridge of the neural plate from the head-fold stage, hence long before closure of the neural tube (Tan and Morriss-Kay, 1986; Nichols, 1981, 1986). These crest cells contribute to the formation of the frontonasal mass, branchial arches, and cranial ganglia in a way that does not significantly differ from what was described in the avian embryo (Tan and Morriss-Kay, 1986; Fukiishi and Morriss-Kay, 1992; Serbedzija et al., 1992). As far as participation of neural crest cells to the frontonasal mass is concerned, Matsuo et al. (1993), using cultured rat embryos, found that both forebrain and midbrain crest cells populate the anterior part of the developing head. Serbedzija et al. (1992) reported a similar behavior by labeling the neural fold by DiI in cultured mouse embryos. Osumi-Yamashita et al. (1994), also using the latter technical approach, confirmed Couly and Le Douarin's (1985, 1987) statement that the anterior neural ridge corresponding to the most rostral region of the prosencephalon does not contribute to the neural crest population but yields the "neural head epithelium," including the nasal placodes, Rathke's pouch, and the oral and nasal epithelium. The full complement of ectomesenchymal cells has reached the mandibular arch by E9 (Lumsden, 1984; Lumsden and Buchanan, 1986). Figure 3.7 shows the fate map and migratory pathways of mouse and rat neural crest cells (see Osumi-Yamashita et al., 1997a, and references therein).

Migration of midbrain neural crest cells was found to be particularly impaired in homozygous rSey embryos in which the Pax6 gene, involved in eye development (Halder et al., 1995), is mutated. DiI labeling showed that crest cells of rSey $^{-/-}$ embryos, emigrating from the midbrain, aggregate at the vicinity of the ophthalmic placode but do not subsequently enter the frontonasal region. Normal midbrain neural crest cells injected orthotopically into the mutant embryo exhibited the same abnormal migration, thus showing that the migration pathway, rather than the crest cells themselves, is responsible for the observed phenotype.

*3.2.3.2 Cranial neural crest and tooth development.* Development of an individual tooth is characterized by a series of reciprocal interactions between the epithelial and the mesenchymal components of the tooth rudiment. Several reviews (e.g., Ruch, 1984; Lumsden, 1987; Thesleff et al., 1995, 1996; Thesleff and Nieminen, 1996) have emphasized the notion that tooth development provides an interesting model system to study tissue interactions during organogenesis.

Participation of mesectoderm to tooth development, already demonstrated in lower vertebrates, was fully confirmed for mammals. Migration studies in rat embryos, cultured for 30–60 hours, allowed the cells of the mandibular tooth-forming region to be traced back to the posterior midbrain neural fold (Imai et al., 1996). By explanting the mandible of these embryos, in organ culture, labeled neural crest-derived cells were seen participating in the anlage of molars that developed in vitro, up to the bud stage.

The question as to whether the determination to form tooth is initially established in the mesenchyme or the ectoderm has not been satisfactorily resolved. In the case of other ectomesodermal appendages, the mesenchyme determines the position of structures whereas the epithelium responds according to its intrinsic specificities (Sengel, 1976).

The fact that tooth patterning is first initiated by the ectoderm has been proposed by Lumsden (see Lumsden, 1988, for a review; see also Mina and Kollar, 1987). Lumsden has explanted the caudal mesencephalic and rostral metencephalic neural folds from 6ss to 12ss mouse embryos and combined them with various regions of embryonic surface ectoderm. The associated tissues were then cultured in the anterior eye chamber of isogenic mice (Lumsden, 1984). Cartilage and bone developed in all types of explants while teeth formed in combinations of cranial neural crest with mandibular arch epithelium but not with limb bud epithelium. Interestingly, teeth formed also in explants where mandibular arch epithelium was associated with trunk neural crest. Therefore, mammalian neural crest has an odontogenic potential that is not restricted to the crest of presumptive tooth-forming region and not even to the cephalic neural crest. It seems therefore that, during normal development, the neural crest cells receive a signal from the oral epithelium which would be the initial promoter of tooth patterning. However, the experiments of Wagner (1949, 1955) described earlier in this chapter (see Section 3.1.2) provided convincing evidence for a primary role of the mesenchyme in the initiation of tooth development. Therefore, in the mouse experiments just described, the neural crest cells, which had already reached the branchial arches at the time of the experiments (E9), might have induced the epithelium to acquire specific tooth-forming capacities before the dissection of the tissues.

Recent advances have been made on the molecular control of tooth patterning. Two different hypotheses have been proposed to account for the arrangement and morphology of teeth in mammals. The field theory assumes that different concentrations of morphogens regulate the differential morphogenesis of initially identical primordia (Butler, 1939). The other hypothesis proposes that the stem cells giving rise to different classes of teeth differ from each other initially (Osborn, 1978). In such a model, designated the clonal model, the neural crest cells have already acquired positional identity at the time they reach their final destination in the jaws. A modern molecular version of the clonal model was proposed by Sharpe (1995) according to which shape and position are specified by the combined action of different homeobox genes expressed in neural crest-derived facial mesenchyme.

As will be discussed below, several transcription factors of the homeobox-containing gene family are expressed in the neural crest cells populating the maxillary and mandibular arches. These genes are respectively expressed in spatially restricted domains prior to overt tooth development. Thus an "odontogenic homeobox code" would be generated that would be critical in patterning tooth development (see discussion below for *Dlx* genes).

Another important notion that emerged from the screening of the molecular basis of tissue interactions in teeth development is that the signaling networks

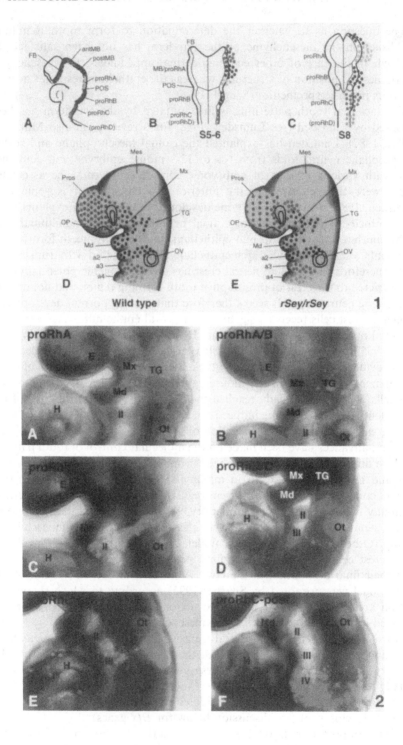

Wild type      rSey/rSey    1

used in this system are similar to those involved in other organs such as the limb and even in some aspects of fly development. Secreted molecules endowed of morphogenetic properties of the Shh, FGF, BMP, and Wnt families and transcription factors of the Msx family have been shown to be involved in the regulation of tooth development. A large amount of work has been produced concerning the epitheliomesenchymal relationships involved in tooth develop-ment but is beyond the scope of this book. The reader is referred to recent reviews devoted to odontogenesis edited by Ruch (1995) and to a review by Thesleff and Sharpe (1997).

**Figure 3.7** Panel 1: Diagrams showing migration patterns of cranial crest cells in mouse and rat embryos. (A, B) Lateral and dorsal views at 5–6ss. (C) Dorsal view at 8ss. (D, E) Lateral view of wild-type and homozygous *rSey* embryos, respectively, at the pharyngula stage. (A) At the time of mammalian cranial crest cell emigration, four morphological units are present in the rostral neural plate, from anterior to posterior: forebrain (FB); midbrain (MB, anterior MB, posterior MB) + presumptive prorhombomere A (proRhA, rostral hindbrain); proRhB (preotic hindbrain); and proRhC + presumptive proRhD (caudal hindbrain); preotic sulcus (POS) is an obvious landmark in the hindbrain. Regions in which DiI labeling of crest cells was performed are shown by different colors. (B) Unlike crest cells in other animals and trunk crest cells, cranial crest cells in mammals emigrate from the neuroepithelium before its closure. At 5–6ss, crest cells begin emigrating from the forebrain and midbrain/proRhA region. (C) At 8ss, zones free of crest cells exist at the boundaries between proRhA/B (preotic sulcus) and proRhB/C, thereby making three streams in the hindbrain region. The forebrain cannot be seen in this view. (D) Normal embryo at the developmental stage in which migration of cranial crest cells is nearly complete. The most anteriorly situated facial primordium is the frontonasal prominence underlying the olfactory placode (OP), to which crest cells from both the forebrain and midbrain migrate. Caudally, the first pharyngeal (branchial) arch appears, later developing into the maxillary (Mx) and mandibular (Md) prominences which are, respectively, the primordia of upper and lower jaws. Situated further caudally are the second, third, and fourth pharyngeal arches (a2, a3, and a4). Crest cells derived from the posterior midbrain and proRhA migrate to the first arch, those from the proRhB to the second arch, and those from proRhC and proRhD to the third and fourth arches, respectively. (E) In homozygous *rSey* embryos, migration of midbrain crest cells into the frontonasal region is specifically impaired, though crest cells from other regions migrate normally. OV, otic vesicle; TG, trigeminal ganglion. (Reproduced, with permission, from Osumi-Yamashita *et al.*, 1997.) Panel 2: Segmental distribution of neural crest cells labeled at proRhA (A), proRhB (C), anterior region of proRhC (E), and posterior region of proRhC (F), as well as at the boundaries between proRhA and proRhB (B), and between proRhB and proRhC (D). Synthesized images of bright-field and corresponding dark-field images of lateral views of whole-mount embryos. Mx, maxillary prominence; TG, trigeminal ganglion; Ot, otic vesicle; II, second pharyngeal arch; III, third pharyngeal arch; IV, fourth pharyngeal arch; H, heart primordium. Bar = 200 μm. (Reproduced, with permission, from Osumi-Yamashita *et al.*, 1996.)

*For a colored version of this figure, see www.cambridge.org/9780521122252.*

### 3.2.4 The fate of the anterior cephalic neural fold and neural plate corresponding to the prosencephalon studied in the avian embryo

The early work on the amphibian embryo established that the forehead neural ridge does not contribute to head cartilage and that bilateral excision of the neural folds at this level results in the absence of telencephalon and of the anterior part of the diencephalon (Hörstadius, 1950). This was interpreted as if most of the anterior neurectodermal ridge becomes incorporated into the brain. In amniotes, neither the fate of the anterior neural fold nor the rostral limit of the neural crest were known, when Le Douarin and coworkers (Couly and Le Douarin, 1985, 1987, 1988; Couly *et al.*, 1993) decided to apply the quail–chick marker system to this problem.

*3.2.4.1 The anterior cephalic neural fold.* For the chick neural folds to be accessible for excision and substitution by their quail counterpart, the operation had to be carried out prior to neural tube closure, i.e., at the neurula stage (from 0 somite, about 1 hour prior to the appearance of the 1–5ss). Fragments of the neural fold, about 150 μm in length and 40 μm in depth, were dissected from quail and chick and the quail fragments were grafted orthotopically into stage-matched chick embryos. A similar paradigm was also applied to the neural plate itself. The surgery did not preclude normal brain development provided that the graft was properly incorporated into the host neural epithelium (Fig. 1.7).

The resulting fate map of the forebrain represented in Fig. 3.8 shows that the anterior neural ridge corresponds, in its medial area, to the presumptive territory of the adenohypophysis which is contiguous to the area of the neural plate yielding the hypothalamus. Thus, the adenohypophyseal–hypothalamic region is located mediorostrally in the neural primordium and just anteriorly to the presumptive neurohypophysis which is itself flanked by the area of the future optic vesicles (Couly and Le Douarin, 1987). The territories from which the telencephalon arises are bilaterally and rostrally located with respect to the above-mentioned structures. They are lined externally by that part of the fold from which develop (1) the epithelium of the olfactory cavities including the sensory olfactory placodes (zone A of Fig. 3.8), and (2) the vestibular epithelium of the nasal cavity, the epidermis of the nasofrontal area, and the beak (including egg tooth) (zone B of Fig. 3.8). Caudally, are located the anlagen of the thalamus (in the neural plate) and epiphysis (laterally), while the corresponding neural fold yields the epidermis covering the forebrain (zone C of Fig. 3.8).

*3.2.4.2 Relationships between the placodal ectoderm and the neural fold.* Interestingly, these quail to chick substitution experiments (Couly and Le Douarin, 1985, 1987, 1988) revealed that the placodal territories of the hypophysis and olfactory epithelium are in continuity with those of the floor of the third ventricle yielding the hypothalamus and with the olfactory bulb, respectively. This early spatial juxtaposition of precursor cells which become parts of integrated functional units (i.e., hypothalamohypophyseal complex, and the peripheral and central olfactory structures) results therefore from the development of a few common neuroepithelial progenitors. One can thus assume that these cells are restricted early in their potentialities to yield related cell types, further cell specifications take place when central (hypothalamus and olfactory bulb) and peripheral (adenohypophysis and olfactory epithelium) subunits disjoin due to morphogenetic movements affecting the corresponding head regions.

In amphibians, several workers have reported that the placodal ectoderm, yielding the anterior cephalic sensory ganglia, originates from a region designated the primitive placodal thickening (Platt, 1896; Brachet, 1907; Knouff, 1927, 1935), located outside the neural fold itself. The future olfactory neurons, in particular, are localized in a region situated on the outer aspect of the primitive neural fold, in apposition to it and close to the anlage of the pituitary (Rohlich, 1929; Carpenter, 1937; Jacobson, 1959; Van Oostrom and Verwoerd, 1972; Klein and Graziadei, 1983). From the study of Couly and Le Douarin (1987, 1988) it is clear that, in the avian embryo, these territories are part of the neural fold itself and are, as previously mentioned, in close spatial association with the prospective region of the CNS to which their derivatives will become functionally connected.

Similar observations were done for the anlage of the trigeminal placode which, in the avian embryo, is located in the neural fold itself at the neurula stage. This was shown in experiments where either the neural fold alone, or a strip of ectoderm excluding the neural fold, were grafted from quail to chick at 3ss at the level of the presumptive mesencephalon (Couly and Le Douarin, 1990). Only in the type of experiments involving the neural fold was the trigeminal ganglion entirely of quail origin. The quail cell types included not only the glial cells and the proximal substance P-containing neurons which are of neural crest origin (Ayer-Le Lièvre and Le Douarin, 1982; D'Amico-Martel and Noden, 1983) but also the large distal neurons which are derived from the placodal ectoderm. In fact, the neural fold of the mesencephalic region contains at the early neurula stage both the neural crest and placodal components of the trigeminal ganglion. Like the olfactory placode, the trigeminal placodal territory migrates distally from the neural primordium during head morphogenesis. It becomes located laterally to the neural tube as shown by D'Amico-Martel and Noden (1983) and later on releases cells which secondarily meet those of neural crest origin arising from the mesencephalon.

These experiments thus revealed that the anterior neural fold has a larger range of developmental potentialities than previously considered. Those include glandular tissues (the adenohypophysis), sensory epithelium (olfactory), placodal ectoderm destined to yield sensory ganglion neurons (trigeminal placode), as well as the superficial ectoderm of the mouth roof and of the face.

*3.2.4.3 The rostral limit of the neural crest and the developmental fate of the diencephalic and mesencephalic neural crest.* The experiments described above showed the rostral limit from which the neural fold cells undergo

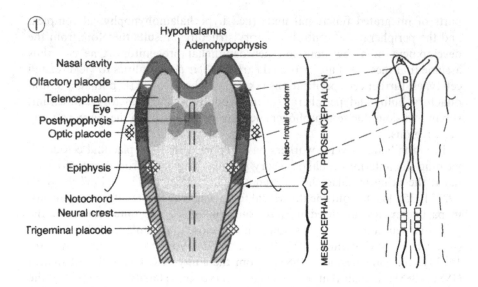

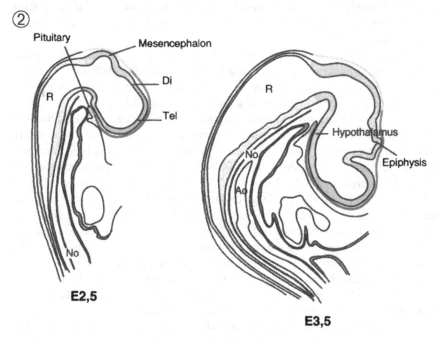

**Figure 3.8** Fate map of the neural plate in an avian embryo at 3ss. (1) The neural plate is represented flat with its limiting anterior and lateral neural folds. This map was deduced from substitution experiments of fragments of the neural fold as drawn in the embryo represented on the right. Other experiments involved grafting of the anterior neural fold which yields Rathke's pouch and of definite regions of the neural plate (see Couly and Le Douarin, 1987, for details). (2) Diagram showing the evolution of the territories represented on the fate map in (1). It appears that the two territories corresponding to the telencephalon join on the dorsal midline and extend more rostrally

the epitheliomesenchymal transition which generates the flux of emigrating crest cells. This limit corresponds to the mid-diencephalon, roughly at the level of the epiphysis primordium: the isotopic graft into a chick embryo of a quail neural fold segment located between 300 and 450 μm from the rostral end of the embryo at 3ss yielded the epiphysis and corresponds to the rostral limit of the neural crest (Couly and Le Douarin, 1987). A similar fragment covering the territory of the diencephalon and anterior half of the mesencephalon (Fig. 3.9) gave rise to the bones of the max-illary, nasal and orbital regions – i.e., premaxillary, maxillary, nasal, and vomer (all membrane bones), interorbital septum (still cartilaginous at E14, the time of chimera analysis), sclerotic, and supraorbital bone. In addition, the frontal bone in its totality, together with the anterior region of the squamosal, the basipresphenoid, the rostrum of the parasphenoid, and the pterygoid, were of graft origin. The sutural spaces between the membra-nous bones (the two frontal bones particularly) were also made up of quail connective tissue. Moreover, these experiments revealed that the caudal boundary of the participation of the neural crest to the floor of the skull lies within the sella turcica and corresponds to the anterior tip of the notochord (Fig. 3.9).

In addition to bones, this segment of neural crest yields the dermis of the scalp and, as mentioned before, the meninges of the forebrain.

Similar experiments involving more posterior mesencephalic fragments of the neural fold showed that the parietal bones and dermis at the same level, the squamosal, the jugal, the quadrate, the pterygoid, the palatine, and most of the skeleton of the lower jaw (Meckel's cartilage, dentary, angular, opercular) (see also Johnston *et al.*, 1974; Le Lièvre and Le Douarin, 1975) were of graft origin. Further, more recent experiments to be described below (see Section 3.5.2) will show the exact contribution of the posterior half of the mesen-cephalic neural crest to the lower jaw.

## 3.3 Relationships between the cephalic paraxial mesoderm and the neural crest-derived mesectoderm

### 3.3.1 The contribution of the paraxial cephalic mesoderm to the skull

In order to delineate the respective contribution of the neural crest and of the paraxial mesoderm to the skull, the quail–chick chimera system was used to

**Figure 3.8** (*cont.*)
than the level of the hypothalamus and of Rathke's pouch (which yields the adenohypophysis). That part of the brain which develops rostrally to the tip of the notochord (No) is later covered by neural crest-derived bones. Ao, dorsal aorta; R, rhombencephalon; Di, diencephalon; Tel, telencephalon.

*For a colored version of this figure, see www.cambridge.org/9780521122252.*

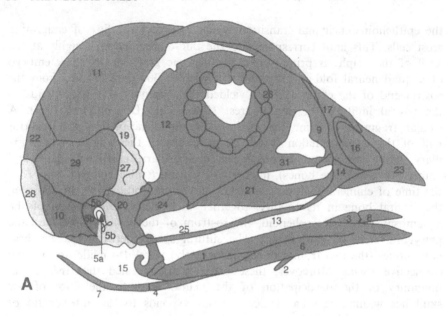

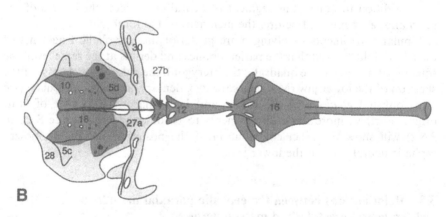

**Figure 3.9** Contribution of the neural crest (red), the cephalic mesenchyme (blue), and the somitic mesoderm (green) to the vertebrate cranium. (A) Lateral view of the cephalic skeleton of a 14-day-old avian embryo. (B) Basal view of the chondrocranium of a 10-day-old embryo. 1, angular; 2, basibranchial; 3, basihyal; 4, ceratobranchial; 5a, columella; 5b, otic capsule; 5c, otic capsule (pars ampullaris); 5d, otic capsule (pars cochlearis); 6, dentary; 7, epibranchial; 8, entoglossum; 9, ethmoid; 10, exoccipital; 11, frontal; 12, interorbital septum; 13, jugal; 14, maxilla; 15, Meckel's cartilage; 16, nasal capsule; 17, nasal; 18, basioccipital; 19, postoccipital; 20, quadrate; 21, palatine; 22, parietal; 23, premaxilla; 24, pterygoid; 25, quadratojugal; 26, scleral ossicles; 27a, basipostsphenoid; 27b, basipresphenoid; 28, supraoccipital; 29, squamosal; 30, orbital capsule; 31, vomer. (Reproduced, with permission of Company of Biologists Ltd, from Couly *et al.*, 1993.)

*For a colored version of this figure, see www.cambridge.org/9780521122252.*

map the cephalic mesoderm in 3ss embryos. At a slightly later stage the fate of the first six somites was also revealed by the same substitution method.

The principle of the experiments used for the cephalic mesoderm consisted of dividing the sheet of mesodermal cells lying on each side of the folding neural plate into medial and lateral parts corresponding respectively to the regions lying on the cephalic vesicles and covering the foregut. At 3ss, the cell density is higher in the lateral than medial moiety and their fate is significantly different (see Fig. 3.10).

The osteogenic potencies were found to be restricted to the medial mesoderm (Fig. 3.11), while most of the cephalic mesoderm yields muscles, the connective component of which is of neural crest origin (Couly *et al.*, 1992; see also Ayer-Le Lièvre and Le Douarin, 1982; Noden, 1982, 1983b, 1986a). The cephalic mesoderm also contributes to the meninges of the midbrain and hindbrain regions. The bones labeled by quail cells in the experiments involving the graft of the medial paraxial mesoderm were the corpus sphenoidalis, the orbitosphenoid, and the otic capsules. No cells of mesodermal origin were ever found in the frontal and parietal bones.

Neither dorsally nor ventrally was the dermis labeled in any of the experiments carried out in this study. The dermis in fact was found to originate entirely from the neural crest. These results differ from those reported by Noden who claimed that somitomeres yield dermis (Noden, 1983a). This discrepancy may have arisen from the fact that this author operated on embryos at later stages (up to 11 somites; i.e., stages HH9 and HH10; see Noden, 1983a), when a few neural crest cells are already adherent to the superficial ectoderm that was involved in these operations. In fact, neural crest cells start to leave the neural fold at the mesencephalic level as early as the 6–7ss (Cochard and Coltey, 1983). Already at 8–9ss, neural crest-derived cells entirely cover the mesoderm dorsally.

### 3.3.2 Analogies and differences between cephalic and truncal paraxial mesoderm

One can conclude that the cephalic mesoderm (considered by Meier (1979, 1981) to form pseudosegments called somitomeres – see discussion in Couly *et al.*, 1992, on this point) has the same developmental potentialities as its truncal equivalent, the somites: they yield skeleton, muscle, and also vascular endothelial cells (see Couly *et al.*, 1995; Christ *et al.*, 1990, 1991). One major difference, however, is that the paraxial mesoderm lining the cephalic vesicles does not yield dermis. The equivalent of the somitic dermatome is therefore absent in the cephalic paraxial mesoderm and the function of extending underneath the ectoderm to form the dermis is fulfilled instead by the neural crest.

Interestingly, the ectodermally derived cephalic dermis has the capability of developing calcified structures (i.e., membrane bones that constitute most of the skull), a property extremely reduced in the mesodermally derived dermis of higher vertebrates.

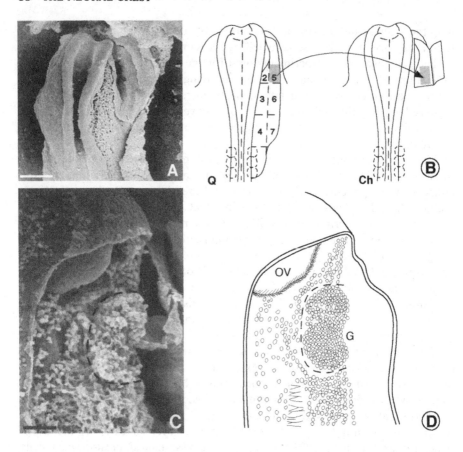

**Figure 3.10** (A) SEM view of a 3-somite quail embryo in which the superficial ectoderm has been removed on the right side. The paraxial mesoderm is visible. Bar = 100 μm. (B) Schematic drawing of the graft of the rostral part of the paraxial cephalic mesoderm from a quail (Q) to a chick (Ch) embryo. The seven types of transplants exchanged between quail and chick embryos are indicated from 1 to 7. (C) SEM view of the graft 5 hours after implantation, with the corresponding drawing in (D). G, graft; OV, optic vesicle. (Reproduced, with permission of Company of Biologists Ltd, in modified form, from Couly *et al.*, 1992.)

The origin of the occipital bone was further documented by grafting individually the first six somites from quail to chick at stages ranging from three somites (for the first three somites) up to six somites (for the next ones). A detailed analysis of the fate of grafted somites can be found in Couly *et al.* (1993). In summary, the first five somites contribute to the occipital bone: the first somite yields the exooccipital, and the second, third, fourth and rostral part of the fifth yield the basioccipital, and condyles. As mentioned before, the paraxial mesoderm forms the supraoccipital. Thus the basioccipital develops from the concentration around the notochord of sclerotomal cells belonging to somites 2–5, and a very small part of somite 1. The complex morphogenesis

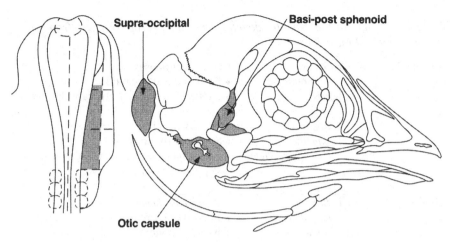

**Figure 3.11** The cephalic paraxial mesoderm of 3–5ss chick embryos is exposed by peeling off the superficial ectoderm, and is divided into seven areas which are individually substituted by their quail counterpart (see Fig. 3.10). The osteogenic potencies of the mesoderm are restricted to the medial mesodermal region which yields the basipostsphenoid, most of the otic capsule and the squamosal bone. Except for the occipital, which is of somitic origin, the rest of the skull and the visceral skeleton are derived from the neural crest.

yielding the occipital bone, which is a highly composite skeletal structure, is largely controlled by *Hox* genes, as evidenced by the fact that targeted mutations of certain *Hox* genes result in abnormalities in the basioccipital bone (Lufkin *et al.*, 1992).

In the occipital region, the dermis is derived from the dermomyotome of the first five somites (Fig. 3.4).

Tongue and cervical muscles are found to be labeled by quail cells following grafts of these anterior somites. Tongue muscles are innervated by the hypoglossal nerves originating from the posterior rhombencephalon corresponding to the level of the first five somites (see Noden, 1983a,b; Lumsden and Keynes, 1989; Keynes and Lumsden, 1990; Couly *et al.*, 1993).

It is important to underline that the analysis carried out in the 1980s and early 1990s by Couly and colleagues have mostly confirmed and refined the results obtained earlier and amply described in the first edition of this book concerning the developmental potentialities and fate of neural crest cells. However, one important discrepancy has emerged with respect to the origin of the skull vault. It had been previously claimed that the frontal bone was of mixed neural crest–mesodermal origin and that the parietal bones were mesodermally derived. This discrepancy is essentially attributable to differences in the stage at which the experiments were performed. The experiments reported by Couly *et al.* (1993) have all been performed at 3ss – i.e., earlier than those of the previous authors (e.g., Le Lièvre, 1978). At 3ss, the head neural fold can be removed and replaced selectively, prior to emigration of neural crest cells. Similarly, the paraxial cephalic mesoderm, which is accessible to surgical

Table 3.4. *Origin of the cephalic skeleton from the neural crest and from the cephalic paraxial mesoderm*

| "Chordal" skeleton | Bones of somitic origin | Cartilaginous bones<br>• Basi- and exo-occipital<br>• Pars canalicularis of otic capsule (partly) |
|---|---|---|
| | Bones of cephalic mesoderm origin | • Supraocciptal<br>• Sphenoid (basipost, orbito-)<br>• Pars canalicularis and cochlearis of otic capsule (partly) |
| "Prechordal" skeleton | Bones of neural crest origin | Cartilaginous bones<br>• Interorbital septum<br>• Basipresphenoid<br>• Sclerotic ossicles<br>• Ethmoid, pterygoid<br>• Meckel's cartilage<br>• Quadratoarticular, hyoid<br>• Pars cochlearis of otic capsule (partly) |

Membranous bones

| – of skull | – of face |
|---|---|
| • Frontal | • Nasal |
| • Parietal | • Maxillar |
| • Squamosal | • Vomer |
| • Columella | • Palatine |
| | • Quadratojugal |
| | • Mandibular |

manipulation at that stage, is not possibly contaminated by mesenchyme of neural crest origin. Moreover, analysis of the chimeras was systematically performed at a developmental stage (E8–14) when the different components of the skull and associated dermis can be clearly identified.

### 3.3.3 The notion of prechordal and chordal skull

As mentioned before, the rostral limit of the mesodermal contribution to the basis of the skull corresponds to the extreme tip of the notochord and lies in the sella turcica, composed of a basipresphenoid of neural crest origin and a basi-postsphenoid which is mesodermal in nature. This leads to the distinction of a "chordal" skull derived from the mesoderm and of a "prechordal" (or "achordal") skull derived from the neural crest, the respective bones of which are listed in Table 3.4.

According to Gans and Northcutt (1983), one can consider that vertebrates have evolved from the basic body plan of the cordates by addition, rostrally to the notochord, of a "new head." The new head developed from the ectoderm, as vertebrates became predators, and involves both sense organs and nervous structures in which the information provided by sense organs is processed.

The results obtained in quail–chick chimeras support this view by showing that the entire achordal skull, including the vault formed by the frontal and parietal bones and the anterior half of the corpus sphenoidalis, is derived from the neural crest. Equally, the jaws, like all the facial skeleton (and odontoblasts of teeth in mammals) which develop ventrally and far from the notochord, are neural crest-derived.

This can be related to the fact that the differentiation of somitic mesenchyme into cartilage is strictly dependent upon its interaction with the notochord and floor plate (Pourquié et al., 1993). Extirpation of both notochord and floor plate in the chick embryo results in the absence of axial skeleton. The effect of floor plate and notochord on the development of the paraxial mesoderm was recently shown to be mediated by the product of the gene Sonic hedgehog (Shh), expressed by both these structures at the critical stages of somitic differentiation (Johnson et al., 1994; Teillet et al., 1998). One can therefore understand that the rostral limit of the notochord and floor plate, lying just at the level of the hypothalamus–adenohypophysis territories from the neurula stage onward, coincides with the limit of that part of the skull which is mesodermally derived.

During the course of evolution, the telencephalon which forms essentially from the lateral regions of the neural plate has been the site of considerable growth compared to the rest of the brain. The cerebral hemispheres have developed in front of the level of the Rathke's pouch to be finally located rostrally and above the diencephalon (Fig. 3.8). Protection to this new part of the brain has been insured by the neural crest cells of the di- and mesencephalic neural fold which have spread over the forebrain and midbrain to construct the skull vault, the optic and nasal capsules, and the facial skeleton which has provided vertebrates with an organ of predation, the jaws.

## 3.4 Segmental nature of the brain primordium: molecular implications for neural crest development

Whether the high level of complexity and diversity which characterizes the vertebrate nervous system is originally established within repeating cell groups, or segments, distributed along the anteroposterior axis, has long been a subject of debate. While segmentation is a developmental strategy encountered in many invertebrates (see e.g., Lawrence, 1981; Weisblat and Shankland, 1985; Akam, 1987; Ingham, 1988, for discussions on this question), its contribution to vertebrate development is less clear and for many years was recognized only for the somitic and nephrogenic mesoderm.

However, the fact that a pattern of repetitive bulges, the neuromeres, can be seen, albeit transiently, in certain regions of the neural tube has long been reported (Von Baer, 1828; Orr, 1887; Neal, 1918). Neuromeres are particularly conspicuous in the hindbrain where they have been designated as rhombomeres (r) and are found in constant number in all vertebrates (r1 to r8) (see reviews of Gräper, 1913, and Vaage, 1969). The developmental significance attributed to rhombomeres gave rise to two opposing views: the first considered them as artifacts resulting from the action of fixing agents or as transient embryonic structures due either to longitudinal compression of the neural tube, to the local strain exerted by nerves, or to more rapid growth than the surrounding tissues. The second view was "the phylogenetic interpretation that neuromeres are visible remnants of a primitive segmentation of the nervous system and consequently reliable cues to the original number of metameres in the vertebrate head" (Neal, 1918). While some authors suggested that a relationship might exist between the development of the cranial nerve nuclei and the initial segmental pattern of rhombomeres (Streeter, 1908; Meek, 1910; Gräper, 1913), others thought that they were unrelated processes (Neal, 1918). Although segmentation of the brain anlage has been evidenced in the hindbrain through modern approaches, molecular and anatomical cues point to the presence of neuromeres in the forebrain and midbrain as well (see Rubenstein *et al.*, 1994, and Shimamura *et al.*, 1997).

As will appear later in this chapter, this question of brain segmentation is highly relevant for our understanding of neural crest development, since segmentation is assorted by differential gene activities, which turned out to be critical for the patterning of neural crest derivatives.

### 3.4.1 The cellular basis of hindbrain segmentation

Thanks to the work of Lumsden and Keynes (1989), this question has been revisited by studying the cellular basis of rhombomere segmentation. These authors reached the conclusion that the neuronal pattern of hindbrain development coincides with its segmentation.

In the chick embryo the constrictions leading to the individualization of the eight rhombomeres are established according to a pattern represented in Fig. 3.12 and extending from stage HH9 (i.e., 6ss) to HH12 (16ss).

The first rhombomere is adjacent to the mesencephalon at the isthmus and the last (r8) is continuous with the spinal cord and extends to somites 1–4 inclusively. It is remarkable that, during their brief existence, most of the basic structures of the hindbrain, mainly those including the delimitation of branchiomotor nerve nuclei, are formed. Identification of specific neuronal populations and their projections during this period reveal that neurogenesis follows a two-segment repeat and that pairs of metameric units also cooperate to generate repeating sequences of cranial branchiomotor nerves (Fig. 3.13). Accumulation of axons are detectable from HH13 in the boundaries separating adjacent rhombomeres. Moreover, each segment is characterized by cell lineage restriction, since the progeny of single cells, labeled after rhombomere formation by intracellular marking with fluorescent dextrans *in ovo*, only seldom

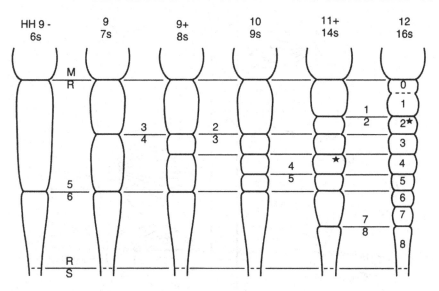

**Figure 3.12.** Diagram showing the emergence of rhombomere pattern in the chick embryo hindbrain between HH9 (6ss) and HH12 (16ss). Each stage has been drawn to the same length and proportions. The rhombencephalic boundary (M/R), like the rhombomere boundaries (5/6, 3/4, etc.), is marked by an external groove. The rhombospinal boundary (R/S, dashed line), lying at the level of the last occipital and first cervical somites, is not marked by any obvious morphological discontinuity. The isthmic rhombomere (O) has uncertain status. The times of first neuronal differentiation at different rhombomere levels are marked by asterisks. (Reproduced, with permission from Lumsden, 1990.)

crossed the limit between two consecutive rhombomeres. Therefore, when they become defined, rhombomeres can be considered as polyclonal lineage-restriction units (Fraser *et al.*, 1990; Lumsden, 1990; see also Birgbauer and Fraser, 1994), a characteristic they share with insect compartments (Garcia-Bellido *et al.*, 1973; Lawrence, 1989).

### 3.4.2   Molecular basis of hindbrain segmentation

An array of genes encoding regulators of transcription have been shown to possess large domains of expression whose anterior limits are situated in the rhombencephalon and coincide with rhombomere boundaries. They constitute a class of vertebrate genes homologous to the homeotic genes discovered in *Drosophila*. A number of other genes have a more restricted expression territory during development, which is limited to one or adjacent rhombomeres. These genes encode various types of proteins: transcription factors, membrane receptors or secreted proteins. Several of them (with the corresponding references) are listed in Fig. 3.14, which provides an idea of the complexity of the regulatory pathways involved in the patterning of the hypobranchial region of

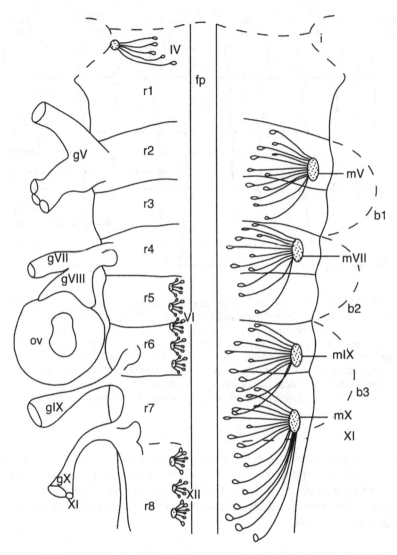

**Figure 3.13** Diagram of a stage-21 chick embryo hindbrain from ventral (pial) aspect, based on neurofilament-stained and axon tracer-injected animals. The relationship of cranial sensory ganglia (gV–gX), branchial motor nuclei (IV, VI, XII), and the combined roots of the sensory and branchial motor nerves (mV–mXI) are shown in relation to the rhombomeres (r1–r8) and arches (b1–b3). ov, otic vesicle; fp, floor plate; i, isthmus or midbrain/hindbrain boundary. (Reproduced, with permission, from Lumsden and Keynes, 1989.)

the embryo. The first genes whose expression was shown to be segmentally regulated in the hindbrain were the *Hox* genes. Their discovery was at the origin of a considerable impetus in the research of the mechanisms controling head and facial development.

*3.4.2.1 Homeotic genes of* Drosophila *led to the discovery of the homeobox selector genes of vertebrates.* Ninety years after the description of the first homeotic mutations by Bateson (1894), a common motif called the homeobox was discovered to be shared by the genes of the *Drosophila* homeotic complexes (*HOM-C*) (McGinnis et al., 1984; Scott and Weiner, 1984).

Modern knowledge about homeotic developmental control genes first came from the genetic analysis carried out by Lewis on the *Bithorax* complex (*BX-C*), which regulates development of the middle and posterior body parts (Lewis, 1978). Another cluster of selector genes which act more anteriorly in the embryo, the *Antennapedia* (*Antp*) complex (*ANT-C*) was later defined by Kaufman and coworkers (see Scott et al., 1983). Thus, the *BX-C* and the *ANT-C* designated together as the *HOM-C* specify much of the body plan corresponding to the "trunk" of the fly.

In the mid-1980s a nucleotide sequence, highly conserved in all *HOM-C* genes and designated the homeobox, was discovered. The homeobox is a DNA segment that encodes a precisely defined protein domain, the homeodomain of 60 amino acids, the three-dimensional structure of which revealed the presence of a helix–turn–helix motif similar to that found in prokaryotic gene regulatory proteins. Thus, the homeoproteins (proteins containing a homeodomain) act as transcription factors regulating the activity of other genes. The homeobox sequences of the *HOM-C* genes present a high level of homology with that of its first discovered "prototype" *Antennapedia*. This led them to be grouped under the term *Antp*-type (or class I) homeobox genes. Divergent homeoboxes have been found in many other transcriptional regulators, namely in genes expressed in the cephalic extremity of the fly such as *empty spiracle* and *orthodenticle*, or in fly developing muscles such as *Msh*.

The role of homeotic genes was already perceived in the 1970s as essential in the process of determination of a given embryonic territory to a definite fate. An important step in our understanding of the genetic control of pattern formation in insects was made when Garcia-Bellido discovered and defined the developmental units he designated compartments. Developmental compartments were defined as precise anatomical regions limited by invariant boundaries and formed by all the descendants of a small set of founder cells (Garcia-Bellido et al., 1973; Lawrence, 1973). Garcia-Bellido (1975) showed that some of the homeotic mutations resulted in transformations of domains corresponding exactly to the compartments they revealed by cell lineage experiments. This led to the assumption that such genes act as controling or selector genes in the founder cells of compartments to determine which part of the body they will construct. Selector genes "act as binary switches with loss of function producing a transformation in which the gene is active and gain of function tending to produce the opposite transformation in the place where the gene is normally inactive" (Lawrence and Morata, 1994). Thus, Garcia-Bellido's hypothesis (1975) linked genetics and cell determination with cell lineage and anatomy.

The first compartments to be defined in the embryo during fly development are the parasegments, comprising the posterior half of one segment and the

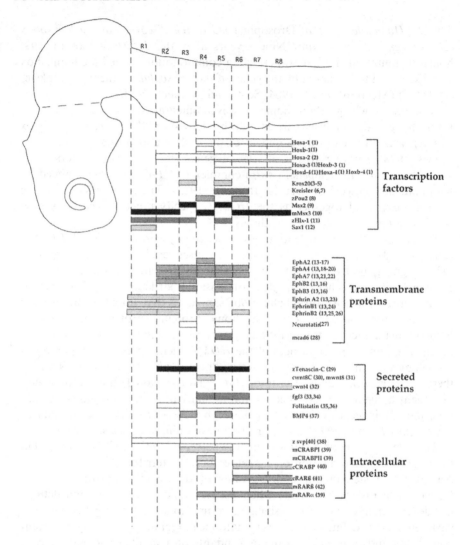

**Figure 3.14** Schematic view of the embryonic head showing segmentation into rhombomeres (R1–R8). A variety of genes are expressed in the developing hindbrain in a segmentally restricted manner, and for a definite period of time during early development of the hindbrain and hypobranchial region. Note that the expression patterns of many of these genes are dynamic. The expression sites indicated here may vary in time. Light tints correspond to a lower level of expression than dark tints. z, m, c before the name of a gene indicates that it has been described in zebrafish, mouse, and chick, respectively. The numbers correspond to the following references: (1) McGinnis and Krumlauf, 1992; (2) Prince and Lumsden, 1994; (3) Wilkinson *et al.*, 1989b; (4) Nieto *et al.*, 1991; (5) Oxtoby and Jowett, 1993; (6) Cordes and Barsh, 1994; (7) Eichmann *et al.*, 1997; (8) Hauptmann and Gerster, 1995; (9) Graham *et al.*, 1993; (10) Shimeld *et al.*, 1996; (11) Fjose *et al.*, 1994; (12) Schubert *et al.*, 1995; (13) Eph Nomenclature Committee, 1997; (14) Ruiz and Robertson, 1994; (15) Ganju *et al.*, 1994; (16) Becker *et al.*, 1994; (17) Weinstein *et al.*, 1996; (18) Nieto *et al.*, 1992; (19) Winning and Sargent, 1994; (2) Bovenkamp and Greer, 1997; (21) Ellis *et al.*,

anterior half of the adjacent segment. Each parasegment is founded by all the cells that lie between the anterior boundaries of adjacent stripes formed by the pair-rule genes *fushi-tarazu* (*ftz*) and *even-skipped* (*eve*).

*3.4.2.2 Cloning of the vertebrate* Hox *genes and their genomic organization.* Thanks to the possibilities offered by recombinant DNA technology, homeobox-containing genes were subsequently isolated not only in insects and phylogenetically related organisms but also in vertebrates. The first vertebrate homeobox gene identified was cloned in *Xenopus* (Carrasco *et al.*, 1984), and was followed by a number of others in mouse, man, and birds (see Duboule, 1992, and references therein). All the vertebrate genes containing a homeobox of the class I type were designated *Hox* genes and considered as homologs to the genes of *Drosophila HOM-C*.

The remarkably constant property of homeotic genes is that, in all animals examined to date, they are clustered, and the order of homologous genes on a given chromosome is conserved. This conservation strongly suggests that the genomic organization has important functional implications.

*Hox* genes are best known in mouse and man where they are distributed on four chromosomes in complexes approximately 120 kb in length in which the genes are all oriented in the same $5'-3'$ direction of transcription (Kessel and Gruss, 1990; Duboule, 1992; McGinnis and Krumlauf, 1992). The present nomenclature adopted in 1992 (see Duboule, 1994) of the 39 mammalian *Hox* genes identified so far is indicated on Fig. 3.15, together with the previous terminology that is used in a series of ante-1994 important reports.

The genomic structural and organizational similarities between vertebrate *Hox* and *Drosophila HOM-C* led to the assumption that the *Hox* gene clusters of vertebrates arose by a two-step process of duplication from a common ancestral cluster (Lewis, 1978). This notion was supported by the discovery that, in several species such as the crustacean *Artemia* (Averof and Akam, 1993), the worm *Caenorhabditis elegans* (Kenyon and Wang, 1991), and some primitive chordates (Pendleton *et al.*, 1993; Holland *et al.*, 1994), a single *Hox/HOM-C*-type cluster also exists.

Alignments of genes in Fig. 3.15, which summarizes the organization and homology between the four *Hox* clusters and the *Drosophila HOM-C* are made on the basis of multiple domains of sequence identity, including the homeodomain itself, as well as on the relative positions of the genes within the

**Figure 3.14** (*cont.*)
1995; (22) Taneja *et al.*, 1996; (23) Cheng and Flanagan, 1994; (24) Flenniken *et al.*, 1996; (25) Bergemann *et al.*, 1995; (26) Smith *et al.*, 1997a; (27) Wijnholds *et al.*, 1995; (28) Inoue *et al.*, 1997; (29) Tongiorgi *et al.*, 1995; (30) Hume and Dodd, 1993; (31) Bouillet *et al.*, 1996; (32) Hollyday *et al.*, 1995; (33) Wilkinson *et al.*, 1988; (34) Mahmood *et al.*, 1996; (35) Albano *et al.*, 1994; (36) Graham and Lumsden, 1996; (37) Graham *et al.*, 1994; (38) Fjose *et al.*, 1995; (39) Ruberte *et al.*, 1992; (40) Maden *et al.*, 1991; (41) Smith and Eichele, 1991; (42) Mendelsohn *et al.*, 1991.

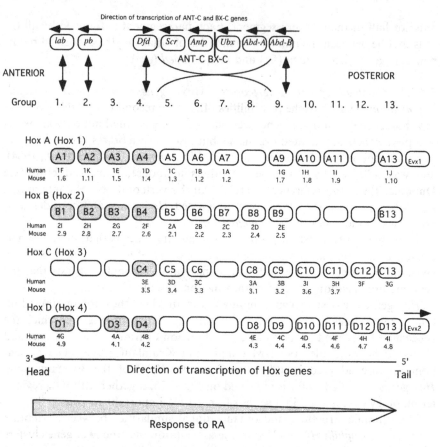

**Figure 3.15** Organization and homology between the four vertebrate *Hox* gene clusters and the *Drosophila HOM-C*. Vertebrate *HoxA–D* and *Drosophila HOM-C* clusters are believed to derive from an archetypal ancestor complex by duplications. The *Drosophila HOM-C* is shown at the top and the vertical bar indicates the junction between *ANT-C* and *BX-C* complexes which are split. Paralogous genes aligned vertically share high levels of homology and are thought to be derived from duplication of the same ancestor gene. From these homologies it is thought that *labial (lab), proboscipedia (pb), Deformed (Dfd)*, and *Adbominal-B (Abd-B)* are related with groups 1, 2, 4, and 9, respectively. Recent nomenclature is shown in the boxes, old nomenclature is listed underneath. In places where no gene is found, an empty box is left. *Hox* genes expressed in the branchial arches are in grey. At the bottom, a thin arrow indicates the direction of transcription. The large arrow indicates the decrease of *Hox* gene sensitivity to retinoic acid (RA). (Modified from McGinnis and Krumlauf, 1992.)

respective complexes. Thus 13 different sets of genes constitute paralogous groups because they share properties while belonging to different clusters.

The strongest homologies are with the *HOM-C* genes *labial* forming group 1, *proboscipedia* (group 2), *Deformed* (group 4), *sex comb reduced (SCr)* (group 5), and *Abdominal B (Abd-B)* (groups 9–13). Group 3 has no *Drosophila* counterpart.

In all the organisms studied so far in this respect it was found that the genes are sequentially expressed in time and space according to their position in the cluster. The genes localized at the 3′ end of the DNA molecule are the first to be expressed in the more anterior region of the embryo. This rule of colinearity holds for vertebrates, and insects, as well as nematodes (see Gaunt *et al.,* 1988; Duboule and Dollé, 1989; Graham *et al.,* 1989).

Another important aspect of colinearity is the posterior prevalence. This was discovered first by Struhl (1983) who showed that, when certain homeotic genes are derepressed, the phenotype is principally determined by the most posteriorly acting protein present. The fact that this was not due only to transcriptional regulation in which the product of posterior genes further repress transcription of anterior ones was demonstrated (Hafen *et al.,* 1984; Struhl and White, 1985). In fact, phenotypic suppression follows a hierarchical rule in which the more posteriorly acting gene products override the more anterior ones (Gibson and Gehring, 1988; Gonzales-Reyes and Morata, 1990; Duboule, 1991). The mechanism, although not yet elucidated, might be based upon competition of the various homeoproteins for binding sites on target genes.

The expression pattern of *Hox* genes was first established in the mouse embryo by *in situ* hybridization. Many *Hox* genes have now been cloned in chick, *Xenopus,* fish, and human.

In mammalian and avian species the expression patterns are generally similar. Only genes of the first four paralogous groups are expressed in the rhombencephalon, and their anteriormost limits of expression are situated at different transverse levels where they coincide with rhombomere boundaries.

The *HoxB* cluster was the first to be studied. The most 3′ gene of this complex, *Hoxb-1,* has an expression domain which first extends up to the limit between rhombomere 3 (r3) and r4. This domain, however, is dynamic, since by day 8.5 postcoitum (dpc) in the mouse (Murphy *et al.,* 1989; Wilkinson *et al.,* 1989a) or stage HH11 in the chick (Sundin and Eichele, 1990), it has receded caudally from r5 and r6 to remain active and even upregulated in r4. The rostral limit of *Hoxb-2* expression is located at the r2/3 boundary, while that of *Hoxb-3* is at the boundary of r4/5 (Wilkinson *et al.,* 1989b).

*Hoxa-2* was first found to have the same anterior limit of expression as *Hoxb-2* in the mouse and to be active more anteriorly in the chick where its expression domain reaches the r1/2 boundary (Prince and Lumsden, 1994). Finally, more careful analysis has shown that, in the mouse also, *Hoxa-2* is expressed in r2 albeit in only a subset of cells (Krumlauf, 1993). *Hoxa-1* (previously *Hox1.6*) has initially the same pattern as *Hoxb-1* but its anteriormost expression (in r4–6) is transient and later on regresses to the r6/7 limit (Lufkin *et al.,* 1991). Other genes fall in general into the expected pattern of colinearity – i.e., the more 3′ a gene lies within a cluster the more anterior its expression domain. Another exception, however, is that the *Hoxc-4* gene has an anterior expression limit one rhombomere more caudal than its paralogs *Hoxa-4,* *Hoxb-4,* and *Hoxd-4* (Fig. 3.16).

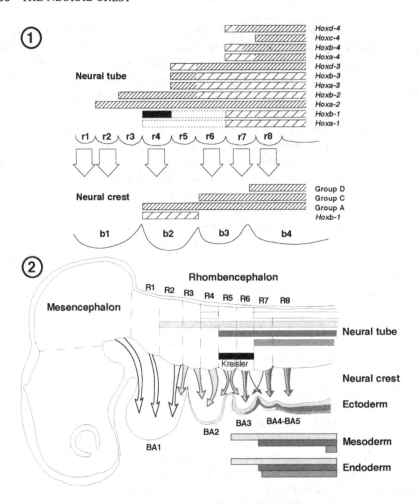

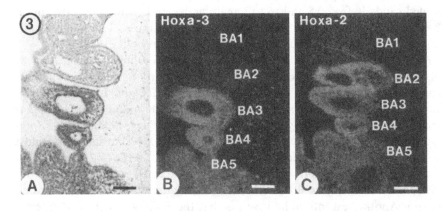

Among the genes of the paralogous group I, *Hoxd-1* shows no expression in the neural tube at the stages analysed (Sundin and Eichele, 1990; Hunt *et al.*, 1991a). Finally, the level of transcripts at a given level and stage of development varies in a rhombomere-specific manner even for paralogous genes, as indicated in Fig. 3.16.

In summary, detailed descriptive studies of *Hox* gene expression along the anteroposterior axis from the rhombencephalic down to the posterior levels have shown that the developing neural tube and axial organs (notochord and paraxial mesoderm) can be divided into successive anteroposterior domains individually characterized by a definite spatiotemporal pattern of *Hox* gene activity. It was then proposed that such patterns define a code designated "Hox code" (Kessel and Gruss, 1991). As a general rule, the neural crest cells express the same Hox code as the transverse level of the neural tube from which they originate (Hunt *et al.*, 1991a,b). One exception is that the neural crest cells exiting from r2 do not express the *Hoxa-2* gene (Prince and Lumsden, 1994).

Several genes other than *Hox* genes turned out to be expressed in domains corresponding to single or adjacent rhombomeres (Fig. 3.14).

*3.4.2.3* Krox-20, Hox *genes and* Eph *receptors are involved in a regulatory cascade controlling hindbrain segmentation. Krox-20* was originally identified as a serum-responsive gene in mouse NIH-3T3 fibroblasts. It encodes a protein with three zinc finger domains similar to those of the transcription factor Sp1. It is therefore considered that the *Krox-20* gene product acts as a transcription factor involved in the resumption of growth in fibroblasts. In the embryo, the *Krox-20* gene is also expressed in the neuroepithelium (except the floor plate) of the hindbrain. The first observations in the mouse showed that by 9.5 dpc two domains of expression corresponding to r3 and r5 were clearly delineated (Wilkinson *et al.*, 1989b) (Fig. 3.17). The same segments also expressed receptor tyrosine kinases (RTK) of the Eph family (see Chapter 2). In fact, several members of the Eph family of RTKs have a segment-restricted pattern of

**Figure 3.16** Summary of *Hox* gene expression in the hindbrain and corresponding neural crest cells. Panel 1: Filled area denotes region of upregulated *Hoxb-1* expression. Dense hatching denotes areas of high-level expression, sparse hatching denotes areas of lower-level expression. Dotted lines show transient gene expression; b1–b4, branchial arches. (Reproduced, with permission of Company of Biologists Ltd, from Prince and Lumsden, 1994.) Panel 2: Schematic representation of the expression of certain *Hox* genes of the first paralogous groups in chick or quail embryos at E3 when the branchial arches (BA) are being colonized by neural crest cells originating from the posterior half of the mesencephalon and the rhombomeres (R1–R8). The arrows indicate the anteroposterior origin of the neural crest cells migrating to each BA. Expression of *Hox* genes is also indicated in the superficial ectoderm. Panel 3: Frontal sections of a chick embryo at E3 showing *Hoxa-3* (A, B) and *Hoxa-2* (C). *In situ* hybridization with the *Hoxa-3* probe seen in bright-field illumination (A). (B, C) Dark-field pictures of *Hoxa-3* (B) and *Hoxa-2* (C). Bar = 100 μm. (Reproduced, with permission of Company of Biologists Ltd, from Couly *et al.*, 1996.)

*For a colored version of this figure, see www.cambridge.org/9780521122252.*

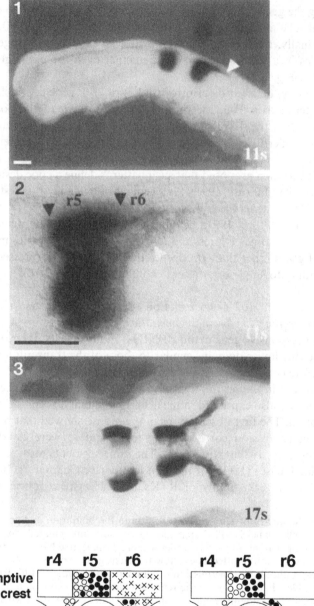

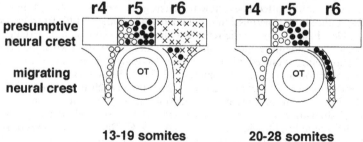

expression in the hindbrain and in the paraxial mesoderm, suggesting that they might be involved in the segmental patterning of the hindbrain.

Early in development of the chick, one of these RTKs, the *Sek-1* gene (now designated *EphA4*; see Becker *et al., 1994*), is expressed in a broad domain and later is upregulated in r3 and r5 together with *Krox-20* gene (Nieto *et al.*, 1992). In the mouse, in contrast, *Krox-20* expression domains appear first anteriorly (r3) and then posteriorly (r5) as narrow stripes of four cells in length that subsequently broaden to 10–12 cells in length. The stripes initially form fuzzy domains which, as they broaden, become defined by sharp limits coinciding with rhombomere boundaries. The dynamics of *Sek-1* and *Krox-20* expression domains were studied at the single-cell resolution level in the mouse and chick by Irving *et al.* (1996). The problem was raised as to how the sharp boundaries between expressing and non-expressing domains are established. Several hypotheses can be proposed. They may result from restriction in cell movements together with upregulation of cell-adhesive differences between cells of two consecutive rhombomeres. Another possibility may be a switch in cell identity (i.e., from an expressing to a non-expressing state and vice versa). If such a mechanism were at least partly to account for the establishment of the limits between positive and negative domains it would indicate that the cells are not irreversibly committed. In such a case, their status with respect to a given segmental identity would be regulated by environmental cues. The role of environmental cues in defining segment identity was actually demonstrated by transposition of the presumptive territories of rhombomeres in the avian embryo (Grapin-Botton *et al.*, 1995, 1997; Itasaki *et al.*, 1996).

*Krox-20* expression also occurs in lower vertebrates: in zebrafish it has been described by Oxtoby and Jowett (1993). In *Xenopus, Krox-20*-expressing cells are precisely aligned with r5. Since the otic vesicle is adjacent to r4 in this species, r5-derived cells are all *Krox-20*-positive and migrate in a direct ventral route to fill the third branchial arch.

Null mutants for the *Krox-20* gene were produced (Schneider-Maunoury *et al.*, 1993). In heterozygous mutant embryos, *lacZ* expression from the disrupted *Krox-20* locus occurs in a pattern identical to *Krox-20* transcripts and only in postotic crest cells. However, in homozygous *Krox-20*$^{-/-}$ mutants, *lacZ* expression is detected in neural crest cells both rostral and caudal to the otic

**Figure 3.17** (1) Expression of *Krox-20* in r3 and r5 and in the corresponding premigratory neural crest in an 11ss chick embryo. (2) Magnification of the region of r5–r6, showing that *Krox-20* is also expressed in the r6 premigratory neural crest. (3) At 17ss the postotic migrating neural crest cells exiting from r5 to r6 clearly express *Krox-20* mRNA. Bars = 100 μm. Arrowheads indicate *Krox-20*-expressing cells in r6. (4) Summary of the migration patterns of *Krox-20*-expressing neural crest cells. Premigrating neural crest in dorsal r5 and r6 is depicted in the upper part, and migrating neural crest cells are shown below. r5 neural crest cells not expressing *Krox-20* are indicated with empty circles, r5 neural crest cells expressing *Krox-20* are indicated with filled circles and r6 neural crest cells are indicated with crosses. OT, otic vesicle. (Reproduced, by permission of Oxford University Press, from Nieto *et al.*, 1995.)

vesicle. In these mice definitive r3 and r5 fail to form and morphological segmentation is disrupted (Schneider-Maunoury *et al.*, 1993; Swiatek and Gridley, 1993). Moreover, disruption of the *Krox-20* gene results in a loss of *Hoxb-2* expression in r3/r5, suggesting that *Hoxb-2* expression is controlled by *Krox-20* (Schneider-Maunoury *et al.*, 1993; Swiatek and Gridley, 1993), a finding that has been confirmed by Sham *et al.* (1993). Thus, a direct coupling seems to occur between *Krox-20* and *Hox* gene expression, and between *Krox-20* expression and segmentation.

Bouillet *et al.* (1995) have cloned a cDNA sequence corresponding to a gene induced in P19 embryonal carcinoma cells by retinoic acid (RA) treatment. This cDNA is identical to a newly reported member of the Eph-family of RTK genes, the MDK1 gene (now called *EphA7*) (Ciossek *et al.*, 1995; Ellis *et al.*, 1995) whose homolog has also been cloned in zebrafish (ZDK1). In the mouse, chicken (Cek11), and human (Hek11) MDK1 starts to be expressed in the hindbrain in a region corresponding to r3 before any sign of segmentation is apparent. In E9.5 embryos, strong expression is seen in r2–3 and r5, and weaker expression in r4 and r6.

Disruption of certain *Hox* genes of the first paralogous groups in the mouse have severe repercussions on the development of hindbrain and its neural crest derivatives (to be described below). In *Hoxa-2$^{-/-}$* mice no expression of MDK1 is found in r3 and an abnormal upregulation of MDK1 is seen in r4 as compared to wild-type. The fact that MDK1 is downregulated in r3 in the absence of the *Hoxa-2* gene product (Taneja *et al.*, 1996) indicates that it lies downstream to *Hox* genes in the regulatory cascade controlling rhombomere development. A hierarchy is thus disclosed concerning the genetic control of rhombomere development with *Krox-20* acting upstream of *Hox* genes which themselves regulate the expression of some *Eph* RTKs. Moreover *Krox-20* was also shown to be involved in the regulation of other secreted proteins like follistatin.

*3.4.2.4 The* MafB *(Kreisler) transcription factor is necessary for the construction of the fifth and sixth hindbrain segments.* The recessive mouse mutation *Kreisler (kr)* was identified in an X-ray mutagenesis screen and the corresponding gene was shown to encode a basic domain leucine zipper (bZIP)-type transcription factor of the Maf family (Cordes and Barsh, 1994) designated as *MafB/kr* (Eichmann *et al.*, 1997). The homozygous *kr/kr* mutant mice are characterized by behavioral abnormalities (head tossing and running in circles, inability to swim), and deafness (Hertwig, 1944).

During embryogenesis *MafB/kr* is expressed in r5 and r6 as well as in the neural crest cells derived from these rhombomeres (Fig. 3.18), and in many other sites such as the kidney. It remains expressed later in development in vestibular and acoustic nuclei and in differentiating neurons of the spinal cord and brainstem as shown in the avian embryo by Eichmann *et al.* (1997).

In *kr/kr* mouse embryos, hindbrain segmentation is strongly impaired. At 9.5 dpc the three first rhombomeres appear normal, but r4–7 are replaced by a morphologically unsegmented neural tube. In addition, the position of the

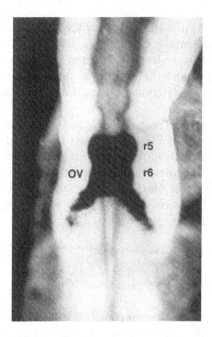

**Figure 3.18** *MafB/kr* expression in a 14ss chick embryo. *MafB/kr* is expressed in rhombomeres r5 and r6 and the associated neural crest cells, migrating caudal to the otic vesicle (OV).

developing otic vesicle is shifted from its normal location near the junction of r5 and r6 to a more lateral position. Frohman *et al.* (1993) and McKay *et al.* (1994) have analyzed the hindbrain of *kr/kr* mutants in terms of expression of *Hox* genes (*Hoxb-1, Hoxb-2, Hoxb-3, Hoxb-4, Hoxa-3, Hoxd-4*), *Krox-20, FGF3*, and cellular retinoic-acid binding protein I (*CRABP-I*). *Krox-20* expression is extinguished in r5 but maintained in r3 (Frohman *et al.*, 1993; McKay *et al.*, 1994). This and other findings are consistent with a loss of r5 and at least part of r6 in *kr/kr* mice. Accordingly, the derivatives of the neural crest originating from these rhombomeres (i.e., the glossopharyngeal ganglion and nerve and the abducens nerve) are missing in the mutant (McKay *et al.*, 1994). It was later shown that *MafB/kr* can directly regulate some of these genes. This transcription factor is able to bind *Hoxb-3* promoter and activate the expression of this gene in r5, and r6 under certain conditions (Manzanares *et al.*, 1997).

An unusual amount of cell death is seen in the unsegmented abnormal neural tube. These observations have been interpreted as if the cells that would normally become specified at an early stage as r5–6, adopt an r4 character instead, producing an excess of r4 cells that are subsequently disposed of by cell death (McKay *et al.*, 1994). The *kreisler* gene seems therefore to be required early for the construction of the fifth and sixth hindbrain segments (see below).

Grapin-Botton *et al.* (1998) have recently shown that expression of *MafB/kr* is subjected to environmental regulatory cues that also affect *Hox* genes. Thus, transplantation of r5–6 at the level of r8 (in the 5ss chick embryo) results in the upregulation of *Hoxb-4* and the downregulation of *MafB/kr*, followed by a complete homeotic transformation of the differentiated neural phenotype of r5–6 derivatives.

This posteriorizing effect can be obtained by grafting either posterior somites or RA-impregnated beads in contact to r5–6 left *in situ*. The effect of the same RA beads and somites on r3–4 is the opposite: instead of down-regulating *MafB/kr*, the expression of this gene is induced in r3–4. It appears, therefore, that expression of *MafB/kr* depends on a definite concentration of a posteriorizing factor that seems to be distributed in a decreasing gradient of concentration along the caudorostral axis.

### 3.4.3  Forebrain and midbrain segmentation

The morphological and histological complexity of the adult forebrain makes it extremely difficult to determine the topological relationships existing between their initial components. Methodological and conceptual advances have recently provided new ways to envisage how the complex structures of the adult brain can be built up.

Cell-marking techniques have provided precise fate maps of the neural primordium (Couly and Le Douarin, 1988; Eagleson *et al.*, 1995), and molecular markers have been discovered which have regionally restricted domains of expression often separated by sharp boundaries, thus providing markers for forebrain subdivisions. These techniques have revitalized the old idea that the forebrain is a neuromeric structure (Bulfone *et al.*, 1993; Figdor and Stern, 1993).

Among the genes expressed in the brain are at least 30 homeobox genes, some of them being related to the genes *Distalless* (*Dll*), *Empty spiracle* (*Ems*) and *Orthodenticle* (*Otd*), that direct the pattern of head development in *Drosophila*. The boundaries of expression of several genes appear either perpendicular or parallel to the initial longitudinal axis of the forebrain. Thus, on the basis of gene expression patterns and of embryological, histological, and morphological criteria, a model has been proposed to account for forebrain organization: the prosomeric model (Rubenstein and Puelles, 1994; Shimamura *et al.*, 1997, and references therein).

The midbrain is characterized by the expression of *Engrailed* genes (*En1* and *En2*) and by a transverse region corresponding to the midbrain–hindbrain junction, endowed with organizing activity (see Le Douarin, 1993; Joyner, 1996, for references).

Mutations of *En1* and *En2* impair the development of the midbrain and rostral rhombencephalon (Joyner *et al.*, 1991; Wurst *et al.*, 1994) but do not seem to affect the development of neural crest derivatives.

## 3.5 Relationships between brain segmentation and the development of the cranial, facial, and visceral skeleton

As mentioned earlier in this chapter, the fate of the cephalic neural crest was established before segmentation of the brain primordium was recognized. When it was established that, at the critical stage of its specification, the brain anlage is divided into metameric units characterized by specific sets of gene activities, it appeared essential to establish with precision what is the contribution of each of these units to the derivatives of the neural crest. The neural derivatives of the neural crest will be discussed in this respect in Chapter 5. In this section the diencephalic, mesencephalic, and rhombomeric origin of the various parts of the head skeleton and of the cephalic muscle-associated connective tissues will be described from experiments whereby segments of the neural fold were labeled at the onset of crest cell emigration. Two methods were used: DiI labeling and the quail–chick chimera system.

### 3.5.1 Evidence for the contribution of each rhombomere to the branchial arches as revealed by DiI labeling

After showing that the formation and disposition of motor neurons in the chick rhombomeres conforms to the rhombomere patterns (Lumsden and Keynes, 1989), Lumsden et al. (1991) used a vital dye (DiI) to follow the migration of the cranial neural crest at stages ranging from 3 to 16 somites. Observations of the embryos during the 2 days following the labeling by intensified video fluorescence microscopy actually revealed the migration streams of cells originating from each rhombomere.

It appeared clearly that the pattern of crest cell emigration from the rhombomeres to the branchial arches largely correlates (with some exceptions, see below) with the segmented disposition of the rhombencephalon, the branchial arches thus appearing as the ventral components of hindbrain segments (see Figs. 3.16 and 3.20).

This being established, the patterns of Hox gene expression in the mouse rhombomeres and branchial arches revealed that the neural crest cells express the same set of Hox genes as the rhombomere from which they originate (Hunt et al., 1991a,b). A few exceptions to this rule exist, however. For example, the neural crest cells which correspond to r2 do not express Hoxa-2 (Prince and Lumsden, 1994; Couly et al., 1996) (Fig. 3.18).

DiI labeling of the neural fold prior to neural crest cell emigration (Lumsden et al., 1991; Sechrist et al., 1993; Birgbauer et al., 1995) or labeling of migrating crest cells by HNK1 Mab or by an antibody against a neurofilament epitope (Sechrist et al., 1995) revealed that at the time of migration the crest cells segregate into three main migratory streams composed of cells originating from r1–2, r4, and r6 which filled branchial arches BA1 (called maxillomandibulary arch for r1–2), BA2 (also called hyoid arch), and BA3 for r4 and r6, respectively. These fluxes of crest cells are in register with the cranial nerve entry/exit points which are located in the same segments. As already reported by previous authors (Anderson and Meier, 1981; Tosney, 1982; Noden, 1988) no neural crest cells are seen migrating laterally to r3 and in the region of the otic vesicle lateral to r5. However, previous quail–chick chimera experiments (Couly and Le Douarin, 1990), and observations of neural crest cell migration using acetylcholinesterase as a marker for premigratory and early migratory neural crest cells, had established that neural crest cells emerged uniformly from the hindbrain neural fold as they do at the other levels of the neuraxis (Cochard and Coltey, 1983).

Graham et al. (1994) attributed the absence of neural crest cells at the r3 and r5 levels to cell death. They reported that the neural folds of r3 and r5 exhibited elevated levels of apoptosis at a stage when neural crest cells should be emigrating from these neuraxial levels. Strong expression of the homeobox containing gene Msx2 preceded the onset of apoptosis. Prior to Msx2 expression, the TGFβ family member BMP4, which by stage HH6 is expressed uniformly in the hindbrain neural fold, shows higher levels of expression first in r3 and then also in r5 by HH9. If r3 and r5 are isolated in culture and therefore removed from the influence of their neighboring even-numbered rhombomeres, they do not show the upregulated expression of BMP4 and Msx2 and produce neural crest cells. In contrast, if reassociated with r4 and r6 both r3 and r5 behave as they do in situ. Moreover, addition of recombinant BMP4 protein to explant cultures of r3 or r5 which produce neural crest cells when isolated, upregulates Msx2 expression. Increase in Msx2 expression is followed by apoptosis of the neural crest cell population.

This segmental apoptotic pattern of neural crest cells along the neuraxis was interpreted as an ontogenetic mechanism through which the streams of neural crest cells which colonize the successive branchial arches become strictly individualized.

Although relevant, the observation and assumptions of Graham et al. (1994) are not as strict as they appeared. First, other series of labeling experiments performed by Fraser's and Bronner-Fraser's groups (Sechrist et al., 1993; Birgbauer et al., 1995) and by Nieto et al. (1995) confirmed that r3 and r5 are not devoid of the capability to yield neural crest cells in vivo. The latter, however, do not follow a dorsoventral migration route from the time they leave the neural primordium but migrate longitudinally first, as can be seen in Fig. 3.19. Moreover, at the level considered (r3 and r5) the majority of crest cell emigration takes place before the period of maximal cell death (11–12ss) (Birgbauer et al., 1995). Finally, labeling of crest cells by the quail–chick chimera system also demonstrated the contribution of r3 and r5 neural crest cells to branchial arches (Couly et al., 1996; Köntges and Lumsden, 1996).

Similar studies have been performed in Danio rerio (Schilling and Kimmel, 1994) at early stages of somitogenesis (12–15 hours of development), when the cephalic neural crest forms two cords of cells laterally to the neural keel down to the otic placode. The cephalic paraxial mesoderm appears as a thin string of cells (one or two cells wide) along the yolk sac, which is continuous caudally with the first somite (Fig. 3.19). Labeling of single cells by vital dye injection

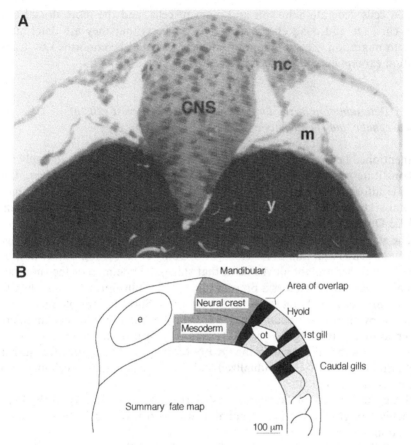

**Figure 3.19** Relative contribution of neural crest and mesoderm to the mesenchyme of the branchial arches in zebrafish embryo. (A) Transverse section through the head of a 12-h zebrafish embryo in a region just caudal to the eye, stained with toluidine blue to reveal general cellular architecture. nc, neural crest; m, cephalic paraxial mesoderm; y, yolk; CNS, central nervous system. Bar = 25 μm. (B) Summary fate map drawn by outlining the areas of neural crest and mesoderm that generate each arch. Black regions denote the overlap between fate map areas. e, eye; ot, otic vesicle. (Reproduced, with permission of Company of Biologists Ltd, from Schilling and Kimmel, 1994.)

led to the establishment of a fate map for both neural crest (Fig. 3.19) and paraxial mesoderm. It was found that single neural crest cells and mesodermal cells produce clones whose extension is restricted to single segments and to single cell types, indicating that, well before rhombomere boundaries are present, the cell movements taking place during branchial arch formation are constrained to a large extent and take place essentially within a given transverse plane. When migration is completed, mixing of cells of neighboring arches is further prevented by the endoderm outpocketings of the branchial pouches. The fate of the cells can be predicted according to their position within the neural crest: the more laterally located cells form neurons, medially located cells generate Schwann and pigment cells, and the more dorsal cells form cartilage and connective tissue. This suggests that they are determined prior to migration. This assumption, however, should be confirmed by transposition experiments before being accepted as fact.

### 3.5.2 Determination of the long-term fate of neural crest cells of mesencephalic and rhombomeric origin

As mentioned before, vital dyes have the advantage of being easy to apply and to constitute a non-invasive way to follow cell migrations. However, they provide information only over a short period of time. This is why the quail–chick marker system was applied to this analysis simultaneously in Lumsden's and Le Douarin's laboratories.

The stages elected for the operation were stages HH8 + to 9 + for Köntges and Lumsden (1996) and 5–6ss (HH8) for Couly et al. (1996). Since rhombomere boundaries are not all visible at that stage, the fate map of the rhombencephalon was constructed as a prerequisite (see Grapin-Botton et al., 1995, for the fate map of rhombomeres in the 5ss embryo). Substitutions of chick rhombomeres by their quail counterpart were carried out using an ocular micrometer so that in the large majority of cases they turned out to be very precise, as attested by the labeling with the QCPN Mab applied 24 hours after surgery when rhombomeres are individualized and when the neural crest cell migration stream is visible (Fig. 1.5).

Since the mesencephalic neural crest participates significantly to the hypobranchial structures, the mesencephalon was also involved in these series of operations.

First, the segmental distribution of crest cells into the maxillary process and the branchial arches already observed with larger grafts of neural tube or neural fold and with DiI (Johnston et al., 1974; Le Lièvre and Le Douarin, 1975; Noden, 1978a,b; Lumsden et al., 1991) was fully confirmed by these experiments. Although r3 and r5 were considered to produce no neural crest cells, their contribution to branchial arches (1 and 2 for r3, and 2 and 3 for r5) was found to be significant (Fig. 3.20). This confirmed the observations made by Sechrist et al. (1993), Birgbauer et al. (1995), and Nieto et al. (1995).

The stability of the quail marker in quail–chick chimeras permitted the determination of the precise contribution of each brain segment to the skeleton, the connective components of the head muscles, the peripheral head ganglia and nerves (see Chapter 4), and finally to the heart (see above) and glandular structures associated with the aortic arches and pharynx (carotid body, ultimobranchial bodies, thyroid, parathyroids, thymus) with a resolution that was not provided by the previously applied experimental designs.

In birds, the craniomandibular articulation includes the quadrate and the pterygoid (also called pterygoquadrate), which articulates to the palatine bone (roof of the palate) anteriorly, to the squamosal posteriorly, to the quadratojugal/jugal bone laterally and to the lower jaw through the articular ventrally (Fig. 3.20). The neural crest from the posterior half of the mesence-phalon yields cells participating to the squamosal bone, the pterygoquadrate, the quadratojugal, the articular, Meckel's cartilage (nearly complete), and all the membrane bones of the mandible. The proximal part of the lower jaw, however, is derived from neural crest cells arising from r1–3 (Couly *et al.*, 1996; Köntges and Lumsden, 1996) (Fig. 3.20).

The columella (avian homolog of the mammalian stapes) is mostly of neural crest origin and made up of cells from r4 (i.e., from BA2, the hyoid arch), with a possible contribution from the paraxial mesoderm (Köntges and Lumsden, 1996).

Neural crest cells derived from r1 are found throughout the quadrate and pterygoid bones, also called pterygoquadrate, whereas midbrain crest cells occupy only its dorsal margin and articulation. Crest cells from r2 are essen-tially located in ventral articulations with the lower jaw. The boundaries between these crest cell populations are not sharp, but r1- and r2-derived cells are always found in the posterior quarter of the jaw in agreement with their localization in the first branchial arch (see Figs. 3.20 and 3.21).

The lower jaw skeleton in the embryo is composed of Meckel's cartilage covered by several membrane bones: the angular, supra-angular, opercular, and dentary. The retroarticular process is formed by calcification of connective tissues taking place during the second half of the incubation period. It can thus be identified only in embryos older than E9–10. These tissues originate from cells belonging to the hyoid arch (i.e., mainly r4 with also a contribution from r3; see Köntges and Lumsden, 1996).

The hyoid cartilage is formed of a series of skeletal elements: entoglossum (or paraglossals), basihyal (basihyoid), basibranchial (or urohyal), ceratobran-chial, and epibranchial (Fig. 3.21), the most anterior of which (entoglossum and basihyal) constitute the tongue skeleton.

The hyoid bone was found to have a highly composite origin with the entoglossum derived from the posterior mesencephalon (Couly *et al.*, 1996) and its other components formed by cells from the successive anteriormost rhombomeres. As a general rule, cells originating from different segmental origins remain coherent and do not mix. This results in sharply defined borders separating the crest cell populations originating either from the mesencephalic or the various rhombomeric levels. These borders do not coincide with ana-tomical entities.

Some slight discrepancies exist between the reports of Couly *et al.* (1996) and Köntges and Lumsden (1996) as to the exact extent of the contribution of each rhombomere to the hyoid bone, but as a whole their interpretations of the segmental contribution of the hindbrain to the hypobranchial structure are similar. These discrepancies may have arisen from variations in the grafting experiments. Thus, for Köntges and Lumsden, r3-, r4-, and r5-derived cells participate in the posterior end of the paraglossals, whereas for Couly *et al.* they do not and are found more posteriorly in the basihyal (r3, r4) and the ceratobranchial (r4–5). Crest cells from r6–7, which colonize the third and fourth branchial arches, form the posterior part of the basihyal for Köntges

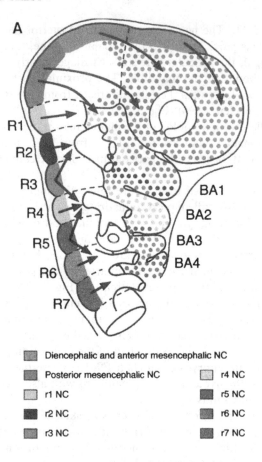

**Figure 3.20** (A) Migration map of cephalic neural crest cells in the avian embryo. The origin of neural crest cells found in the nasofrontal and periocular mass and in the branchial arches is color-coded. Anterior mesencephalon contributes to the nasofrontal and periocular mass. Posterior mesencephalon also participates in these structures, but in addition populates the anterodistal part of BA1. The complementary portion of BA1 derives from R1/R2 together with a small contribution of R3. The major contribution to BA2 comes from R4. Neural crest cells arising from R3 and R5 split into strains participating, respectively, to two adjacent arches: R3 cells migrate to BA1 and BA2; R5 cells migrate to BA2 and BA3. R6 and R7 derived cells migrate to BA3 and BA4. (B) Color-coded (see A) fate map of the neural crest issued from the prosencephalon, mesencephalon, and rhombomeres 1–4. The bones, cartilages, and muscles of the jaw are numbered in the upper panel. 1, articular; 2, quadrate; 3, quadratojugal; 4, pterygoid; 5, palatine; 6, squamosal; 7, jugal; 8, Meckel's cartilage; 9, supra-angular; 10, angular; 11, dentary; 12, opercular; 13, splenial; 14, columella; 15, depressor mandibulae; 16, pterygoideus; 17, pterygoquadrate; 18, retroarticular process. a–f represent respectively the contribution of anterior mesencephalic (a), posterior mesencephalic (b), r1 (c), r2 (d), r3 (e), and r4 (f) neural crest to the facial skeleton and the connective tissue of the mandibular retractor muscle in f.

*For a colored version of this figure, see www.cambridge.org/9780521122252.*

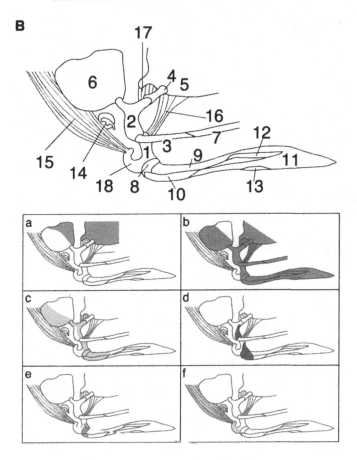

**Figure 3.20** (*cont.*)

and Lumsden (1996), whereas for Couly *et al.* (1996) they are found more posteriorly in the basibranchial.

Both groups of authors agree that r6 forms most of the epibranchial and ceratobranchials (see Fig. 3.21) and that in this case the r6–7 populations of neural crest cells mix. Couly *et al.* (1996) show also that cells from r8 participate in the distal part of the basibranchial.

The articular bone has two processes which serve as sites of attachment for muscles (pterygoideus and depressor mandibulae). Köntges and Lumsden (1996) noticed that in this case as for the other bones and associated muscles the connective tissues of these muscles have the same transverse origin as the bones to which they are attached (see Fig. 3.22B, F). For these authors, this is the case also in the viscerocranium, for all branchial and hypoglossal (tongue) muscles. This means that each rhombomeric or mesencephalic crest population remains coherent and that their segmental origin is maintained throughout

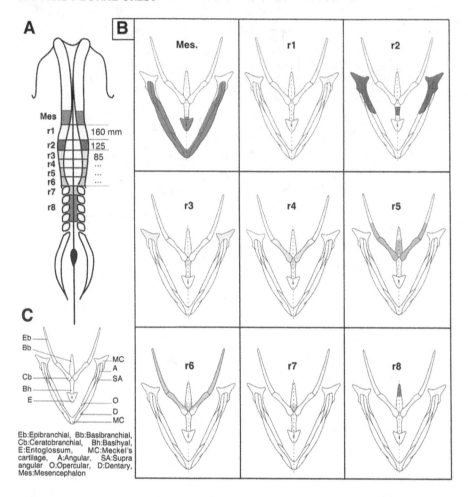

**Figure 3.21** Respective origins of the lower jaw and hyoid bone from midbrain and hindbrain neural crest. (A, B) Results of experiment whereby the chick neural folds corresponding to each individual rhombomere and to the posterior mesencephalic half, were substituted by their quail counterpart at 5ss. (C) Lower jaw and hyoid bones with the corresponding legends. (Reproduced, with permission of Company of Biologists Ltd, from Couly *et al.*, 1996.)

*For a colored version of this figure, see www.cambridge.org/9780521122252.*

ontogeny, a fact which is proposed to account for the permanence of the cranial skeletomuscular pattern throughout vertebrate evolution (Köntges and Lumsden, 1996). The fact that the connective tissue component of the muscles and their insertion points on the bone are strictly ontogenetically related demonstrates a key role for connective tissue in patterning the muscles in the head as already proposed by Noden (1983a, 1986b) and evidenced for the limb muscles by Chevallier and Kieny (1982).

Moreover, this suggests that the neural crest has a special function in patterning the skeletomuscular connectivity. This underlying pattern, which could only be revealed by the ability of the quail–chick system to disclose the long-term fate of single rhombomere-derived cells, also provides an answer to the old question of craniofacial pattern formation (see Noden, 1986b; McClearn and Noden, 1988) as to how branchial and hypoglossal muscles become anchored to specific regions of the neurocranium and splanchnocranium.

The investigations described above have also pointed out the fact that the muscles whose connective tissue belongs to a given segmental unit are innervated by nerves arising from the same rhombomeres. For example, the mandibular attachment sites of the lower jaw for the jaw-closing adductor muscles are derived from r2 and r3, and these muscles are innervated by the trigeminal neurons of r2–3 (Köntges and Lumsden, 1996). Similarly, the depressor mandibulae whose connective tissue is of hyoid arch (BA2) origin is innervated by facial motor neurons lying in r4 and r5.

This review of the origin and fate of the neural crest cells shows that the classical nomenclature of branchial arches as maxillary and mandibular for BA1 and hyoid for BA2 has become obsolete. In fact, the hyoid bone does not originate from the so called hyoid arch but from cells of BA1, BA2, BA3, and BA4–5 as well. Moreover, the mandible is not derived exclusively from BA1 since the retroarticular process originates from cells which have initially migrated to BA2 (r4 cells).

### 3.5.3 *Regeneration capabilities of the cephalic neural crest*

Ablation of embryonic territories in "regulative embryos" such as those of vertebrates is generally followed by more or less complete regeneration of the excised region together with its reorientiation toward a novel fate in accordance with that of the missing cells. The regulative capacities of the embryonic cells limiting the excision depend upon the nature and extension of the lesion as well as on the developmental stage at which the operation is performed. This has been well illustrated in the past for the neural crest, the ablation of which has yielded variable results ranging from no effect to more or less extended deficiencies (references in Couly *et al.*, 1993) (e.g., Stone, 1929; Hammond and Yntema, 1947, 1964; Le Lièvre, 1974; Langille and Hall, 1988a,b).

In more recent experiments (Scherson *et al.*, 1993; Hunt *et al.*, 1995; Sechrist *et al.*, 1995), the neural fold either alone or together with larger or lesser portions of the lateral wall of the neural tube at the rhombencephalon level was excised uni- or bilaterally in 4–12ss chick embryos. No malformations consecutive to the ablation were reported to occur in neural tube (brainstem) and neural crest derivatives. The sectioned neural tube was labeled by DiI on its dorsal surface and labeled cells were described as migrating and following "normal migratory routes" (Scherson *et al.*, 1993). This would mean that the neural tube, even from its most ventral levels, is able to regenerate not only the missing tube but also a neural crest. It was later proposed (Sechrist *et al.*, 1995) that regeneration of the neural crest takes place only if apposition of the remaining neuroepithelium with the surface ectoderm occurs. This contact is followed by upregulation of the *Slug* gene which normally precedes neural crest cell emigration. In this second report from Bronner-Fraser's group, it is mentioned that regulation of neural crest cells occurs essentially at 3–5ss and slowly declines after this stage. Moreover, in this series of experiments, the extent of neural tube ablation seems to bring about a limitation in the capacity of the neuroepithelium to reform a neural crest since the regulative response decreases with the depth of the ablation. In a more recent article, Buxton *et al.* (1997) found that the contact between the surface section of the neural tube with the superficial ectoderm was not able to induce the expression of *Slug*. In contrast, annealing of the neural tube by fusion of the two sectioned surfaces on the midline was shown to be critical in the upregulation of *Slug* gene expression in the dorsal part of the truncated neural tube. Notable is the fact that *Pax3* is first expressed dorsally and then downregulated when *Slug* transcripts start to accumulate. According to Buxton *et al.* (1997), *BMP4* is not reexpressed at the site of the excision and would therefore belong to regulatory pathways distinct from *Pax3* and *Slug*.

The regeneration capacity of the cephalic neural crest after its partial removal at 5–6ss was confirmed in general terms by Couly *et al.* (1996). Thus, the excision of the tip of the neural fold over several consecutive rhombomeres was followed by the perfect regulation of deficiencies. However, when the excision included also the whole mesencephalic territory, large deficiencies in skeletal and connective facial structures were observed. Moreover, excision of either the alar plate or of the basal plate was not followed by regulation of the corresponding brainstem and mesencephalic structures and large regions of the brain were missing in such embryos (Etchevers and Le Douarin, unpublished results).

The source of the cells able to compensate for the removed neural folds was systematically investigated using the quail–chick marker system (Couly *et al.*, 1996). When the neural fold was removed unilaterally, the contralateral neural fold rapidly flowed over the area of excision and yielded crest cells bilaterally in sufficient number to constitute normal structures on both sides. This was confirmed by the experiment represented in Fig. 3.22, where the neural folds are excised in a chick embryo on both sides and replaced by the equivalent neural fold from a quail embryo on one side only. A few hours later, the quail neural crest is seen migrating bilaterally (Fig. 1.4).

The capacity of the ventral neural tube to regenerate a neural crest was tested by replacing the ventral half of the chick neural tube by its quail counterpart, as indicated in Fig. 3.23. Although great care was taken so that the superficial ectoderm would contact the graft, no quail neural crest cells ever

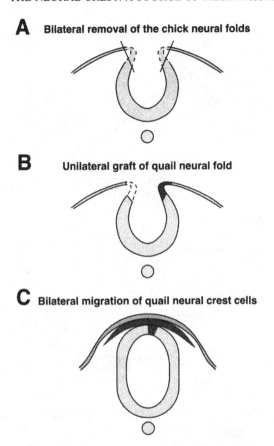

**A** Bilateral removal of the chick neural folds

**B** Unilateral graft of quail neural fold

**C** Bilateral migration of quail neural crest cells

**Figure 3.22** Diagram showing an experiment designed to explore the capacity of the contralateral neural fold to regulate the deficiencies due to the unilateral excision of the neural fold. After bilateral removal of the chick neural folds (A) followed by unilateral graft of the same structure excised from a quail (B), quail neural crest cells migrate on both sides of the chimera (C).

exited from the surface of excision. Couly *et al.* (1996) further demonstrated that the source of the regenerating cells was the neural crest itself limiting the excised territory. This was shown by grafting quail neural fold at these levels and demonstrating that quail neural crest cells filled the gap corresponding to the excision by undergoing a caudorostral and rostrocaudal longitudinal migration (Fig. 3.24).

Different experimental designs were used to compare the extent of regulation of the neural fold of the chick left *in situ* to that of the grafted quail neural fold. It was found that the regeneration capacity was identical for the grafted quail or the endogenous chick neural fold and depended only on the level of the excision. This definitively rules out the improbable possibility that introduction of quail grafts might have biased the results observed. Finally, the regeneration of neural crest cells by the neural fold bordering the excised area was also

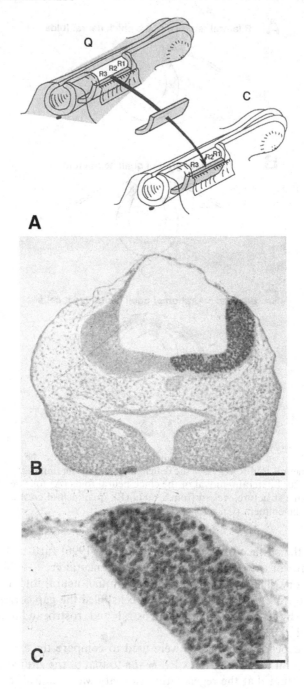

**Figure 3.23** (A) Schematic drawing showing the graft of the ventral half of a quail (Q) neural tube into a chick (C) embryo at 5ss at the level of R1–R3. The dorsal ectoderm is put back to its normal position after graft completion. (B, C) Transverse sections at the rhombencephalic level of a 30ss chimera. The neural epithelium was treated with

observed by Saldivar *et al.* (1997). In conclusion, the divergent interpretations given to the striking phenomenon of neural crest regeneration show once more that the use of precise and stable cell markers is necessary for following highly dynamic phenomena.

In the fate map established by grafting fragments of the quail neural fold into chick embryos from the midmesencephalic level to r8, the contribution of the posterior half of the mesencephalon and of each rhombomere to the facial skull could be determined (Figs. 3.20, 3.21) (Couly *et al.*, 1996). The regulation experiments just described revealed that the neural crest cells migrating along the anteroposterior axis to compensate the deficiencies resulting from the ablation of the neural folds in r1 to r7 included, expressed the *Hox* code corresponding to their level origin. Neural crest cells were therefore not influenced by the environment of the branchial arches in their capacity to express *Hox* genes.

This stability in the *Hox* code of neural crest cells was confirmed when the neural fold was transposed either anteriorly or posteriorly. Moreover, these experiments showed that expression of *Hox* genes in the branchial arch ectoderm is independent from that of the neural crest cells which colonize this arch. Thus, when the neural fold of r1–2 was transposed to the r4–6 level, the neural crest cells emigrated into branchial arch 2 instead of branchial arch 1. Although they failed to express *Hoxa-2*, the surface ectoderm of branchial arch 2 expressed *Hoxa-2* normally. This shows that the neural crest cells do not induce the superficial branchial arch ectoderm to express a given *Hox* code as hypothesized by Hunt *et al.* (1991a).

### 3.5.4 Patterning capacities of the neural crest

Already in 1946, Hörstadius and Sellman raised the question as to whether a prepattern of its mesectodermal derivative neural crest is established in the cephalic neural crest. In an interesting experiment, they lifted off the neural plate and ridge of the head region in *Amblystoma mexicanum* at the neurula stage and replaced it after a 180° rotation. Many anomalies of the cranial skeleton resulted from the operation and the transplanted neural crest populations formed skeletal structures more appropriate in shape to their original location along the neuraxis than to their new position in the host.

In the chick embryo, heterotopic transposition of fragments of the neural primordium along the neuraxis (Le Douarin and Teillet, 1974) led to the notion

**Figure 3.23** (*cont.*)
pancreatin before grafting to remove contaminating mesodermal cells. The grafted half neural tube of the quail is perfectly integrated in the chick embryo. The graft clearly did not yield neural crest cells since no quail cells are seen in the chimera apart from the grafted neuroepithelium itself. Note the close contact between the graft and the host's superficial ectoderm (in B and C). Quail cells are evidenced by the QCPN Mab. Bar = 90 μm in B and 20 μm in C. (Reproduced, with permission of Company of Biologists Ltd, from Couly *et al.*, 1996.)

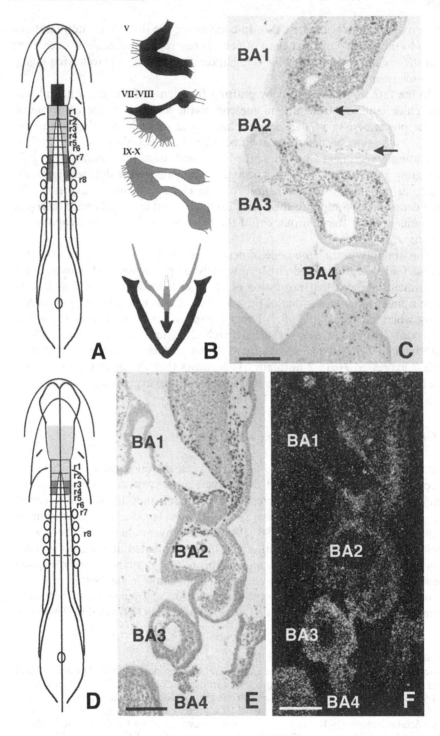

that the population of neural crest cells exiting from a given level of the neuraxis is not committed to a definite fate. The latter is acquired in response to signals from the environment of the target tissues that the neural crest cells colonize. This was particularly well documented for the derivatives of the neural crest belonging to the PNS and to endocrine tissues. For example, neural crest cells arising from the posterior level of the rhombencephalon, which normally constitute most of the enteric nervous system and do not reach the suprarenal gland, can do so and can even differentiate into normal medullary cords of chromaffin cells if they are transplanted to the brachial level (somites 18–24). Inversely, neural crest cells from the brachial level, transplanted to the posterior rhombencephalon, will colonize the gut and differentiate into enteric plexuses (see Chapter 5). Similar experiments involving the mesencephalic neural crest have shown, however, that one particular neural crest-derived lineage is endowed with a certain level of commitment. Thus, the mesencephalic neural crest transplanted at the truncal level yields ectopic pieces of cartilage (e.g., rods of cartilage were seen in the dermis or in the kidney at the graft site). Moreover, Noden (1983b) transplanted the mesencephalic neural fold (which normally yields cells which colonize BA1 and participate in the formation of the lower jaw) to the hindbrain (about the otic level of the rhombencephalon) and found that, in this ectopic position, the transplanted crest cells change the fate of the second (so-called hyoid) arch into that of a first arch. In fact, a second set of jaw skeletal tissues including Meckel's cartilage developed while the endogenous hyoid bone was reduced. This indicated that, in contrast to the plasticity of its neural derivatives, the mesectodermal derivatives of the neural crest are endowed with intrinsic patterning activity. Even transplanted ectopically, at the hindbrain level, the mesencephalic neural crest cells develop into a lower jaw as they would have done if left *in situ*.

Recent experiments carried out by Couly *et al.* (1998), in which definite fragments of the 5–6ss quail mesencephalic neural fold and associated dorsal neural tube were transplanted at the level of r4–6 of stage-matched chick,

**Figure 3.24** (A) Demonstration that the neural crest regenerates from the neural folds limiting the excised territory. The neural fold was bilaterally excised in a 5ss chick embryo from the midmesencephalic level to somite 3. Excision was followed by the bilaterally orthotopic graft of fragments of the quail neural fold at level r7–r8a (anterior) and corresponding to the posterior mesencephalon. The territory devoid of neural fold (r1–r6 included). (B) Schematic representation of quail cell distribution in the cranial ganglia of nerves V, VII–VIII, IX–X, and in the hypobranchial skeleton. The screen indicates the origin of the regenerating cells. (C) QCPN Mab staining of an operated E3 chick embryo revealed that branchial arches 1–4 (BA1–4) are colonized by quail cells. Note that invading cells have not yet completely filled up BA2 (arrows). (D) Schematic drawing of an experiment in which the neural folds were removed bilaterally from the diencephalic/mesencephalic constriction down to r5. Excision was followed by the orthotopic bilateral graft of quail neural fold fragments corresponding to r3/4. The result is seen after QCPN Mab staining at E3, on (E) and after (F) *in situ* hybridization with a *Hoxa*-2 probe. Bar = 100 μm. (Reproduced, with permission of Company of Biologists Ltd, from Couly *et al.*, 1996.)

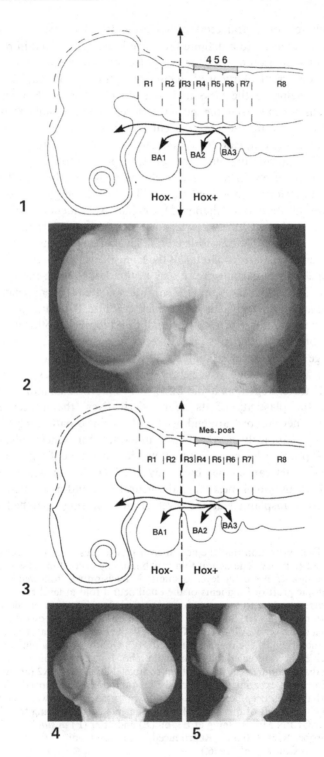

confirmed Noden's findings and resulted also in the duplication of some skeletal features of the lower jaw. In this experimental situation, the mesencephalic crest cells failed to express any *Hox* genes. As far as their "Hox code" is concerned, the crest cells colonizing BA2 in these circumstances were therefore in the same situation as those of mice embryos in which the *Hoxa-2* gene was disrupted (Gendron-Maguire *et al.*, 1993; Rijli *et al.*, 1993). As will be discussed below, $Hoxa\text{-}2^{-/-}$ mice also exhibit duplication of the lower jaw. These transposition experiments thus produce a loss of function of *Hoxa-2* which is restricted to crest cells of the second branchial arch.

Couly *et al.* (1998) also selectively transplanted the neural fold from the mesencephalon to the r4–6 level and subsequently found ectopic rods of quail cartilage, but these were not properly organized into recognizable lower jaw structures. It appears, therefore, that the presence of the neural tube neuroepithelium from the mesencephalic level is critical for the patterning capacities of the neural crest to be fully expressed.

It was then tempting to manipulate the midhindbrain neural crest so that the reverse situation would be generated: i.e., colonization of BA1 by crest cells expressing *Hoxa-2*. The experiments described above, showing that regeneration of neural crest cells takes place by longitudinal migration of crest cells along the anteroposterior axis from the neural folds limiting the excised territory, provided a cue for devising the following experimental paradigm: the idea was to lead neural crest cells from a posterior origin, thus expressing *Hox* genes, to colonize BA1 while avoiding *Hox* gene-negative cells to reach it.

First, it was shown that, at 5–6ss, the resection of the neural fold including the presumptive mesencephalon, plus the first three rhombomeres (r1–r3), is followed by a regeneration from caudal to rostral from the territory corresponding to r4–6. These r4–6 neural crest cells express the *Hoxa-2* gene and colonize BA1. In contrast, the diencephalic neural crest cells, that are left *in situ*, migrate essentially to the frontal and nasal area and do not contribute to BA1 (Fig. 3.24). If, in contrast, the resection does not include the anterior half of the mesencephalic neural fold, BA1 is essentially colonized by crest cells of mesencephalic origin that do not express *Hoxa-2* and develop into a normal jaw. The experimental designs represented in Figs 3.24 and 3.25 result in the colonization of BA1 by *Hoxa-2*-expressing quail cells. In the chimeric embryos resulting from this operation the lower jaw was absent and the BA1 quail cells were eventually found exclusively in neural derivatives. It thus appears that this *Hoxa-2* gain-of-function experiment has a morphogenetic outcome opposite to

**Figure 3.25** (1) Schematic drawing showing that the neural fold has been removed from the level of the diencephalic–mesencephalic constriction down to r6 inclusive (the operation is carried out at 5ss). Quail neural fold fragments are implanted bilaterally at r4–r6 level. The regenerating neural crest cells fill up BA2, and BA1, and to some extent the maxillary bud. Thus, *Hoxa-2* cells colonize BA1. (2) At E4 the operated embryos lack a lower jaw. (3) The same operation is performed except that the grafted quail neural folds are of posterior mesencephalic origin. (4) Normal head of an E4 chick. (5) The operated embryo develops a normal lower jaw.

the loss of function obtained in mice by targeting the *Hoxa-2* gene (Gendron-Maguire *et al.*, 1993; Rijli *et al.*, 1993) and in chicken by transposition of neural folds from rostral to caudal so that not expressing neural crest cells *Hoxa-2* colonize BA2 (Fig. 3.25).

Strikingly, if in the experimental design represented in Fig. 3.25 the endogenous r4–6 neural fold is replaced by a segment of mesencephalic crest, then mesencephalic crest cells (expressing no *Hox* genes) colonize BA1 and a normal jaw entirely of graft origin develops.

In conlusion, the interpretation according to which *Hoxa-2* acts as a selector gene which inhibits the development of a lower jaw type skeleton (Rijli *et al.*, 1993) (see discussion in Section 3.6.1.2) is corroborated by these experiments.

Moreover, while confirming the seminal observation made by Noden (1983b), from which it was perceived, for the first time, that the neural crest cells themselves possess a certain amount of patterning information, these experiments raise the following points:

1. The midbrain neural crest cells keep their capacity to differentiate into cartilaginous structures in ectopic positions whether the latter are truncal or rhombencephalic.
2. Patterning of skeletal structures typical of a lower jaw occurs only if mesencephalic neural crest cells migrate into BA1 or BA2 and not posteriorly.
3. Patterning of lower jaw skeletal structures in the BA2 context occurs only if the transplanted neural fold is associated with its mesencephalic neuroepithelium. Transplantation of the neural fold alone merely results in the development of cartilage rods.
4. Transposed fragments of mesencephalic and rhombencephalic neural fold along the anteroposterior axis (whether or not the graft includes the adjacent fragment of the neural tube) yield neural crest cells that keep their endogenous Hox code.

## 3.6   Genetic analysis of the development of the mesectodermal derivatives of the neural crest: roles of *Hox* genes and retinoic acid (RA)

### 3.6.1   Disruption of certain Hox genes of the first paralogous groups interferes with the development of neural crest-derived branchial arches structures

*3.6.1.1   Disruption and overexpression of* Hoxa-1 *gene* (Hox1-6). In the mouse, *Hoxa-1* expression in the prospective r4–6 region of the hindbrain is only transient (at 7.5 to 8.5 dpc) (see Fig. 3.16) and precedes the restriction of *Hoxb-1* to r4, and of *Krox-20* to r3 and r5 and the overt hindbrain segmentation.

*Hoxa-1*$^{-/-}$ mouse embryos have only five rhombomere-like structures (r1–r5) (Lufkin *et al.*, 1991; Chisaka *et al.*, 1992; Mark *et al.*, 1993; Carpenter *et al.*, 1993). The first three rhombomeres (r1–r3) seem normal in size and also in the expression pattern of several markers such as CRABP-I, *Hoxa-2*, *Hoxb-2*, and *Krox-20* (Dollé *et al.*, 1993). In contrast, r4 and r5 are strongly reduced and fused in a single "segment-like structure" which seems to incorporate r6. This is suggested by the absence of *Krox-20* expression normally present in r5 and by the juxtaposition of a short expression domain of *Hoxb-1* (corresponding to an r4 reduced in size) with the expression domain of *Hoxa-3* and *Hoxb-3* characteristic of normal r6. Juxtaposition of r4 and r6 does not lead to the formation of a rhombomere boundary as shown by Guthrie and Lumsden (1991) in another experimental paradigm.

In *Hoxa-1* mutants the motor neurons of the facial nerve (derived from r4 and r5) are scarce and those of the abducens nerve are absent. Interestingly, in *Krox-20*$^{-/-}$ mice and in the *kreisler* mutation characterized by a severe reduction or absence of r5, agenesis of the abducens nerve motor nucleus is also observed (Frohman *et al.*, 1993; Schneider-Maunoury *et al.*, 1993).

In *Hoxa-1*$^{-/-}$ mice, the motor nuclei of nerves V and IX (glossopharyngeal), respectively derived from r1–3 and from r6–7, are not affected.

The absence of defects in regions caudal to r6 in *Hoxa-1*$^{-/-}$ mice indicate that the functional domain of this gene corresponds to its more rostral domain of expression and does not involve the mesectodermal derivatives.

Overexpression of *Hoxa-1* has been obtained in the zebrafish by injecting synthetic RNA into the fertilized egg (Alexandre *et al.*, 1996). This resulted in the abnormal development of the anterior hindbrain, and the duplication of Mauthner neurons in r2. *Krox-20* expression was increased in r3 and not in r5, in contrast to the situation in the mouse where overexpressing *Hoxa-1* had no effect on *Krox-20* (Zhang *et al.*, 1994b). In these experimental fish embryos r3 was enlarged and the first pharyngeal arch skeleton (i.e., Meckel's cartilage and palatoquadrate) failed to form. The second arch-derived ceratohyals (also called ceratobranchials) were thickened and partially bifurcated. It seems, therefore, that the neural crest cells that normally form the first pharyngeal arch have to some extent been respecified since in the fish the crest cells which populate the first branchial arch originate predominantly from r1–3, whereas those populating the second arch originate from r3–5 (Schilling and Kimmel, 1994). Fusion of the first two branchial arches in the *Hoxa-1*-overexpressing embryos seems a likely mechanism to account for this phenotype. This was not observed in the *Hoxa-1* transgenic mice in which *Krox-20* expression was normal. However, transformation of r2 to r4 was found in both the mammalian and fish models. It is interesting to note that the phenotype generated by treating early zebrafish gastrulas with a pulse of RA (Hill *et al.*, 1995) is strikingly similar to that generated by overexpression of *Hoxa-1*. This must be related to the fact that treatment with RA causes an ectopic anterior expression of *Hoxa-1*, thus showing a functional link between RA and this *Hox* gene (Holder and Hill, 1991; Hill *et al.*, 1995).

*3.6.1.2   Disruption of* Hoxa-2(Hox1-11) *gene in the mouse results in spectacular malformations of the mesectodermal derivatives of the anteriormost branchial arches.* In *Hoxa-2$^{-/-}$* fetuses, analysed at 17.5 or 18.5 dpc (Gendron-Maguire *et al.*, 1993; Rijli *et al.*, 1993) the skeletal elements derived from the second branchial arch are lacking (Fig. 3.26). Such is the case of the stapes (S), the styloid bone (SY) and the lesser horn of the hyoid bone (LH). They had, instead, abnormal skeletal pieces membranous and endochondral in origin which consitute a duplication of proximal bones normally present in the first mandibular arch: a supernumerary malleus (M2) fused to an ectopic truncated Meckel's cartilage (M2) and extra incus (I2), tympanic (T2), and squamosal (SQ2) bones. This caudal set of first arch structures are positioned in such a way that, together, they form a mirror image of their orthotopic counterpart. In addition, the *Hoxa-2$^{-/-}$* mutants possess a rod-like cartilage fused rostrally to the alisphenoid (homologous to the reptilian epipterigoid) and to the supernumerary incus (homologous to the reptilian quadrate). This supernumerary element, absent in normal mice, was homologized to the reptilian upper jaw cartilage pterygoquadrate (PQ) on the basis of its anatomical relationships with the alisphenoid and the incus. In addition, the epithelial auditory meatus normally derived from the first branchial arch was duplicated.

The molecular identity of the rhombomeres and of the neural crest they produce was tested in *Hoxa-2$^{-/-}$* embryos by looking at the expression of genes such as *Hoxb-1*, *Hoxb-2* (the paralogs of *Hoxa-2*) *Hoxa-3*, *Krox-20*, *goosecoid* (*Gsc*), and CRABPI. Rhombomeres 2–6 were present in the mutant embryo and showed a normal expression pattern for these markers (Gendron-Maguire *et al.*, 1993; Rijli *et al.*, 1993). Thus *Krox-20* was exclusively expressed in r3 and r5 whereas *Hoxb-1* was restricted to r4 in 9.5 dpc *Hoxa-2$^{-/-}$* mice. In the neural crest, *Hoxb-2* was expressed in the cells destined to form the acousticofacial ganglia and in the mesenchyme of the second and third branchial arches, as in normal embryos.

*Goosecoid* homeobox gene transcripts, which are restricted to the caudal mesenchyme of the first branchial arch and the most rostral portion of the second arch at E10.5 (Gaunt *et al.*, 1993), were not altered in *Hoxa-2* mutants (Rijli *et al.*, 1993).

All these observations justify the interpretation according to which removal of *Hoxa-2* activity during the specification of the branchial arches reveals the existence of a skeletogenic ancestral pattern program common to the neural crest cells which colonize the two first branchial arches, namely the posterior mesencephalic crest cells and those which arise from r1–4 inclusively. This pattern is manifested by the development of atavistic bones present in reptiles.

Evolution from reptiles to mammals has involved changes of the jaw joint that have been well identified in the Therapsids which are considered as the reptilian progenitors of mammals (reviewed in Mark *et al.*, 1995). In therapsids the lower jaw is formed by the dentary bone and several postdentary elements, including the articular (see Fig. 3.27) which is derived from the caudal portion of Meckel's cartilage. Articulation of the lower with the upper jaw is mediated by the articular bone which contacts the quadrate. During evolution of

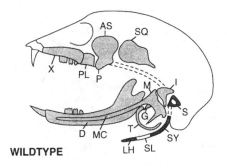

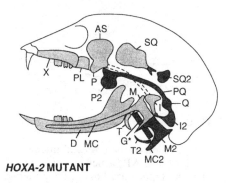

**Figure 3.26** Skeletal elements derived from BA1 ▨ and Ba2 ■ in the wild-type and *Hoxa-2* null mutant mouse fetuses. In the wild-type, the stapes (S), styloid bone (SY), stylohyoid ligament (SL), and lesser horn of the hyoid bone (LH) all derive from the cartilage of the second arch (Reichert's cartilage). In the *Hoxa-2* mutant, several bones of the lower jaw are duplicated. The duplicated bones are thought to originate from the hyoid (BA2) arch which, in the absence of Hoxa-2 activity, develops according to a BA1 pattern. The duplicated bones are: P and P2, pterygoid and extrapterygoid bones; I and I2, incus and extraincus; MC and MC2, Meckel's cartilage and extra-Meckel's cartilage; M and M2, malleus and extramalleus; T and T2, tympanic and extratympanic; SQ and SQ2, squamosal and extrasquamosal; G and G*, gonial and extragonial. Other bones are: Q, quadrate; AS, alisphenoid; D, dentary; PL, palatine; X, maxillary; PQ, pterygoquadrate: the latter present in *Hoxa-2*$^{-/-}$ mice, represents the reemergence of an atavistic reptilian character. (Reproduced, with permission, from Rijli *et al.*, 1993. Copyright Cell Press.)

therapsids to mammals, the dentary-bone has enlarged caudally to meet the squamosal, thus establishing the dentary-squamosal or temporomandibular joint (see reviews in Olson, 1959; Allin, 1975; De Beer, 1985; Walker, 1987; Mark *et al.*, 1995). Moreover, elements of the primary jaw joint were reorganized to become the three-ossicular chain of the middle ear. Thus, the articular and quadrate bones and the articuloquadrate joint were annexed to the middle ear to form the malleus, the incus and the malleoincudal joint that became located between the tympanic membrane and the stapes. The gonial and angular bones became part of the adult malleus and tympanic bone, respectively, while other reptilian bones have disappeared. Thus, the spenial and coronoid

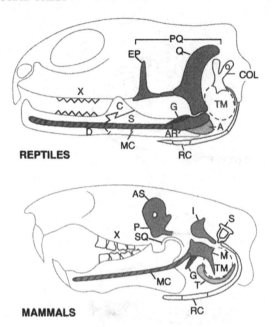

**Figure 3.27** Schematic representation of the jaws and middle ears of reptilian and mammalian fetuses. A, angular bone; AR, articular bone; AS, alisphenoid bone; C, coronoid bone; COL, columella; D, dentary bone; EP, epipterygoid bone; G, gonial (prearticular) bone; I, incus; M, malleus; MC, Meckel's cartilage; P, pterygoid bone (i.e., the pterygoid process of the alisphenoid bone); PQ, pterygoquadrate cartilage; Q, quadrate bone; RC, Reichert's (second arch) cartilage; S, stapes; SQ, squamosal bone (note that although not represented here, the squamosal bone also exists in reptiles); T, tympanic bone; TM, tympanic membrane; X, maxillary bone. (Reproduced, with permission, from Mark *et al.*, 1995.)

have no mammalian equivalent, and the quadrate bone was significantly reduced in size during the therapsid phase of mammalian evolution (Allin, 1975) (Fig. 3.27).

In the mouse first arch this ground pattern is modified by a *Hox*-independent but RA-dependent process (see below), whereas *Hoxa-2* acts as a selector gene in the second arch. In the absence of *Hoxa-2* activity, second arch neural crest cells produce a virtually complete first arch-type set of bones. However, the specification of hyoid arch neural crest cells in this case corresponds to an atavistic ground pattern intermediate between the reptilian and mammalian type of differentiation as shown by the presence of a pterygoquadrate (a reptilian primitive character) and of an incus (resulting from the transformation of the reptilian quadrate seen in mammals).

The fact that the first two mammalian branchial arches have an identical ground pattern has to be related to their derivation from the branchial basket of their jawless agnathan ancestors formed by identical cartilages associated with the gills (Langille and Hall, 1988a,b, and references therein). This interpretation, proposed by Rijli *et al.* (1993) and Gendron-Maguire *et al.* (1993), is supported by the fact that, as shown in quail–chick chimeras, when *Hoxa-2* expressing neural crest cells colonize BA1 the jaw skeleton fails to develop. (Couly *et al.*, 1998) (see Section 3.5.4 above.).

*3.6.1.3 Null mutants of* Hoxa-3 (Hox1-5) *and the Di George syndrome.* As mentioned above, members of the third paralogous group of *Hox* genes have their anterior limit of expression in the neural tube at the boundary between r4 and r5. Thus *Hoxa-3* and *Hoxb-3* are expressed not only in the neural tube but also in neural crest cells emanating from r5–r8 and also in the third and fourth branchial arches endoderm and ectoderm (see Fig. 3.16).

*Hoxa-3* null mutants produced by Lufkin *et al.* (1991) and by Chisaka and Capecchi (1991) have an array of abnormalities in several pharyngeal neural crest derivatives. These defects include malformations of cartilages and bones of the jaw, disorganization of the throat musculature, of the heart and great vessels, deletion of the thymus and parathyroids, and various degrees of thyroid hypoplasia. In addition, abnormalities of the trachea are also seen in *Hoxa-3$^{-/-}$* mice.

These malformations are strikingly similar to those recorded in the congenital human Di George syndrome (Di George, 1968; Burke *et al.*, 1987; Pueblitz *et al.*, 1993) characterized in the haploid state (diploid mutants are lethal) by hypoparathyroidism, thymic, and thyroid hypoplasia, cardiac and large aortic trunk deficiencies, plus craniofacial malformations.

Genesis of the thymic and thyroid malformations in the *Hoxa-3$^{-/-}$* mice embryos was carefully investigated by Manley and Capecchi (1995). Neural crest cells were labeled by various methods: (1) using either DiI as proposed by Serbedzija *et al.* (1992), or an antibody to CRABP-I, known to be expressed at E9.5 in neural crest cells migrating from r2 to r6 into the pharyngeal arches (Maden *et al.*, 1992); (2) applying an antibody to *Hoxb-1* to label r4 neural crest cells; (3) applying *in situ* hybridization with the *Hoxa-3* mutant allele inserted in place of the endogenous *Hoxa-3* gene; this construct produces a fusion transcript initiated from the *Hoxa-3* promoter. Probes against *Hoxa-2*, *Hoxb-3*, and *Hoxd-3* were also used on these mice.

All these methods provided convergent results showing that migration of neural crest cells to BA3 and BA4 was not affected in the mutant either quantitatively or temporally. This means that *Hoxa-3* is not required for the expression of *Hoxa-2* or of other genes of the third paralogous group. Neither is it required for the proper neural crest cells migration to the branchial arches.

One clear effect of *Hoxa-3* deficiency on further development of these neural crest cells was the observed differences in *Pax-1* gene expression in mutants and wild-type embryos. *Pax-1* expression begins at E9.5 in the third pharyngeal pouch and continues throughout thymus development (Deutsch *et al.*, 1988). At E9.5 *Pax-1* expression is about the same in normal and mutant mice. From E10.5, there is a marked decrease in *Pax-1* expression in the third branchial arch of *Hox-3$^{-/-}$* embryos whereas it remains at normal levels in the other sites of expression of *Pax-1* (e.g. first and second branchial arches and somites). FGF3 normally present in the endoderm of the third pharyngeal pouch is not affected in the mutants and neither is *Hoxa-2* expression. It appears therefore that *Hoxa-3* is required for *Pax-1* expression in the third branchial pouch. Since *Pax-1*$^{-/-}$ mice show thymus hypoplasia (Dietrich and Gruss, 1995), it seems that *Hoxa-3* is upstream of *Pax-1* in the regulatory pathway involved in thymus development.

The abnormalities of the thyroid gland and calcitonin cells are variable in the homozygous *Hoxa-3* null mutant. They affect endodermally derived follicular cells and neural crest-derived calcitonin-producing cells of neural crest origin. As shown by Fontaine *et al.* (1973), the precursors or calcitonin-producing cells initially home to the ultimobranchial (UB) bodies derived from the fourth pharyngeal pouch endoderm. The UB bodies thereafter fuse with the thyroid gland thus allowing calcitonin-producing cells to colonize the thyroid and to become the so-called parafollicular "clear" cells (see Chapter 5, Section 5.8.3). Thyroid hypoplasia is a general rule in *Hoxa-3*$^{-/-}$ mice with a number of thyroid follicles severely reduced. Fusion of the UB bodies and thyroid anlagen failed to occur in several cases, leading to the formation of small vesicular structures containing calcitonin-producing cells as observed in the Di George syndrome in humans (Di George, 1968).

The similarity between the *Hoxa-3* null mutant phenotype and that of humans affected by the Di George syndrome is certainly remarkable. However, the latter is autosomal dominant and associated in certain cases with deletions and translocations involving chromosome 22, whereas the anomalies resulting from *Hoxa-3* loss of function are autosomal recessive and *Hoxa-3* maps to chromosome 7. It should be noted that most patients with the Di George syndrome have a normal karyotype and may present point mutations in the *Hoxa-3* gene. The similarity of the two phenotypes suggests that the two genes belong to a common developmental molecular pathway.

### 3.6.2 Role of RA in the development of hindbrain and of its neural crest derivatives

It has long been known that retinoids (vitamin A derivatives) are essential for normal development, growth, vision, maintenance of several tissues, reproduction, and survival (Wolbach and Howe, 1925; Wilson *et al.*, 1953; for reviews see Sporn *et al.*, 1984; Livrea and Packer, 1993). Vitamin A, all-*trans*-retinol of dietary origin, is converted into its carboxylic acid derivative retinoic acid (RA) which exists in two forms: all-*trans*-retinoic acid and 9-*cis*-retinoic acid. It was rapidly shown that RA was the biologically active form of the vitamin since most of the effects (except for vision; Wald, 1968) due to vitamin A deficiency in fetuses and in young and adult animals can be prevented and reversed by RA administration. On the other hand, excess of vitamin A or RA administered during development has dramatic teratogenic effects (see Morriss-Kay, 1992, and references therein), and local administration of retinoids in certain embryonic areas such as the developing limb bud has spectacular morphogenetic consequences (see Tabin, 1991, for a review). This has led to the notion that RA could be an important signaling molecule acting as a morphogen and playing a critical role during organogenesis.

This assumption is supported by the fact that RA has been detected endogenously in the embryos of all the major vertebrate groups where it is distributed in different embryonic fields. It is present during gastrulation in zebrafish (Costaridis et al., 1996) and Xenopus (Chen et al., 1992; Kraft et al., 1994). Moreover, RA is synthesized from retinol in Hensen's node of the mouse (Hogan et al., 1992). Later in development, RA has been found to be distributed in the developing CNS of 10.5 dpc mouse embryos in concentrations decreasing from posterior to anterior so that in the forebrain and midbrain only very low levels of RA are observed whereas the hindbrain contains slightly higher levels, the highest concentrations being found in the spinal cord (Horton and Maden, 1995).

Figure 3.28 (Lohnes et al., 1995) illustrates the major components of the retinoid-signaling pathway.

Apart from retinol and its derivatives, various cellular proteins endowed with binding properties for these products have been discovered and their role elucidated. First, the role of the cellular retinol-binding proteins (CRBPI and II) is probably to store retinoids and regulate free levels of biologically active retinoids in a given cell (reviewed in Livrea and Packer, 1993). The second class of proteins binding retinoids are encoded by two multigene families, the retinoic acid receptors (RAR$\alpha$, $\beta$, $\gamma$), and the retinoid X receptors (RXR$\alpha$, $\beta$, $\gamma$). Both families encode ligand-inducible *trans*-regulators belonging to the nuclear receptor multigene family (reviewed in Kastner et al., 1994; Petkovitch, 1992; Kliewer et al., 1994). RARs appear in different isoforms (RAR$\alpha$1 and 2, RAR$\beta$1-4, RAR$\gamma$1 and 2; see Lohnes et al., 1995, for references) which are derived by differential promoter usage and alternative splicing. Furthermore, the expression of the second of each RAR isoform (i.e. RAR$\alpha$2, $\beta$2, $\gamma$2) is modulated by RA through a specific DNA sequence designated RARE (retinoic acid responsive enhancer) present in the promoter region of each of these genes.

The notion has progressively emerged that the morphogenetic effects of RA are linked to those of *Hox* genes. RA receptor genes as well as *Hox* genes encode transcriptional factors that belong to an integrated functional network implicated in patterning the skeleton of the limb, of the vertebral column, and of the head and teeth.

*3.6.2.1 Relationships between retinoids and* Hox *gene expression.* Embryological, cell culture, and transgenic experiments led to the notion that the most anteriorly expressed *Hox* genes are *in vivo* mediators of RA morphogenetic activity.

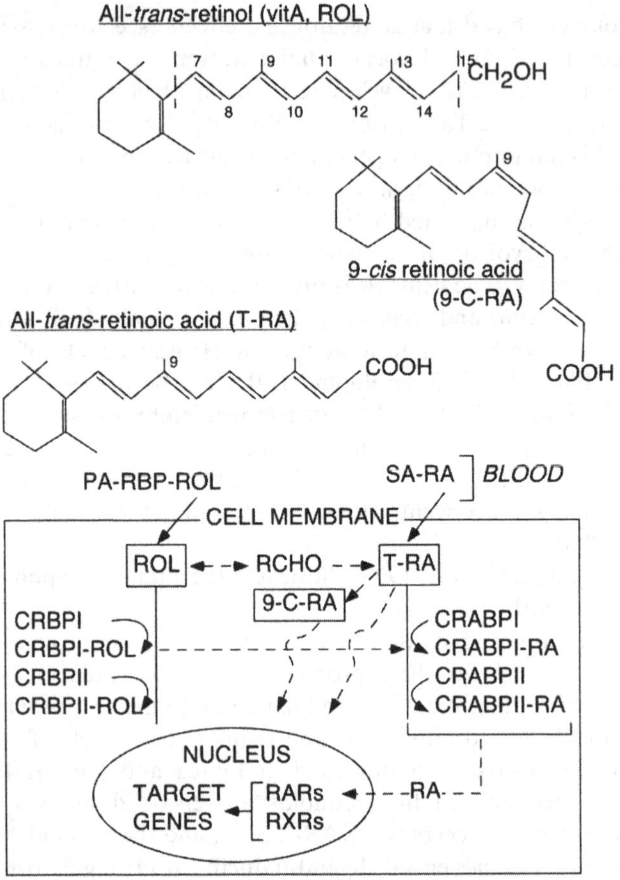

**Figure 3.28** Schematic representation of the major components of the retinoid-signaling pathway. CRBP, cellular retinol-binding protein; CRABP, cellular retinoic acid-binding protein; RBP, retinol-binding protein; RAR, retinoic acid receptor; RXR, retinoic X receptor. (Reprinted from Lohnes *et al.*, 1995. Copyright (1995), with permission from Elsevier Science.)

A. RA AS A REGULATOR OF *Hox* GENE EXPRESSION IN CULTURED CELLS. Soon after homeobox-containing genes were discovered in vertebrates, the pioneering work of Gruss and his colleagues showed that expression of *Hox* genes can be influenced by all-*trans*-RA (e.g., Colberg-Poley *et al.*, 1985; Breier *et al.*, 1986). Murine F9 teratocarcinoma cells can be cultured virtually indefinitely without differentiating. If treated with RA, they differentiate into non-tumorigenic parietal endodermal cells and during this process almost all of their *Hox* genes are activated. One of the early response genes to RA was later identified as being *Hoxa-1* (formerly *Hox*1.6; LaRosa and Gudas, 1988), and a RARE element identical to that found in the *RAR-β* gene (De Thé *et al.*, 1990; Hoffmann *et al.*, 1990; Sucov *et al.*, 1990) was further identified in its 3' region (Langstone and Gudas, 1992). Similar RAREs were also found in *Hoxb-1* (Marshall *et al.*, 1994; Studer *et al.*, 1994) and *Hoxd-4* (Popperl and Featherstone, 1993).

Boncinelli and coworkers conducted a thorough study of *Hox* gene induction by RA in teratocarcinoma cells (see Boncinelli *et al.*, 1991).

The human embryonal carcinoma (EC) cell lines NT2/D1 and tera2/clone13 which can be induced by retinoids to differentiate into a variety of cell types including neurons (see Andrews, 1988, and references therein) were used to test the effect of RA on the induction of *Hox* genes (Andrews, 1984, 1988; Andrews *et al.*, 1986). *Hox* genes were shown to be differentially activated sequentially by RA in a concentration-dependent manner starting from the most 3' (e.g. *Hoxb-1*) up to the most 5' (e.g., *Hoxb-9*) gene.

Thus concentrations of RA required to induce *Hoxb-1* at peak levels is as low as 10 nM whereas 1 μM is necessary for *Hoxb-5–9* gene induction. Activation is sequential and strictly colinear with the 3'-5' linear arrangement of *Hox* genes on the chromosome. Moreover the 3' located *Hox* genes respond faster to RA induction than more 5' located ones. RA induction of the most 5' paralog groups of *Hox* genes does not occur and the boundary between responding and non-responding genes corresponds to the ninth paralogous group (Mavilio, 1993, and references therein).

Thus *Hox* genes expressed in the hindbrain which belong to the more 3' located paralogous groups are more susceptible to RA influence. This led to the reasonable assumption that endogenous levels of RA are critical in the expression of these genes and therefore on hindbrain morphogenesis.

B. VITAMIN A DEFICIENCY AND EXCESS OF VITAMIN A BOTH PERTURB HINDBRAIN SEGMENTATION AT MORPHOLOGICAL AND MOLECULAR LEVELS. Vitamin A (retinol) is an essential component of the vertebrate diet and deprivation as well as excess of vitamin A during pregnancy is followed by congenital malformations that affect the development of many structures, including the eye, genitourinary tract, branchial arch derivatives, heart, lung, and kidneys.

Experimental data on the relationships between vitamin A deficiency or excess of vitamin A and hindbrain segmentation have been produced recently on both avian and mammalian embryos.

A vitamin A-deficient diet applied to quails results in embryonic malformations characterized by abnormalities in nervous system development concerning pattern specification, neurite outgrowth, and neural crest derivatives. The neural tube of RA-deficient embryos had thinner walls than that of normal embryos, while anti-neurofilament (anti-NF) antibody staining revealed fewer neurons expressing this protein. No neurons extented neurites outside the neural tube. While patterning of the neural tube along the dorsoventral axis seemed to be unaffected (on the criteria provided by several dorsoventral markers), its anteroposterior patterning was highly abnormal in the hindbrain region. The most affected region corresponds to the myelencephalon since the features to be described below indicate that r4–8 might be absent as if they had failed to be specified during CNS development. The other alternative is that cells corresponding to the r4–8 region of the neural tube may still be present without any visible signs of segmentation at either the morphological or molecular levels.

The trigeminal neural crest pattern and the expression of *Hoxa-2* were normal in the anterior rhombencephalon suggesting that r1–3 develop normally in vitamin A-deficient embryos, except, however, that *Krox-20* gene expression was not detected in r3.

More posteriorly, the transverse stripe of *Hoxb-1* expression, a prominent marker for r4, was not present. Transcripts of *Krox-20* were missing in r5, *FGF3* expression did not occur in r4–6, and the region of *Hoxb-4* expression corresponding to r7–8 was absent.

In RA-deficient embryos the migration pattern of neural crest cells into the branchial arches was severely affected. The streams of neural crest cells exiting from the more anterior rhombomeres were normal but branchial arches 2 and 3 were not discernible because the branchial grooves did not form and HNK1-positive neural crest cells were absent in this region. Moreover, at the postotic level, the cells from r5 to r7, which normally colonize branchial arch 3, were totally absent (Maden *et al.*, 1996). Internally, the aortic arches were fused to form one large blood vessel. There was extensive cell death in neural crest cells along the whole neural axis so that the head was only sparsely populated with mesectodermal cells and the dorsal root ganglia (DRG) failed to differentiate and to express neurofilament proteins.

These observations, in agreement with *in vitro* studies, strongly suggest that RA supports the survival and proliferation of crest-derived cells (and particularly neuronal precursors; Henion and Weston, 1994), and interacts with the genes of the apoptotic pathway (Maden *et al.*, 1996).

It is noteworthy that vitamin A deficiency does not significantly affect the development of branchial arch 1 which is colonized by cells that do not express *Hox* genes (see above).

Excess of RA applied to the mouse embryo results in changes of the expression pattern of 3′ *Hox* genes and dysmorphogenesis in the hindbrain and its neural crest derivatives (Morriss, 1972; Morriss and Thorogood, 1978; Webster *et al.*, 1986). *Hoxb-1*, *Hoxb-2*, *Krox-20*, *En1-2*, and *RAR-β* were among the genes whose expression pattern is altered by RA (Morriss-Kay *et al.*, 1991; Holder and Hill, 1991; Conlon and Rossant, 1992; Marshall *et al.*, 1992).

Thus, by treating pregnant mice on days 7–9 of gestation with a single teratogenic dose of RA (20 mg/kg), Conlon and Rossant (1992) observed, 4 hours after the onset of the treatment, that the anterior limits of expression of *Hoxb-1*, *Hoxb-2*, and *RAR-β* were shifted rostrally, whereas, in agreement with the results obtained *in vitro* with the NT2/D1 cell line (see Mavilio, 1993, and references therein), *Hoxb-9* did not respond to RA treatment. The anterior site of expression of *Krox-20* corresponding to r3 was abolished. In contrast to the effect of RA on *Hoxb-1*, *Hoxb-2*, and *Krox-20*, the anteriorization of *RAR-β* expression was only transient. In the embryos examined 20 hours after RA administration, the *RAR-β* limit of expression resumed its normal level. As a whole, RA administered early in development (at gastrulation stage) induces transformation of r2/3 into an r4/5 identity. Expression of *Hoxb-1* and of *Krox-20* in the neural crest was also affected by early RA treatment since it was shifted anteriorly in the crest exiting from r2 and r3 as it is in the neural tube at this level.

By constructing transgenic mice in which the regulatory region of *Hoxb-1* was linked to the *LacZ* reporter gene, Marshall *et al.* (1992) saw that, in normal E10.5 mice, the transgene was expressed at high levels in r4 and in the facial motor nerve VII whose axons exit from r4 and originate from motor neurons located in r4 and r5 while innervating the second branchial arch. After a single injection of RA at 7.5 dpc (20 mg/kg), the transgene expression limit was shifted rostrally and later on formed a transverse band of strong expression at the level of r2 in addition to the normal r4 expression site. Moreover, cranial nerve V, which exits from r2, abnormally expressed the transgene, and the distribution of motorneurones in r2 and r3 was abnormal and reminiscent of that seen in r4–5.

In summary, RA changes the identity of r2/3 which are "posteriorized" but does not affect r4/5 (Morriss-Kay *et al.*, 1991; Conlon and Rossant, 1992; Marshall *et al.*, 1992).

The effect of exogenously applied RA on chick embryos has been examined by Sundin and Eichele (1992). When applied at the primitive-streak stage (gastrulation), RA strongly perturbs the development of the neural anlage: the primordia of the forebrain and eye are smaller; midbrain and anterior hindbrain are fused in a single vesicle reduced in size; a massive cluster of neural crest cells accumulates anteriorly to the otic vesicle representing the primordium of the acousticofacial ganglion which may result from the aggregation of the normally r4-derived ganglion to an extra structure originating from r2/3. This would result from an r2/3 to r4/5 conversion, also observed in the mouse (Sundin and Eichele, 1992). Such dysmorphogenesis is consecutive only to an early applied treatment and does not follow RA administration at the neurula stage. As far as *Hoxb-1* is concerned, modifications of its expression pattern following early RA administration in the chick are similar to those observed in the mouse.

This general posteriorization of hindbrain specific pattern and hypotrophy of forebrain and midbrain structures was also found to occur in *Xenopus* embryos treated with excess of RA at late blastula and gastrula stages (Durston *et al.*, 1989; Sive *et al.*, 1990; Papalopulu *et al.*, 1991; Ruiz i Altaba and Jessell, 1991). As in the mouse, sensory and motor axons of cranial nerves V and VII form a single root. The typical rhombomeric segments are abnormal and *Krox-20* is expressed in a single large stripe.

In zebrafish treated with RA at the midgastrula stage, lesions in midbrain and rostral hindbrain were regularly obtained with the disappearance of expression of the *engrailed* gene, which is a specific marker of mesencephalic and r1 territories. The posterior limit of the effect was located at the r4/5 boundary and the partial transformation of r2 to r4 occurred (Holder and Hill, 1991; Hill *et al.*, 1995).

It is thus striking that excess of RA affects anterior rhombomeres by a process of posteriorization (Morriss-Kay *et al.*, 1991; Leonard *et al.*, 1995). In contrast, RA deficiency causes abnormalities in posterior rhombomeres, the development of which thus appears to require appropriate doses of this morphogen. Moreover, experiments carried out in zebrafish, in which overexpression of *Hoxa-1* reproduces a phenotype strikingly similar to that obtained with excess RA, suggest that *Hoxa-1* could be a target for exogenously applied RA (Alexandre *et al.*, 1996).

*3.6.2.2 Sites where vitamin A derivatives are present in the embryo.* The striking effects of either excess or deprivation of RA on morphogenesis raised the problem of the presence and production of endogenous retinoids.

Direct measurements of endogenous retinoids were carried out by Durston *et al.* (1989) in *Xenopus* embryos. They found all-*trans*-RA at the gastrula and neurula stage in the range of $10^{-7}$ M, which was somewhat higher than the values determined in the chick limb bud by Thaller and Eichele (1987). Durston *et al.* (1989) also showed that the minimal doses of exogenous RA capable of producing respecification in anterior axial structures are in the same range. In contrast, higher doses, well above physiological concentration, led to teratogenic effects such as anterior truncations.

A series of elegant experiments evaluated the presence of RA in the embryo. These experiments relied upon the use of transgenic mice where a *LacZ* reporter gene was placed downstream to a RARE-containing regulatory sequence (Mendelsohn *et al.*, 1991; Reynolds *et al.*, 1991; Rossant *et al.*, 1991; Balkan *et al.*, 1992; Zimmer and Zimmer, 1992). The idea was that endogenous retinoids activate the transcription of the construct which can therefore be detected histochemically, thus revealing the embryonic area where retinoids are present. The period of interest is when retinoids most strikingly affect development – i.e., between 7.5 (head-fold stage) and 9 dpc.

Mendelsohn *et al.* (1991) found expression of the transgene restricted to that part of the body going from the anterior/posterior hindbrain limit down to the posterior neuropore (corresponding to the trunk). Expression was seen in the neurectoderm and mesoderm. Thus neither forebrain, midbrain, anterior hindbrain, nor posterior trunk contained retinoids able, in this situation, to trigger expression of the transgene.

The transgene used by Rossant *et al.* (1991) consisted of three copies of a 34-bp oligonucleotide encoding an RAR-$\beta$ located upstream of a hsp-68 promoter linked to the *LacZ* coding region. By day 8.5, the anterior limit of *LacZ* expression was in the preotic sulcus, and it was seen in all three germ layers including endoderm. All studies, using minimal promoters containing RAR-$\beta$ regulatory regions or RAREs (Balkan *et al.*, 1992) reached the same conclusions: at these critical stages of organogenesis, known to respond to variations of RA concentrations, neither anterior nor posterior regions of the embryo contain endogenous RA at levels detectable by this transgenic method.

Interestingly, if these transgenic animals were exposed to excess of RA, the domain of expression of the transgene expanded both anteriorly and posteriorly, showing the capacity of applied RA to induce the endogenous *RAR-β* gene (Reynolds *et al.*, 1991; Rossant *et al.*, 1991; Balkan *et al.*, 1992; Zimmer and Zimmer, 1992).

Another method for evincing the presence of endogenous retinoids in embryonic tissues consisted of stably transfecting F9 cells with an RA-driven *LacZ* reporter gene (Wagner *et al.*, 1992). If RA-containing embryonic tissues are placed on top of such F9 cell cultures they will switch on *LacZ* production. Tissues from the anterior CNS, such as dorsal diencephalon, were found to release very little RA in contrast to the spinal cord which turned out to be an effective source. As a whole these studies show that the early embryo certainly contains endogenous RA, which is essentially localized from the posterior hindbrain down to the posterior neuropore. The problem was then raised of where retinol of dietary origin is converted to the active RA compounds.

*3.6.2.3 Synthesis of RA takes place in Hensen's node and primitive streak.* Hornbruch and Wolpert (1986) carried out an experiment strongly suggesting the presence of significant concentrations of RA in Hensen's node of the chick. They implanted this region of the embryo into limb buds and found that it induced pattern duplications similar to those obtained with implants of the zone of polarizing activity (ZPA) or RA-releasing beads. Hogan *et al.* (1992) used mouse node and found that the ability to induce digit duplication increased from 6.5 to 7.75 dpc. In contrast, tissues anterior to the node were ineffective. Moreover, quantitative data on RA production were provided by Hogan *et al.* (1992). They measured conversion of radiolabeled retinol into RA and found it to be four times higher in the node than in the primitive streak and 12 times higher in the node than in anterior tissues at 7.5 dpc.

The presence of RA in Hensen's node and surrounding tissues was also demonstrated in chick embryos. Thus, Wagner *et al.* (1992) found that, at the primitive streak stage in the chick, Hensen's node contained 20 times more RA than the rest of the area pellucida and that RA level continued to increase considerably in the following stages. It appears, therefore, that Hensen's node is a high point of RA concentration and possibly the source of the RA gradient postulated by Kessel and Gruss (1991).

*3.6.2.4 RA receptors and CRBP, CRABP expression during development.* At the time when it was found that RA causes pattern changes during limb development (Tickle *et al.*, 1982; Summerbell, 1983), the only protein known to interact with RA was cellular RA-binding protein I (CRABP-I). Later, a second cellular RA-binding protein (CRABP-II) was discovered (Giguère *et al.*, 1990). In the late 1980s, the discovery of a nuclear receptor for RA (Giguère *et al.*, 1987; Petkovich *et al.*, 1987) was decisive in our understanding of how the retinoid signal is transduced and exerts so many diverse effects. Several types of RA receptors have been identified. RARα, β, and γ and the retinoid X receptors, RXRα, β, and γ are ligand-inducible *trans*-regulators which modulate the transcription of target genes by interacting with *cis*-acting DNA response elements in the promoter region of target genes. RARs are efficiently activated by either *all*-trans or 9-*cis* RA, whereas RXRs are efficiently activated only by 9-*cis*-RA. Isoforms of RAR, such as RARα1, RARβ2, are generated through the differential usage of two promoters and/or alternative splicing (reviewed in Chambon, 1994). RA receptors are active in a dimeric form either of two RARs or of one RAR combined with an RXR.

Cellular retinol-binding proteins (CRBP) have also been identified and it has been of interest to screen various embryonic stages to find out where and when the different proteins related to RA metabolism are expressed during development. A large body of data was accumulated on this subject in the late 1980s and early 1990s (for references see, e.g., Hofmann and Eichele, 1994). Only information revelant for the purpose of this chapter will be reviewed hereafter.

At presomitic stages in the mouse (Conlon and Rossant, 1992) and in the chick (Smith and Eichele, 1991) RARβ was found to be expressed by the three germ layers, but more in the paraxial and lateral regions of the embryo than in the midline. The anterior limit of expression reaches the presumptive level of the posterior hindbrain. At the early somite stages of the mouse (8 dpc) RARα is uniformly transcribed, RARβ anterior boundary is still located at the hindbrain level (Ruberte *et al.*, 1991; Conlon and Rossant, 1992), RARγ transcripts are present in the caudal region of the embryo and excluded from somites but present in the frontonasal process and in branchial arches during and after neural crest cells migration (Ruberte *et al.*, 1990). CRABP-I is present in the hindbrain neuroepithelium up to the preotic sulcus, in the mesoderm lateral to the midbrain and hindbrain, in the lateral plate but not in the primitive streak (Ruberte *et al.*, 1991). CRABP-II is present in the neural ectoderm at all axial levels (Ruberte *et al.*, 1992). CRBP is expressed in ectoderm, mesoderm at the streak level and in the hindbrain neurectoderm.

In conclusion, at the time of axis formation, RARs and retinoid-binding proteins are expressed, often in distinct patterns (see Fig. 3.14). Interestingly, CRABP-I is expressed mainly in regions lateral to the streak, a region which, as reported below, has been found to be a site of RA synthesis (Hogan *et al.*, 1992).

*3.6.2.5 Morphological effects of targeted mutations of RA-binding proteins in the mouse.* Null mutant mice lacking functional RA receptors have been generated by homologous recombination. Spectacular defects, which are similar to those resulting from deprivation of RA during pregnancy, have been observed only in fetuses in which two receptor genes were mutated. This, however, occurred only for certain gene combinations. Thus, although not viable, since they all die within a few hours after birth, double null mutant mice lacking both the *RARα* and *RARβ2* genes ($RAR\alpha^{-/-}/\beta2^{-/-}$ mutants) or both the *RARβ2* and *RARγ* genes ($RAR\beta2^{-/-}/\gamma^{-/-}$ mutants) (Lohnes *et al.*, 1994; Mendelsohn *et al.*, 1994c) did not show deficiencies in the mesectoderm-derived craniofacial structures. In particular, patterning of the dentition and that of the cusps appeared normal in all of these *RAR* double mutants at 18.5 dpc, which corresponds to the last developmental stage that can be analysed.

In contrast, almost all the derivatives of cranial mesectoderm were severely affected in $RAR\alpha^{-/-}/\gamma^{-/-}$ embryos and fetuses (Lohnes *et al.*, 1994; Mendelsohn *et al.*, 1994b). In these double null mutants at 18.5 dpc, all the structures derived from the frontonasal process were partially or completely absent. Thus, the frontal, nasal, premaxillary, ethmoid, comprising the nasal capsule, and the nasal septum and presphenoid bones were missing for the most part (Fig. 3.29). Moreover, the upper incisors were bilaterally absent in the $RAR\alpha^{-/-}/\gamma^{-/-}$ fetuses analyzed at 18.5 dpc. In addition, the skeletal derivatives of the second and third pharyngeal arches either were not identifiable (e.g., stapes) or appeared severely malformed. In contrast, the derivatives of the mandibular arch were present and nearly normal in $RAR\alpha^{-/-}/\gamma^{-/-}$ double mutants. Such is the case for the dentary bone, the temporomandibular joint, Meckel's cartilage, malleus, and tympanic bones. Moreover, the spatial arrangement and the shape of the lower incisors and of the lower and upper molars were apparently normal in all $RAR\alpha^{-/-}/\gamma^{-/-}$ fetuses.

$RAR\alpha$ and $RAR\gamma$ are both strongly expressed in the frontonasal process and in branchial arches during and after neural crest cell migration (Ruberte *et al.*, 1990). Since, the agenesis of midfacial structures is preceded by increased cell death in the frontonasal mesectoderm at 10.5 dpc, it is likely that RA is normally required for the survival of postmigratory neural crest cells in this location (Lohnes *et al.*, 1994). In the $RAR$ double mutants, a connection exists between the epipterygoid/alisphenoid and quadrate/incus, respectively – i.e., the rostral and caudal elements derived, during evolution, from the ancestral (reptilian) pterygoquadrate cartilage. The cartilaginous rod linking the incus and the alisphenoid bone is interpreted as a reemergence of a connection lost during therapsids to mammals transition (Lohnes *et al.*, 1994).

None of these craniofacial defects are observed in either $RAR\alpha$ or $RAR\gamma$ single mutants, indicating that these receptors are functionally redundant for the development of the frontonasal mesectodermal structures.

$RAR/RXR$ heterodimers bind *in vitro* much more efficiently to *cis*-acting DNA response elements than $RAR$ or $RXR$ homodimers (Chambon, 1994). Accordingly, the double null mutants $RAR\gamma^{-/-}/RXR\alpha^{-/-}$ are embryonic lethal, in contrast to $RAR\alpha^{-/-}/RAR\gamma^{-/-}$ which can survive until birth (Kastner *et al.*, 1994). $RAR\gamma^{-/-}/RXR\alpha^{-/-}$ fetuses examined at 14.5–15.5 dpc exhibited an apparently normal arrangement and shape of the teeth anlagen.

The fact that mice and chicks, in which *Hoxa-2* expression is altered (see above) as well as $RAR$ mice mutants, exhibit such atavistic changes (i.e., "the reappearance of a lost character typical of remote ancestors and not seen in the parents or recent ancestors of this organism"; Hall, 1984) indicates that, even if they are not expressed, these atavistic features are still potentially present in the ontogenetic program of the neural crest cells in higher vertebrates. Moreover, these results also show that RA has been involved in the regulatory mechanisms responsible for these evolutionary changes (Mark *et al.*, 1995). It is remarkable to observe that transcription factors whose function is controlled by RA have not been identified in insects, suggesting that this regulatory pathway controlling *Hox* gene expression has appeared late during evolution.

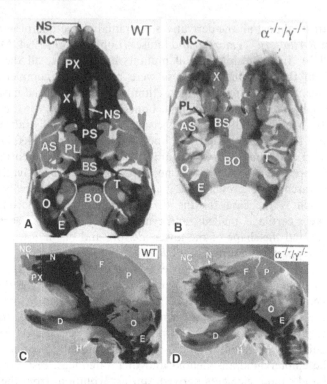

**Figure 3.29** Craniofacial skeletal features of 18.5 dpc wild-type (WT) and RAR double $\alpha^{-/-}/\gamma^{-/-}$ mutant fetuses. Comparison of ventral (A, B) and lateral (C, D) views of the skull and of the dentary bone between wild-type (A, C) and RAR$\alpha^{-/-}/\gamma^{-/-}$ mutant (B, D) fetuses. In the mutant, note the nearly complete absence of the nasal capsule (NC), the complete agenesis of the nasal septum (NS) and of the premaxillary (PX) and presphenoid (PS) bones, the wide median cleft in the basisphenoid bone (BS), and the aplasia of the hyoid bone (H). AS, alisphenoid bone; BO, basioccipital bone; BS, basisphenoid bone; D, mandibular (dentary) bone; E, exoccipital bone; F, frontal bone; N, nasal bone; NC, nasal capsule; NS, nasal septum; O, otic capsule; P, parietal bone; PL, palatine bone; PS, presphenoid bone; PX, incisive (premaxillary) bone; T, tympanic bone; X, maxillary bone. (Reproduced, with permission of Company of Biologists Ltd, from Lohnes *et al.*, 1995.)

*For a colored version of this figure, see www.cambridge.org/9780521122252.*

## 3.7 A number of transcription factors are involved in the ontogeny of the mesectodermal derivatives of the neural crest

As well as *Hox* genes, *Krox-20*, *MafB/kr*, certain Eph-RTKs, and a variety of proteins binding RA whose roles in the development of hindbrain and neural crest derivatives have been discussed earlier in this chapter, data have been gathered on many other genetic pathways that are involved in the development of the mesectodermal derivatives of the neural crest. Some of the genes involved are transcription factors that have been identified owing to their homology to *Drosophila* gene sequences. Others contain sequence-specific DNA-binding proteins that have been identified in the context of studies on mammalian transcriptional regulation.

### 3.7.1  Otx *genes*

In *Drosophila*, two genes that contain homeobox domains are expressed in the anteriormost part of the head: *empty spiracle* (*ems*) and *orthodenticle* (*otd*) (Finkelstein and Perrimon, 1991). The vertebrate homologs of these genes were first isolated in the mouse (Simeone *et al.*, 1992). The mouse cognates of *otd* are *Otx1* and *Otx2*, and those of *ems* are *Emx1* and *Emx2*. They have nested patterns of expression in the rostral brain of the developing embryos so that various levels of the prosencephalon and mesencephalon are characterized by a combined gene expression pattern, suggesting that forebrain and midbrain patterning might be regulated by a genetic code, as are hindbrain and spinal cord by the Hox code (e.g., Puelles and Rubenstein, 1993). Among *Otx* and *Emx* genes, *Otx2* is already expressed in the entire epiblast at the primitive streak stage. During the midstreak stage, its expression domain becomes restricted to the presumptive territory of the rostral head where it has been proposed the future rostralmost part of the brain is specified (Simeone *et al.*, 1993; Ang *et al.*, 1994).

Otx2 mutant mice have been generated (Matsuo *et al.*, 1995; Acampora *et al.*, 1995). Homozygous null embryos died before 10 dpc. They exhibited a truncated body axis where territories corresponding to the rostral head were lacking as well as heart and foregut. In contrast, trunk structures such as neural plate and somites were present. Thus the *Brachyury* gene was normally expressed but extended to the anterior end of the embryo, whereas *En2* transcripts which characterize the midbrain were absent. The presence of *Krox-20* at the anterior extremity of the embryo indicates that the neuraxis of $Otx2^{-/-}$ embryos is truncated at the level of r3.

Heterozygous $Otx^{+/-}$ embryos showed haplotype insufficiency characterized by otocephaly with a loss of lower jaw and eyes, but no defects were found in the palate, maxillar, and middle ear. *Otx2* heterozygous embryos were, however, not all identical, their phenotypes ranging from slight to severe abnormalities where most of the head did not develop. Interestingly, the affected regions coincide with the most caudal and rostral areas where *Otx2* is expressed and where *Otx1*, *Emx1*, and *Emx2* transcripts are not present. As mentioned above, for *Hox* genes, loss-of-function mutations result in the abnormal development of structures derived from their most rostral region of expression. *Otx* and *Emx* genes have nested expression domains, with boundaries at both rostral and caudal ends. It is interesting to see that, in the head, abnormalities induced by loss of function of the most widely spread of these genes lie in the areas where *Otx2* is the only one expressed and therefore where no compensation by *Otx1*, *Emx1* or *Emx2* can occur. The *odt* gene of *Drosophila* functions as a gap and a homeotic gene (Finkelstein and Perrimon, 1991). In vertebrates, its cognate gene *Otx2* deficiency does not produce any homeotic transformation but rather acts like a gap gene since it is followed by atrophies and deletions.

### 3.7.2  Mhox *gene*

*Mhox* was originally identified in the mouse as a nuclear factor that bound the enhancer in the muscle creatine kinase gene (Cserjesi *et al.*, 1992). Its human homolog *Phox* was cloned by Grueneberg *et al.* (1992). *Mhox* homeodomain is most closely related to that of the paired family of homeobox-containing genes although it lacks a paired domain. By 9.5 dpc *Mhox* is expressed strongly in mesenchymal cells of both neural crest and mesoderm origin, in the head, the visceral arches, and also in limb bud and somites (Cserjesi *et al.*, 1992).

Null mutation of *Mhox* gene induces multiple craniofacial defects resulting from the deletion or malformation of specific skeletal elements derived from the neural crest cells colonizing BA1, and from the cephalic mesoderm. The squamosal bone and the ascending lamina of the alisphenoid bone are absent, as well as the zygomatic bone, the gonial bone, the tympanic ring, and the supraoccipital bone. At the base of the skull, the pterygoid bone is hypoplastic and the palatal processes of the palatal and maxillary bones are missing. The mandible and ear ossicles of the *Mhox* mutants also present several defects: hypoplasia of the coronoid, condylar, and angular processes of the mandible, and shortening of the dentary bone. However, the alveolar processes of the mandible and tooth buds are normal in *Mhox* mutants. As in other mutants affecting branchial arch development, disruption of *Mhox* gene results in cranial and skeletal components that are morphologically similar to those of the mammal-like reptiles.

Interestingly, the defects in neural crest derivatives on these mutants were not due to disturbances in cell migration which seems to occur normally. Moreover, the neural derivatives of the cephalic neural crest were not affected.

### 3.7.3  Goosecoid *gene*

*Goosecoid* (*Gsc*) is a homeobox-containing gene discovered as a gene of Spemann's organizer (Blumberg *et al.*, 1991). Two phases of *Gsc* expression have been identified: one during gastrulation (6.4–6.7 dpc in the mouse) (Blum *et al.*, 1992) and another later (from 9.5 dpc) in specific structures such as those forming some components of the head, the limbs, and the ventrolateral body wall (Gaunt *et al.*, 1993). The *Gsc*-expressing cells of the early embryo are fated to form the head process which later give rise to the anterior notocord and endoderm, and to the head mesoderm. At E10.5 *Gsc* is expressed in BA1 and in the anterior one-third of BA2 with a sharp posterior boundary. *Gsc* transcripts generally lie medial to those of *Dlx1* (see below), the two genes occupying distinct complementary domains. Expression persists as these tissues undergo morphogenesis (Gaunt *et al.*, 1993). A related homeobox-containing gene, pituitary homeobox-1, *ptx*, is expressed in the developing mandible and nasal cavities in E11–13 mice in a complementary pattern to *Gsc* (Lanctot *et al.*, 1997). Another gene related to the *Drosophila* gene *bicoid* has been cloned in humans, mouse, and chick, and designated *Rieg, Brx1*, or *Otlx2* (Semina *et al.*, 1996). *Rieg* is expressed in the maxillary and mandibulary epithelia before tooth formation and requires mesenchymal signals for its maintenance (Mucchielli *et al.*, 1997). In humans, this gene is mutated in Rieger's syndrome, which is characterized by mild facial abnormalities and odontogenic defects (Semina *et al.*, 1996).

A null mutation of *Gsc* has been produced in the mouse. Although the gene is expressed at gastrulation stage (by 6.4–6.7 dpc), $Gsc^{-/-}$ mice develop until birth but die soon after parturition with multiple craniofacial defects (Rivera-Pérez *et al.*, 1995; Yamada *et al.*, 1995). In the skull, the orbital processes of the maxillary and frontal bones that support the eye are reduced. The tympanic ring is absent. The mutants lack the anlagen of the turbinal bones and the ventrolateral walls of the nasal cavity. The glandular mucous epithelium, which normally covers the nasal sinuses, is mostly absent in the mutants, although the nasal septum and the vomeronasal organs and cartilages are present. The nasal septum fails to fuse with the palate. Within the middle ear the malleus is malformed but the incus and stapes are normal. The palatine, maxillary, alisphenoid, and pterygoid are also abnormal in morphology and the mandible is shortened.

It should be pointed out that these craniofacial defects are accompanied by rib fusion in about 35% of the mutants. In contrast, although the *Gsc* gene is strongly expressed in the developing limbs, no skeletal abnormalities were detected in the limbs (Rivera-Pérez *et al.*, 1995), showing that the effect of loss of function of this gene does not correlate perfectly with its expression pattern.

### 3.7.4 Dlx *genes*

*Dlx* genes are the vertebrate homologs of the *Drosophila* homeobox containing gene *Distalless* (*Dll*) which is required for the development of the terminal regions of the segmented appendages (legs, antennas, mouthparts). A number of *Dlx* genes have been cloned in the rat, mouse, man, chick, frog, newt, and zebrafish (see Nakamura *et al.*, 1996, for details and references).

There are at least six *Dlx* genes (*Dlx-1, -2, -3, -5, -6, -7*) that are expressed in spatially restricted patterns in craniofacial mesenchyme and ectoderm of vertebrates (Qiu *et al.*, 1997). The homeodomain sequences of the murine Dlx proteins are nearly identical (Porteus *et al.*, 1991; Price *et al.*, 1991; Robinson *et al.*, 1991; Simeone *et al.*, 1994), suggesting that they may bind to similar target nucleotide sequences. *Dlx-1, -2, -5*, and *-6* are expressed in the first and second branchial arches. In the mouse at E10.5, *Dlx-1*, and *-2* are expressed in both proximal and distal regions of BA1 and BA2, whereas *Dlx-3*, *-5*, and *-6* expression is restricted to the more distal domains (Qiu *et al.*, 1997).

*Dlx-3* is also expressed in distal regions of BA1 and BA2 (Bulfone *et al.*, 1993; Robinson and Mahon, 1994). This suggests that different *Dlx* genes control the pattern of development in different regions of the visceral arches.

All known *Dlx* genes are expressed during tooth development (Porteus *et al.*, 1991; Robinson *et al.*, 1991; Dollé *et al.*, 1992; Bulfone *et al.*, 1993; Robinson and Mahon, 1994; Weiss *et al.*, 1994, 1995; Thomas *et al.*, 1995; Nakamura *et al.*, 1996). Mutations have been made in *Dlx-1* and *Dlx-2*, and double mutant mice have been generated (Qiu *et al.*, 1995, 1997).

The phenotypes of these mutants indicate that *Dlx* genes are invovled in proximodistal patterning of the two first branchial arches. *Dlx-2* mutants are characterized by the absence of most of the alisphenoid, and the presence of a supernumerary cartilage proposed to be the homolog of the pterygoquadrate of therapsids. This has already been found in loss-of-function mutations of *Hoxa-2* (Gendron-Maguire *et al.*, 1993; Rijli *et al.*, 1993; Mark *et al.*, 1995) and *RAR* double mutants (Lohnes *et al.*, 1994).

The skull is also abnormal in the mutant where the squamosal and jugal bones are replaced by four bones whose positions or shapes suggest that they can be assimilated to the four dermal bones (squamosal, quadratojugal, postorbital, jugal) of the reptiles and mammal-like reptiles. As mentioned before, *Dlx-1* and *-2* have similar expression patterns. *Dlx-1* is also required for splanchnocranial development but its mutation does not affect skull bones (Qiu *et al.*, 1997). Interestingly *Dlx-1* and *-2* mutants have unique abnormalities, the most striking of which is the absence of molar on the maxillary bone. The fact that *Dlx-3*, *-5*, and *-6* are expressed in the distal arch regions suggests that they may be functionally redundant for *Dlx-1* and *-2*. This could account for the absence of malformation in the distal arch region in the above mentioned mutants.

The fact that *Dlx-1* and *-2* have a role in odontogenic patterning supports the notion put forward by Sharpe (1995) that the molecular mechanisms controlling dental patterning (e.g., tooth shape and position) might be controlled by a combinatorial code of homeobox gene expression in the ectomesenchyme: the "odontogenic homeobox code" (see Section 3.2.3.2). Interestingly, these observations suggest that the development of molars on the lower jaw obey different control mechanisms than the maxillary molars.

### 3.7.5  Pax *genes*

*Pax* genes were isolated on the basis of sequence homology to the *Drosophila* segmentation genes *pair-ruled* (see Gruss and Walther, 1992, for a review). The mouse *Pax* gene family consists of nine members all containing the paired box. *Pax* genes can be subdivided into subgroups which share common expression domains. *Pax3* and *Pax7* form such a paralogous group. *Pax3* function is revealed by the spontaneous *Splotch* mutation, while *Pax7* has been mutated by homologous recombination by Mansouri *et al.* (1996a,b).

The expression domains of *Pax3* and *Pax7* in the cephalic neural crest derivatives have been established in the mouse. *Pax7* is expressed in the nasal pit and olfactory epithelium and in the medial region of the frontonasal mass including the nasal capsule at E13.5. *Pax3* expression extends over both the medial and lateral parts of this region. Furthermore, *Pax3* is expressed in the mandible. A construct was prepared where the β-galactosidase gene was inserted in frame to the *Pax7* sequences. This *Pax7* mutated gene was introduced into the mouse germ line and *Pax7* expression pattern was revealed with great precision in animals heterozygous for this mutation. Expression of *Pax7* is detected from E8.5 until E10 in r1, 3 and 5. Moreover, the streams of neural crest cells exiting from the forebrain (around the optic vesicles), and from r1, r3, and r5 were easily identifiable through their *Pax7* gene expression.

*Pax7* null mice die 3 weeks after birth while $Pax7^{+/-}$ heterozygotes are normal. The defects caused by loss of function of this gene concern the morphogenesis of cephalic neural crest derivatives. $Pax7^{-/-}$ mice do not exhibit evident anomalies of the brain even in the domains where the gene is strongly expressed, and the neural derivatives of the cephalic neural crest are normal. These $Pax7^{-/-}$ mice present a reduced maxilla; the inferior lateral part of the nasal capsule is absent, and the tubules of serous glands associated with lateral wall of the middle meatus and with the nasal septum are reduced in number.

*Pax3* complement the lack of *Pax7* function in the neural tube and in the dorsal somitic derivatives (dermomyotome) as well. Normal expression of *Pax7*, however, does not compensate for all the functions of *Pax3* in the neural tube in *Splotch* mutant. This may be either because the expression patterns of these two genes do not completely overlap (thus *Pax7* is not expressed in the dorsal neural tube) or because *Pax7* is expressed later than *Pax3*. Waardenburg's syndrome type 1 in humans, characterized by deafness and craniofacial defects (Tassabehji *et al.*, 1992), is related to mutation of the *Pax3* gene.

*Pax6* gene is expressed in the anterior midbrain crest cells and in the rostral head ectoderm and forebrain. The *small-eye* (*Sey*) mutant mice and rats present a mutated *Pax6* gene (Hill *et al.*, 1991; Matsuo *et al.*, 1993). These animals have, besides abnormalities in eye development, an ectopic cartilage rod which is thought to substitute for the lateral nasal capsule.

Mutation of *Pax6* causes aniridia in man, characterized by a partial or complete absence of the iris (Nelson *et al.*, 1984; Ton *et al.*, 1991).

### 3.7.6  The Cart-1 homeobox gene

The *Cart-1* gene (for cartilage homeoprotein 1) was isolated by RT-PCR (reverse transcriptase–polymerase chain reaction) from a rat chondrosarcoma tumor cell line. It encodes a paired-class homeoprotein. Its transcripts are present in rib chondrocytes, and in various precartilages and cartilages. Low levels of *Cart-1* transcripts are also found in testes but not in any other tissues. Head and branchial arch mesenchyme of neural crest origin express *Cart-1* from E8.5 onward in the mouse. Transcripts are also found in the limb bud and in various cartilage primordia at the time of mesenchymal cell condensation (Zhao *et al.*, 1994).

*Cart-1*$^{-/-}$ mice die soon after birth with severe craniofacial defects. At early stages (E9) the head mesenchyme, normally surrounding the forebrain and frontonasal areas, is missing in the mutants, whereas the head mesenchyme associated with midbrain and hindbrain seems normally developed. The neural folds do not close at the midbrain level. Apparently the loss of function of *Cart-1* gene causes apoptosis of anteriormost head mesenchymal cells. The skull defects consist of acrania: the cranial vault is absent, only small portions in parietal and frontal bones develop at their lateral- and basalmost levels. Only remnants of the supraoccipital, squamosal, palatine, and alisphenoid are present, and the basipresphenoid bone at the base of the skull is missing. The nasal cartilages are abnormal: the lamina cribrosa and turbinate cartilages are absent and the nasal capsule malformed. No defects are seen in the neural derivatives of the cephalic neural crest.

### 3.7.7  Msx *genes*

*Msx-1* and *Msx-2* genes (originally designated *Hox7* and *Hox8*, respectively, in the mouse) were isolated due to the sequence homology of their homeobox with that of the *Drosophila* gene *Msh* (for muscle segment homeobox) in mouse (Hill *et al.*, 1989; Robert *et al.*, 1989; Monaghan *et al.*, 1991), quail (Takahashi and Le Douarin, 1990), chick (Coelho *et al.*, 1991), and man (Monaghan *et al.*, 1991; Takahashi *et al.*, 1996). Expression patterns were found to be similar in all vertebrates investigated and include a range of neural crest-derived mesenchymal tissues. At day 10 of gestation in the mouse, the medial and lateral nasal processes, the maxillary and mandibulary buds express both *Msx-1* and *Msx-2* genes. By day 12, expression of *Msx-1* in the teeth is restricted to the mesenchyme surrounding the developing tooth germs in both the maxillary and mandibulary processes. Expression of *Msx* genes in the dental papilla is maximal at the cap stage of development and progressively declines at the bell stage prior to differentiation of odontoblasts and ameloblasts.

At the bell stage of tooth development, *Msx-2* expression switches tissue layers, disappearing from the differentiating epithelial ameloblasts and turning on in the odontoblasts. *Msx-1* is expressed in the mesenchyme of the dental papilla and follicles at all stages; the reciprocity of expression suggests an interactive role between *Msx-1*, *Msx-2*, and other genes in regulating epitheliomesenchymal interactions during tooth development (MacKenzie *et al.*, 1992). *Msx-1* and *Msx-2* are expressed in the mesenchyme of the calvarium and in the meninges of the telencephalon shown to be of neural crest origin (Takahashi and Le Douarin, 1990; Couly *et al.*, 1993). It is also present in Rathke's pouch, the developing choroid plexuses, the meninges, the dorsal neural tube, the dorsomedial superficial ectoderm along the whole neural axis and in the dorsal mesenchyme of somitic origin destined to form the spinous process of the vertebra. *Msx-1* and *Msx-2* expression can also be detected in the progress zone of the developing limb bud. Expression is seen in the external ear, the forming eye, the nasal pits, and the forming Jacobson's organs. Dynamic expression of both genes is also observed in hair follicles, feather buds, and cardiac cushions (Hill *et al.*, 1989; Robert *et al.*, 1989; MacKenzie *et al.*, 1991a,b, 1992; Takahashi and Le Douarin, 1990; Takahashi *et al.*, 1992; Monsoro-Burq *et al.*, 1994). Expression of *Msx-2* together with differentiation of bones in the mandible of the E4 chick embryo was shown to be under the control of a signal of ectodermal origin (Takahashi *et al.*, 1991).

*Msx-1* null mutant mice have been generated (Satokata and Maas, 1994). They manifest a cleft secondary palate and a shortening of the maxilla. In addition, teeth do not develop beyond the bud stage. Deficiency of the medial portions of the frontal bones gives rise to an enlarged fontanelle. In the middle ear, while the incus and stapes are normal, the malleus is reduced. No abnormalities are seen in limb, heart, or ciliary body. This relatively discrete phenotype is probably due to a possible redundancy of *Msx-2* which is still acting in the mutants. Haplodeficiency of *MSX-1* gene in humans results in the loss of premolar and molar teeth (Vastardis *et al.*, 1996). Transgenic mice overexpressing the *Msx-2* gene exhibit a phenotype also found in humans and known as the Boston-type craniosynostosis which carries a mutation in one copy of *MSX-2*. This mutation increases the stability of binding of the protein to the *MSX-2*-binding site (TAATG) and therefore acts in a way similar to the overexpression of the gene in transgenic mice (Liu *et al.*, 1995, 1996; Ma *et al.*, 1996a; reviewed in Wilkie, 1997).

No mutants of *Msx-2* have so far been produced.

### 3.7.8 Twist *gene*

*Twist* was initially identified in *Drosophila* as one of the zygotic genes required for dorsoventral patterning during embryogenesis (Simpson, 1983; Nüsslein-Volhard *et al.*, 1984). It encodes a basic helix–loop–helix (bHLH) transcription factor (Thisse *et al.*, 1988; Murre *et al.*, 1989). Vertebrate homologs of *twist* have been isolated (Chen and Behringher, 1995, and references therein). At the neurula stage abundant *twist* transcripts are detected in the somites and in the neural crest-derived cranial mesenchyme in *Xenopus* (*Xtwi*) (Hopwood *et al.*, 1989). In the mouse the expression pattern includes the paraxial mesoderm, the somatopleure, the limb buds, the branchial arch mesectoderm (Ang and Rossant, 1994), and the primary osteoblastic cells of the calvaria in the newborn (Murray *et al.*, 1992). It seems therefore that, in amphibians as in mammals, *twist* regulates genes involved in the development of both mesoderm and mesectoderm.

*Twist*$^{-/-}$ mutants exhibit a failure of neural tube closure in the cranial region and have defects in head mesenchyme branchial arches, somites, and limb buds. Heterozygous mice for the *twist* deletion allele are normal. While in normal embryos neural tube closure involves fusion of neural folds in multiple sites in the midbrain and forebrain regions from E8.5, in *twist*$^{-/-}$ mice the neural folds never initiate fusion. Local hemorrhages occur in the cranial area and the neural tube remains open rostrally down to the level of r4, thus leading to exencephaly.

The head mesenchyme of E8.5 *twist*$^{-/-}$ embryo is abnormal with rounded cells lacking intercellular contacts and expanded extracellular space at the forebrain and midbrain regions. At this level the blood vessels are dilated and extravascular blood cells are abundant. The same disorganization of the mesenchyme reaches the branchial arches at E9.

Cranial ganglia contain many pycnotic nuclei from E9.5 onwards. Examination of neural crest-derived mesenchyme, at migration stage, was carried out by using the specific marker $AP_2$ (see below). It turned out that in *twist* null mutants neural crest cells showed essentially the same pattern of migration as in normal embryos but $AP_2$ expression was weaker in the forebrain-derived mesectoderm.

Construction of chimeric embryos by injection of *twist*$^{-/-}$ ES cells into wild-type blastocysts homozygous for the ROSA26 gene trap insertion that expresses $\beta$-galactosidase ubiquitously (Friedrich and Soriano, 1991), showed that the defect in forebrain and midbrain neural tube closure was directly related to the presence of *twist*$^{-/-}$ and the absence of normal mesectodermal cells in the cranial neural crest. In contrast, the contribution of *twist* null cells to neurectoderm and surface ectoderm which do not express *twist* did not correlate with the development of the cranial neural tube defects.

These results indicate that the mesenchyme of neural crest origin which arises from the diencephalon and mesencephalon plays an active role in the closure of the anterior neural tube destined to give rise to the forebrain and midbrain.

In certain chimeras, head mesectoderm was of mixed mutant and wild-type origin. In such cases, mutant cells formed patches of rounded cells which remained apart from normal cells and did not establish cell–cell contacts like normal mesenchymal cells do. It seems, therefore, that adhesive properties are altered in *twist*$^{-/-}$ cranial mesectodermal cells (Howard *et al.*, 1997).

Possible targets of *twist* are therefore genes encoding surface molecules, cytoskeletal proteins, and extracellular matrix components. Interestingly, in *Drosophila*, the *twist* mutation inhibits mesoderm formation by altering the changes in cellular morphology which are linked to the morphogenetic movements of gastrulation (Leptin and Grunewald, 1990). Thus *twist* alters cell morphology and behavior in flies and mice, showing that it has conserved the same primary function throughout evolution.

### 3.7.9 AP₂ *gene*

The transcription factor gene $AP_2$ was first identified from HeLa cells and was shown to stimulate RNA polymerase II transcription of test promoters *in vitro* in an $AP_2$-binding site-dependent manner.

$AP_2$ was found to be expressed at E8.5 in the mouse in the head mesenchyme and surface ectoderm as well as in the neural folds in the trunk (i.e., in the precursors of neural crest cells). At E9.5 the head and branchial arch mesenchyme was still strongly expressing $AP_2$ as well as the surface ectoderm (Mitchell *et al.*, 1991). Cranial sensory ganglia and DRG were clearly labeled at E10.5 while expression in head and branchial arch ectomesenchyme persisted at that stage and later on. In the neural tube a lateral column, dorsal to the sulcus limitans corresponding to the dorsal entry zone of the sensory axons, was strongly labeled along the whole hindbrain and spinal cord by *in situ* hybridization revealing $AP_2$ transcripts. At E11.5 $AP_2$ expression was also seen in sympathetic ganglia.

$AP_2^{-/-}$ mice have been generated (Schorle *et al.*, 1996). They died perinatally with severe dysmorphogenesis of the face, skull, and cranial ganglia. The head skeleton was acranic with all major skull bone of neural crest origin reduced to remnants. The forebrain was everted with the ventricular epithelium of the cerebral hemispheres facing outward. Medial nasal and mandibular rudiments did not fuse on the ventral midline, resulting in full midline facial clefting. Meckel's and hyoid cartilage primordia were evident but displaced and reduced. The defect in cranial morphogenesis was not due to a loss of expression of *Pax3* or *twist* since transcripts of these genes were found in the neural crest cells that migrated early to the facial rudiments.

It appears therefore that $AP_2$ is an important component in the gene-expression program that regulates craniofacial development.

## 3.8 Roles of various growth factors in the development of the mesectodermal derivatives of the neural crest

A number of studies have been devoted to the tissue interactions involved in the development of the mesectodermal derivatives of the neural crest. They have been reviewed by Hall (1983, 1987). These studies have been particularly extensive concerning tooth development (see, e.g., Thesleff and Sharpe, 1997, and references therein). In this chapter, only some recent findings concerning the factors and their cognate receptors that have been shown to play a critical role in the development of the facial structures will be considered.

A number of secreted factors are produced by either the epithelial or the mesenchymal components of the branchial arches and facial rudiments. Such is the case for members of the fibroblast growth factor (FGF), transforming growth factor $\beta$ (TGF$\beta$), and Wingless (Wnt) gene families. Other signaling molecules such as endothelin-1, PDGF$\alpha$, epidermal growth factor (EGF),

TGFα, and membrane-bound ligands such as *Jagged-1* (or *Serrate-1*) have also been implicated in facial development.

### 3.8.1 The endothelin-1 receptor-A pathway

Endothelins (EDN) consist of three closely related small peptide ligands (EDN1, -2, -3) that bind to one or both of the G protein-coupled endothelin receptors, EDNRA and EDNRB (Arai *et al.*, 1990; Sakurai *et al.*, 1990; Yanagisawa, 1994). Recent evidence shows that endothelins and their receptors are required for the development of specific subsets of neural crest derivatives: EDN3-4 or EDNRB-deficient mice develop white spotted coats and megacolon due to perturbations in the development of melanocytes and enteric nervous system (see Chapters 5 and 6), while disruption of the EDN1 and EDNRA genes causes severe malformations of the pharyngeal arch-derived structures and the heart (Kurihara *et al.*, 1994; Clouthier *et al.*, 1998).

The cephalic neural crest cells start expressing the EDNRA gene as they start migrating. At E3 in quail and chick and from E9 onward in the mouse, EDNRA mRNA is detectable in the ectomesenchyme of the head and branchial arches. In contrast, EDN1 message is confined to the ectodermal and endodermal epithelia, to the central core of mesodermal cells of the branchial arches, and to the ectoderm of the rostral cephalic area (Clouthier *et al.*, 1998; Nataf *et al.*, 1998a,b).

EDNRA is also expressed in the myocardium, while EDN1 mRNA is found in the endothelial layer of the heart and of the arch vessels from E9.5 in the mouse. The phenotypes induced in mice by the targeted disruption of the EDNRA and EDN1 genes are very similar. They mimick the human conditions collectively termed CATCH22 or velocardiofacial syndrome, which includes severe craniofacial deformities and defects of the cardiovascular outflow. One of the striking features of the EDNRA-deficient mice is the complete absence of Meckel's cartilage, the strong reduction of the tongue, and the aberrant development of the middle ear. This suggests that paracrine interactions between crest-derived cells and both ectoderm and mesoderm are essential in forming the skeleton and connective tissue of the head. Moreover, expression of *gsc* gene is absent in EDNRA-deficient mice, suggesting that this transcription factor might be one of the downstream signals triggered by EDNRA activation (Kurihara *et al.*, 1994; Clouthier *et al.*, 1998).

As will be discussed in Chapter 6, although EDN1 can interact with both EDNRA and EDNRB with high affinities, the EDN1/EDNRA axis does not overlap with the EDN3/EDNRB pathway, since in EDN1 and EDNRA mutants the enteric nervous system and the melanocytes were not affected. Moreover, the defects observed in these mutants are not seen in EDN3- and EDNRB-deficient animals (Baynash *et al.*, 1994; Hosoda *et al.*, 1994).

Both END-1 and END-3 peptides are produced by proteolysis of larger proteins (big endothelin) by a specific enzyme designated endothelin converting enzyme-1 (ECE-1). The targeted null mutation of ECE-1 has been recently produced (Yanagisawa *et al.*, 1998), and the ECE-1$^{-/-}$ embryos exhibited craniofacial and cardiac abnormalities virtually identical to the defects seen in EDN1- and EDNRA-deficient embryos. In addition they also showed the developmental defects characteristic of EDN3$^{-/-}$ and EDNRB$^{-/-}$ animals. This means the ECE-1 is an activating protease for both big EDN1 and EDN3 *in vivo*. Mutations in ECE-1 in humans were shown to cause developmental defects, such as Hirshsprung's disease (see Chapter 5), and the velocardiofacial syndrome and related neurocristopathies.

### 3.8.2 *Other growth factors involved in mesectoderm development*

Members of the FGF family involved in the development of the facial primordia, are *FGF2* which is expressed throughout the epithelium of the facial primordia, and *FGF4* and *FGF8* expressed in restricted regions of this epithelium (Crossley and Martin, 1995; Wall and Hoggan, 1995; Vogel *et al.*, 1996; Barlow and Francis-West, 1997; Helms *et al.*, 1997; Richman *et al.*, 1997).

Many members of the TGF$\beta$ family and their receptors are expressed in the developing facial primordia, and their role in patterning the facial structures via epitheliomesenchymal interactions, cell proliferation and chondrogenesis has been demonstrated in several systems (reviews by Kingsley, 1994; Hogan, 1996). TGF$\beta$1, 2 and 3 are all expressed in early facial mesenchyme. The mice in which the TGF$\beta$1 gene has been mutated died at E10 before the onset of facial development (Dickson *et al.*, 1995), but in the TGF$\beta$2 mutant defects of maxillary and mandibular development, resulting in cleft palate (Sanford *et al.*, 1997), have been recorded. TGF$\beta$3 seems to be required for the fusion of the palatal shelves (Kaartinen *et al.*, 1995), a fact consistent with its high level of expression at the palatal interface.

Expression of members of the bone morphogenetic proteins (BMPs, a subdivision of the TGF$\beta$ family) has been studied in detail in the facial primordia both in chick and mouse embryos (Francis-West *et al.*, 1994; Bennett *et al.*, 1995; Wall and Hogan, 1995; Barlow and Francis-West, 1997).

BMP4 is expressed in restricted domains of the developing facial rudiments of the chick embryo. BMP4 expression in the epithelium is often associated with mesenchymal areas in which *Bmp2, Bmp7, Msx1*, and *Msx2* transcripts are present (Francis-West *et al.*, 1994; Wall and Hogan, 1995; Barlow and Francis-West, 1997). Haploinsufficiency of *Bmp4* in C57Bl/6 mice results, in certain cases (12% of heterozygous), in shortening of frontal and nasal bones, while application of BMP2 and BMP4 activates *Msx1* and *Msx2* gene expression, increases cell proliferation, and can result in the bifurcation of skeletal structures (Barlow and Francis-West, 1997).

The null mutation of BMP7, also called osteogenic protein 1 (OP1), has been reported to induce eye defects, probably by disrupting the process leading to lens induction. Moreover, abnormalities were regularly observed in the skull of these mutants, such as smaller basisphenoid bones and marked reduction of the pterygoid bone (Luo *et al.*, 1995).

Another member of the TFGβ family, *Activin bA,* was shown to be expressed in the mesenchyme of the facial and hypobranchial rudiments in mice and rat (Feijen *et al.*, 1994; Roberts and Barth, 1994). Targeted mutations of either the *Activin bA* gene or of its receptor ActRII (Matzuk *et al.*, 1995b) result in cleft palate and loss of incisors and lower molars (Matzuk *et al.*, 1995a). Similar palate abnormalities are observed in *Follistatin*$^{-/-}$ mice. The *sonic hedgehog (Shh)* signaling pathway, which plays an essential role in patterning the neural tube, the paraxial mesoderm and the limb, is also implicated in facial morphogenesis. *Shh* gene is expressed in the epithelia of the facial primordia (Marti *et al.*, 1995; Wall and Hogan, 1995; Barlow and Francis-West, 1997; Helms *et al.*, 1997). The *Shh*-expressing epithelium of the facial processes were shown to have patterning activity since they are able to induce digit duplications when grafted to the anterior region of the limb (Helms *et al.*, 1997). In *Shh*$^{-/-}$ mice, the branchial arches arc relatively normal at E9.5, but the skeletal structures of the face are absent or extremely reduced (Chiang *et al.*, 1996). Mutations of the transmembrane receptor Patched (Ptc) can also induce facial abnormalities (homologs of the *Drosophila* gene *cubitus interruptus*).

Some of the transcription factors of the *Gli* family involved in the *Shh* transduction pathway are expressed in the neural crest cells migrating in the embryonic face of the mouse (Walterhouse *et al.*, 1993; Hui *et al.*, 1994). Null mutations of *Gli2* and *Gli3* result in facial abnormalities (Mo *et al.*, 1997). Namely, *Gli2*$^{-/-}$ mice, show a truncation of the distal part of the maxilla and mandible together with the absence of incisors (Mo *et al.*, 1997). In *Gli3*$^{-/-}$ mice, in contrast, the maxillary region is enlarged. A functional redundancy between *Gli2* and *Gli3* must exist since double *Gli2* and *Gli3* mutants have a more severe phenotype with additional bones being hypoplastic.

In humans, Grieg's syndrome, characterized by a wide forehead and nasal bridge, has been related to the partial loss of *GLI3* function (Vortkamp *et al.*, 1991).

Mouse *Serrate-1* (also called *Jagged-1*), a membrane ligand of the Notch receptor family, is expressed in the developing maxillary and mandibular rudiments (Mitsiadis *et al.*, 1997). In humans, haploinsufficiency of *Jagged-1* causes Allagile's syndrome, characterized by defects in mandible, forehead, and nose development (Li *et al.*, 1997; Oda *et al.*, 1997).

# 4

# From the Neural Crest to the Ganglia of the Peripheral Nervous System: The Sensory Ganglia

## 4.1 General considerations

The sensory ganglia of the peripheral nervous system (PNS) transmit information from peripheral targets to higher somatosensory areas in the spinal cord and brain. They include the dorsal root ganglia (DRG) organized as bilateral metameric units along the spinal cord, and the ganglia located along the path of cranial nerves. In development, sensory neurons originate from progenitors that migrate from the neural crest and certain ectodermal placodes to the homing sites where they differentiate. Nascent sensory ganglia are colonized by subsets of neural and glial progenitors with heterogeneous developmental potentialities. Knowledge of the state of commitment of neural crest precursors invokes a critical role for the local environment encountered along the migratory routes and at the target sites in regulating neural crest development into ganglionic derivatives. The pathways and mechanisms of neural crest cell migration that lead to the formation of segmentally organized ganglia, as well as the factors that regulate the differentiation of progenitor cells into neurons and satellite cells, have been the subject of intensive research during the past 10 years and will be discussed in this chapter.

Upon differentiation, sensory neurons initially extend two axonal processes that grow in opposite directions from the cell bodies to reach peripheral and central target fields. The innervation of the targets is executed with exquisite precision, raising the possibility that sensory neurons become specified at early stages prior to innervation. In spite of this apparent exactitude, many more neurons send their processes to the targets than ultimately survive. This process of target innervation is associated with a period of massive neuronal death. A similar, but quantitatively more dramatic, process of sensory ganglion hypoplasia is observed when ablating the target limb bud. Observation of normal neuronal death that follows target innervation and of experimentally induced death caused by limb ablation led to a general paradigm of target dependence for neuronal survival (see Hamburger, 1992). Thus, both normal and experimental cell death were proposed to be caused by competition among axons for limited accessibility to, or availability of, target-derived survival molecules. Identification of the process of "naturally occurring cell death" as part of the normal developmental program of the embryo led to the subsequent search for trophic agents that would sustain the neurons alive. Research in this direction led to the recognition that nerve growth factor (NGF) acts as a maintenance factor for sensory and sympathetic neurons during programmed neuronal death (reviewed by Levi-Montalcini, 1987), thereby opening a new era in developmental neurobiology.

Further to the discovery of NGF, additional members of what would become the "neurotrophin family" of growth factors were discovered. This is presently a very well-known family of molecules that includes, in addition to NGF, brain-derived neurotrophic factor (BDNF), neurotrophin-3 (NT-3), neurotrophin-4 (NT-4), also named NT-4/5 or NT-5, and the more recently discovered fish-specific NT-6. The neurotrophins exert their activities through a small family of receptors, the trk tyrosine protein kinases. The first trk receptor was described in 1986 as the product of the *trk* oncogene, a chimeric oncoprotein found in human colon carcinoma (Martin-Zanca *et al.*, 1986, 1989). However, it was not until 1991 that the physiological role of the trk protein kinase was elucidated as being the receptor for NGF. In subsequent studies, two additional and highly related molecules joined the family: trkB as the receptor for BDNF and NT-4, and trkC, the primary receptor for NT-3 (reviewed by Barbacid, 1994). The neurotrophins, their receptors, and a variety of responsive cells have been characterized in detail. Localization studies at the mRNA and protein levels have confirmed their expression in the target tissues in agreement with the proposed role as target-derived factors, but have also revealed novel sites of synthesis such as sensory neurons and motorneurons, classically considered as targets for neurotrophin activity. These patterns of synthesis suggested that there may be a compartmentalization in the effects of target-derived compared to locally produced neurotrophins (Schecterson and Bothwell, 1992; Pruginin-Bluger *et al.*, 1997). Indeed, local factors deriving from the neurons were recently shown to play short-range paracrine and autocrine activities within the ganglionic environment.

Research during the past 10 years has revealed that the function of neurotrophins in development is not limited to the period of programmed neuronal death but starts at much earlier stages during ontogeny. A requirement for neural tube signals in the development of neural crest cells into DRG was first reported by Kalcheim and Le Douarin (1986). In subsequent studies, it became evident that BDNF and NT-3 exert distinct effects on neural crest cells both in the embryo and in culture. Ever since, new effects of neurotrophins were uncovered in the early embryo, both on neural and on mesodermal progenitor cells.

In this section, we provide an overview of the development of sensory ganglia, with an emphasis on their origin, organization, and control of cell number mainly by the neurotrophins and also by other factors. We shall concentrate on four distinct phases: segmental migration of neural crest progenitors, differentiation into sensory neurons, factors affecting early maturation before innervation, and trophic interactions during programmed neuronal death.

## 4.2  How is the segmental distribution of spinal ganglia established?

Two pioneering works by Lehman (1927) and Detwiler (1937b) have shown that removal or addition of one or more somites to the axis reduced or increased, respectively, the number of sensory ganglia. These findings provided the first indication that the development of the DRG is intimately linked to the segmentation of the mesoderm. In fact, initial emigration of neural crest progenitors from the neural tube is even along the trunk axis and it is only following somite dissociation into dermomyotome and sclerotome that segmental migration of neural crest cells becomes apparent (Weston, 1963; Tosney, 1978). By using the HNK-1 monoclonal antibody, it was clearly demonstrated that migration of neural crest cells proceeds selectively in the rostral somitic halves and in the spaces separating two consecutive somites (see Chapter 2 for further discussion).

To test whether this segmental migration is driven by intrinsic properties of the crest cells themselves or by somitic signals, two key experiments were performed. First, fragments of the neural tube were rotated along the rostrocaudal axis prior to the onset of neural crest cell migration in the chick embryo. This procedure did not alter the normal pattern of crest cell migration, demonstrating that segmentation is not an intrinsic property of the trunk neural crest. In contrast, reverting the rostrocaudal polarity of individual somites inversed the polarity of the ganglia whose progenitors kept migrating into the original rostral somitic domain that was experimentally brought to a caudal position (Keynes and Stern, 1984).

The exclusive permissive nature of the rostral half of the sclerotome was further tested as follows: three to five consecutive somites were removed in a chick embryo and replaced by a series of either rostral or caudal epithelial somitic halves. In the first case, the operation resulted in the continuous and non-segmented migration of crest cells into the grafted mesoderm and the consequent formation of unsegmented ganglia (Fig. 4.1B). Conversely, grafting of multiple caudal-half somites in tandem prevented the migration of crest cells and gave rise to an area with small ganglia dorsally located with respect to the somites (Kalcheim and Teillet, 1989). Much like the ganglion progenitors, peripheral nerves also grew exclusively into the rostral domain of each segment and behaved in a similar manner as neural crest cells in the experimental

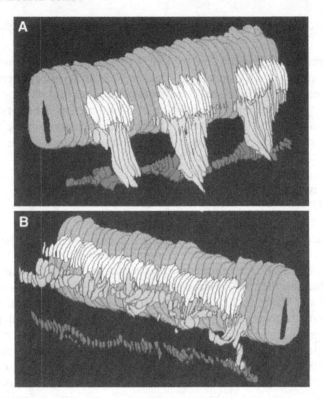

**Figure 4.1** The segmental pattern of peripheral ganglia and nerves depends upon somite integrity. Three-dimensional reconstruction of the neural tube and peripheral nervous structures in an E4.5 embryo receiving an implant of rostral somitic halves. (A) Control side showing three normal DRG, ventral roots, and primary sympathetic ganglia. (B) Continuous DRG polyganglion that resulted from constructing a paraxial mesoderm with only rostral halves of epithelial somites. Whereas three motor nerves condense from axons growing out of the neural tube on the unoperated side (A), motor axons emerging from the neural tube in B do not form individual nerves. Likewise, the primary sympathetic ganglia fail to condense into distinct swellings and remain unsegmented along the operated region.

embryos (Fig. 4.1). Altogether, these results demonstrate that the segmental pattern of neural crest migration and subsequent DRG formation is subservient to somitic mesoderm segmentation, and more specifically to differences between at least permissive (and perhaps also attractive) qualities of rostral half-somite cells and inhibitory properties of the caudal somitic domains. Intensive research is now aimed at elucidating the molecular basis of somite polarity and its implications for neural segmentation (further discussed in Chapter 2).

Weston (1963), Tosney (1978), and Teillet *et al.* (1987) have noticed that in intact embryos, as well as in quail–chick chimeras with grafted neural primordia, neural crest cells exit the neural tube evenly along the trunk neural

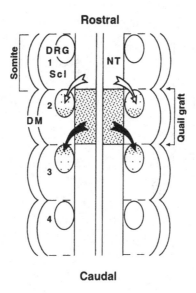

**Figure 4.2** The segmental origin of neural crest cells that colonize the DRG: fate of labeled cells in chick embryos grafted with a one-somite-long segment of quail neural primordium inserted between two successive intersomitic spaces. DRG 1, located rostral to the implant, is exclusively composed of host cells. DRG 2 and 3 are chimeric. The quail cells of DRG 2, which faces the graft, are preferentially localized in the caudal part of the ganglion, whereas those of DRG 3 are mainly found in its rostral portion. The arrows indicate the suggested direction that neural crest cells take from the neural primordium opposite the caudal somitic half to populate the rostral part of the DRG of the subsequent somitic level. DM, dermomyotome; NT, neural tube; Scl, sclerotome. (See Teillet *et al.*, 1987, for further details; reproduced with permission.)

axis and "wait" for the epithelial somites to dissociate into dermomyotome and sclerotome to start their segmental migration. This observation raised the question of the fate of those cells emerging from the neural tube opposite each caudal somitic domain given that this part of the somite does not support crest cell migration. This question was answered in a work by Teillet *et al.* (1987) who addressed the segmental origin and migratory pathways followed by neural crest cells that give rise to DRG. To this end, quail neural primordia, one or two segments long, were precisely implanted into chick hosts between two successive intersomitic spaces or two midsomitic transverse levels and the fate of cells originating at the limits of the grafts was assessed (Figs 4.2 and 4.3). Teillet *et al.* found that, once facing the somite-derived sclerotome, crest cells bearing the quail marker that originate opposite rostral and caudal somitic halves converge to migrate in a polarized fashion into the rostral domain of each sclerotome. Those cells arising opposite the caudal sclerotome, hence being unable to migrate further, relocate into rostral areas to join their siblings by a short longitudinal migration which does not exceed one and a half segments in length. As a result of this polarized migration, each DRG forms in the rostral domain of the

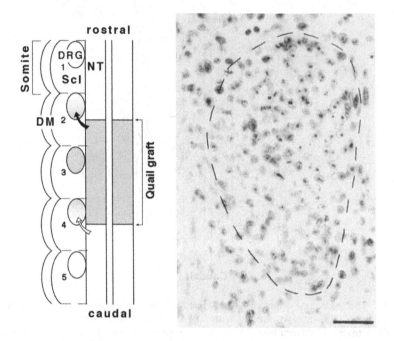

**Figure 4.3** The segmental origin of neural crest cells that colonize the DRG: fate of labeled cells in chick embryos grafted with two-somite-long segment of quail neural primordium ending at midsomitic regions. Left: Diagram showing the distribution of quail cells in the DRG with respect to the grafted segment of neural primordium. DRG 3 is entirely composed of neural crest cells of the quail type, whereas DRG 2 and 4 are chimeric. DRG 2, which faces a neural tube (NT) of host type, contains quail cells preferentially segregated to its caudal half, while DRG 4 (see photomicrograph to the right) faces a tube with quail phenotype and contains quail cells preferentially localized in its rostral half. The arrows in the diagram indicate the suggested direction that neural crest cells take from levels opposite the caudal half of the somite to populate the caudal halves of the respective DRG. (See Teillet *et al.*, 1987, for further details; reproduced with permission.)

somite but is colonized by cells arising opposite both rostral and caudal domains of the corresponding segment and by crest cells that originally exit opposite the caudal half of the preceding somite (summarized in Fig. 4.4). Another interesting observation made in the course of this work was that neural crest cells derived from rostral or caudal portions of the neural tube relative to the somites remain segregated along the rostrocaudal extent of each DRG, in a manner that corresponds to their origin along the neur-axis (Fig. 4.3). This segregation persists throughout development of the ganglia perhaps because, at the end of the migratory process, progenitor cells that had migrated close to each other (as a consequence of arising from the same segmental level) tend to associate with their next neighbors, by virtue of cell adhesion mechanisms that become active at gangliogenesis.

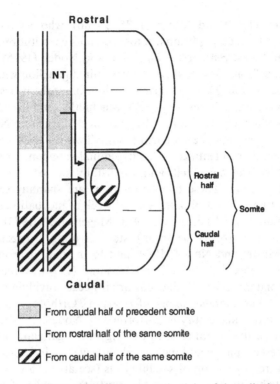

**Figure 4.4** Schematic model illustrating the segmental origin of the cells that constitute the DRG. Each DRG forms in the rostral part of the somite and is colonized by neural crest cells migrating longitudinally and laterally from the part of the neural tube (NT) facing the same somite and the preceding one. Neural crest cells arising opposite the caudal half of the corresponding somite contribute to about 50% of the ganglion, localizing predominantly in its caudal half Cells from the caudal half of the preceding somite populate the rostral 20% of the ganglion, and cells arising at the rostral half of the corresponding segment populate the remaining 30% of the ganglion. (Reproduced, with permission, from Teillet *et al.*, 1987.)

## 4.3  Embryonic origin and cellular heterogeneity of cranial and spinal ganglia

### 4.3.1  Embryonic origin of cranial sensory ganglia

After the pioneer work of His (1868) and until appropriate techniques were applied to the problem, the embryonic origin of the sensory ganglia of the head remained a controversial matter. From observations carried out on the chick embryo, His considered that they were exclusively of neural crest origin, but several other investigators also attributed a role in their histogenesis to the ectodermal placodes (Beard, 1888; Disse, 1897; Landacre, 1910; Landacre and McLellan, 1912; Coghill, 1916; Van Campenhout, 1937; Tello, 1946; Ortmann, 1948).

The dual origin of the trigeminal ganglion in the chick was clearly demonstrated by Hamburger (1961) and subsequently by Johnston (1966), Johnston and Hazelton (1972) and Noden (1975, 1978b), who established the relative share of neural crest and placodal ectoderm in its constitution. Using [³H]TdR-labeled neural crest cells, Johnston (1966) and Noden (1975) noticed aggregation of crest cells at the site where the trigeminal ganglion will form as early as 3 days of incubation. The respective contribution of the neural crest and of the placodes to the cranial sensory ganglia was finally established through the use of the quail–chick chimera system by D'Amico-Martel and Noden (1980), and by Ayer-Le Lièvre and Le Douarin (1982). Two different neuronal cell types can subsequently be identified in the trigeminal ganglion: large neurons, which are derived from placodal cells and are distally located, and smaller proximal ones of crest origin. These two populations are spatially separated and can easily be distinguished during the second week of incubation in the chick (Gait and Farbman, 1973; Ciani *et al.*, 1973; Meyer *et al.*, 1973). The cells of the mediodorsal area (of crest origin) stop dividing between days 5 and 7 (D'Amico-Martel and Noden, 1980) and form small neurons whose average volume, established by Ebendal and Hedlund (1974), is 1080 μm³. The more distal ventrolateral cells (of placodal origin) cease dividing earlier (at 3 and 4 days) and have an average volume of 1800 μm³. Orthotopic transplants of quail neural crest into chick embryos, performed by Noden (1978b), showed that the contribution of the neural crest to the trigeminal ganglion originates from the posterior mesencephalic and metencephalic levels, which provide the ganglion with the entire population of satellite cells (see also below).

The contribution of placodal ectoderm to the ganglia of cranial nerves was greatly clarified by the experimental work of Yntema (1937, 1943) on *Amblystoma*. Areas of ectoderm including the placodes, stained with Nile blue, were implanted orthotopically and their fate followed up to the early stages of ganglion formation. In the chick, following the contribution of Yntema (1944), a methodical study by Narayanan and Narayanan (1980), using the quail–chick chimera system, clearly evaluated the respective contributions of the neural crest and the placodal ectoderm to the glossopharyngeal–vagal complex, formed by the root ganglia of cranial nerves IX and X. The more rostral portion of this complex is the superior ganglion of the glossopharyngeal nerve, and the more caudal portion is the jugular ganglion of the vagus nerve. Each cranial nerve also has a ganglion generally referred to as the trunk ganglion located at a certain distance from the brain. Such are the petrosal ganglion of the glossopharyngeal nerve (IX) and the nodose ganglion of the vagus nerve (X).

The results of this study point clearly to a purely neural crest origin for the root ganglia of cranial nerves IX and X, and a placodal origin for all the neurons of the trunk ganglia of these two nerves. However, in both the trunk and the root ganglia, Schwann cells lining the nerves and ganglionic satellite cells are derived from the neural crest. Observations by Ayer-Le Lièvre and Le Douarin (1982) are in full agreement with the conclusions of Narayanan and Narayanan (1980). Table 4.1 and Fig. 4.5 summarize the present state of knowledge on the embryonic origin of the cranial ganglia.

Table 4.1. *Origin of cranial sensory ganglia*

|  |  | Neurons | | Neuroglia | |
| --- | --- | --- | --- | --- | --- |
| Ganglion | Origin | Placode | Neural crest | Placode | Neural crest |
| Vth nerve (trigeminal) | Trigeminal ganglion | + | + | – | + |
| VIth nerve (facial) | Root ganglion | – | + | – | + |
|  | Geniculate ganglion | + | – | – | + |
| IXth nerve (glossopharyngeal) | Superior ganglion | – | + | – | + |
|  | Petrosum ganglion | + | – | – | + |
| Xth nerve (vagus) | Jugular ganglion | – | + | – | + |
|  | Nodosum ganglion | + | – | – | + |

*Note*: Neurons are derived from ectodermal placodes or from the neural crest, with the exception of the trigeminal ganglion where neurons are of mixed placodal and crest origin. The neuroglia is in all cases exclusively derived from the neural crest.

In more recent studies, the precise segmental origin of the cranial sensory ganglia was reevaluated. This was assessed through both DiI labeling (Lumsden *et al.*, 1991) and quail–chick grafting procedures (Couly *et al.*, unpublished results, summarized in Fig. 4.6). These authors have mapped the contribution of the mesencephalic and of the rhombomeric (r) crest to the cranial ganglia at the 5 somite stage (5ss). It was found that the trigeminal ganglion originates from the neural fold of the caudal half of the mesencephalon and from rhombomeres r1, r2, and r3, which provide both the placodal ectoderm and the neural crest cells which contribute to the distal and proximal regions of the ganglion (Fig. 4.6). As shown by Couly and Le Douarin (1990), the presumptive territory of the trigeminal placode is located in the external aspect of the mesencephalic neural fold at the early 0–3ss. Later on, it migrates laterally and can be selectively removed in 10–12ss chick embryos and replaced by its quail counterpart (D'Amico-Martel and Noden, 1980). This results in the formation of trigeminal ganglia in which the large sensory distal neurons are of quail origin. In contrast, substitution of the chick by the quail neural crest at the level corresponding to the mesencephalon and to r1 and r2, yields chimeric ganglia in which the non-neuronal cells are all of quail type as well as the small, proximal, substance P-containing neurons (Ayer-Le Lièvre and Le Douarin, 1982).

Grafts of the caudal half of the mesencephalic neural fold yield the Schwann cells of the various branches of cranial nerve III (oculomotor), nerve IV (trochlear) and nerve V. The neural participation of r1 concerned the maxillary and mandibulary part of the trigeminal ganglion plus a small region in the ophthalmic branch, the Schwann cells of the maxillary and mandibulary branches of the trigeminal nerves, and of the IV (trochlear) and VI (external oculomotor and abducens) nerves.

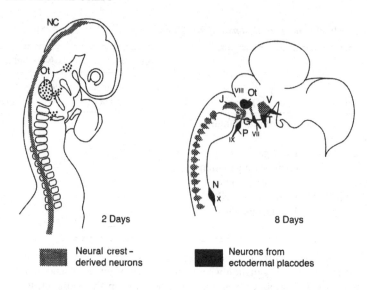

**Figure 4.5** The origin of cells in cranial sensory ganglia. Left panel: ■, neural crest (NC); ▦, location of the ectodermal placodes which contribute to the sensory ganglia. Four placodes located on the branchial arches contribute to (right panel) the trigeminal (T), geniculate (G), petrosum (P), and nodose (N) ganglia. The otic placode (Ot) forms the otic ganglion. Right panel: Disposition of the ganglia of the cranial nerves. ■ The superior (S) and jugular (J) ganglia (like the DRG) contain neurons that originate exclusively from the neural crest. In all ganglia the glial cells are neural crest-derived.

The neural crest cells from r2 migrate to the first branchial arch. Their fate is essentially neurogenic. They contribute to the maxillary and mandibulary part of the trigeminal ganglion. The rostral part of the acousticofacial (VII and VIII nerves) and the geniculate ganglia also contained quail cells. Likewise, Schwann cells of the maxillary and mandibulary branches of the trigeminal nerve and of nerve VII were mainly of quail origin.

Cells originating from the r3 neural fold colonized the posterior region of the first and the second branchial arches. They are later on found in the maxillomandibular part of the trigeminal ganglion, of the acousticofacial ganglion, and as Schwann cells of the corresponding nerves. Some contribution to the jugular ganglion was also found in the embryos subjected to r3 neural fold grafts. Grafts of r4 neural fold contributed to the acousticofacial and geniculate ganglia and to the proximal part of the jugular superior ganglion together with Schwann cells lining nerves VII, VIII, IX, and X.

### 4.3.2   The mesencephalic nucleus of the trigeminal nerve

Although it does not belong to the PNS, the mesencephalic nucleus of the trigeminal nerve deserves a special mention in this context since its origin

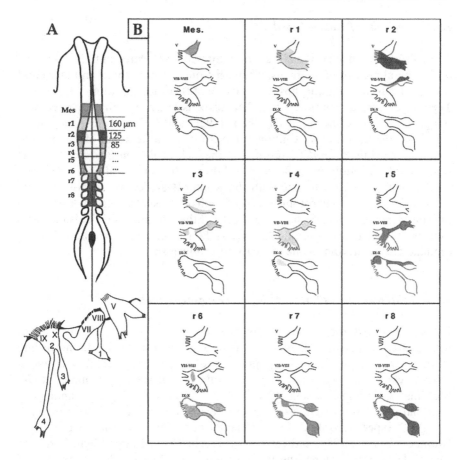

**Figure 4.6** The contribution of mesencephalic- and rhombencephalic-level neural crest to the formation of cranial sensory ganglia. Isotopic and isochronic grafts of quail neural primordium into chick hosts at 5ss. (A) Color-coded scheme illustrating the different transplanted regions. (B) Colonization patterns of different portions of the cranial ganglia by the grafted fragments of neural primordia. Mes, mesencephalon; r, rhombomere; V, trigeminal ganglion; VII, facial ganglion; VIII, vestibuloacoustic ganglion; IX, glossopharyngeal ganglion; X, vagal ganglion; 1, geniculate ganglion; 2, superior jugular ganglion; 3, petrosal ganglion; 4, nodose ganglion.

*For a colored version of this figure, see www.cambridge.org/9780521122252.*

from the neural crest has been demonstrated (Narayanan and Narayanan, 1978a). This was acheived by heterospecific grafting of the mesencephalic neural crest between quail and chick embryos and by showing that the precursor cells of the nucleus migrate from the crest towards the ventricular surface, independently of the simultaneous outward migration of the cells of the tectum originating from the proliferative activity of the neuroepithelium. This seems to be a unique case where migration of two cell types in opposite directions has been clearly demonstrated.

### 4.3.3 The cells of Rohon-Béard

The cells of Rohon-Béard are large transient neurons located in the dorsal part of the spinal cord of amphibians. They exist only is the embryo and larva and disappear during metamorphosis. The cause of their degeneration is unknown. They degenerate at the time when their place is taken by spinal root ganglia (Hughes, 1957). According to Du Shane (1938) and Chibon (1966) they originate from the neural crest. Chibon demonstrated that removal of the crest at the trunk level in *Pleurodeles* resulted in the lack of Rohon-Béard cells in the spinal cord. In addition, following orthotopic bilateral grafts of [$^3$H]TdR-labeled neural crest in the trunk all the Rohon-Béard cells were found to be labeled. In contrast, heterotopic grafts of cephalic crest at the trunk level led to the complete absence of Rohon-Béard cells at the operation site. This indicates that these large neurons are derived from the trunk neural crest and that the cephalic crest is devoid of the capacity to produce such cells.

### 4.3.4 Different types of neurons in spinal sensory ganglia

Developmental studies in avian species reveal that primary sensory neurons within DRG can also be divided into two major neuronal subtypes according to their size and birthdates: a population of large ventrolateral cells and a second class of small diameter dorsomedial neurons (Pannese, 1974; McMillan-Carr and Simpson, 1978; Hamburger *et al.*, 1981). Two analogous neuronal populations were also shown to exist in mammalian DRG: the "A" and "B" types according to the classification by electron microscopists (Rambourg *et al.*, 1983, and references therein), or the large "light" and the small "dark" populations, the latter containing a higher relative amount of granular endoplasmic reticulum and ribosomes (Lawson *et al.*, 1974). More importantly, these two morphologically distinct subclasses of neurons were shown to mediate transmission of different sensory modalities. The small diameter neurons send their central afferents to laminae I and II of the dorsal horn and transmit nociceptive and thermoceptive sensations, whereas the large diameter cells end up within the deeper laminae of the dorsal horn of the spinal cord and transmit mechanoreceptive and proprioceptive stimuli (Koerber and Mendell, 1992; Snider and Wright, 1996).

Birthdate studies performed in chick embryos by McMillan-Carr and Simpson (1978) reveal that large-scale neuronal production in DRG occurs between E4.5 and E6.5 for the lateroventral population and between E4.5 and E7.5 for the dorsomedial neurons. In the quail embryo, the birthdate patterns are very similar to the chick, yet neurons are born about 12 hours earlier (Schweizer *et al.*, 1983). A similar temporal sequence is followed by these two neuronal subsets during the period of programmed neuronal death. A peak of apoptosis is found in the lateroventral region of the chick DRG at E5.5 and only at E9.5 in the mediodorsal area (McMillan-Carr and Simpson, 1978). Thus, in general, the timing of birth, differentiation and death of the large neurons precedes that of the small-diameter neuronal population. These two morphologically and functionally different major neuronal subsets also differ in additional parameters that include expression of phenotypic markers and responsiveness to distinct growth factors.

### 4.3.5 Differential expression of markers to subsets of sensory neurons

The expression of several markers by sensory neurons also appears to be related to their embryonic origin. Cell-surface complex carbohydrate structures carried on lipid and protein cores, such as globoseries oligosaccharides, appear to be associated predominantly with the large light neuronal population in rat DRG and with placodally derived cranial sensory neurons. Conversely, the distribution of lactoseries carbohydrates in DRG and cranial sensory ganglia is consistent with a preferential association with small dark DRG neurons and neural crest-derived cranial sensory neurons (Dodd and Jessell, 1985). These markers are also selective for chick DRG subsets in which the sugar epitope is carried only by glycolipids (Ernsberger and Rohrer, 1988; Scott, 1993). In the quail embryo, however, these markers are not restricted to specific neuronal subsets, as the lactoseries carbohydrates are expressed by most neurons in DRG with a stronger intensity on the ventrolateral ones (Sieber-Blum, 1989c).

The functional classification of these two major neuronal subgroups was at least partially related to the presence of a number of neuroactive peptides in both embryonic and adult DRG, notably substance P, calcitonin gene-related peptide (CGRP), and somatostatin (Hökfelt et al., 1975a,b,c, 1976; Fontaine-Pérus et al., 1985; Price, 1985; New and Mudge, 1986). In avian DRG, substance P immunoreactivity first appears on E5 and its expression is confined to the small-diameter dorsomedial population of DRG neurons that innervate laminae I and II of the spinal cord (Fontaine-Pérus et al., 1985; New and Mudge, 1986). Most substance P-immunoreactive neurons are born between E3 and E6 in lumbosacral DRG (Scott et al., 1990). In a study addressing the origin of the neurons expresssing substance P in cranial ganglia, it was found that, in general, the sensory neurons that express substance P immunoreactivity are derived from the neural crest, with the exception of few neurons in the nodose ganglia which express the marker and which are of placodal origin (Fontaine-Pérus et al., 1985).

In rats, selective subsets of the smaller-diameter DRG neurons were also shown to express substance P-immunoreactive peptide, in an almost non-overlapping pattern when compared with the expression of somatostatin and fluoride-resistant acid phosphatase (Hökfelt et al., 1976; Nagy and Hunt, 1982; Maubert et al., 1992). In contrast to rodents, avian embryonic DRG express little somatostatin, but when these neurons were cultured in vitro in the presence of non-neuronal cells or of medium conditioned by these cells, the expression of somatostatin was shown to be greatly increased (Mudge, 1981).

The markers that characterize the lateroventral population of neurons are far less numerous. In addition to the complex type of globoseries carbohydrates, several forms of neurofilament proteins are expresssed by these neurons, as revealed by specific antibody staining (Price, 1985; Lawson *et al.*, 1993). Furthermore, Sommer *et al.* (1985) have subclassified the large lateroventral (or A) population of murine DRG into three distinct subclasses according to the relative activity of the enzyme carbonic anhydrase. The A$\alpha$ neurons were shown to display large clumps of Nissl substance and moderate activity of the enzyme carbonic anhydrase. The A$\beta$ class displayed small clusters of Nissl substance and moderate enzymatic activity. Finally, subclass A$\gamma$ showed the most intense enzymatic activity. Most notably, the variety of markers that characterize the two major neuronal subpopulations within the DRG, of which only a part is discussed here, suggests that individual neurons express combinations of the above markers. It would be interesting to determine whether this combination of markers is stable for a given set of neurons or can be subjected to functional variations. Although still unknown, the combinatorial expression of two or more of such molecules by individual neurons could define even smaller subpopulations with putative functional significance.

### 4.3.6 Differential response of sensory neurons to growth factors

The different neuronal subtypes defined above also respond to different growth factors. NGF, the prototypic maintenance factor for sensory and sympathetic neurons discovered by Levi-Montalcini (see Levi-Montalcini, 1987), is able to support the survival of small diameter cranial sensory neurons that derive exclusively from the neural crest (Davies and Lindsay, 1985; see also Table 4.2). Also in DRG, where both large and small neurons derive from the neural crest, the activity of NGF appears restricted to the small diameter neurons (Davies and Lindsay, 1985; Gaese *et al.*, 1994), that were shown in separate studies to express substance P immunoreactivity and transmit putative nociceptive sensations (see above). As already discussed, these neurons are also the latest to be born, suggesting that NGF acts on an ontogenetically young neuronal population. It was recently proposed by Barde (1994) that the relationship between embryonic origin, size, birthdate, and differential reactivity to growth factors may have a possible evolutionary relevance. Barde has proposed that NGF evolved from an ancient BDNF or NT-3-like gene based on the following facts. Firstly, whereas NT-3 and BDNF are already detected in cartilaginous fishes, NGF appears only in bony fishes. Secondly, sequence comparisons between BDNF and NGF in different species highlight the relative conservation of the BDNF gene as compared to a significant variability in the NGF sequence, indicating that the "older" BDNF gene may have reached an optimized structure earlier than NGF. The "late" appearance of NGF is probably related with the observation that it acts on phylogenetically recent components of the PNS that also differentiate relatively late in ontogeny. Such is the case for the mediodorsal neurons of the DRG and the neurons of the sympathetic

Table 4.2. *Neuronal responsiveness to neurotrophic factors*

| | Embryonic origin | NGF[1] | BDNF[2] | NT-3[3] | NT-4[4] | NT-5[5] | NT-6[6] | CNTF[7] | LIF[8] | bFGF[9] |
|---|---|---|---|---|---|---|---|---|---|---|
| DRG[a] | NC sensory | + | + | + | + | + at early stages | + | + | + | + |
| Jugular[b] | NC sensory | + | + | + | + | + at early stages | | + | + | |
| TMN[c] | NC sensory | -, + | + | + | | - | | | | |
| Trigeminal[d] | NC (MD), placode (LV) sensory | -VL, + MD | + mainly VL neurons | +/- VL | + | + MD, at early stage - LV | | + | | |
| Nodose[e] | Placode sensory | - | + | + | +/-, + | -, + | - | -, + | + | |
| Petrosal[f] | Placode sensory | - | + | | | | | | | |
| Geniculate[g] | Placode sensory | - | + | | | | | | | |
| Vestibular[h] | Placode sensory | - | + | | | | | | | |
| Sympathetic ganglia[i] | NC | + | - | -, ?, + | - | + | + | + | -, survival +, differ. to cholinergic | + |
| Ciliary[j] | NC parasympathetic | - | - | -? | | - | | + | | + |

*Note:*
+, indicates effects on survival and/or differentiation; +/-, small but significant effects; -, no effect; empty cells stand for not determined.

*References:*
a1. Davies and Lindsay, 1985; Davies et al., 1986; Diamond et al., 1992; Hamburger et al., 1981; Hory-Lee et al., 1993; Lindsay, 1988; Ruit et al., 1992.
a2. Hofer and Barde, 1988; Hory-Lee et al., 1993; Leibrock et al., 1989; Lindsay, 1988.
a3. Hohn et al., 1990; Hory-Lee et al., 1993; Maisonpierre et al., 1990b; Ockel et al., 1996; Rosenthal et al., 1990.
a4. Hallbrook et al., 1991.
a5. Berkemeier et al., 1991.
a6. Gotz et al., 1994.
a7. Lin et al., 1990; Manthorpe et al., 1982.
a8. Murphy et al., 1991, 1993, 1994.
a9. Eckenstein et al., 1990.
b1. Davies and Lindsay, 1985; Davies et al., 1993b.
b2. Davies et al., 1986.
b5. Davies et al., 1993a.
c1. Davies et al., 1987; Lindsay, 1988; Paul and Davies, 1995.
c2. Davies et al., 1986; Lindsay, 1988.

c3. Hohn et al., 1990.
c5. Davies et al., 1993a.
d1. Buchman and Davies, 1993; Davies et al., 1993b; Davies and Lumsden, 1984; Dimberg et al., 1987.
d2. Davies et al., 1986.
d3. Hohn et al., 1990.
d4. Ibanez et al., 1993.
d5. Davies et al., 1993a.
d7. Manthorpe et al., 1982.
e1. Davies and Lindsay, 1985; Katz et al., 1990; Lindsay and Rohrer, 1985; Rosenthal et al., 1990.
e2. Davies et al., 1995; Hohn et al., 1990; Lindsay et al., 1985; Maisonpierre et al., 1990b.
e3. Davies et al., 1995 (Nt-3 with activity on trk C-/- mutants); Hohn et al., 1990; Lindsay et al., 1985; Rosenthal et al., 1990.
e4. Hallbrook et al., 1991; Thaler et al., 1994.
e5. Berkemeier et al., 1991; Davies et al., 1993a.
e6. Gotz et al., 1994.
e7. Barbin et al., 1984; Manthorpe et al., 1982; Thaler et al., 1994.
e8. Thaler et al., 1994.
f1. Davies and Lindsay, 1985.
f2. Davies et al., 1986.

g1. Davies and Lindsay, 1985.
g2. Davies et al., 1986.
h1. Davies and Lindsay, 1985.
i1. Campenot et al., 1991; Levi-Montalcini, 1987; Ruit et al., 1990.
i2. Lindsay et al., 1985; Maisonpierre et al., 1990b.
i3. Davies et al., 1995 (NT-3 with effect in trkC-/- mutants); Dechant et al., 1993b (survival of E7 but not of E11 neurons); Hohn et al., 1990; Rosenthal et al., 1990.
i4. Hallbrook et al., 1991.
i5. Berkemeier et al., 1991.
i6. Gotz et al., 1994.
i7. Lin et al., 1990; Rao et al., 1990, 1992a, b, c.
i8. Yamamori et al., 1989.
i9. Eckenstein et al., 1990.
j1. Eckenstein et al., 1990.
j2. Lindsay et al., 1985; Maisonpierre et al., 1990b.
j3. Ernfors et al., 1994b; Hohn et al., 1990; Maisonpierre et al., 1990b.
j5. Berkemeier et al., 1991.
j6. Gotz et al., 1994.
j7. Barbin et al., 1984; Lin et al., 1990.
j8. Rao et al., 1990.
j9. Eckenstein et al., 1990; Unsicker et al., 1987.

ganglia, which appear relatively late during vertebrate evolution (see Barde, 1994, and references therein). The findings that the early phases of development of sensory neurons of the DRG depend upon BDNF and NT-3 signaling but not upon NGF, are consistent with the proposed notion of a time sequence in responsiveness to the different neurotrophins during embryonic ontogeny (see detailed discussion in the next section).

Scott (1993) has investigated the trophic requirements of chick DRG neurons bearing the AC4/SSEA-1 epitope that recognizes lactoseries glycoconjugates. The immunoreactive neurons, of small and medium size, innervate both skin and muscle targets and are localized in the ventrolateral region of the ganglia of young embryos, but in hatchlings they become evident in the dorsomedial part of the DRG. These AC4/SSEA-1-positive neurons are supported *in vitro* primarily by BDNF, to a lesser extent by NT-3, but not by NGF. Furthermore, it was clearly established that the large neurons which transmit directly proprioceptive information from muscle spindles and Golgi tendon organs to spinal motor neurons depend for their survival *in vitro* primarily upon NT-3 (Hory-Lee *et al.*, 1993; Oakley *et al.*, 1995). These results were fully confirmed by phenotypic analysis of mutant mice carrying deletions either in the NT-3 or in the *trkC* genes (see Section 4.4.5).

In summary, increasing evidence relates the embryonic origin of distinct sensory neurons with specific morphological, biochemical and functional features, suggesting that both time- and space-related parameters (localization of the neuron within the ganglion, innervation patterns) play a critical role in the regulation of sensory development and function.

## 4.4 Control of cell number in developing ganglia

### 4.4.1 Control of the development of sensory neural crest progenitors by the microenvironment

*4.4.1.1 The neural crest–neural tube–somite complex.* Neural crest cells that give rise to DRG remain in the vicinity of the neural tube throughout migration and during condensation into ganglia, a process that occurs within the rostral half of each somite (Fig. 4.7). Based on these topographical relationships, the assumption was tested that both the CNS primordium and the somites are the sources of specific signals controling the shape and size of the organizing ganglia. A means of analyzing this hypothesis consisted of experimentally changing the topographical relationships between the neural crest cells and the neural tube, or the neural crest cells and the adjacent somites.

The possibility that the neural tube plays a role in the development of sensory ganglia was first suggested by the observation that sensory neurons differentiate only in proximity to the CNS primordium (Le Lièvre *et al.*, 1980; Schweizer *et al.*, 1983; Le Douarin, 1986). Subsequently, this notion was tested in a more direct manner by performing neural tube ablations soon after the migration of neural crest progenitors. Neuralectomy caused the death of DRG

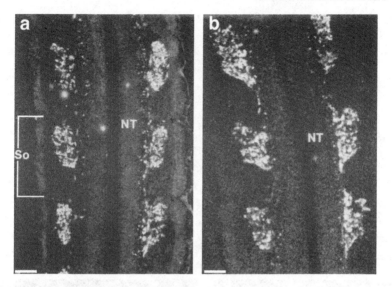

**Figure 4.7** The segmental arrangement of DRG and peripheral nerves. Immuno-fluorescent staining with the HNK-1 antibody of sections from a 30-somite chick embryo at the level of somites 21–23. (a) Frontal section at a dorsal level showing the neural crest-derived cells forming the DRG on both sides of the neural tube (NT) and occupying the rostral part of each somite (So). (b) Frontal section through the same embryo at the level of the ventral neural tube depicting HNK-1-positive Schwann cells metamerically organized within the rostral sclerotomes and lining along the ventral root fibers. Bar = 50 μm. (Reproduced, with permission, from Kalcheim and Le Douarin, 1986.)

cells, but had also an adverse effect on the somites which died within the 24–48-hour period following ablation (Teillet and Le Douarin, 1983). A less-invasive manipulation was performed afterwards that consisted in implanting into chick embryos at 30ss an impermeable silicon membrane between the neural tube and the recently migrated neural crest cells along a length of five consecutive somites (somitic level 20–25; Kalcheim and Le Douarin, 1986; Fig. 4.8). Separation of the crest cells from the neural tube at this specific stage resulted in the selective death of the separated cells located in the prospective DRG region and consequently in a total absence of DRG (Fig. 4.9a). In contrast to the neural tube ablation, this operation caused no apparent impairment in the development of crest-derived sympathoblasts and Schwann cells located further ventrad. Neither did it affect the survival of the mesodermal somites or their derivatives with which the crest cells remained in contact. This require-ment for neural tube-derived signals is only transient because lack of CNS input at slightly later stages, when young sensory neurons begin to establish contacts with their targets, only partially affected ganglion development (Yip and Johnson, 1984; Kalcheim and Le Douarin, 1986).

The results of the membrane implants showed that the neural tube is neces-sary for the survival of sensory ganglion progenitors, and that an interaction

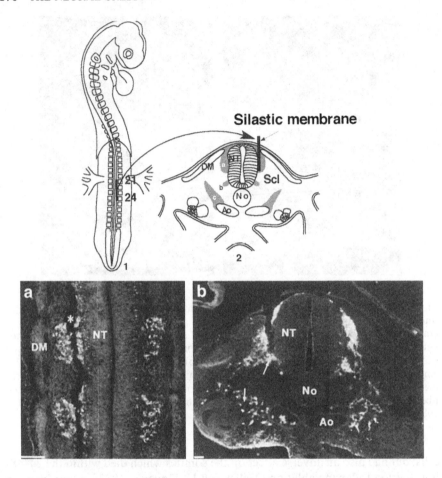

**Figure 4.8** Separation of the DRG anlage from the neural tube by means of an impermeable silastic membrane. A unilateral incision between the neural tube (NT) and the adjacent somites is performed at the level of somites 21–24 of a 30ss chick embryo. The silastic membrane is thus introduced into the slit and protrudes dorsally to compensate for embryonic growth. The photomicrographs represent frontal (a) and transverse (b) sections of operated embryos fixed 2 hours after implantation. The operation attained only the DRG while leaving intact the Schwann cells and sympathetic ganglia (arrows). Ao, dorsal aorta; DM, dermomyotome; No, notochord. Bar = 50 μm. (Reproduced, with permission, from Kalcheim and Le Douarin, 1986.)

with the somitic environment was not enough to rescue the cells from death. The question was then raised whether the somites have any effect on the regulation of neural crest cell number, in addition to their role in ganglion segmentation. Results from experiments in which the rostrocaudal composition of the somites was altered by replacing normal somites with multiple rostral or caudal somitic halves shed light on this issue. Altering the metameric organization of DRG and sympathetic ganglia also had profound effects on their size (Kalcheim and Teillet, 1989; Goldstein *et al.*, 1990; Goldstein and Kalcheim,

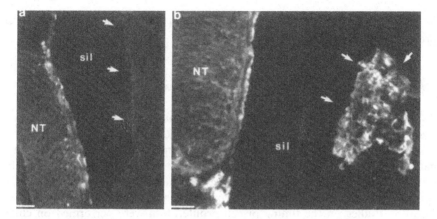

**Figure 4.9** Increased survival time of neural crest-derived cells on the somitic side of a silastic membrane preadsorbed with a neural tube extract. HNK-1 immunolabeling of two sections from E4.5 chick embryos. (a) After implantation of a collagen-treated silastic (sil) membrane (control). Note the absence of HNK-1 stained cells distal to the implant (arrows) and the few immunofluorescent cells adjacent to the tube surface. (b) After implantation of a neural tube extract-treated silastic membrane. Many fluorescent cells are evident on the somitic side of the membrane (arrows) (see Kalcheim and Le Douarin, 1986). Similar results were obtained after preadsorbing silastic membranes with FGF2, BDNF or NT-3. Bars = 25 μm. (Reproduced, with permission, from Kalcheim and Le Douarin, 1986.)

1991). Grafting of multiple rostral half-somites in place of normal segments led to the formation of unsegmented DRG (polyganglia) whose volume and cell number was significantly greater than the sum of the ganglia located in the intact contralateral side. One mechanism that accounts for increased ganglion size is enhanced proliferative activity of the cells in the polyganglia (Goldstein *et al.*, 1990). This stimulation of the proliferation of neural crest cells developing within a mesoderm of rostral type may result either from a direct mitogenic effect of rostral somitic cells, or from a neural tube-derived mitogen whose accessibility to the crest progenitors has increased in view of the continuous area of contact between the crest cells and the neural tube. Subsequent studies performed *in vitro* have confirmed the notion that a neural tube-derived factor is able to stimulate the proliferation of a subset of neural crest cells. Moreover, this mitogenic effect was significantly higher when the factor was applied to mixed cultures of dissociated neural crest cells migrating within the somites (Kalcheim *et al.*, 1992).

In summary, results of the studies modifying in a controlled manner the environment in which the peripheral ganglia develop show that initial DRG size is regulated both by the neural tube and by the rostral part of each somite. These results raised many questions concerning the identity of the factors that mediate the effects of the neural tube, the cellular processes they influence, the state of committment of the responsive crest cells, and the mechanisms whereby the somites modulate factor activity. The molecular aspects of these processes are discussed in the next section.

### 4.4.2 Trophic control of neural crest development

*4.4.2.1 Multiple factors promote the development of neural crest cells into sensory neurons: the neurotrophins and basic FGF (FGF2).* Among the processes that contribute to the regulation of the initial number of cells within a ganglion are the initial amount of migrating cells, the proliferation and the differentiation of the neural crest progenitors into neurons and satellite cells. Mitosis and differentiation lead to the production of large numbers of neurons. Detailed studies on the timing of cell proliferation were performed on chick embryos, where elevated proliferative activity was found to occur through days 4 and 5 of incubation (E4-5), followed by a significant drop thereafter. Withdrawal from the cell cycle is followed by differentiation of sensory neurons. McMillan-Carr and Simpson (1978) have reported that large-scale neuronal production occurs in chick DRG between 4.5 and 6.5 days in the lateroventral region and between 4.5 and 7.5 days in the mediodorsal area. Thus, most sensory neurons are likely to arise from multipotent precursors that continue dividing following colonization of the ganglionic anlage. It is interesting to mention in this respect that, unlike the situation in the embryo, when premigratory or migrating neural crest cells are cultured under serum-free conditions, a subpopulation of cells becomes readily postmitotic and differentiates into sensory-like neurons (Ziller *et al.*, 1987; Sieber-Blum, 1989c; see also Chapter 7). This experimental paradigm highlights possible differences inherent to the crest population, thus suggesting that distinct mechanisms control the development of progenitor cells. Only recently have we begun unraveling some of the factors that operate on neural precursors.

A. *IN VIVO* APPROACH. As a result of the finding that sensory ganglion progenitors require specific neural tube-derived signals (see previous section), a search began for molecules that mediate these early effects. Impermeable membranes that served as barriers to disconnect the neural crest cells from the neural tube were implanted following preadsorption of embryonic extracts or purified factors. Neural crest cells that form the DRG could be temporarily rescued provided the membranes were pretreated with a neural tube extract, but not a liver extract, or with purified neurotrophins such as BDNF or NT-3, but not NGF (see Fig. 4.9b; Kalcheim and Le Douarin, 1986; Kalcheim *et al.*, 1987; Brill and Kalcheim, unpublished results). Also the basic form of fibroblast growth factor (FGF2) was effective to a similar extent compared with BDNF. Interestingly, no synergistic effect was obtained when cotreating the membranes with both FGF2 and BDNF compared with each factor separately, suggesting that the two molecules may act on overlapping subsets of crest precursors (Kalcheim, 1989). These results showed for the first time that, much like differentiated neurons, neural crest precursors are programmed to die unless provided with specific survival cues. Moreover, they also revealed an early role for specific neurotrophins, originally found to prevent neuronal death and promote neurite extension of several peripheral and central neuronal types (Barde, 1989). Likewise, FGF2 was classically considered to trigger the proliferation of a wide variety of mesodermal cells and also to promote neuronal survival (Gospodarowicz, 1991).

Subsequent studies have confirmed that the neural tube is an early source of these factors. NT-3 and FGF2-like immunoreactive proteins and mRNA were detected in the avian CNS primordium (Kalcheim and Neufeld, 1990; Pinco *et al.*, 1993; Yao *et al.*, 1994). Furthermore, *in situ* hybridization has revealed a dynamic pattern of expression of BDNF transcripts in the avian neural tube of developing chick embryos. Expression of BDNF gene was already observed in 18ss embryos throughout the entire cross-sectional plane of the neural tube at thoracic levels. Slightly later, expression of BDNF became confined to the dorsal part of the neural tube at the levels where neural crest formation and migration were in progress (Fig. 4.10, see also Kahane *et al.*, 1996).

*In vivo* studies have also shown that the combined presence of the neural tube, and of a mesoderm permissive for neural crest migration, stimulates the proliferation of avian neural crest cells that constitute the DRG (Goldstein *et al.*, 1990). Further experiments have shown that this *in vivo* activity can be mimicked *in vitro* upon treatment of neural crest cells with NT-3. NT-3 is a mitogen for cultured avian crest cells, as it stimulates the proportion of cells encountered in the S-phase of the mitotic cycle, in addition to stimulating an increase in total cell number (survival effect) (Kalcheim *et al.*, 1992; Pinco *et al.*, 1993). Consistent with this finding, NT-3 was found to keep rat sensory precursors of the DRG in a proliferative state by inhibiting their differentiation into neurons (Memberg and Hall, 1995). It is interesting to point out in this context that retinoic acid (RA) was also found to stimulate the proliferation and differentiation of distinct subsets of neuronal progenitor cells (Henion and Weston, 1994; Dupin and Le Douarin, 1995; Rockwood and Maxwell, 1996). Since RA was found to induce the expression of the trk and p75 neurotrophin receptors in neural crest-derived neuroblastoma cells (Haskell *et al.*, 1987; Kaplan *et al.*, 1993) and in sympathetic neurons (Kobayashi *et al.*, 1994c; Rodriguez-Tebar and Rohrer, 1991), it would be relevant to test whether RA acts by upregulating neurotrophin dependence of progenitors of sensory and sympathetic neurons.

B. *IN VITRO* APPROACH. Additional effects displayed by the neurotrophins and by FGF2 on neural crest cells were characterized to a great extent by means of *in vitro* approaches. BDNF (Kalcheim and Gendreau, 1988), NT-3 (Pinco *et al.*, 1993), and FGF (Brill *et al.*, 1992) were found to stimulate the differentia-tion of a subset of neural crest progenitors into neurons expressing substance P and A2B5 immunoreactivities. The ability of all three factors to stimulate neurogenesis of the neural crest progenitors raises the possibility of redundancy in the response of individual neural crest cells to the various factors, much like

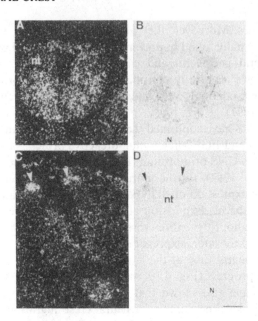

**Figure 4.10** Expression of BDNF mRNA in the neural tube of early chick embryos. (A, B) Dark-field and bright-field images, respectively, of a transverse section through somite 17 of a 24ss embryo showing BDNF mRNA signal distributed throughout the neural tube (nt). (C, D) Dark-field and bright-field images, respectively, of a transverse section through the thoracic region of a 36ss embryo showing the expression of BDNF mRNA restricted to the dorsal part of the neural tube (arrowheads). N, notochord; nt, neural tube. Bar = 50 μm. (Reproduced, with permission, from Kahane *et al.*, 1996. Copyright © 1996. Reprinted by permission of John Wiley & Sons, Inc.)

the combined action of BDNF and FGF2 *in vivo* on survival of DRG precursors (for further discussion see Section 4.4.6).

Two issues related to the possible mechanisms of action of these factors are worth discussing. The first concerns the state of commitment of the responsive cells. Distinct sensory neurons can differentiate both from dividing multipotential cells as well as from early committed progenitors (see Chapter 7). The experiments outlined above made use of culture conditions devoid of serum that facilitated the differentiation of precursors within 15–20 hours of explantation from a postmitotic progenitor subset (Ziller *et al.*, 1983, 1987). These progenitors are, therefore, likely to belong to an early specified neural crest subset. The mechanism whereby exogenous factors influence such progenitors must be by activating, rather than by inducing, a program for survival or differentiation. Furthermore, mitotically active neural crest precursors also have the ability to respond to NT-3 (Kalcheim *et al.*, 1992), and to FGF2 (Bannerman and Pleasure, 1993; Kalcheim, 1989). In contrast to the early postmitotic precursors, this cell population is likely to contain multipotent cells able to develop into sensory neurons and also into other derivatives.

FGF2 and NT-3 have, therefore, more than one type of activity on distinct avian neural crest cells. In other words, neural crest progenitors, which progressively change their developmental potentialities, may remain responsive to a given growth factor throughout ontogeny. The nature of their response to this environmental signal will, however, differ according to their actual state of commitment.

*4.4.2.2 Primary neuronal differentiation involves cell-mediated environmental stimuli.* The second issue concerns the observation that primary neuronal development from neural crest progenitors requires a cell-mediated process. Growth factors such as NT-3, BDNF, and FGF2 have little or no effect on neuronal number when added to pure neural crest cultures (they have, however, significant effects on the non-neuronal cells; see Kalcheim, 1989; Brill *et al.*, 1992; Kalcheim *et al.*, 1992). In contrast, when neural crest cells are grown on mesodermal cells expressing these molecules, or on mesodermal cells to which these factors are exogenously added, neuronal development is greatly stimulated (see Fig. 4.11; see also Pinco *et al.*, 1993, for NT-3; Kalcheim and Gendreau, 1988, for BDNF; Brill *et al.*, 1992, for FGF2). Likewise, the mitogenic effect of NT-3 on neural crest cells is further enhanced in cultures of somitic cells that contain migrating neural crest progenitors (Kalcheim *et al.*, 1992). Several mechanisms could account for these observations. Somitic mesodermal cells are likely mediators in the presentation of the growth factors to the responding cells. Thus, FGF2 could specifically bind to somite-associated heparan sulfate proteoglycans, molecules to which the FGFs show high affinity (Saksela and Rifkin, 1990; Bashkin *et al.*, 1992). In fact, heparin and heparan sulfate proteoglycans were implicated in the modulation of FGF activity by at least three mechanisms: (1) by providing a matrix-bound or a cell-surface reservoir of FGF for responsive cells; (2) by being essential cofactors for the interaction of FGF with their tyrosine kinase receptors; and (3) by regulating the secretion and internalization of the FGFs (reviewed in Rapraeger *et al.*, 1994).

Likewise, the response of neural crest cells to neurotrophins could be enhanced by the low-affinity p75 receptor expressed both on crest cells and on somitic cells (Wheeler and Bothwell, 1992). It has been clearly documented that, in several systems, the low-affinity p75 receptor, common to all neurotrophins, cooperates with the tyrosine kinase receptors that are more specific for each neurotrophin to create high-affinity sites, thus modulating the responsiveness to this family of growth factors (reviewed in Chao, 1992, 1994).

Moreover, the glycoprotein laminin, present in the basement membrane that surrounds the neural tube and in the dermomyotome/sclerotome interface (Rogers *et al.*, 1986), could act as a specific attachment molecule for the above factors, thereby increasing their concentration in a local manner. Two observations support this possibility. First, adsorption of BDNF or FGF2 together with laminin on silastic membranes prior to implantation between neural tube and somites was necessary for rescuing neural crest cells that

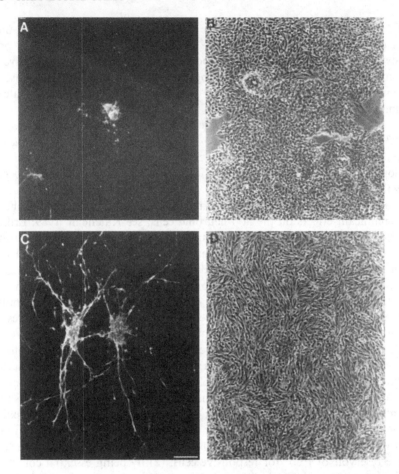

**Figure 4.11** NT-3 stimulates neuronal differentiation of neural crest progenitors: the effect of NT-3-producing cells on neurite outgrowth from neural crest clusters. Two-day-old neural crest clusters were cultured on top of a monolayer of (A, B) control chinese hamster ovary (CHO) cells, or (C, D) CHO cells that produce recombinant NT-3. At the end of 24 hours of incubation in serum-free medium, cocultures were fixed and stained with the HNK-1 antibody. Note in (C) the presence of two HNK-1-immunoreactive clusters showing extensive neurite outgrowth. In contrast, only few crest cells differentiate into neurons on the control CHO line. (B, D) Phase contrast. Bar = 120 μm. (Reproduced, with permission, from Pinco *et al.*, 1993. Copyright © 1993. Reprinted by permission of John Wiley & Sons, Inc.)

were disconnected from the neural tube (Kalcheim *et al.*, 1987; Kalcheim, 1989). Second, when dissociated mixed somite–neural crest cells were cultured on a laminin substrate, a ten-fold less concentration of NT-3 was required to obtain a similar stimulation in the proportion of proliferating neural crest cells when compared to cells seeded on a fibronectin substrate (Kalcheim *et al.*, 1992).

*4.4.2.3 Neural crest cells express receptors for neurotrophins and FGF2.* Validation of a model invoking the activities of BDNF, NT-3, and FGF2 on neural crest cells requires that the respective receptors be expressed on these progenitors. The observations that all growth factors are active at nanomolar concentrations (50 pg/ml to 1 ng/ml) suggest that their activities are indeed mediated via appropriate high-affinity binding sites (see Fig. 4.14). Evidence derived from RT-PCR analysis and *in situ* hybridization confirmed this view. Heuer *et al.* (1990) have shown that subpopulations of migrating neural crest cells express the avian counterpart of the human FGF receptor (flg). The most selective receptor for BDNF, trkB (Glass *et al.*, 1991; Klein *et al.*, 1991; Soppet *et al.*, 1991; Squinto *et al.*, 1991), is expressed on avian neural crest cells (Yao *et al.*, 1994) and on chick DRG cells from E3.5 (Dechant *et al.*, 1993a). The most selective receptor for NT-3, trkC (Lamballe *et al.*, 1991, 1993, 1994), is synthesized by neural crest cells already at premigratory stages (Fig. 4.12; and see Kahane and Kalcheim, 1994) and continues to be expressed on crest cell subsets migrating in the ventral pathway that leads to the formation of DRG (Fig. 4.12; and see Kahane and Kalcheim, 1994; Tessarollo *et al.*, 1993; Williams *et al.*, 1993; Yao *et al.*, 1994). More recently, Henion *et al.* (1995) have demonstrated that neural crest progenitors migrating in the ventral pathway express the full length isoform of trkC which is able to transduce a biological signal. This observation is of great importance, as truncated isoforms lacking the cytoplasmic domain of the receptor have been identified primarily on non-neuronal cells (reviewed by Barbacid 1995; see also Lamballe *et al.*, 1993; Tsoulfas *et al.*, 1993; Valenzuela *et al.*, 1993). The function(s) of these isoforms is still elusive, although Biffo *et al.* (1995) have suggested that, at slightly later stages, the truncated form of the trkB receptor may serve to restrict the availability of its cognate ligand by prompting its internalization.

Interestingly, in all the experiments outlined above, as well as in *"in vivo"* studies, NGF was without activity on neural crest cells (Kalcheim and Le Douarin, 1986; Kalcheim *et al.*, 1987). In agreement with this result, no NGF binding to high-affinity sites could be revealed on migrating neural crest cells (Speight *et al.*, 1993). Specific high-affinity sites appear on a population of neuron-like cells that express tyrosine hydroxylase only after 1–2 weeks in culture, provided they are treated with NGF (Bernd, 1987, 1988).

### 4.4.3 Early stages of sensory ganglion development

*4.4.3.1 Variations in initial DRG size along the neuraxis.* DRG display regional variations in cell number along the rostrocaudal axis. It was known for a long time that limb-innervating ganglia are much larger than their non-limb-innervating counterparts because a comparably lower percentage of neuronal death takes place in the former (Hamburger and Levi-Montalcini, 1949). Moreover, it was recently shown that brachial-level ganglia are larger than cervical ones already from the stage of gangliogenesis and prior to the onset

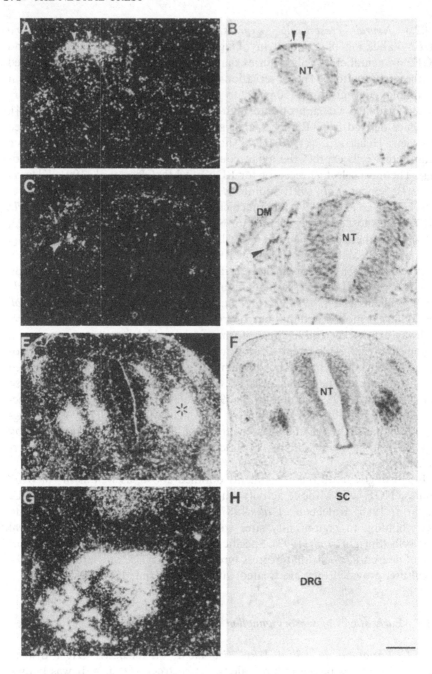

**Figure 4.12** Expression of the *trkC* gene in subsets of neural crest cells and DRG. (A, B) Transverse section through a 22ss embryo at the level of somite 17 showing restriction of *trkC* signal to the dorsal part of the neural tube that contains presumptive premigratory crest cells (arrowheads). (C, D) Transverse section through the rostral somitic half of a 28ss embryo showing *trkC*-positive neural crest cells within the sclerotome. Sclerotomal

of programmed cell death (Goldstein, 1993). These early differences are not due to different rates of DRG cell proliferation in the two regions, but appear to be related to size differences of the rostral sclerotomes in the two areas. Since the rostral sclerotome in brachial areas is about twice the length of the cervical-level sclerotome (Goldstein, 1993), a comparably larger number of neural crest cells is likely to invade the former. In support of this notion, grafting of brachial-level somites into cervical areas of the axis, or of cervical-level somites into wing-innervating areas, results in the formation of DRG with a size appropriate to the grafted segments (Goldstein *et al.*, 1995). So, axial-level differences in the number of neural crest cells able to migrate into a given anlage may be an early determinant of initial ganglion size.

Another striking example of variation in early DRG development along the rostrocaudal axis is the transient lifespan of the rostralmost ganglia in the embryos of *Amniotes*. The ganglia that develop at the level of the occipital and of the first one or two cervical segments, called Froriep's ganglia, disappear early in development despite their apparent normal formation and segmentation when compared to the permanent DRG (Lim *et al.*, 1987). Rosen *et al.* (1996) have shown that, in the chick, DRG corresponding to the level of the second cervical vertebra (C2) forms by 50 hours of incubation and has the same volume as the permanent DRG C5 and C6 by E2.5, but starts degenerating by E3, until its complete disappearance by E10 (Fig. 4.13). The cellular mechanisms accounting for these size differences are both decreased proliferation of the precursors cells as well as a more significant extent of cell death at parallel stages. As the rostral somitic tissue where these ganglia develop has been implicated in modulating proliferation and differentiation of the neural crest precursors that colonize the DRG (Goldstein *et al.*, 1990; Kalcheim *et al.*, 1992), the possibility exists that molecular differences between the somites account for the different properties observed. In line with such a suggestion is the finding that overexpression of the caudally expressed *Hoxb-8* gene, involved along with other members of the *Hox* family in determining positional properties of body segments, results in prolonged life of the murine Froriep's C2 DRG (Charite *et al.*, 1994). This was interpreted as a "posterior transformation" making the C2 ganglion similar to DRG in more caudal segments. These differences between the occipital somites and the more caudal somites might be involved in determining their differential fates. While most somites along the axis give rise to the vertebrae of the spine, the rostralmost four pairs of somites fuse to form the basi- and exo-occipital chondrocranium, and somite

**Figure 4.12** (*cont.*)
cells show no expression of *trkC*. (E, F) Transverse section through an E5 embryo showing intense and homogeneous distribution of *trkC* transcripts in the DRG(*). Note also that, at this stage, the ventral region of the neural tube containing spinal motor neurons lacks *trkC* mRNA. (G, H) DRG of an E8 embryo. *trkC* mRNA is localized to specific neurons in the lateroventral area of the DRG, and shows a diffuse distribution to mediodorsal cells. DM, dermomyotome; NT, neural tube; SC, spinal cord. Bar = 50 μm. (See Kahane and Kalcheim, 1994.)

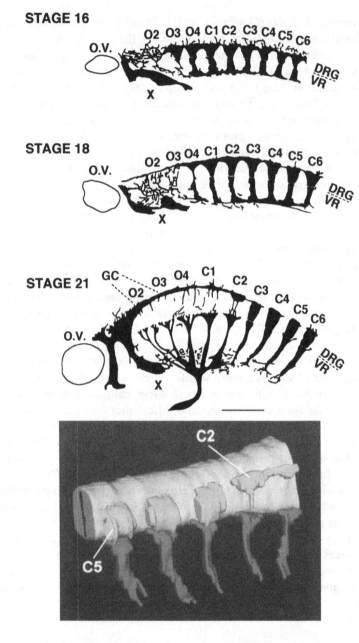

**Figure 4.13** Morphology of the transient Froriep's and neighboring conventional ganglia at early stages of development. Upper panel: Camera lucida drawings of chick embryos stained with the HNK-1 antibody to delineate neural crest and nervous structures. The C2 DRG is normal at stages 16 and 18, but reduced at stage 21. X, vagus nerve; O, occipital segments; C, cervical segments; VR, vertical root; O.V., otic vesicle. Lower panel: A three-dimensional reconstruction of neural structures of a serially sectioned stage 23 embryo at the transition between the hindbrain and spinal cord. The unusual shape and smaller girth of the C2 ganglion compared to normal ganglia such as

pairs 5 and 6 contribute to the formation of the highly specialized atlas and axis (Couly *et al.*, 1993). These dramatic changes in the morphology of the rostral-most six somitic pairs may be the immediate cause for the degeneration of the Froriep's ganglia.

*4.4.3.2 Factors affecting the development of sensory neuroblasts.* Factors that affect the differentiation of neural crest progenitors also act on young neuroblasts to stimulate their maturation prior to the onset of programmed neuronal death. Both BDNF and NT-3 were found to accelerate a maturational change in the shape of young sensory neuroblasts explanted from E4.5 avian DRG (Wright *et al.*, 1992). The active factors may be provided by the adjacent spinal cord where they are present at these stages. In addition, sensory ganglia have been shown to synthesize BDNF, suggesting that BDNF may act by an auto-crine route. This view is supported by an experiment in which application of antisense BDNF oligonucleotides was shown to inhibit neuronal maturation (Wright *et al.*, 1992).

A role for NT-3 during the early stages of avian sensory ganglion development was further confirmed by treating embryos with neutralizing antibodies to this neurotrophin (Gaese *et al.*, 1994). Hybridoma cells secreting antibodies to NT-3 were placed onto the chorioallantoic membrane of 3-day-old quail embryos. These cells proliferate in the embryo and continue secreting the antibodies. In anti-NT-3-treated embryos, the number of neurons counted in the E6 DRG was reduced by 35% compared with control embryos. A similar treatment with anti-NGF-secreting cells showed only a 5% reduction in neuronal number at this age. These observations demonstrate that young DRG neurons require NT-3 but are largely NGF-independent. In contrast, a 30% reduction in the number of DRG neurons was obtained on E11 upon NGF deprivation, pointing to a selective effect of NGF on the survival of neural crest-derived neurons during the period of programmed cell death. In agreement with these results, a complementary approach, that consisted of *in ovo* neutralization of the trkC receptor protein from E2.75 until E4.5 or E7.5, resulted in a severe reduction in the number of DRG cells. Interestingly, the young trkC-expressing neurons that were affected by trkC deprivation were found to become dependent for survival upon either NT-3 or NGF after target innervation (Lefcort *et al.*, 1996). Taken together, these results are consistent with the notion that multiple neuronal progenitors within the DRG display an early response to NT-3, and that some of them will switch their dependence to NGF at later stages. In contrast to the situation observed in the DRG,

**Figure 4.13** (*cont.*)
C5 is clearly observed, and the remnant of its degenerating root is still present.
Rostrally, C2 is continuous with the remnants of the other Froriep's ganglia, known as the "ganglion crest." Bar = 100 μm. (Reproduced, with permission, from Rosen *et al.*, 1996. Copyright © 1996. Reprinted by permission of John Wiley & Sons, Inc.)

placode-derived neurons of the nodose ganglia displayed a selective dependence upon NT-3 both before and during programmed neuronal death, but remained refractory to NGF throughout ontogeny (Gaese *et al.*, 1994).

The multifunctional nature of NT-3 activity before target innervation is further illustrated by results of mutant mouse embryos lacking NT-3 gene activity. Fariñas *et al.* (1996) have found that the decreased number of DRG neurons observed in the mutants at birth results, not only from reduced survival during programmed cell death (see section 4.4.5), but also from events taking place at earlier stages. These include elevated apoptosis of newly born postmitotic neurons, and a premature differentiation of sensory neurons which, under normal conditions, continue proliferating before neurogenesis. These observations contrast with findings by Elshamy and Ernfors (1996), who reported that, in the absence of NT-3, proliferating sensory progenitors undergo increased apoptosis.

Another factor that plays diverse effects throughout DRG development is the cytokine leukemia inhibitory factor (LIF). LIF was shown to stimulate the generation of sensory neurons from murine neural crest cells cultured *in vitro* (Murphy *et al.*, 1991, 1993). In cultures derived from recently formed DRG, LIF stimulates the conversion from a neurofilament-negative phenotype into a neurofilament-positive phenotype in sensory neuroblasts, but NGF has no activity prior to a stage in which full neuronal differentiation has been reached (Murphy *et al.*, 1991, 1993).

### 4.4.4 *Neurotrophic activities during the period of normal neuronal death*

4.4.4.1 *The neurotrophic hypothesis.* The occurrence of cell death in many developing systems of the vertebrate embryo at specific times during normal development was already noticed by German embryologists around the 1920s (Gluksmann, 1951; see Raff, 1992, and references therein). Concerning the nervous system, it is only since the pioneering studies of Hamburger and Levi-Montalcini (1949) that it became formally accepted that the development of the nervous system involves, not only cell division and neuronal differentiation, but also death by a normal process later defined as apoptosis (reviewed by Hamburger, 1992, and Oppenheim *et al.*, 1992b). Neuronal death is a particularly dramatic event as it affects a very significant, though quantitatively variable number of neurons, depending on the system and axial level considered.

Observation of neuronal death during normal development, and, more importantly, the results of experiments in which manipulation of the amount of target tissue was found to modulate the size of the neuronal populations, led Hamburger to postulate a general paradigm for the effects of the target on neuronal survival that is formulated in "the neurotrophic model" (reviewed by Hamburger, 1992). This model predicts a need for the establishment of a match between the number of neurons in a given system and the size of the target they innervate. Following an initial phase of neuronal overproduction in the embryo, the process of target innervation causes a subsequent selection of neuronal populations. The extent of neuronal death would then be determined by a mechanism of competition for properties inherent to the target, such as size, limiting amounts of specific survival factors, or, as more recently proposed, restricted accessibility to potential sources of factor whose amounts are not necessarily limiting (Oppenheim, 1989; Landmesser, 1992). Cell–cell (i.e., neuron–target) interactions would then play a critical role in the survival of vertebrate neurons and in the final shaping of synaptic connections (see also Changeux and Danchin, 1976).

*4.4.4.2 Characteristics of programmed cell death.* The fact that the apoptotic process affects predictable numbers of cells at specific developmental stages, suggested that it reflects a cell intrinsic program inherent to each cell (for review see Johnson and Deckwerth, 1993). That cell death can be prevented by inhibitors of RNA and protein synthesis proved that apoptosis is regulated by active mechanisms which require *de novo* protein synthesis and expression of new genes (Martin *et al.*, 1992). Thus, instead of passive death due to the absence of survival cues, the apoptotic process is now considered to occur by an active "suicide program." According to this view, the mode of action of neurotrophic molecules consists of inhibiting the cascade of events that leads to cell death. Many genes involved in this suicide program are known. Genetic analysis in the nematode *Caenorhabditis elegans* has identified two of them, *ced-3* and *ced-4*, which participate in the stereotypic death of more than 10% of the total number of somatic cells of the larva (Ellis and Horvitz, 1986). More recently, the mammalian homologue of *ced-3* has been cloned and identified as interleukin 1$\beta$ converting enzyme (Gagliardini *et al.*, 1994). Overexpression of this gene product in DRG neurons induces them to die. Moreover, the activity of the interleukin 1$\beta$ converting enzyme is suppressed either by the cowpox virus *crmA* gene or by the product of the *bcl-2* proto-oncogene, previously shown to inhibit death of certain hemopoietic cells deprived of cytokines (Hockenbery *et al.*, 1990; Nunez *et al.*, 1990), and of cranial sensory neurons deprived of neurotrophins (Allsopp *et al.*, 1993). Both *bcl-2* and *crmA* are equivalent in their ability to rescue sensory neurons from death and are likely to act through the same pathway, as no additive survival was obtained upon coexpression of the two genes in DRG neurons (Gagliardini *et al.*, 1994). In contrast, parasympathetic ciliary neurons that are refractory to neurotrophins and depend upon CNTF for survival *in vitro* (Table 4.2) are not rescued by *bcl-2*, thus showing the presence of more than one intracellular pathway for apoptosis (Allsopp *et al.*, 1993). Therefore, during normal development, neuronal survival induced by neurotrophic factors may be accounted for by the activation of both *bcl-2*-sensitive and -insensitive mechanisms. Conversely, the absence of such factors may lead to the activation of suicide genes of the *ced* family as well as of additional, still unknown, death gene products.

### 4.4.5 Target-derived survival factors

*4.4.5.1 Nerve growth factor (NGF).* The classical pathway for mediating neurotrophic effects involves the synthesis and secretion of a neurotrophic factor by the target cells of the responsive neuron, and its uptake by a receptor-mediated mechanism followed by retrograde axonal transport to the neuronal soma. NGF, discovered in the 1950s, represents the first molecular confirmation of the neurotrophic model. In the PNS, it was shown to promote survival and neurite outgrowth of sympathetic and neural crest-derived sensory neurons, but not of neurons derived from the ectodermal placodes (Table 4.2). Then it was established that NGF is synthesized in the target fields innervated by responsive neurons from the time of innervation onward (Davies *et al.*, 1987; Rohrer *et al.*, 1988a). Moreover, the amount of NGF produced was correlated with the density of target innervation (Korsching and Thoenen, 1983a; Heumann *et al.*, 1984; Shelton and Reichardt, 1984). NGF was also shown to be transported retrogradely to the neuronal soma (Hendry *et al.*, 1974; Johnson *et al.*, 1978; DiStefano *et al.*, 1992), where it exerts its functions (Levi-Montalcini, 1987; Barde, 1989; Thoenen, 1991). Furthermore, target-derived NGF is able to bind both trkA and the low-affinity p75 receptor (Fig. 4.14; reviewed by Barbacid, 1994; Chao, 1994).

Following the leading experiments of Levi-Montalcini and colleagues, showing that injection of NGF into embryos rescues neurons that are normally lost during development, the physiological role of NGF in neuronal survival was further substantiated by treating embryos with anti-NGF antibodies. Such treatment led to the selective loss of NGF-dependent neuronal populations (both immunosympathectomy and sensory hypoplasia). In the early experiments of NGF neutralization, the specificity of the antisera employed was unknown in relation to the neurotrophins discovered afterwards (Johnson *et al.*, 1986; Rohrer *et al.*, 1988b). Yet, the results obtained in these experiments are in good agreement with those of later experiments in which antibodies that do not cross-react with BDNF or NT-3 were employed (Ruit *et al.*, 1992). To identify which classes of DRG neurons depend upon NGF, antibody treatment was combined with staining of central projections with the carbocyanine dye, DiI. These assays revealed a striking selectivity to NGF of small-diameter sensory neurons that express trkA, project to laminae I and II of the spinal cord, and transmit nociceptive and thermoceptive stimuli (Carroll *et al.*, 1992; Ruit *et al.*, 1992). The reader is also referred to results discussed in Section 4.4.3.

As anticipated by previous studies based on the use of antibodies to NGF and pharmacological approaches, gene-targeting experiments confirmed the specific role of NGF and trkA on survival of small peptidergic DRG neurons serving both nociceptive and thermoceptive functions (Table 4.3). This was further validated by results of behavioral studies performed on heterozygous animals that survive for longer times than the null mutants. NGF +/− and trkA +/− mice exhibited decreased responses to pain-inducing stimuli when compared to wild-type animals (Smeyne *et al.*, 1994; Crowley *et al.*, 1994). It is

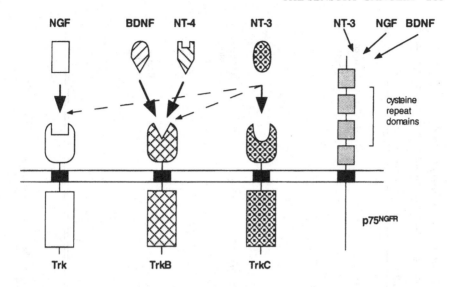

**Figure 4.14** Two types of receptors have been identified for the neurotrophins. The trk proto-oncogene product was found to encode a family of tyrosine kinase receptors comprising three members: Trk (or TrkA), TrkB, and TrkC. NGF has a high binding specificity for TrkA. Both BDNF and NT-4 preferentially bind to TrkB, and NT-3 to TrkC (indicated by thick arrows). Although the cellular context apparently restricts the neurotrophins to act via their preferred receptors, pharmacological doses of neurotrophins, or overexpression of the ligands or their receptors in cell lines, can result in an interaction between some of the factors and their non-preferred receptors (broken arrows). Only the catalytic receptor isoforms are represented in the diagram. Neurotrophins also bind with lower affinity to the p75 NGF receptor. (Reviewed by Chao, 1994.)

interesting to compare the phenotype of trkA and of the low-affinity NGF receptor p75 null mutant mice. In trkA mutants, specific neuronal subsets die massively during development, whereas, in the p75 null mice, the DRG and trigeminal ganglia show a more subtle phenotype that includes some neuronal loss *in vivo* and a reduced sensitivity to NGF *in vitro*, suggesting that the p75 receptor is involved in modulating NGF activity (Table 4.3; Davies *et al.*, 1993b; Lee *et al.*, 1994).

*4.4.5.2 Brain-derived neurotrophic factor.* Discovery and characterization of the prototypic neurotrophin NGF was followed by the findings that medium conditioned by cultured glioma cells as well as mammalian brain extracts contain molecule(s) with activity on survival of a subset of sensory neurons which did not respond to NGF (Barde *et al.*, 1978, 1980). This line of experiments, initiated by Thoenen and his colleagues, led to the subsequent purification of BDNF from pig brain. BDNF was thus considered as a neurotrophic factor present in central targets with activity on neurons located either in the CNS, or in sensory ganglia (Barde, 1989; Leibrock *et al.*, 1989). Several studies have shown that BDNF binds preferentially to the trkB receptor (Fig. 4.14),

Table 4.3. *Percent reduction in the number of neurons in homozygous mutant mice compared to wild-type mice*[a]

| | NGF | NT-3 | BDNF | NT-4 | BDNF/NT-4 | trkA | trkC | trkB | trkA/trkB | trkA/trkC | trkB/trkC | p75 |
|---|---|---|---|---|---|---|---|---|---|---|---|---|
| *Sensory ganglia* | | | | | | | | | | | | |
| Trigeminal | | 61[1], 64[14], 65[15] | 44[9], 27[13] | 98[6] | 70[6] | 70–90[8] | | 60[7] | | | | Reduced response to NGF[11] |
| Geniculate | | 25[1] | 48[6], 40 (vol)[13] | 50[6] | 94[6] | | | | | | | |
| Vestibular | | ns[1] | 76[6], 82[9], 87 (vol)[13] | 21[6] | 82[6] | | 16[2] | 56[2] | | | 58[3] | |
| Cochlear | | 85[1] | ns (vol)[13] | | | | 51[2] | 15[2] | | | 61[3] | |
| Superior jugular | ns[12] (vol) | ns[1] | | | | | | | | | | |
| Petrosal nodose | | 30[1] | 43[13], 61[5], 57[6], 66[9] | 56[5], 59[6] | 79[5], 90[6] | | | | | | | |
| TMN | | 56[1], 52[14] | 48[9], 39[13] | | | | | | | | | |
| DRG | 70[12] | 62 (C1)[1], 78 (L5)[1], 55[14], 58[15] neuroblast loss[17] | 34[13], 30[9] | 14[6] | | 70–90[8], 73[3], small diameter | 17[3], 19[4] | 20[3], 30[7] | 78[3] | 93[3] | 41[3] | Reduced response to NGF[10], reduced cutaneous innervation[16] |
| *Sympathetic ganglia* | | | | | | | | | | | | |
| SCG | 81[12] | 48[1], 53[14], reduction of neuroblasts[17] | ns[9], ns[13] | ns[6] | ns[6] | Significant reduction | | | | | | Reduced response to NGF[16] |
| *Motor nuclei* | | | | | | | | | | | | |
| Trigeminal | | ns[1] | ns[5,6,9,13] | ns[6] | ns[6] | | | | | | | |
| Facial | | ns[1], [14] | ns[5] | ns[5], 8[6] | ns[5], 11[6] | | 69[7] | | | | | |
| Spinal MNs | ns[15] | ns[5,13] | | | | 31.6[4], in motor axons | 35[7] (L2–L5) | | | | | |

*Note:*
[a] Prepared with the help of I. Silos Santiago. ns, not significant.

*References:*
[1]Fariñas, 1994; [2]Schimmang et al., 1995; [3]Minichiello et al., 1995; [4]Klein et al., 1994; [5]Conover et al., 1995; [6]Liu et al., 1995; [7]Klein et al., 1993; [8]Smeyne et al., 1994; [9]Ernfors et al., 1994a; [10]Lee et al., 1994a; [11]Davies et al., 1993b; [12]Crowley et al., 1994; [13]Jones et al., 1994; [14]Ernfors et al., 1994b; [15]Tojo et al., 1995; [16]Lee et al., 1994; [17]Elshamy and Ernfors, 1996; [18]Elshamy et al., 1996.

and has a highly specific effect *in vitro* on definite subpopulations of spinal and cranial sensory neurons, but has no effect on sympathetic or parasympathetic cells (Table 4.2). Moreover, the phenotypes of BDNF and trkB null mutant embryos are in agreement with data stemming from the previous pharmacological studies. Thus, cranial sensory neurons of the trigeminal, nodose, geniculate, and trigeminal mesencephalic nucleus are significantly affected in both mutants, so are some DRG neurons, while sympathetic ganglia are not affected (Table 4.3). The most dramatic effect obtained in these mutants refers to the loss of vestibular neurons and of a subset of cochlear neurons responsible for the innervation of the outer hair cells (Table 4.3). Behavioral analysis of trkB mutants was not possible because of their early postnatal lethality, but the BDNF null mice live for a longer time and reveal a phenotype of spinning, head bobbing and hindlimb extension during attempted locomotion, all features characteristic of abnormalities in the function of the vestibular system. Interestingly, trkC mutants reveal a more severe loss in cochlear neurons than in vestibular ones, and in double trkB/trkC null mutant mice most cochlear as well as vestibular neurons degenerate at some point in development (Table 4.3; also see Schimmang *et al.*, 1995).

*4.4.5.3 Neurotrophin-3.* The identification of highly conserved regions in the protein sequences of NGF and BDNF led to the discovery of additional members of the family: NT-3 (Leibrock *et al.*, 1989), NT-4 (also known as NT-4/5 or NT-5), and more recently fish-specific NT-6 (Table 4.2). Various lines of evidence support a role for NT-3 as a neurotrophic factor for sensory neurons. *In vitro* experiments have revealed an effect of NT-3 on survival of neurons derived from both the neural crest and the ectodermal placodes, such as DRG (reviewed by Barde, 1989), nodose (Hohn *et al.*, 1990; Dechant *et al.*, 1993b), cochlear (Pirvola *et al.*, 1992), and early trigeminal ganglion neurons (Buchman and Davies, 1993). Subsequent studies in avian embryos further defined the identity of the responsive populations among the DRG neurons. Hory-Lee *et al.* (1993) have elegantly shown that large-diameter sensory neurons with proprioceptive function that innervate muscles are the most sensitive to this neurotrophin *in vitro*. Moreover, in support of a role for NT-3 as a target-derived factor is the fact that neutralization of muscle-derived NT-3 in the embryo during the period of normal cell death results in loss of the 1A muscle afferent fibers while the cutaneous innervation remains intact (Oakley *et al.*, 1995). Nevertheless, in the adult mouse, NT-3 was shown to be essential for the maintenance of specific cutaneous afferents that subserve fine tactile discrimination (Airaksinen *et al.*, 1996). Furthermore, treatment of embryos with hybridoma cells secreting function-blocking NT-3 antibodies showed as well a further decrease in neuronal numbers in DRG beyond that observed in control embryos during the phase of normal cell death (Gaese *et al.*, 1994). Conversely, when exogenous NT-3 was administered between E6 and E9, a 30% increase in neuronal numbers was counted (Ockel *et al.*, 1996). Like NGF and BDNF, NT-3 may access the neuronal soma by retrograde transport from the peripheral targets (DiStefano *et al.*, 1992).

Finally, results from mice lacking NT-3 fully confirmed its role on survival of limb proprioceptive afferents (Table 4.3; Ernfors *et al.*, 1994b; Fariñas *et al.*, 1994; Tojo *et al.*, 1996). Thus, mutant embryos lack muscle spindles and Golgi tendon organs, and display severe movement defects in their limbs. Moreover, the number of muscle spindles in mice heterozygous for the mutation is half of that in the wild-type mice, indicating that NT-3 availability is limited (Ernfors *et al.*, 1994b). The deficits observed were correlated with loss of large-diameter DRG neurons that express parvalbumin and carbonic anhydrase (Ernfors *et al.*, 1994b), are localized in the lateroventral portion of the DRG, and synthesize trkC (Mu *et al.*, 1993; Tessarollo *et al.*, 1993; Kahane and Kalcheim, 1994; Zhang *et al.*, 1994a). Indeed, a marked reduction in the expression of the trkC receptor was detected in trigeminal ganglia and DRG in mice homozygous for the NT-3 mutation (Tojo *et al.*, 1995). Consequently, disruption of the trkC gene resulted in the elimination of 1A muscle afferents and in behavioral deficits similar to, but less severe than, those observed in the embryos homozygous for the NT-3 mutation (Klein *et al.*, 1994).

Altogether, these convergent lines of evidence support the contention that, in addition to other functions played earlier in development, NT-3 regulates survival of sensory neuronal subsets during the period of normal cell death, and that its activity on 1A muscle afferents is at least partly mediated through the trkC gene product. It has nevertheless to be mentioned that additional receptors may also contribute to mediating some effects of NT-3, since the number of neurons missing in the NT-3 mutants is much higher than that seen in the trkC null mutant mice in peripheral ganglia such as the DRG and the cochlear ganglia. Consistent with such a notion, it was recently shown that NT-3 can promote the *in vitro* survival of sensory and sympathetic neurons isolated from embryos that are homozygous for a null mutation in the trkC gene (Davies *et al.*, 1995). However, this responsiveness is lost in trkC and trkA, or trkC and trkB, double mutants (Davies *et al.*, 1995), suggesting that, at least under experimental conditions (lack of the preferred receptor), NT-3 can also signal through trkA and trkB. Most interestingly, the opposite occurs in the CNS where the number of spinal motor axons is reduced by 30% in the trkC mutants, but is not changed in the NT-3 knockout mice (Table 4.3), suggesting that absence of NT-3 input to spinal motor neurons can be compensated for by other molecules.

*4.4.5.4 Other neurotrophic factors: CNTF and FGF.* Two additional families of factors are well known for their effects on survival of sensory neurons *in vitro*: ciliary neurotrophic factor (CNTF), and members of the FGF family. CNTF was so named for its ability to rescue cultured parasympathetic ciliary neurons (Barbin *et al.*, 1984; see also detailed discussion of CNTF in Chapter 5). The spectrum of CNTF activity on cultured neurons appears, however, to be much broader (Table 4.2). Mammalian CNTF promotes the survival of more than 40% of E10 and E11 DRG neurons, but it has no activity on E8 DRG. This factor was also shown to rescue cultured chick nodose and trigeminal neurons excised from E8 and E10 embryos, respectively (Manthorpe *et al.*,

1982). In spite of its effects *in vitro*, administration of CNTF to chick embryos *in ovo* between E5 and E9 was without effect on the survival of DRG or nodose ganglion neurons (Oppenheim *et al.*, 1991). These results raise the question whether CNTF plays any meaningful role during embryonic development. In contrast to its still uncertain function in embryonic life, a well-established effect of CNTF is to rescue motor neurons during adulthood and after lesion to the motor system (Sendtner *et al.*, 1990, 1992; Masu *et al.*, 1993; Kato and Lindsay, 1994; Tan *et al.*, 1996; Aebischer *et al.*, 1996). In line with these observations, retrograde transport to motor neurons as well as to sensory neurons was shown to be increased after peripheral nerve injury (Curtis *et al.*, 1993). It still remains to be tested whether CNTF has any effect on the adult sensory system.

Nishi and Berg (1981) have characterized a growth-promoting activity (GPA) in chicken eyes similar to mammalian CNTF. Purification of GPA was accomplished several years later using chick sciatic nerves as a source (Leung *et al.*, 1992), and it was shown that GPA shares 47% sequence homology with CNTF and has similar biological activities (Leung *et al.*, 1992). More recently, isolation of a chick cDNA coding for a GPA receptor revealed that this receptor is 70% identical to the human CNTF receptor $\alpha$, and mediates activities of both GPA and CNTF on survival of cultured ciliary, sensory, sympathetic, and motor neurons (Heller *et al.*, 1995). Thus, GPA and CNTF share many functional similarities, suggesting the possibility that GPA is the avian homolog of mammalian CNTF. However, these two factors have different biochemical properties with respect to their interaction with high-affinity receptors on sympathetic neurons (Heller *et al.*, 1993), and with respect to their ability to be secreted in a biologically active form. Whereas GPA was shown to be secreted from the producing cells, CNTF is apparently unable to do so (Leung *et al.*, 1992).

In contrast to the neurotrophins and to CNTF, FGF1, and FGF2, the first characterized members of the growing family of FGFs that were implicated in development of the nervous system, may not act as classical target-derived growth factors because they do not undergo retrograde axonal transport. Moreover, neither of these factors is secreted from the cells in a conventional way as they both lack the N-terminal signal sequence required for secretion. Nevertheless, FGF2 was shown to use the secretion machinery of proteoglycans in order to be coreleased along with these molecules, and be subsequently expressed on the cell surface (reviewed by Rapraeger *et al.*, 1994). Thus, these factors are more likely to act as short-range signals to influence nearby neurons. Another possibility is that they play intracrine activities, an example of which is the mitogenic activity of both FGF1 and FGF2 that takes place upon translocation of the factor into the cell nucleus (reviewed by Eckenstein, 1994). In the PNS, FGF2 was shown to exert differentiative and survival actions on cultured sensory and sympathetic neurons (Table 4.2; see also Unsicker *et al.*, 1987; Eckenstein *et al.*, 1990, 1991), but it was ineffective on the survival of these neurons during programmed cell death *in vivo* (Oppenheim *et al.*, 1992a). Moreover, FGF2 is a potent mitogen for cultured rat Schwann cells (Davis and Stroobant, 1990), and can rescue Schwann cell progenitors isolated from peripheral nerves of young rat embryos (Jessen *et al.*, 1994).

In conclusion, the results available show that cranial and spinal sensory neurons are affected by a variety of factors belonging to different families and acting via different mechanisms. Of the three families discussed above, only the neurotrophins were convincingly shown to be involved in sensory neuron survival during the restricted time period of normal neuronal death in the embryo and *in vitro*. CNTF and FGF were shown to be active on neuronal survival in culture, but without effect *in vivo* during programmed cell death. Their exact role therefore remains elusive. It is possible that, in the embryo, CNTF and FGF affect neuronal properties other than survival. Cooperative activity of these factors with specific neurotrophins may exist, or they may serve as important signals for differentiation or maintenance before and after this restricted time window.

### 4.4.6  Developmental sequence and overlapping activities of neurotrophins during development of sensory neurons

*4.4.6.1  Target-independent survival of neuroblasts.* The target dependence for survival of sensory neurons develops as a dynamic series of events. Neural crest precursors transiently depend upon neurotrophins derived from the neural tube (see previous sections). Following differentiation, but before target innervation, sensory neurons were reported to undergo a phase in which their survival *in vitro* is independent of exogenous neurotrophins. The duration of this phase was proposed to be correlated with the distance from the developing ganglion to its central or peripheral targets. This is based on the observations that neurons with more distant targets survive longer in culture before becoming neurotrophin-dependent. The time of onset of neurotrophin dependence is the shortest for the vestibuloacoustic neurons, followed by the geniculate, petrosal, and finally nodose neurons whose targets are the most distant. Such a mechanism was suggested to ensure neuronal survival during axonal growth (Davies, 1994, and references therein). Nevertheless, since BDNF and NT-3 receptors are continuously expressed throughout early development of sensory ganglia on a majority of the ganglionic cells, it is likely that, during this particular time window, neurotrophins have effects other than survival. Alternatively, young neuroblasts sending their axons to the periphery may depend upon intrinsic neurotrophins for survival, which act either by autocrine or paracrine routes (Davies and Wright, 1995).

*4.4.6.2  Redundant activities of growth factors.* Neural crest progenitors and postmitotic neurons following target innervation, display overlapping responses to various molecules. This notion is illustrated by the finding that BDNF, NT-3 or FGF2 promote neuronal differentiation of a subset of early postmitotic neural crest progenitors (see previous sections). In addition, both

BDNF and NT-3 stimulate maturation of young neurons explanted from E4.5 chicken DRG (Wright *et al.*, 1992), or from E10–11 mouse trigeminal ganglia (Buchman and Davies, 1993). Moreover, individual E6 DRG neurons expressing substance P immunoreactivity respond to both NGF and BDNF. This early redundancy might explain why mouse embryos lacking the NT-3 gene show an apparently normal migration and condensation of DRG progenitors, and no difference is recorded in cell number between mutants and wild-type embryos at 9 and 10 days of development (Fariñas *et al.*, 1996).

Redundancy was also revealed in embryos with targeted deletions in the neurotrophin genes. For example, analysis of DRG in neonatal animals revealed that homozygous embryos lacking the NT-3 gene show the absence of 55-80% of the normal complement of ganglionic neurons, whereas BDNF, NGF, and NT-4 mutants reveal a lack of 30, 70, and 14% of DRG neurons, respectively (see Table 4.3 and references therein). These results suggest that a proportion of the missing cells must respond to more than one neurotrophin. Unfortunately, analysis at late stages does not provide information as to the exact time when these overlapping activities are critical for neuronal survival. The exact sequence of activities initiated by a given molecule or combination of factors will be eventually disclosed when a detailed temporal analysis of the effect of single and combined gene deletions is accomplished.

Patterns of mRNA expression of the different trks in adult rat DRG also provided a means of following the response to their respective ligands. Analysis of the distribution of mRNAs for trkA receptors, performed in rat DRG, has revealed that the percentage of trkA-expressing neurons decreases from embryonic life to adulthood to about half of the embryonic values (70% of DRG neurons). These observations suggest that some DRG neurons lose their responsiveness to NGF some time during development (McMahon *et al.*, 1994). This work revealed that no overlap exists between trkA- and trkC-expressing neurons. By contrast, the majority of cells expressing trkB, including visceral, cutaneous, and muscle afferents, express one or two other receptor types, demonstrating that, even in adults, neuronal populations able to respond to multiple factors are present (McMahon *et al.*, 1994).

Thus, in general, individual progenitor cells might already express a wide repertoire of functional receptors. This response to multiple factors is likely to decline upon maturation, rendering functional groups of differentiated neurons preferentially responsive to a given growth signal (Lindsay *et al.*, 1985). It must then be assumed that one role played by local signals encountered during gangliogenesis, differentiation, and target innervation is to refine this repertoire by stabilizing the expression and function of a given receptor type while down-regulating "less relevant" receptor species.

*4.4.6.3 Sequential activities of growth factors.* Simultaneously with a progressive and partial loss of redundancy in the responsiveness to combinations of neurotrophins, some sensory neurons undergo a switch in neurotrophin dependence. These dynamic events are further illustrated with the following examples. Mouse trigeminal neurons display a transitory survival response to both BDNF and NT-3 during the earliest stages following target field innervation. This response is lost as neurons become NGF-dependent shortly before neuronal death begins in the ganglion (Buchman and Davies, 1993). Likewise, early survival of the proprioceptive sensory neurons of the trigeminal mesencephalic nucleus is supported by either BDNF or NT-3 (Davies *et al.*, 1986). Only at later stages, when neuronal death is well under way, can the two neuronal populations listed above be rescued by CNTF (Allsopp *et al.*, 1993). In the mouse, a factor that affects DRG development is LIF. LIF is required as a primary factor to affect the differentiation of immature murine neuroblasts into neurofilament-positive neurons, whereas their subsequent survival can be supported by either LIF or NGF. At this stage, each molecule was shown to act on non-overlapping populations (Murphy *et al.*, 1993). Consistent with the switch in factor dependence, the expression pattern of the respective receptors changes, but the mechanisms regulating these changes remain to be clarified (Paul and Davies, 1995; see also Birren *et al.*, 1993, and DiCicco-Bloom *et al.*, 1993, for sympathetic neurons). Perhaps the first molecule in the cascade modulates responsiveness to those factors that affect progressively more mature cells, by upregulating the expression of their cognate receptors.

### 4.4.7 Modulation of responsiveness to neurotrophins by the neurofibromin gene product

Von Recklinghausen's neurofibromatosis is a human disease characterized by the appearance of benign and malignant tumors of neural crest derivatives, including melanocytes and peripheral ganglion components (Riccardi, 1991; Gutmann and Collins, 1992). The neurofibromatosis type 1 gene has been cloned. Its protein product, neurofibromin, was characterized as a 250-kDa protein expressed in specific regions during early development and becoming ubiquitously expressed during late fetal life. Neurofibromin immunoreactivity has been found in migrating neural crest cells at both trunk and cranial levels, and in crest-derived tissues such as the DRG and peripheral nerves (Stocker *et al.*, 1995, and see references therein). In the DRG and CNS, increases in mRNA and neurofibromin coincide with the period of neuronal differentiation (Huynh *et al.*, 1994). Neurofibromin contains a 350-amino acid domain that has homology to mammalian and yeast GTPase-activating proteins that function as negative regulators of the p21 ras oncoprotein (Ballester *et al.*, 1990; Buchberg *et al.*, 1990). In addition, neurofibromin can regulate ras in a GTPase-independent fashion. Regulation by neurofibromin of the intracellular cascade involving ras is of interest in view of the involvement of the ras pathway in neurotrophin signaling by the trk receptors in both sensory and sympathetic neurons (Ng and Shooter, 1993; Borasio *et al.*, 1993, and references therein).

Based on this correlation, Vogel *et al.* (1995) tested the hypothesis that the neurofibromin gene is involved in the development of neurotrophin responsiveness. To this end, they exploited the availability of mouse embryos lacking the neurofibromin gene. Embryos homozygous for a null allele of the neurofibromin 1 gene exhibit abnormalities in neural crest-derived tissues, most notably hyperplasia of the sympathetic ganglia, and die by E14.5. Moreover, the heterozygotes show higher propensity for tumor development when compared to the wild-types (Brannan *et al.*, 1994; Jacks *et al.*, 1994). Cultures of dissociated DRG from E13.5 mutant and wild-type embryos were established and treated with NGF. Interestingly, whereas the majority of normal neurons die whithin 48 hours after seeding in the absence of the factor, mutant neurons survived for 48 hours and longer under the same conditions. Although both normal and mutant DRG neurons are able to respond to NGF, the maximal survival response of the mutant neurons is obtained with a ten-fold lower concentration of added factor. Likewise, E12 DRG neurons are normally able to survive 24 hours in culture before becoming dependent upon NGF for survival. By contrast, the mutant neurons did not develop any dependence upon NGF and continued surviving for days in its absence. These results suggest that sensory neurons, from neurofibromin–/– DRG can survive and differentiate in the virtual absence of exogenous factors. Similar results were obtained for trigeminal and nodose sensory neurons, and for sympathetic superior cervical ganglion cells. The possibility was then tested that survival in the absence of exogenous neurotrophins results from dependence upon endogenous factors produced by the neurons themselves. This possibility was ruled out because addition to wild-type neurons of medium conditioned over the mutant neurons failed to support their survival. Such a negative answer does not exclude, however, the possibilities that the neuronal factors are very diluted in the conditioned medium, or that the factor has decreased stability after a conditioning period of a few days. A more definitive answer to this question would have been provided upon treatment of the cultures of mutant DRG neurons with antisense oligonucleotides to the neurotrophins and/or with specific antibodies, paradigms used in similar studies attempting to test autocrine or paracrine activities (see next section). In any case, it is clear that lack of the neurofibromin 1 gene, a tumor suppressor gene, results in increased neuronal survival *in vitro* of sensory and sympathetic neuronal populations, features that may underline tumor development. These *in vitro* findings are consistent with the hyperplasia of sympathetic ganglia found in the mutant embryos lacking neurofibromin gene activity. Cell counts of sympathetic ganglia performed in the mutant mice revealed a 2–3.3-fold increase in cell number and an elevated mitotic index when compared to the wild-type ganglia. In contrast, no difference could be found between normal and mutant embryos in the number of DRG, nodose, and trigeminal ganglion neurons, unlike the situation *in vitro*. Perhaps abnormalities in neuronal survival in the mutant sensory ganglia become apparent at older embryonic stages, but these cannot be tested because of embryonic mortality (Brannan *et al.*, 1994).

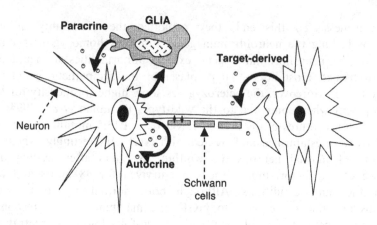

**Figure 4.15** Identified mechanisms that mediate the development and maintenance of neuronal and glial cells. Neuronal survival is promoted by target-derived factors that arrive in the cell body by retrograde transport. In addition, autocrine mechanisms were implicated in the maturation and survival of immature neuroblasts and adult neurons, respectively. Reciprocal paracrine interactions between neurons and glia (ganglionic satellite and Schwann cells) were also shown to affect the concerned cell types. Of the mentioned activities, distinct neurotrophins were shown to mediate target-derived neuronal survival, autocrine effects, and a paracrine action of sensory neurons on glial cell differentiation. (See text for further details.)

### 4.4.8 Evidence for autocrine and paracrine effects of neurotrophins

In line with a target-derived nature of neurotrophin activity, their synthesis in the central and peripheral target fields innervated by responsive peripheral neurons was demonstrated and discussed earlier in this chapter. More recently, *in situ* hybridization revealed that BDNF and NT-3 mRNAs are also synthesized within the rodent DRG (Schecterson and Bothwell, 1992). Moreover, BDNF is produced in avian DRG from E4.5 onward, as shown by PCR (polymerase chain reaction) studies (Wright *et al.*, 1992), and *in situ* hybridization (Pruginin-Bluger *et al.*, 1997). These observations suggest that neurotrophins may also have local roles within the ganglionic environment to promote either neuronal or glial development (Fig. 4.15).

Local autocrine activities for BDNF on target neuronal cells were suggested to occur in two separate instances. First, during a short time window following neuronal differentiation when neurons are believed to survive independently of their targets. This is based on the observation that treatment of neurons with antisense oligonucleotides to BDNF prevented neuronal maturation (Wright *et al.*, 1992). The second case concerns adult rat DRG neurons which have the property of surviving in culture without added factors, even as single neurons, suggesting that, at adulthood, autocrine mechanisms can account for survival. Antisense oligonucleotides corresponding to the C-terminal of the mature protein were used to inhibit BDNF translation both in mass cultures and in clonal neuronal cultures. About 35% of the neurons died in both types of experiments and could be rescued by addition of exogenous factor (Acheson *et al.*, 1995).

In another study, it was found that NGF upregulates the expression of the L1 cell adhesion molecule in early postnatal mouse Schwann cells, and that treatment of the cells with antibodies against NGF abolishes the effect observed (Seilhemer and Schachner, 1987). This activity can also be accounted for by local interactions, and even by an autocrine mechanism since Schwann cells are known to produce NGF.

Since the process of gliogenesis within peripheral ganglia follows in a general temporal sequence that of neurogenesis and because neuron–glia interactions are essential for the development (proliferation, differentiation, survival, etc.) of the latter cell type (Holton and Weston, 1982a,b; Rudel and Rohrer, 1992; Shah *et al.*, 1994), the possibility was then examined that neuron-derived BDNF can mediate some parameters related to this interaction. One of the earliest markers that define the glial lineage in avian embryos is the Schwann cell myelin protein (SMP) antigen (Dulac *et al.*, 1988). The acquisition of the SMP–positive phenotype was proposed to be part of a cell autonomous pathway of differentiation common to all PNS glial lineages because this marker is expressed *in vitro* by all types of peripheral glia (Le Douarin, 1993). Nevertheless, the expression of SMP is tightly regulated by environmental cues. For instance, the microenvironments of the DRG and the gut were shown to repress transcription of the SMP gene (Dulac and Le Douarin, 1991; Cameron-Curry *et al.*, 1993). In striking contrast, activation of the SMP marker was observed upon dissociation of DRG into single cells (Cameron-Curry *et al.*, 1993). Moreover, coculturing neurons with glial cells greatly stimulates the number of glial cells that express SMP (Cameron-Curry *et al.*, 1993, Pruginin-Bluger *et al.*, 1997). Several lines of evidence support the contention that BDNF mediates the paracrine effect of the neurons on the expression of SMP by glial cells in culture. First, treatment of DRG cultures containing both neurons and non-neuronal cells with immunoadhesins bearing the extracellular domain of the trkB receptor that bind to the factor with high affinity and sequester it in a selective manner abolishes this stimulation. Second, the effect of the trkB fusion proteins can be completely prevented if the cultures are supplemented with an excess of exogenous BDNF. Third, neither trkC nor trkA immunoadhesins that selectively bind to NT-3 and NGF, respectively, have an effect on the proportion of glial cells that acquire the SMP marker. Fourth, soluble BDNF mimics the effect of the neurons on the expression of SMP when added to purified glial cells, but has no effect on the proliferation or survival of these cells (Pruginin-Bluger *et al.*, 1997). These results provide the first evidence for a paracrine effect of neuron-derived BDNF on the generation of new SMP-immunoreactive glial cells from SMP-negative cells. While BDNF affects Schwann cells and satellite ganglionic cells in culture, previous studies showing that the DRG environment represses SMP expression by satellite cells *in vivo* raise the possibility that the final phenotypic identity of a given cell is the result of an interaction between positive and negative cues. What factor(s) then repress expression of SMP by satellite cells *in vivo*? In contrast to BDNF, it was found that NT-3 inhibits the acquisition of the SMP marker by glial cells. NT-3 is poorly synthesized in the avian ganglia, but is expressed there at the protein level perhaps due to retrograde transport from the target (Pinco *et al.*, 1993). Thus, the interplay between local and retrograde cues present in neurons can modulate the development of specific phenotypic features in satellite cells. Most interestingly, it was found that the p75 common neurotrophin receptor is involved in mediating the antagonistic activities of BDNF and NT-3, suggesting that this receptor is able to mediate a biological response in glial cells, and that p75 is able to discriminate between different members of the neurotrophin family (see also Carter *et al.*, 1996).

Altogether, growing evidence substantiates the notion that, in addition to being target-derived neurotrophic factors, the neurotrophins, widely expressed in different embryonic tissues, act in a compartmentalized manner near their sites of synthesis to affect developmental processes in neural and non-neural cell types (for review see Brill *et al.*, 1995; Kalcheim, 1996, 1998).

# 5

# The Autonomic Nervous System and the Endocrine Cells of Neural Crest Origin

## 5.1   General considerations

The autonomic nervous system (ANS) derives entirely from the neural crest. It consists of the catecholaminergic and non-catecholaminergic cells of the sympathetic nervous system, including both sympathetic ganglia and adrenal chromaffin cells, the cholinergic and non-cholinergic parasympathetic ganglia, and the enteric ganglia. The ANS provides motor innervation to smooth muscles and viscera. Visceral sensory endings are also present in most internal organs from which myelinated and unmyelinated afferent fibers ascend to the spinal cord and brain through various visceral nerves, one of the most important being the vagus nerve which extends over most of the gut. Visceral motor centers are located in the spinal cord, brainstem, and higher centers of most vertebrates. Visceral motor neurons within the CNS are connected to postganglionic neurons with which they form at least a two-neuron visceral motor pathway for the reflex control of all smooth muscles, of the cardiac muscle, and of glandular epithelia.

The sympathetic branch of the ANS includes preganglionic neurons located in the intermediolateral column of the spinal cord along thoracolumbar segments of the axis. Fibers from preganglionic neurons synapse with second-order neurons located in the paravertebral and prevertebral sympathetic ganglia. The parasympathetic division consists of preganglionic fibers found in cranial nerves III, VII, IX, and X, as well as in sacral nerves that synapse on ganglia localized close to or within the target organ. As a general rule, the preganglionic myelinated axons are therefore much longer, and unmyelinated postganglionic fibers shorter, in the parasympathetic system than in the sympathetic system. The enteric branch of the ANS is the largest and most complicated division of the PNS. Langley (1921) considered the enteric nervous system as a separate division of the ANS. He thought that many of the neurons were not connected to the CNS at all. In fact, its component neurons are organized into defined plexuses, and mediate reflex activities even in the absence of central inputs. These neuronal microcircuits contain primary afferent neurons, interneurons, and motor neurons, expressing a rich diversity of chemical messengers comparable to that of the brain. Thus, the enteric nervous system is developmentally and functionally truly autonomic.

Much is known about the neurotransmitters and modulators that mediate activity in the functional ANS. During the past 10 years we have witnessed significant progress in deciphering the mechanisms that govern the differentiation of neural crest precursors into the different components of the ANS, namely the roles played by combined and sequential activities of identified factors. Furthermore, null mutations in the genes coding for these factors or for their receptors have been obtained and proved to be extremely valuable, not only for the understanding of normal development, but also as experimental models for investigating the etiology of specific diseases. These aspects of ANS development will be the focus of the present chapter.

## 5.2 Origin and migratory pathways followed by distinct autonomic progenitors

The embryonic origin of autonomic ganglioblasts was the subject of intense research from almost the beginning of this century. Although controversial for a while, it is now well established that these cells derive exclusively from the neural crest. These important discoveries, however, led to controversial results as to the axial level of origin of the crest progenitors that give rise to the various derivatives (Uchida, 1927; Van Campenhout, 1931, 1932; Yntema and Hammond, 1945, 1947, 1954, 1955; Kuntz, 1953; Andrew, 1964, 1969, 1970, 1971). A reinvestigation of this question was thus undertaken using the quail–chick chimera system (Le Douarin and Teillet, 1971, 1973). Isotopic and isochronic grafts of small fragments of quail neural primordium (corresponding to a length of four to seven segments) were systematically carried out into chick hosts (and vice versa) along the entire length of the neural axis. Figure 5.1 illustrates the levels of the axis that give rise to the sympathetic ganglia, adrenomedullary cells, and to the enteric nervous system.

### 5.2.1 The sympathetic ganglia

*5.2.1.1 Embryonic origin.* Results based on experiments with quail–chick chimeras have demonstrated that avian sympathetic ganglia arise from the part of the neural crest located caudal to the fifth pair of somites; those sympathoblasts originating from the level of somites 5–10 mainly contribute to the formation of the superior cervical ganglion (SCG), but also colonize the pair of

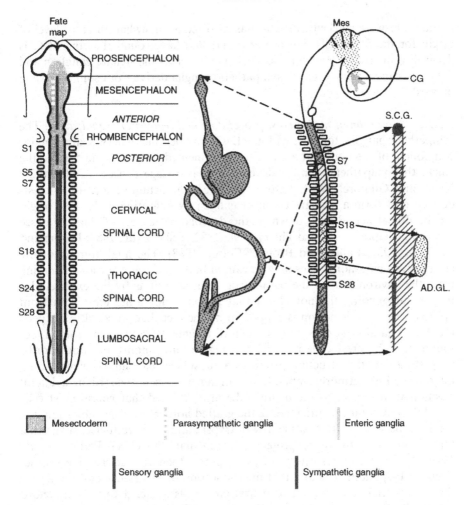

**Figure 5.1** The origin of autonomic ganglia, enteric ganglia, and adrenomedullary cells. The neural crest caudal to the level of the fifth somite pair gives rise to the ganglia of the sympathetic chain. The adrenomedullary cells originate from the neural crest between somite levels 18 and 24. The vagal neural crest (somites 1–7) gives rise to the enteric ganglia of the preumbilical region, the ganglia of the postumbilical region originating from both the vagal and the lumbosacral neural crest (see text for further details regarding mammals; see also Fig. 5.9). The ganglion of Remak (RG) is derived from the lumbosacral neural crest (posterior to level of somite 28). The ciliary ganglion (CG) is derived from the mesencephalic crest (Mes). AD.GL., adrenal gland; S.C.G., superior cervical ganglion.

*For a colored version of this figure, see www.cambridge.org/9780521122252.*

sympathetic ganglia located immediately caudal to it (Le Douarin and Teillet, 1973). It has also been proposed that the rat SCG derives from the thoracic neural crest cells that migrate first ventrally and then rostrally along the dorsal aorta (Rubin, 1985). A recent study, in which premigratory mouse neural crest populations were marked with DiI, has reported a somewhat different level of origin for the SCG cells. Durbec *et al.* (1996a) have found that the SCG is derived from crest cells of somitic levels 1–5, whereas the stellate ganglion and the more caudal paravertebral sympathetic ganglia derive from somitic level 6 onward.

*5.2.1.2 Migration of neural crest progenitors and segmental organization.* The sympathetic ganglia, together with the DRG and peripheral nerves, are a major component of the PNS that are distinctly segmented (Figs. 5.2 and 5.3). In the chick, the sympathetic ganglia develop in a two-stage process (Tello, 1925; Kirby and Gilmore, 1976). Before reaching their definitive locations, crest-derived cells form a primary chain of ganglia by migrating from their origin on the dorsal neural tube down to the dorsolateral aspect of the aorta (see details in Chapter 2; see also Thiery *et al.*, 1982; Teillet *et al.*, 1987; Loring and Erickson, 1987; Lallier and Bronner-Fraser, 1988). The final homing of the crest cells to para-aortic sites may be caused both by attractive cues emanating from the environment of the para-aortic area, as well as by the extracellular matrix surrounding the notochord that prevents migrating crest cells from approaching it. This notion is supported by the fact that, in coculture experiments of neural crest cells with notochord fragments, Newgreen *et al.* (1986) found that the crest cells avoided the region surrounding the notochords, suggesting that this structure produces a substance that inhibits neural crest migration. Furthermore, implantation of an ectopic notochord in a lateral position with respect to the neural tube also revealed that neural crest cells avoid the environment surrounding the grafted notochord. This inhibition was found to be sensitive to treatment with trypsin and chondroitinase, suggesting that chondroitin-containing proteoglycans mediate this effect (Pettway *et al.*, 1990). More recently, it was shown that paranotochordal sclerotome synthesizes the F-spondin gene and that microinjection of neutralizing antibodies to the protein enable migration of neural crest cells adjacent to the notochord (Debby-Brafman *et al.*, 1999; see also Chapter 2).

From the initially continuous rostrocaudal distribution of neural crest cells adjacent to the aorta, first observed at E3.5 (stage HH22; Hamburger and Hamilton, 1951), distinct swellings form by E4–4.5 that constitute the primary sympathetic ganglia (Tello, 1925; Kirby and Gilmore, 1976; Lallier and Bronner-Fraser, 1988). The secondary, or definitive, sympathetic chains of the chick arise from a secondary dorsolateral migration of cells from the primary chain on E5 (Tello, 1925; Kirby and Gilmore, 1976). In addition to their migration in the transverse plane, sympathoblasts migrate longitudinally two somites rostrally and three somites caudally of their segmental level of origin, so that individual ganglia are composed of crest cells that arise from up to six consecutive segments (Yip, 1986).

Analysis of the pattern of segmentation of sympathetic ganglia revealed that, similar to DRG and peripheral nerves, the locations of primary and secondary sympathetic ganglia are related to each rostral somitic domain (Fig. 5.3; see also Fig. 5.1). Moreover, sympathetic ganglia require the normal

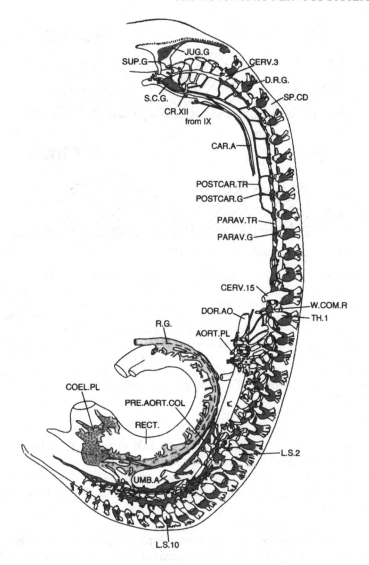

**Figure 5.2** The organization of the peripheral ganglia and plexuses of the 8-day-old chick embryo. AORT.PL, aortic plexus; CAR.A, carotid artery; CERV.3, cervical nerve 3; CERV.15, cervical nerve 15; COEL.PL, coeliac plexus; CR.XII (from IX), cranial nerve; DOR.AO, dorsal aorta; D.R.G., dorsal root ganglion; R.G., Remak ganglion; JUG.G, jugular ganglion; L.S.2, lumbosacral nerve 2; L.S.10, lumbosacral nerve 10; PARAV.G, paravertebral ganglion; PARAV.TR, paravertebral trunk; POSTCAR.G, postcarotid ganglion; POSTCAR.TR, postcarotid trunk; PRE.AORT.COL, preaortic column; RECT., rectum; S.C.G., superior cervical ganglion; SP.CD, spinal cord; SUP.G, superior ganglion; TH.1, thoracic nerve 1; UMB.A, umbilical artery; W.COM.R, white communicating ramus. (Modified from Yntema and Hammond, 1945, and Hammond and Yntema, 1947.)

*For a colored version of this figure, see www.cambridge.org/9780521122252.*

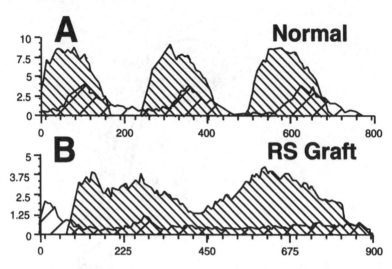

**Figure 5.3** Relationship between the primary sympathetic chain and DRG along the neuraxis. (A) A normal chick embryo at day 4 of development. (B) A 4-day-old chick embryo that received a graft of multiple rostral half somites from a quail donor on day 2. The cross-sectional areas of the sympathetic chain comprising three sympathetic ganglia (SG) and three contiguous DRG on one side of a normal embryo are shown in A. The sympathetic ganglia reach their peak diameter slightly rostrally to the DRG. In the presence of a rostral type of mesoderm both the DRG and the SG are continuous and lack segmentation. DRG area plotted with thin lines, SG area with thick lines. (Reproduced, with permission of Company of Biologists Ltd, from Goldstein and Kalcheim, 1991.)

rostrocaudal alternation of the somitic mesoderm for acquisition of proper segmental morphogenesis because, in the presence of a mesoderm composed of only rostral somitic halves, segmentation of the sympathetic chains is prevented (Fig. 5.3; see also Goldstein and Kalcheim, 1991; Kalcheim and Goldstein, 1991).

*5.2.1.3 Relationships between the sympathetic ganglia and the preganglionic neurons of the spinal cord.* The basic metameric pattern of the peripheral sympathetic ganglia can be traced back to the spinal cord with respect to the distribution of the sympathetic preganglionic neurons located in the intermediolateral column. These preganglionic neurons were shown in both rat and chick embryos to be organized into discrete segmental units. Mapping of projections has shown that individual sympathetic preganglionic neurons located in rostral positions within a given segment project only rostrally within the sympathetic chain, and those with a more caudal position project caudally, even though the spinal segment in which they reside provides an innervation to both rostral and caudal ganglia. Thus, the position of the preganglionic neurons in the spinal cord is related to the direction in which they project in the sympathetic chain (Forehand et al., 1994). Yip (1996) has also shown that grafting of cervical somites into thoracic levels, or deleting single somites from the axis, causes an alteration of the normal pattern of preganglionic fiber projections to the sympathetic ganglia. Taken together, these findings reveal a segmental organization of preganglionic sympathetic neurons in the developing spinal cord, and a modulation by axial-specific cues of their axonal projections once they exit the CNS.

### 5.2.2  The adrenal medulla

The cells producing the hormone adrenaline have for long been recognized as neural crest derivatives. In the avian embryo, the adrenal medulla derives from the neural crest of somitic levels 18–24, termed the "adrenomedullary level of the crest" (Fig. 5.1; Le Douarin and Teillet, 1971; Teillet and Le Douarin, 1974). The adrenal is composed of two main populations: adrenal chromaffin cells (pheochromocytes), and small intensely fluorescent (SIF) cells which are mainly found within sympathetic ganglia. Adrenal chromaffin cells have small cell bodies usually 20 μm in diameter, and have few or no processes. The cell bodies are packed with large (100–150 nm) dense-cored amine-storing vesicles, and they release their catecholamines (adrenaline and noradrenaline – norepinephrine) directly into the circulation (see Doupe et al., 1985a, and references therein). SIF cells are intermediate in morphology between chromaffin cells and sympathetic neurons (reviewed by Eranko, 1978). SIF cells resemble adrenal chromaffin cells in that they have small cell bodies packed with large catecholamine-storage granules. They also resemble sympathetic neurons in their ability to grow short neurites and make synapses (see Doupe et al., 1985b, and references therein).

In avian embryos, adrenal chromaffin cells and SIF cells are thought to arise from a population of neural crest cells that initially localize in the primary sympathetic chains of the embryo (see Cochard et al., 1979; Anderson and Axel, 1986). It has been suggested that, when the primary sympathoblasts remigrate in a dorsal direction to form the definitive sympathetic ganglia, other progenitors migrate ventrally to contribute chromaffin cells to the developing adrenal gland and the aortic plexus (Fernholm, 1971; Polak et al., 1971; see also Fig. 5.2).

### 5.2.3  The parasympathetic ganglia

These ganglia arise from discrete levels of the axis, mostly from cranial and sacral regions of the neural crest. The cranial ganglia include the ciliary, otic, submandibular, lingual, ethnoid, and sphenopalatine (Narayanan and Narayanan, 1978b; Noden, 1978b). Among these, the ciliary ganglion deserves special attention. Two populations of neurons differentiate in these ganglia, the large ciliary neurons that innervate the constrictor muscle fibers of the iris and the ciliary body, and the smaller choroid neurons that innervate smooth muscle cells associated with blood vessels in the choroid layer (Marwitt et al., 1971).

The ciliary ganglion of avian species has been a useful model system both for embryological as well as already classical physiological studies, notably those pioneered by Pilar (see, e.g., Landmesser and Pilar, 1972, 1974). The origin of the neurons and glia of these ganglia from the mesencephalic crest was first determined by Hammond and Yntema (1958) based on extirpation experiments. This was later confirmed by Narayanan and Narayanan (1978b) through the use of the quail–chick chimera system.

### 5.2.4    The enteric nervous system

*5.2.4.1    Enteric ganglia and plexuses.* Neurons and glia of the enteric nervous system (ENS) are organized into two main plexuses: the myenteric plexus of Auerbach and the submucosal plexus of Meissner. Precise localization of the levels of origin and colonization of the crest was performed in a series of studies based on the use of the quail–chick chimera system by Le Douarin and Teillet (1973), who established that the crest cells that colonize the gut emigrate from two regions of the neuraxis. Neural crest cells arising at the vagal level (somites 1–7) colonize the entire length of the gut, whereas those arising at the sacral level (caudal to somite 28) contribute only to the innervation of the postumbilical portion of the bowel (see Figs 5.1 and 5.9).

Recently, the migration of neural crest cells to the gut was investigated in the mouse embryo. Murine premigratory neural crest cells were labeled *ex utero* with the fluorescent tracer DiI along restricted vagal segments. When somitic levels 1–5 were marked, the progeny of the labeled crest cells was found along the entire gut. When somites 6 and 7 were marked, labeled cells were found only in the foregut derivatives. These results suggest that, while the innervation of the foregut is contributed by the entire vagal region of the neuraxis, the innervation of the midgut and hindgut derives from the anterior vagal area (Durbec *et al.*, 1996b; see also below for sacral contribution).

The observation made in quail–chick chimeras that the sacral neural crest colonizes the postumbilical level of the gut, was subsequently confirmed using three additional techniques. Sacral neural crest cells were tagged prior to the onset of migration, either with the fluorescent tracer DiI, or with the replication-deficient retrovirus LZ10 previously used for lineage studies (Galileo *et al.*, 1990; Pomeranz *et al.*, 1991a). The third technique employed was to follow the distribution of sacral-level neural crest cells with the NC-1/HNK-1 antibody (Pomeranz and Gershon, 1990). Regardless of the nature of the marker, labeled cells were observed to settle exclusively in the postumbilical portion of the gut, including the ganglion of Remak. These results were further confirmed and extended to the murine bowel (Serbedzija *et al.*, 1991).

In avians, migration of neural crest cells could be precisely followed by labeling the cells with the HNK-1 antibody or by the quail–chick marker. Emigration from the dorsal neural tube in the vagal region takes place between 9 and 15ss. Vagal-level crest cells first migrate dorsoventrally between the ectoderm and paraxial mesoderm until they arrive at the primitive gut wall, where they associate with the splanchnopleural mesenchyme. Then, the crest cells migrate in a caudal direction and colonize the entire length of the primitive gut, a process that is completed by E8 with the arrival of the progenitors to the hindgut (Tucker *et al.*, 1986).

In mice, the foregut becomes colonized as early as at E9 (Rothman and Gershon, 1982), and migration progresses distally until E13.5 when the entire gut is colonized (Kapur *et al.*, 1992). At the time they first arrive in the murine bowel, these crest-derived cells express no neuronal markers. The wave of neuronal differentiation of the enteric progenitors follows, in a rostral to caudal gradient, the initial colonization of the bowel. Myenteric neurons expressing catecholaminergic traits and neurofilament immunoreactivity first become evident in the foregut at E9.5–10 and appear at progressively more caudal levels until E14.5, when they are seen in the distal portion of the rectum (Teitelman *et al.*, 1981; Baetge and Gershon, 1989).

After homing at distinct levels of the gut, crest-derived cells differentiate into the ganglia of the Auerbach and Meissner plexuses. Whereas the neurogenic ability of neural crest cells arising in the vagal level of the neuraxis has been fully established, the question remained open as to the ability of sacral-level crest to give rise to enteric neurons. To address this question, several investigators isolated segments of hindgut during a time window that follows the colonization by sacral crest cells but precedes innervation by cells arising in vagal regions. These explants were grown as chorioallantoic membrane grafts. Three of the studies, in which the presence of neurons was assessed either by morphology (Allan and Newgreen, 1980), acetylcholinesterase activity (Smith *et al.*, 1977), or with the E/C8 Mab (Ciment and Weston, 1983), failed to reveal neuronal differentiation of the sacral neural crest cells. In a later study, however, the presence of neurons expressing neurofilament immunoreactivity and of glial cells expressing GFAP (glial fibillary acidic protein) could be observed among the NC1/HNK1-positive crest cell population, both in organotypic cultures and in grafts performed on the chorioallantoic membrane of chick embryos (Pomeranz and Gershon, 1990). These results suggested that the avian sacral neural crest may have neurogenic potential, at least *in vitro*.

Recently, *in vivo* experiments (Burns and Le Douarin, 1998) utilizing the quail–chick grafting technique have provided evidence that the sacral neural crest actually contributes to the neural population of the gut. When the sacral neural crest from quail was grafted into chick, quail cells were observed (using the quail cell-specific antibody, QCPN) within the chick hindgut from E7 onwards (Fig. 5.4). These cells colonized the gut in a rostral direction, ascending to the umbilicus, and were particularly numerous in the distal hindgut. Some of these graft-derived cells were shown to be neuronal, as double labeling of individual cells was observed with QCPN and with a human anti-neuronal autoantibody (ANNA-1) which recognizes neuron-specific RNA-binding proteins of the Hu family (Fairman *et al.*, 1995). Further, double labeling was also observed with QCPN and the NADPH-diaphorase staining technique, which has been shown to label cells containing neural nitric oxide synthase (NOS)

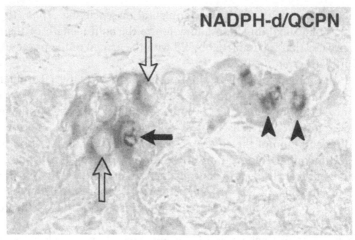

Figure 5.4 The sacral-level neural crest contributes to intestinal neurons. NADPH-diaphorase histochemistry. Myenteric ganglion. NADPH-diaphorase stained the cytoplasm of neurons violet (open arrows), while the nuclei of quail cells were stained brown (arrowheads). Double-labeled neurons possessed violet cytoplasm with brown nuclei (arrow). (Reproduced, with permission of Company of Biologists Ltd, from Burns and Le Douarin, 1998.)

*For a colored version of this figure, see www.cambridge.org/9780521122252.*

(Ward *et al.*, 1992; Young *et al.*, 1992). These antibody labeling experiments clearly demonstrate that the sacral neural crest contributes to the neuronal population of the enteric nervous system.

Abnormalities in the colonization of the distal colon by neural crest cells can occur, which result in a lack of innervation (Hirschsprung's disease in humans). Due to the clinical and scientific importance of this disease, and the experimental models available for the aneural megacolon phenotype, we shall deal with this issue separately at the end of this chapter.

*5.2.4.2   The ganglion of Remak.* The ganglion of Remak is unique to avian species and arises from the lumbosacral neural crest. The formation of this ganglion was extensively studied by Teillet (1978). It appears at stage 24HH in the chick, and at stage 18 of Zacchei (1961) in the quail, at the level of the rectum as an accumulation of crest cells in the dorsal mesentery. These cells migrate further rostrally within the dorsal mesentery to reach the level of the hepatic and pancreatic ducts. Ganglion cells thus become distributed along the entire length of the colon and ileum. Pomeranz and Gershon (1990) have further distinguished between two continuous streams of HNK1/NC1 immunostained cells extending from the sacral crest to the hindgut in chick embryos. They have observed that, by E4, the first stream of cells gives rise to the ganglion of Remak, and, between E4 and E7, the second stream of cells ascends within the colorectum to the umbilicus. Notably, the first group coexpressed immunoreactivity both for the general crest marker HNK-1/NC1 and for a neurofilament-associated protein revealed by the E/C8 antibody, whereas the second group of cells was devoid of the neural marker during migration.

*5.2.4.3   The origin of the interstitial cells of Cajal.* The interstitial cells of Cajal (ICC) were first described a century ago by Ramon y Cajal as aggregates of spindle-shaped cells with few processes that are distributed in various parts of the gastrointestinal tract in close association with both muscles and nerves. These cells were implicated in the regulation of intestinal contractions serving as the "pacemaker" of the gut (reviewed by Thuneberg, 1989). Their identity as neuronal or non-neuronal cells remained, however, a subject of debate. While some authors, including Cajal, considered them to be primitive nerve cells, others proposed that they were glial in nature (Stohr, 1955), or even mesenchymal in nature based on their ultrastructural characteristics (Gabella, 1989). Since all neurons and glia of the enteric nervous system originate from the neural crest, confirming or invalidating a neural crest origin for these cells would be informative for understanding their characteristic structure and physiology. An important finding in this direction was that ICC express the c-kit protein, a tyrosine kinase that acts as the receptor for stem cell (steel) factor (Ward *et al.*, 1994; Torihashi *et al.*, 1995). The importance of the c-kit receptor for the survival and/or function of these cells was shown by postnatal injection of neutralizing c-kit antibodies, which significantly perturbed gut motility in the mouse (Maeda *et al.*, 1992). In addition, the number of interstitial cells was greatly reduced in mice carrying a point mutation in the c-*kit* gene, enabling their physiological roles to be determined in different regions of the gastrointestinal tract (Ward *et al.*, 1994; Huizinga *et al.*, 1995; Burns *et al.*, 1996). The possibility of using c-kit as a marker for these cells has recently enabled their embryonic origin to be studied.

As the ICC cell type also exists in the chicken gut (Gabella, 1989), Lecoin *et al.* (1996) constructed quail–chick chimeras in which the vagal neural primordium of chick embryos was replaced by its quail counterpart to produce a neural crest-derived enteric innervation of quail type. Embryo sections were then subjected to hybridization *in situ* with a c-*kit* probe that reacts with the quail gene and the identity of the c-*kit*-positive cells as being of the quail or chick type was analyzed. All the c-*kit*-positive cells were identified to be of the chick type, implying that they do not derive from the neural crest. This observation was confirmed by grafting aneural gut segments onto the chorioallantoic membrane, which revealed the development of c-*kit*-positive cells possessing the typical ultrastructural characteristics of interstitial cells even in the absence of neural crest-derived ganglia (Lecoin *et al.*, 1996). Similarly, aneural segments of mouse embryonic fetus were grafted under the renal capsule of adult mice, and checked for the appearance of c-kit-immunoreactive protein. In spite of the lack of neurons in these preparations, c-*kit*-positive cells developed in the explants (Young *et al.*, 1996). These results clearly demonstrate that the ICC are derived from the gut mesenchyme.

## 5.3 Transmitter phenotypes and markers of developing autonomic sublineages

During embryonic development of the ANS, neural crest cells acquire particular differentiated traits that can be revealed both *in vivo* and in tissue culture. These traits include the acquisition of the capacity to metabolize specific neurotransmitter compounds. An interesting question, in this regard, is whether the same mechanisms that are responsible for the specification of a particular kind of neuron also govern the neuron's choice of neurotransmitter phenotype. This issue is particularly relevant in light of three main observations. First, although most sympathetic postganglionic neurons use catecholamines to mediate neurotransmission and the parasympathetic neurons are mostly cholinergic, the reverse situation has also been well documented in several cases (Sjokvist, 1963a,b; Ehinger and Falck, 1970; Furness and Costa, 1971). Second, in addition to these major neurotransmitters, the presence of a variety of neuropeptides has been identified in both systems. Specific sets of sympathetic neurons contain one or more peptides that colocalize with the more classical transmitters and are believed to act as cotransmitters or neuromodulators (Hökfelt *et al.*, 1980; Lundberg *et al.*, 1980, 1982a,b; Lundberg and Hökfelt, 1986). Third, a shift in the pattern of synthesis of specific neurotransmitters has been thoroughly documented to occur in autonomic neurons, both *in vitro* and *in vivo*, in response to particular environmental signals (Schotzinger *et al.*, 1994). These observations suggest that determination of the phenotypic identity of a neuron is a multistep process, resulting not only from lineage restrictions, or from the response of respective progenitor cells to an initial set of environmental cues, but also from complex interactions that take place throughout the period of innervation and mature function.

### 5.3.1 Transmitter phenotypes of embryonic sympathetic neurons

One of the characteristics of the adrenergic cells is that they display enzymatic activities which are related to synthesis and degradation of catecholamines. They also have the capacity to take up, store, and release catecholamines. Shortly after the neural crest cells stop migrating and coalesce in the vicinity of the aorta, all the sympathetic neuroblasts express catecholaminergic traits. *In vitro*, a subset of crest cells also acquires the capacity to synthesize, store, take-up, and release catecholamines (Enemar *et al.*, 1965; Iversen, 1967; Kirby and Gilmore, 1976; Allan and Newgreen, 1977; Cochard *et al.*, 1978; Rothman *et al.*, 1978). Expression of the catecholaminergic phenotype does not, however, define a particular sublineage, because many autonomic cells express this marker transiently during development. Three major classes of crest-derived cells express the catecholaminergic phenotype: principal sympathetic neurons characterized by small (50 nm) dense-cored vesicles, small intensely fluorescent (SIF) cells, and chromaffin cells of the adrenal medulla. Besides this similarity, these cells exhibit significant differences in their structure, response to growth factors and hormones, and function (reviewed by Anderson, 1993a).

Sympathetic neuroblasts also contain neuropeptides. Neuropeptide Y (NPY) and vasoactive intestinal polypeptide (VIP) have an early onset of appearance that begins at neurogenesis in rodent and avian embryos (Garcia-Arraras et al., 1987, 1992; Tyrrell and Landis, 1994; Hall and MacPhedran, 1995). Whereas NPY is expressed in most tyrosine hydroxylase (TH)-positive cells in both the SCG and stellate ganglia of rat embryos, VIP-immunoreactive protein is expressed in about 30% of the stellate ganglion cells and is absent in the SCG (Tyrrell and Landis, 1994). Another neuropeptide, somatostatin (Garcia-Arraras et al., 1984), also begins to be expressed during embryonic life. In contrast, the expression of enkephalin and calcitonin gene-related peptide (CGRP) in sympathetic neurons only occurs postnatally (reviewed in Rao and Landis, 1993).

### 5.3.2 Modulation of the catecholaminergic phenotype during development

Catecholamine-containing cells have also been found in some populations of rodent embryonic DRG (Katz et al., 1983; Jonakait et al., 1984; Katz and Erb, 1990; Katz, 1991). This catecholaminergic phenotype, which is transient in the sensory ganglia of the rat, can be prevented from extinction if the ganglionic cells are cultured at low, but not at high, density. These observations were taken to imply that the degree of cell aggregation influences phenotypic expression (Katz, 1991). Furthermore, the ganglionic non-neuronal cells were found to inhibit expression of TH in high-density cultures, and this effect was found to be partially mediated by LIF (Fan and Katz, 1993). In this respect, it is worth mentioning that avian DRG, which do not express catecholaminergic markers in the embryo, can be induced to do so when cells are dispersed in culture. Moreover, the number of TH-positive cells can be upregulated in these cultures by insulin and insulin-like growth factors (Xue et al., 1987, 1992). It can be then concluded that the microenvironment of sensory ganglia, in particular that conferred by the non-neuronal component, regulates the expression of catecholaminergic traits in sensory neurons. Whereas this negative regulation is partial and stage-dependent in rodents, it is complete in avians.

Most interestingly, a transient expression of catecholamines was also reported to occur in the gut of mice and rat embryos. These transient catecholaminergic cells express characteristic neuronal traits which include the presence of growth-associated protein-43 (GAP-43), neurofilament proteins, microtubule-associated proteins MAP 2 and 5, peripherin, NCAM (neural cell adhesion molecule), acetylcholinesterase, and the low-affinity NGF receptor p75 (Baetge and Gershon, 1989). Since these cells retain the ability to proliferate, they are considered to be progenitor cells rather than fully differentiated neurons. While the catecholaminergic phenotype of these progenitors is retained in vitro, it disappears progressively in the environment of the gut where these cells give rise to non-catecholaminergic enteric neurons that express either serotonin, substance P, or NPY (reviewed in Gershon *et al.*, 1993). In analogy with the situation described above for the DRG, the adrenergic phenotype is never expressed *in vivo* in the avian gut environment (Smith *et al.*, 1977). Nevertheless, a subpopulation of neural crest-derived cells that colonized the bowel retains the capacity to express TH upon disso- ciation and growth either in mass or in clonal cultures (Pomeranz *et al.*, 1993; Sextier-Sainte-Claire Deville *et al.*, 1994), or to colonize the sympathetic ganglia and express their adrenergic traits upon retrotransplantation into the migratory pathways of young hosts (Rothman *et al.*, 1990).

An analogous situation occurs for neural crest cells which express catechol- amines and TH protein on their way to colonize the adrenal gland. Upon differentiation into chromaffin cells, they acquire the ability to synthesize phe- nylethanolamine-*N*-methyltransferase (PNMT) under the influence of gluco- corticoids produced by the adrenal cortex (Seidl and Unsicker, 1989; see also review by Beato, 1989). Simultaneously with the upregulation of PNMT expression, glucocorticoids were shown *in vitro* to repress the expression of neuronal-specific genes in primary sympathetic neurons (Anderson and Axel, 1986; Anderson and Michelson, 1989; Vogel and Weston, 1990; Anderson *et al.*, 1991), and in the chromaffin-derived PC12 cell line. These neuron-specific genes include peripherin, an intermediate filament protein present in mammals but not in birds (Leonard *et al.*, 1987), SCG10 (Stein *et al.*, 1988), and GAP-43 (Federoff *et al.*, 1988).

In addition to catecholamines, a number of markers are expressed both by sympathetic neuroblasts, by adrenal chromaffin cells, and by SIF cells. Such are SA1-5 (Carnahan and Patterson, 1991a,b; Anderson *et al.*, 1991; Sejvar *et al.*, 1993), and the CC1 and CC4 (Schwarting *et al.*, 1992) epitopes, all defined by Mabs. Other markers, however, are specific for distinct cell types from the onset of their expression, such as the antigen recognized by the B2 antibody that reacts exclusively with sympathetic neurons (Anderson and Axel, 1986), and the expression of peanut agglutinin binding sites present only on neuroen- docrine derivatives (Katz *et al.*, 1995). A relationship between sympathetic and enteric neuroblasts was suggested by the fact that both cell types express in common the SA1-5 antigens and TH immunoreactivity. Enteric neuroblasts lose the two categories of markers upon terminal differentiation and acquire instead the B2 antigen, which characterizes committed sympathetic neuroblasts (Gershon *et al.*, 1993, and references therein). The B2 antigen is subsequently lost, and the transiently catecholaminergic cells go on to acquire their final, non-catecholaminergic phenotype.

### 5.3.3 Neurotransmitter plasticity in the autonomic nervous system

The expression of a neurotransmitter type is subject to changes during on- togeny as well as in response to experimental conditions (see Rao and Landis, 1993). Some sympathetic neurons whose initial transmitter phenotype is noradrenergic, show a developmentally regulated switch and become cholinergic neurons. The occurrence of this catecholaminergic to cholinergic shift was demonstrated in a compelling way in the early 1970s, in a series of elegant studies carried out *in vitro* on neurons of the SCG. When grown in the absence of non-neuronal cells, SCG neurons synthesize, store, and release norepinephrine (noradrenaline) as do their normal counterparts *in vivo*. Coculturing the SCG neurons with non-neuronal cells such as glia, skeletal myotubes, or cardiac myocytes, results in a dramatic increase in acetylcholine synthesis at the expense of noradrenaline and in the formation of functional synapses (O'League *et al.*, 1974; Furshpan *et al.*, 1976). This effect on neurotransmitter switch could be mimicked by treatment with medium conditioned by the non-neuronal C6 cells, showing that the activity is mediated by diffusible signals (Patterson *et al.*, 1975; Landis, 1976; Patterson and Chun, 1977). Most importantly, a simultaneous increase in acetylcholine, and a decrease in norepinephrine synthesis, were shown to occur in individual cultured cells upon treatment with a heart cell conditioned medium, thus reflecting a true phenotypic switch rather than a selection due to preferential survival of specific neuronal subsets (Reichardt and Patterson, 1977). The cholinergic differentiation factor from heart cells was subsequently purified and cloned by Yamamori *et al.* (1989). This molecule turned out to be identical to LIF (Table 4.2), a protein endowed with pleiotropic functions since it stimulates hemopoietic cell differentiation and inhibits differentiation of embryonic stem cells. Besides LIF, two other factors are known that induce cholinergic function in cultured sympathetic neurons. CNTF, identified for its ability to support the survival of ciliary neurons, induces choline acetyltransferase and inhibits TH activity (Saadat *et al.*, 1989; Table 4.1). In addition, a membrane-associated neurotransmitter-stimulating factor (MANS) was partially purified from spinal cord and ganglionic membranes, and has a function similar to that of LIF and CNTF in promoting the survival of parasympathetic ciliary neurons (Wong and Kessler, 1987). The question then arises whether these factors mediate the choice of neurotransmitter phenotype *in vivo*.

The best studied example of phenotypic plasticity of neurotransmitter type *in vivo* are the sympathetic neurons that innervate the sweat glands in the rat footpad. These neurons are originally catecholaminergic, but were shown to switch their transmitter phenotype to become cholinergic when their axons contact their target tissue during the second postnatal week (Landis and Keefe, 1983; Leblanc and Landis, 1986; Stevens and Landis, 1987, 1990). Cross-innervation studies performed *in vivo* by Landis and colleagues have shown that this switch is induced by factors produced by the target tissue itself. If neurons that normally innervate the sweat glands are experimentally led to innervate the parotid glands whose innervation is noradrenergic, these neurons fail to undergo the normal noradrenergic–cholinergic switch (Schotzinger and Landis, 1990). Furthermore, normal noradrenergic neurons that innervate hairy skin become cholinergic if forced to innervate the sweat glands (Schotzinger and Landis, 1988). This catecholaminergic–cholinergic switch is also accompanied by a change in the pattern of expression of distinct neuropeptides. Neurons innervating the sweat glands first express NPY when they are noradrenergic and then switch to VIP and CGRP expression when they become cholinergic. This shift in neuropeptide synthesis was also shown to be target-dependent (Landis et al., 1988; Schotzinger and Landis, 1988; Stevens and Landis, 1990). These observations suggest that the target tissue is important, not only for retrograde control of cell number, but is also required for specifying the final choice of neurotransmitter identity and function.

Footpad extracts contain factors that induce choline acetyltransferase and VIP, but reduce catecholamines and NPY content in cultured sympathetic neurons (Rao and Landis, 1990; Rohrer, 1992). Rao et al. (1992a,b) raised the question as to whether LIF, CNTF, or MANS was responsible for the cholinergizing activities emanating from the footpad extracts. Despite the biochemical similarity of the footpad activity with CNTF, neither CNTF mRNA nor protein could be detected in the sweat glands or in the Schwann cells associated with them. Furthermore, the footpad activity is biochemically distinct from LIF, and LIF antibodies failed to block the inducing activity present in the footpad extract. Finally, MANS failed to induce cholinergic activity in sympathetic neurons. Thus, the identity of the cholinergic differentiation factors that mediate the phenotypic switch occurring upon innervation of the sweat glands remains to be determined.

An analogous switch of neurotransmitter phenotypes has also been shown to occur in cultured avian sympathetic neurons. Both FGF1 and FGF2 induce choline acetyltransferase activity in these neurons and cause a concomitant decrease in TH. Since the time course of FGF activity is slow on both enzymes, it was proposed that the factors act by inducing de novo gene expression rather than by activation of a preexisting pool of enzyme (Zurn, 1992). VIP, which is often associated with a cholinergic phenotype, was also shown to increase the cholinergic properties of sympathetic neurons via activation of cAMP. Moreover, VIP also causes an initial rise in TH activity, followed by a secondary decline. Therefore, FGF and VIP appear to act by different mechanisms since FGF does not alter cAMP activity and has no stimulatory effect on the levels of TH (Berreta and Zurn, 1991). Since these studies were carried out with populations of chick sympathetic neurons, they did not address the question of whether the observed effects reflect a true shift in transmitter phenotype of single cells. Another study of quail sympathetic neurons, however, has clearly shown that, whereas the great majority of sympathetic neurons express TH immunoreactivity, about half of them also express cholinergic traits (Barbu et al., 1992). These data, showing that some sympathetic neurons simultaneously express adrenergic and cholinergic phenotypes, suggest that a phenotypic switch is also likely to occur in avian sympathetic neurons. This double phenotypic identity observed in defined culture conditions might, therefore, be equivalent to a transition state during the shift between phenotypes, similar to that observed in the rat sympathetic neurons (Potter et al., 1986).

Contrary to the noradrenergic–cholinergic switch that takes place during development of certain sympathetic neurons, no such change is known to occur normally in the reverse direction. Precursors of presumptive cholinergic neurons that derive from the neural crest, such as ciliary ganglion neurons, synthesize acetylcholine as early as the neural fold stage and during their migration; however, they do not express catecholamines or TH (Smith *et al.*, 1979). The question was raised as to whether cholinergic neurons are able to display plasticity in neurotransmitter phenotype when experimentally placed in an environment that stimulates adrenergic differentiation. To this end, cholinergic ciliary ganglia from E4–6 quail embryos were back-transplanted into chick hosts at E2. Ganglia were grafted between the neural tube and somites at the adrenomedullary level of the axis. Analysis of the distribution of ganglion-derived cells was performed after gangliogenesis had taken place in the host. It revealed the presence of sympathetic neurons expressing adrenergic traits and bearing the quail marker (Le Douarin *et al.*, 1978). These results were the first evidence to suggest a plastic behavior of cholinergic ganglia in an adrenergic environment. These observations, however, did not make it possible to determine whether the cholinergic neurons themselves switch phenotype or whether non-neuronal cells of the ganglia become catecholaminergic. Subsequent experiments by Dupin (1984) revealed that the non-neuronal population of the transplanted ciliary ganglia succeeded in differentiating into sympathetic neurons expressing catecholaminergic traits in the host ganglia, whereas the neurons did not survive.

In a later study, Coulombe and Bronner-Fraser (1986) developed conditions that enabled neurons to survive. The conditions made it possible to trace potential phenotypic shift in the cholinergic neurons themselves. E6.5 quail ciliary ganglion neurons were retrogradely labeled with fluorescent latex microspheres that were injected into their target sites. About 20% of the neurons in the ganglia and none of the glial cells were labeled. Ganglia were then excised, dissociated, and injected into the ventral pathway of migration in the trunk region of chick hosts. Four to five days after injection, embryos were examined for cells containing both fluorescent microspheres and catecholamines. A certain proportion of the labeled neurons was incorporated into host sympathetic ganglia and was shown to express catecholamines. These observations led to the conclusion that neurotransmitter plasticity also occurs in cholinergic neurons, at least under experimental conditions.

## 5.4   Development of autonomic cells from the neural crest

### 5.4.1   *Factors that affect initial development of adrenergic cells*

The onset of adrenergic differentiation of neural crest precursors depends upon interactions of these cells with structures bordering the migratory routes taken by the autonomic progenitors. These progenitors migrate through the rostral sclerotome and intersomitic spaces to home at para-aortic sites next to, but not in contact with, the notochord and ventralmost part of the neural tube. The importance of the microenvironment of the migratory routes in the development of the sympathetic derivatives was demonstrated in avian embryos by ablating either the notochord or the neural tube. These experiments revealed that the formation of sympathetic ganglia requires the presence of the notochord and/or the neural tube. In fact, ablation of both axial organs induces the death of all neural crest cells present in the paraxial mesoderm, the cells of which are also subjected to apoptosis (Teillet and Le Douarin, 1983; Rong *et al.*, 1992). Furthermore, after rotating the neural tube along its dorsoventral axis, the expression of catecholamine histofluorescence in condensing sympathetic ganglia was prevented, suggesting that the ability of the neural tube to regulate the sympathetic phenotype is restricted to its ventral part (Stern *et al.*, 1991). Removal of the notochord also resulted in an inhibition of expression of the paired homeodomain protein Phox2, and of the zinc finger protein, GATA-2, two transcription factors that characterize early sympathetic ontogeny. Similarly, the rate-limiting enzyme of catecholamine metabolism, TH, is no longer expressed after notochord removal (Groves *et al.*, 1995). These observations indicate that signals imparted by the notochord/ventral neural tube are required for the early phases of sympathetic ganglion development. In line with the above observations, brain extracts or medium conditioned by cultured neural tubes were shown to stimulate sympathetic differentiation (Howard and Bronner-Fraser, 1985, 1986). Moreover, conditioned medium of cultured notochords had no effect (Howard and Bronner-Fraser, 1986). Moreover, addition of the notochords to neural crest cells associated with gut tissue strongly promoted adrenergic differentiation (Rothman *et al.*, 1993b). These findings suggest that molecules that associate with the cell surface or with the extracellular matrix must be present along the migratory routes of the neural crest in order to obtain normal sympathetic neurons. Indeed, overlaying a reconstituted basement membrane-like matrix on cultured neural crest was shown to promote strong adrenergic differentiation (Maxwell and Forbes, 1988, 1990; Forbes *et al.*, 1993).

Signals originating from the axial structures may act either directly on neural crest cells, or indirectly via the sclerotomal cells which constitute the actual substrate on which sympathetic ganglion progenitors migrate. An indirect action is consistent with the results of recent experiments showing that ectopic notochord grafts induce a ventral fate in the paraxial mesoderm, leading to the differentiation of sclerotome at the expense of dermomyotome (Pourquié *et al.*, 1993). This ventralizing effect of the notochord was found to be mediated by the protein encoded by the vertebrate homolog of the *Drosophila hedgehog* gene, designated *sonic hedgehog* (*Shh*) or *vertebrate hedgehog* (*Vhh*) (Fan and Tessier-Lavigne, 1994; Johnson *et al.*, 1994; Fan *et al.*, 1995). The question is raised as to whether *Shh* might be involved in promoting the appearance of the adrenergic phenotype in neural crest cells. The possibility that this secreted protein exerts a direct effect on neural crest cells was examined *in vitro*. Modest adrenergic differentiation was obtained upon addition of the N-terminal recombinant peptide to secondary avian neural crest cultures (Reissmann *et al.*, 1996). In striking contrast, a dramatic stimulation of the development of TH-positive cells was obtained with specific members of the TGFβ family, whose expression was shown, both in *Drosophila* and vertebrates, to be induced by Shh (O'Farrell, 1994; Perrimon, 1995). These proteins include TGF$\beta$1 itself (Howard and Gershon, 1993), BMP-4 and BMP-2, and also BMP-7 (Varley et al., 1995; Reissmann et al., 1996; Varley and Maxwell, 1996). Glial cell line-derived neurotrophic factor (GDNF), another member of the TGF$\beta$ superfamily, was found to promote adrenergic differentiation of cultured mouse neural crest cells (Maxwell et al., 1996), but not of their avian counterparts (Reissmann et al., 1996). Furthermore, localization studies have shown that the dorsal aortic tissue synthesizes BMP-4 and BMP-7 already prior to the onset of expression of TH mRNA by the homing sympathetic progenitors (Reissmann et al., 1996). Taken together, these results raise the hypothesis that notochord-derived SHH triggers a cascade of inductive changes that initially affect the sclerotome, and subsequently the more ventrally located paraaortic region. Analysis of the pattern of expression of TGF-$\beta$-related factors and of adrenergic development in embryos lacking *Shh* gene function will be helpful for critically testing this notion, because, as expected, these embryos are defective in the formation of sclerotomal derivatives such as the vertebral column and ribs (Chiang et al., 1996).

It has long been established that chick embryo extract is a potent inducer of adrenergic differentiation in primary cultures of quail truncal neural crest cells (Ziller et al., 1987). The exact nature of the factors involved in this effect has not been clarified so far. *In vitro* analysis of adrenergic cell differentiation in culture has brought about cues concerning this problem. For instance, catecholamines were found to stimulate adrenergic differentiation by regulating the transcription of the TH gene, through activation of $\beta$-adrenergic receptors and cAMP production (Dupin et al., 1993). This is coherent with the fact that the notochord synthesizes and releases catecholamines at the critical stage of sympathetic ganglion development (Dupin et al., 1993). Along the same line, norepinephrine was suggested to modulate the expression of the adrenergic phenotype in quail neural crest cells grown in clonal cultures because treatment of crest cell cultures with norepinephrine uptake inhibitors blocked the production of the enzymes TH and dopamine $\beta$-hydroxylase (DBH) by the cultured cells (Sieber-Blum, 1989c). Other factors, such as insulin and insulin-like growth factor-1 (Nataf and Monier, 1992; Xue et al., 1992), and glucocorticoids (Smith and Fauquet, 1984), were reported to stimulate *in vitro* the number of cells that differentiate along the adrenergic pathway at a stage when neural crest cells have already colonized the sclerotome.

Two independent studies, carried out *in vitro*, have shown that retinoic acid (RA) is also a likely candidate for mediating early differentiation of sympathetic cells (Dupin and Le Douarin, 1995; Rockwood and Maxwell, 1996). Dupin and Le Douarin showed that RA strongly stimulates the differentiation of adrenergic TH-positive cells when it is applied during the first 3 days of the culture of trunk neural crest cells, a period of time which just precedes the onset of TH synthesis *in vivo*. In contrast, if RA is applied later (from day 3 to day 6 of culture), no effect on adrenergic differentiation is observed while the number of melanocytes is greatly increased. Moreover, treatment of single-cell cultures with RA resulted in the induction of adrenergic traits in multipotent crest precursors (Dupin and Le Douarin, 1995). The study performed by Rockwood and Maxwell (1996) pointed instead to a mitogenic effect of RA on adrenergic cells that already express TH immunoreactivity.

The *in vitro* effects of RA are likely to be of significance in the embryo, since retinoids are present in areas such as the ventral neural tube known to influence sympathetic development (Wagner *et al.*, 1990, 1992). In addition, RA receptors are expressed by migrating neural crest cells and their PNS derivatives (Perez-Castro *et al.*, 1989; Maden *et al.*, 1991; Ruberte *et al.*, 1993, and references therein).

### 5.4.2 Control of the proliferation of sympathetic neuroblasts and chromaffin cells

5.4.2.1 *Sympathetic neuroblasts.* Ziller *et al.* (1987) have clearly shown that the *in vitro* development of adrenergic cells expressing TH immunoreactivity involves extensive proliferation of the neural crest progenitors and growth conditions that are provided by culture medium supplemented with 10% chick embryo extract (Fig. 5.5). RA, which was proposed to act as a mitogen for cultured sympathetic progenitors (see previous section), is the only factor that has been unequivocally identified to promote the proliferation of neural crest cells that acquire TH. Although NT-3 has been shown to stimulate proliferation and survival of avian neural crest cells grown in defined medium (Kalcheim *et al.*, 1992), lack of terminal sympathetic differentiation under these conditions made it impossible to determine whether the responsive population contains progenitors with the ability to become adrenergic cells.

An interesting feature of sympathetic progenitors is their continued proliferation *in vivo* even after differentiation into cells expressing either TH or DBH, catecholamines or somatostatin (Rothman *et al.*, 1978, 1980; Rohrer and Thoenen, 1987; Ziller *et al.*, 1987). The proportion of mitotically active cells that express the above markers declines with time in the embryo as well as in culture (Cohen, 1974; Rothman *et al.*, 1978, 1980; Maxwell and Sietz, 1985). Most notably, sympathetic neuroblasts are able to extend neurites before and during cell division both in culture and *in vivo*. The pattern of *in vitro* neurite growth (number, branching, position, length) of the parental neuroblasts was shown to be preserved by their progeny. Whereas 18% of the neurites remained intact as the soma divided, 82% of the processes retracted into the cell body during late "prophase" of the cell cycle and regenerated within minutes after completion of cytokinesis (Wolf *et al.*, 1996). *In vivo*, up to 15% of dividing sympathetic neuroblasts from E16 rat SCG revealed the presence of efferents in the internal and external carotid nerve, hundreds of micrometers away from the cell soma (Fig. 5.6; DiCicco-Bloom, personal communication).

What factors do affect sympathetic neuroblast proliferation? Treatment with epidermal growth factor (EGF) or with insulin was shown to stimulate thymidine (TdR) incorporation into nuclei and cell division in culture (DiCicco-Bloom *et al.*, 1990, 1993). In a later study, the effect of factors on

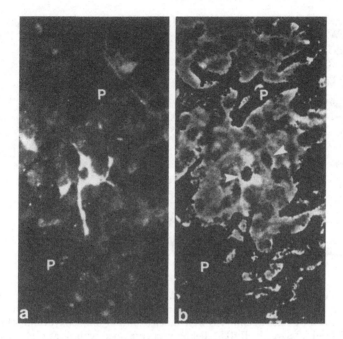

**Figure 5.5** Adrenergic cells expressing tyrosine hydroxylase immunoreactivity differentiate from cycling progenitors. (a) Tyrosine hydroxylase (TH) immunofluorescent cell in a 6-day-old culture of neural crest cells derived from total neuron primordium and grown in medium supplemented with serum and 10% chick embryo extract. (b) The same field where fluorescence is superimposed upon a brightfield image. The nucleus of the TH-positive cell is labeled with [$^3$H] TdR (arrowhead). Other TH-negative cells have also incorporated the label. P, pigment cell. (Reproduced, with permission, from Ziller *et al.*, 1987.)

process-bearing neuroblasts was tested, and it was found that the most remarkable effect on cell proliferation is obtained upon treatment with a combination of insulin, NT-3, and the PACAP peptide. This combination drove almost 95% of the cells into the S-phase compared to 38% in controls (Wolf *et al.*, 1996).

The proliferation of sympathetic neuroblasts is also subjected to a negative regulation. CNTF was found to decrease proliferation of TH-immunoreactive cells derived from E7 chicken sympathetic ganglia. Concomitant with this effect, a dramatic increase in VIP expression was observed to occur in 50–60% of the cells. These results suggest that CNTF stimulates the differentiation of avian sympathoblasts by repressing the mitotic cycle (Ernsberger *et al.*, 1989). Likewise, retroviral transfer of the v-*src* oncogene into proliferating E6 sympathoblasts was shown to downregulate proliferation and to stimulate instead the differentiation and long-term survival of the infected cells. In contrast, retroviral infection with v-*myc* maintained high levels of cell proliferation and an immature phenotype (Haltmeier and Rohrer, 1990), invoking the possibility of opposite roles for these neural oncogenes in this system.

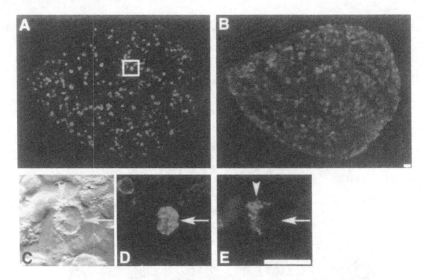

**Figure 5.6** Mitotically active superior cervical ganglion (SCG) neuroblasts possess axons projecting into efferent nerves. To identify dividing neuroblasts bearing axonal projections *in vivo*, E16.5 rat embryos were labeled with bromodeoxyuridine (BrdU) for 1 hour and fixed afterwards. BrdU labeling was combined with retrograde rhodamine dextran tracing as follows: SCG was dissected, the efferent internal and external carotid nerves were sectioned with dextran-coated knives 1 mm from the ganglion, and the explant was incubated for 4 hours in culture to allow retrograde transport. Confocal images of 8-μm sections indicate BrdU nuclear labeling in A and dextran labeling in B. Higher magnification of the boxed region in A illustrates a neuroblast in C by DIC optics that colocalizes nuclear BrdU (arrow, D) and retrograde cytoplasmic tracer (arrowhead, E). Bar = 25 μm. (Kindly provided by E. DiCicco-Bloom.)

*For a colored version of this figure, see www.cambridge.org/9780521122252.*

#### 5.4.2.2 Chromaffin cells.

Adrenal chromaffin cells proliferate extensively before birth, and this process essentially continues throughout life, albeit at a slower pace (Malvadi *et al.*, 1968; Tischler *et al.*, 1989). Both FGF2 and NGF induce cell proliferation of neonatal rat chromaffin cells (Lillien and Claude, 1985; Claude *et al.*, 1988). In addition to its mitogenic action, FGF2 also stimulates neuronal differentiation of chromaffin cells, while later causing them to become dependent on NGF for survival. This differentiative effect of FGF is specifically suppressed by dexamethasone, while its mitogenic action is not, suggesting that a complementary action of FGF2 and steroids could maintain the expansion of a population of cycling cells while inhibiting their differentiation into neurons in the adrenal environment (Stemple *et al.*, 1988).

A decrease in proliferation occurs in the rat adrenal medulla during the first postnatal weeks. This reduction was shown to correlate with the innervation of the adrenal medulla, suggesting that functional activity mediated by neurotransmitters downregulates chromaffin cell proliferation (Slotkin *et al.*, 1980). This notion was examined in culture, where depolarization of the chromaffin cell membranes was achieved by raising the concentration of KCl to 20 mM. These conditions significantly decreased the mitogenic effect elicited by NGF. Furthermore, specific activation of calcium influx blocked cell proliferation, suggesting that the effect is mediated by the activation of voltage-gated calcium channels. Direct activation of adenylate cyclase also inhibited the response to NGF. In contrast, depolarization of cultures treated with phorbol ester was ineffective in blocking proliferation, indicating that the proliferation of chromaffin cells does not require the activation of protein kinase C (Tischler *et al.*, 1994). Thus, multiple signaling mechanisms, tuned by local glucocorticoids, growth factors, and by neural activity are likely to regulate mitogenic responses in chromaffin cells.

### 5.4.3 Sequential and synergistic activities of growth factors on the differentiation of sympathetic neuroblasts

*5.4.3.1 Neurotrophin-3.* Treatment with NT-3, but not with NGF, enhanced the survival of mitotically active sympathetic neuroblasts prepared from E15.5 rat SCG, or from the paravertebral sympathetic chains of E14.5 rat embryos (Birren *et al.*, 1993; DiCicco-Bloom *et al.*, 1993). These results provide support for the general belief that sympathetic progenitor cells are not responsive to NGF, and that this neurotrophin acts on the survival of postmitotic sympathetic neurons. These results were confirmed *in vivo* by the phenotype of mice lacking NT-3 gene function. Null NT-3 mice display a reduced number of proliferating neuroblasts when compared to their normal counterparts. This effect, observed between E11 and E17 in the mutant mice, was suggested to result from increased apoptosis of the proliferating cell population (Elshamy *et al.*, 1996).

Furthermore, treatment of cultured sympathetic cells with doses of NT-3 higher than required for supporting their survival (more than 10 ng/ml) was shown to induce mitotic arrest and expression of trkA mRNA, resulting in responsiveness to NGF, a characteristic which normally appears later in development. In fact, several antimitotic agents, including aphidicolin and mitomycin C, were shown to induce trkA better than high doses of NT-3, suggesting that synthesis of trkA is primarily induced by withdrawal from the cell cycle (Verdi and Anderson, 1994). Interestingly, NGF itself was reported to up-regulate the synthesis of p75 NTR, the common neurotrophin receptor, thus completing the molecular repertoire required for maintaining a full responsive state to NGF (Verdi and Anderson, 1994). Consistent with the above results, the level of expression of trkC mRNA was found to decrease in rat ganglia, while that of trkA mRNA increased between E14.5 and the time of birth (Birren *et al.*, 1993; DiCicco-Bloom *et al.*, 1993; Verdi and Anderson, 1994). In avian embryos trkC mRNA was also found to be expresssed in primary sympathetic ganglia but not in cells that constitute the secondary, definitive chains which are, instead, responsive to NGF (Kahane and Kalcheim, 1994). These observations suggest that the switch in trophic dependence observed *in vitro* is likely to reflect a normal sequence evolving during development. Thus, sympathetic precursor cells undergo a stage-dependent change in their needs for exogenous factors, which ultimately leads them to respond to NGF.

What is the source of NT-3 that affects sympathetic neuroblasts prior to target innervation? Verdi *et al.* (1996) have reported that NT-3 is synthesized by the non-neuronal cells adjacent to the sympathetic ganglia of E13.5 and E14.5 mouse embryos, but not by the sympathetic neuroblasts. Moreover, *in vitro*, the levels of NT-3 mRNA in non-neuronal cells can be upregulated by the neuroblasts themselves, by CNTF, PDGF, or by neuregulins such as glial growth factor-2 (GGF-2). To further test the effect of GGF-2 on NT-3 production, cultures composed of both neuroblasts and non-neuronal cells were treated with blocking antibodies to GGF-2. This treatment was shown to attenuate the effect of the neuroblasts on NT-3 production by the non-neuronal cells, suggesting a role for endogenous GGF-2 in the synthesis of NT-3. The notion was then put forward that reciprocal cell–cell interactions exist, whereby neuroblast-derived neuregulins promote NT-3 production by neighboring non-neuronal cells, and that these cells in turn promote neuroblast survival and further differentiation into a NGF-dependent neuron (Verdi *et al.*, 1996).

*5.4.3.2   FGF and CNTF.* An additional *in vitro* system, by means of which it was attempted to characterize the factors involved in neuronal development, is an immortalized cell line developed by Birren and Anderson (1990). Embryonic rat adrenomedullary cells expressing the HNK-1 marker were isolated and immortalized with the v-*myc* oncogene to produce a stable cell line, MAH. The MAH cells were shown to have retained the ability to differentiate into sympathetic-like neurons. They do so under FGF2 treatment, which also has a mitogenic action on this cell line, as it has on primary chromaffin cells (Stemple *et al.*, 1988). Although FGF induces an initial neuronal differentiation response in the majority of MAH cells, only a minute proportion of the FGF-treated cells gain NGF responsiveness and undergo terminal differentiation into postmitotic neurons. Therefore, additional signals were postulated to be required for full differentiation to take place. Ip *et al.* (1994) found that CNTF, a factor that inhibits the proliferation of sympathetic progenitors (see previous section), and triggers an intracellular pathway different from that of both FGF and NGF, can collaborate with FGF to promote the terminal differentiation of MAH cells into an NGF-dependent state that closely resembles that of postmitotic sympathetic neurons. Furthermore, NGF dependence of MAH cells could be experimentally induced by overexpressing a trkA receptor cDNA construct followed by treatment with NGF (Verdi *et al.*, 1994). These results indicate that signal transduction elicited by the recombinant trkA receptor can override the need for an initial treatment with FGF and CNTF. The possible involvement of NT-3, that was found critical during early stages for survival of proliferating primary precursor cells and for induction of trkA mRNA, was not tested in this cell line.

Primary chromaffin cells obtained from adult rats were considered to represent another model system to study signals involved in neuronal differentiation. In this system, the NGF-dependent switch into sympathetic neurons has been known for a long time to be potentiated by medium conditioned by heart cells (Doupe *et al.*, 1985a). Since this medium was found to contain LIF, a cholinergic-inducing factor which closely resembles CNTF in its spectrum of responsive populations (Yamamori *et al.*, 1989; see also Section 5.3.3 and Table 5.2), and in elements of the signal transduction cascade (Ip and Yancopoulos, 1994, 1996), the possibility was then tested whether CNTF or LIF mediates the chromaffin cell–neuron conversion promoted by the heart cell conditioned medium. It was found that CNTF or LIF in combination with FGF2 enhances the ability of NGF to induce transdifferentiation of postnatal chromaffin cells into neurons (Ip *et al.*, 1994). The synergistic effects of the growth factors observed both in the immortalized MAH cell line and in chromaffin cells have been found difficult to demonstrate for the embryonic primary sympathetic precursors. Therefore, the validity of these models to gain insight into the signals that regulate the initial steps of differentiation of the primary sympathetic neuroblasts remains uncertain.

*5.4.3.3   GDNF and c-ret.* New insights from embryonic mutant mice lacking the tyrosine kinase receptor gene *ret* have revealed the complete absence of SCG in 12.5-day-old embryos (E12.5), but no abnormalities could be detected in the sympathetic ganglia located more caudally along the axis. These results clearly show that the SCG has certain distinct requirements for differentiation and/or survival from the sympathetic ganglia located more caudally in the chain. Since the neural crest progenitors were found to migrate normally to the ganglionic rudiment in the null mutants, it was proposed that ret activity is necessary either for the survival or for the differentiation of the sympathetic progenitors that form the SCG (Table 5.1; Durbec *et al.*, 1996a). Glial cell line derived neurotrophic factor (GDNF) was recently found to act as a ligand of ret (Treanor *et al.*, 1996; Trupp *et al.*, 1996), but a mutation in this gene caused only a limited reduction in the volume of the SCG (20%, Sanchez *et al.*, 1996; 35%, Moore *et al.*, 1996) when compared to the dramatic phenotype apparent in the ret mutation. These results suggest: firstly, that the GDNF/ret complex participates in the early development of the rostralmost sympathetic ganglion, the SCG, and, secondly, that the ret receptor can be activated by factors other than GDNF. The identity of these putative molecules and the possibility that they act on distinct ganglionic subpopulations remain to be established.

### 5.4.4   Requirements for chromaffin cell differentiation

When neural crest progenitors colonize the adrenal anlagen, they become associated with mesodermal cells that constitute the adrenal cortex. In the environment of the adrenal gland, precursor cells that synthesize noradrenaline convert into adrenaline-producing endocrine chromaffin cells. The enzyme responsible for this conversion is PNMT. The induction of PNMT is believed to be triggered specifically by glucocorticoid hormones produced by the adrenal cortex.

In fact, a pivotal role for glucocorticoid hormones in the normal function of the adult adrenal gland was already suggested years ago in classical experiments by Wurtman and Axelrod (1965), Margolis et al. (1966), and Wurtman and Axelrod (1966). They showed that hypophysectomy of adult rats results in a dramatic decrease in the activity of the adrenal enzyme PNMT, and that replacement therapy with the synthetic hormone dexamethasone can prevent this loss. Furthermore, experimental manipulations that raise the levels of prenatal glucocorticoids in the embryo cause a concomitant increase in PNMT activity and adrenaline content in newborn adrenal glands (Parker and Noble, 1967).

Seidl and Unsicker (1989) precisely defined the onset of differentiation of chromaffin cells in rat embryos. While adrenaline and PNMT activity were present in the E17 adrenal glands, they were undetectable at E16. This shift occurring between E16 and E17 was found to parallel a rise in the levels of adrenal and plasma glucocorticoids, and in the expression of glucocorticoid receptors. Culturing cells from E16 rat adrenals in the absence of adrenal cortical input revealed no expression of either adrenaline or PNMT. Treatment of these cultures with glucocorticoids induced adrenaline biosynthesis, and also promoted the survival of the young chromaffin cells. These results suggested an influence of glucocorticoids on the differentation of endocrine cells at a time following the initial establishment of the receptor machinery that confers responsiveness to glucocorticoid signaling. In another study (Anderson and Axel, 1986), some response to dexamethasone could be detected as early as E14.5, suggesting that functional receptors are already present at this age on a subpopulation of precursor cells. Alternatively, it is possible that the culture conditions had a stimulatory effect on the synthesis of glucocorticoid receptors in response to the ligand.

While glucocorticoids exert a positive influence on the expression of adrenal chromaffin-specific markers, they were shown to repress in vitro the expression of several neuron-specific proteins, including peripherin, SCG10 and GAP-43 in PC12 cells, and to inhibit neuronal differentiation induced by NGF or FGF (reviewed in Anderson, 1993a). Moreover, the extinction of neuronal markers in vivo precedes the onset of PNMT expression in the chromaffin progenitors (Anderson and Axel, 1986; Vogel and Weston, 1990). Based on these results, Anderson (1993b) has proposed that chromaffin cell development arises as a result of two sequential and interdependent steps. First neuronal commitment is inhibited. This inhibition is a prerequisite for a later secondary decision to express PNMT and to differentiate into a chromaffin phenotype.

The role of glucocorticoid receptors in the development of the chromaffin cells was subjected to further examination following the production of mouse embryos carrying a targeted disruption of the glucocorticoid receptor gene (Cole et al., 1995). Wild-type mice with normal adrenal glands showed strong immunostaining for synaptophysin, TH, and PNMT. In contrast, mice homozygous for the glycocorticoid receptor mutation showed enlarged adrenal glands with no detectable PNMT or adrenaline at any stage examined (from E13.5, shortly after colonization of the adrenal medulla, until early postnatal stages). Cole *et al.* (1995) have assigned the lack of characteristic chromaffin markers to the virtual absence of TH-positive adrenergic cells. In a later study, Unsicker and colleagues found no decrease in the number of cells immunoreactive for the enzyme TH in the homozygous mutants when compared to wild-type mice. Moreover, they found that the adrenomedullary cells that developed in the mutants are true chromaffin cells as they have characteristic dense-cored vesicles that distinguish them from sympathetic neurons (Unsicker, personal communication). As a result of the absence of PNMT expression, however, chromaffin cells in the mutant embryos express noradrenergic features but do not synthesize adrenaline. The results of mutant mice are therefore consistent with the notion that glucocorticoid signaling is important for modulating the expression of PNMT and the consequent acquisition of adrenergic properties, but is not required for regulation of other processes such as cell survival or proliferation, and for the differentiation of another important feature typical of chromaffin cells, the presence of large chromaffin granules.

## 5.5 The relationship between sympathetic, adrenomedullary, and enteric progenitor cells

Sympathetic neurons, adrenal chromaffin cells and catecholaminergic SIF cells were proposed to arise from neural crest cells that initially localize in the primary sympathetic chains of the embryo (Rau and Johnson, 1923; Willier, 1930; Cochard *et al.*, 1979; Anderson and Axel, 1986). These progenitors express a variety of neuron-specific markers. Moreover, chromaffin and SIF cells are able to convert phenotypically into sympathetic neurons. These initial findings led Anderson and Axel (1986) to propose the notion that sympathetic neurons and chromaffin cells develop from a common progenitor cell, the sympathoadrenal precursor, that segregates from the neural crest "as a separate lineage," and which is restricted in its developmental potential to give rise to these cell types. A decade later, many results have accumulated that allow a reevaluation of that hypothesis; they are reviewed below.

It should be stressed, at this point, that postulating a true lineage relationship between distinct cell types requires a direct demonstration of a common ancestor. This can only be achieved by clonal analysis with appropriate lineage tracers (see Chapter 7). Such experiments were not performed for the putative sympathoadrenal precursors, perhaps because of technical difficulties involved in marking individual neural crest cells. Thus, although the idea of the sympathoadrenal precursor is consistent with the proven existence in the neural crest lineage of restricted intermediate progenitors, a formal proof for a lineage relation of the autonomic precursors is still lacking.

### 5.5.1 Expression in common of neuronal and chromaffin-specific markers

Progenitors of sympathetic neurons and adrenomedullary cells coexpress markers specific to each cell type before final diversification occurs. In the rat embryo, these cells are first recognizable at E11.5 at para-aortic locations, and express neuronal markers such as TH, catecholamine storage, release, and high-affinity uptake mechanisms (Grillo, 1966; Gabella, 1976), SCG10, and neurofilament proteins as well as the chromaffin-specific markers SA1–5 (Carnahan and Patterson, 1991a,b). Subsequent to E11.5, the expression of the chromaffin markers is extinguished in the sympathetic ganglia but is retained in the adrenal medulla. Concomitant with the loss of SA1–5, sympathetic neuroblasts acquire the B2 neuronal marker. However, B2-positive neuroblasts are unable to respond to signals that induce chromaffin cell differentiation and thereby do not convert into chromaffin cells. Thus, appearance of this marker was interpreted to be the expression of a commitment to a neuronal fate (Anderson et al., 1991). During chromaffin cell differentiation, the reverse situation occurs. The acquisition of endocrine characteristics and of PNMT activity leading to adrenaline production is a late event that follows the loss of neuronal traits (Anderson and Axel, 1986; Vogel and Weston, 1990; see also Patterson, 1990). It should be noted that the transition from one cell type to another is not readily complete since some B2-positive cells appear in the adrenal gland, and SA1-containing cells can be detected in sympathetic ganglia.

### 5.5.2 The problem of a common lineage for adrenomedullary cells, sympathetic progenitors, and the enteric neurons

The results discussed above based on the analysis of marker expression suggest that the neural crest diversification pathway includes a subset of progenitors giving rise, at least, to sympathetic neurons and chromaffin cells. On the one hand, the observations that not all cells show the transient coexpression of dual markers and that other cells express the markers in inappropriate locations, raise the possibility that subsets of both sympathetic neurons as well as chromaffin cells arise from precursors other than the proposed sympathoadrenal progenitor. On the other hand, this line of evidence does not rule out the possibility that sympathoadrenal cells have additional fates during normal development. For instance, Durbec et al. (1996a) have followed the migration and fate of populations of neural crest cells from levels posterior to somite 6, and found that these cells give rise not only to sympathetic ganglia (with the exception of the SCG) and adrenomedullary cells, but also to a significant proportion of the foregut enteric nervous system. The existence of a close relationship between sympathoadrenal and enteric precursors was previously suggested by Carnahan et al. (1991), based on previous observations that migrating enteric neuroblasts transiently express several markers also found in the sympathetic and adrenal precursors (Gershon *et al.*, 1984; Baetge and Gershon, 1989; Baetge *et al.*, 1990). Thus, because it is also capable of generating a subset of enteric ganglia, the putative sympathoadrenal progenitor could be considered to be, at least, a tripotent intermediate progenitor.

Another finding of extreme interest is that the murine SCG, along with the enteric nervous system at midgut and hindgut levels, are formed from crest cells of somitic levels 1–5 of the axis, and they degenerate in the absence of ret receptor activity, whereas the sympathetic ganglia posterior to somite 6, the chromaffin cells, and the foregut enteric ganglia do not depend upon *ret* gene function (Schuchardt *et al.*, 1994; Durbec *et al.*, 1996a). These results suggest that not all the enteric neurons are lineally related to the sympathoadrenal cells (see below). Furthermore, they stress the involvement of axial-specific cues in the segregation of neural crest-derived cells. Along this line, it was never precisely determined whether, in the avian embryo, the location of progenitors with postulated sympathoadrenal characteristics is restricted to somitic levels 18–24 of the axis, that give rise to the adrenomedullary cells, or whether progenitors coexpressing sympathetic and chromaffin-specific markers extend along the entire sympathetic chain including the SCG.

*5.5.2.1   The MASH-1 transcription factor.* A most productive approach for investigating the mechanisms that underlie the development of the nervous system in higher vertebrates consists in searching for putative developmental control genes based on their homology with invertebrate systems, where a function for these gene products has already been established. In *Drosophila*, the *achaete-scute* complex of basic helix–loop–helix transcription factors was found to be necessary for the determination of particular neuronal subsets (reviewed by Ghysen and Dambly-Chaudière, 1988). With the idea of identifying homologous genes involved in the determination of the sympathoadrenal lineage in mammals, Johnson *et al.* (1990) have performed a PCR amplification with *achaete-scute* gene degenerate primers, using as a template cDNA from the MAH cell line (Birren and Anderson, 1990; see previous sections). This search led to the cloning of the rat *achaete-scute* homologue MASH-1.

Subsequent antibody staining and *in situ* hybridization revealed that *Mash-1* is transiently expressed in neural crest cells after they colonize the anlage of the sympathetic ganglia. Double labeling immunocytochemistry with MASH-1 antibodies and either TH or SCG10 antibodies revealed that *Mash-1* expression precedes that of the other markers. MASH-1 was also detected in the adrenal medulla, in subsets of enteric precursors, and in parasympathetic neurons located in the mesenchyme surrounding major blood vessels. By contrast, *Mash-1* expression was not detected in sensory ganglia, Schwann cells, or putative melanoblasts, suggesting that, in the PNS, this transcription factor is a specific marker for autonomic progenitors (Lo *et al.*, 1991b; Guillemot and Joyner, 1993). A similar spatiotemporal pattern of expression was reported for *Cash-1*, the avian homologue of *Mash-1*, during development of the sympathetic ganglia in chick embryos (Jasoni *et al.*, 1994; Ernsberger *et al.*, 1995).

These interesting localization patterns raised the question as to whether *Mash-1* is involved either in determining the segregation of a putative sympathoadrenal lineage from other neural crest derivatives, or is acting at later stages of the cascade leading to differentiation of neuronal and/or endocrine derivatives. It was therefore of interest to establish what is the state of commitment of the MASH-1-positive precursors, and if they are restricted to become neurons or have the ability to develop both into neuronal as well as glial cell types. To answer these questions, Guillemot *et al.* (1993) generated mutant mice carrying a deletion in the *Mash-1* gene. *Mash*$^{-/-}$ mice revealed no impairment in the homing of neural crest cells to para-aortic sites, but showed severe deficits in subsequent development of the sympathetic ganglia. A lack of staining for multiple markers (TH, SCG10) was observed, suggesting that lack of MASH-1 blocked cell differentiation, or, alternatively, that the cells were actually missing in the mutants. In contrast to neurons, ganglionic glial cells were apparently unaffected as determined by expression of the glial marker F-spondin. These results raise the question of whether *Mash-1* is expressed by precursors able to develop both into neurons and glia but is important only for neurogenesis, or, alternatively, whether *Mash-1* is expressed by already committed neuroblasts.

In the gut, MASH-1 is expressed by a subset of transient catecholaminergic cells. Mutation in the *Mash-1* locus, which eliminates sympathetic neurons, was also shown to prevent the development of enteric serotoninergic neurons. In contrast, another identified population of enteric neurons that expresses CGRP developed normally (Blaugrund *et al.*, 1996). This suggests that *Mash-1* is required for the development (differentiation or survival) of the population of catecholaminergic progenitors which later on gives rise to serotoninergic neurons in the gut, but not for lineage determination, as originally postulated (Anderson, 1993b).

In further support of such an interpretation, it was found that the adrenomedullary cells thought to arise from MASH-1-positive sympathoadrenal precursors were not affected by the lack of this factor. The adrenal medullae of mutant mice were only slightly smaller than the normal wild-type organs, and their component chromaffin cells exhibited PNMT activity. Two alternative possibilities might account for the results observed. First, another source of chromaffin precursors never expressing *Mash-1* exists in the embryo. If this were true, then only subsets of chromaffin cells could be postulated to derive from a common sympathoadrenal precursor. An alternative view could be that chromaffin cells develop entirely from a separate progenitor sublineage, a feature that does not necessarily disclaim the ability of chromaffin cells to transdifferentiate into neurons. If *Mash-1* were indeed critical for neurogenesis, then chromaffin cells from *Mash-1* null mutants would be expected to fail to convert into sympathetic neurons even if given the appropriate conditions (CNTF, FGF, and NGF), in contrast to the observed conversion of chromaffin cells to neurons that can be triggered in normal chromaffin cells (see below).

*5.5.2.2 The conversion of chromaffin cells to sympathetic neurons.* A second line of evidence that led to the postulation of the existence of a restricted sympathoadrenal precursor is that chromaffin cells and SIF cells have the ability to interconvert phenotypically into sympathetic neurons if supplied with appropriate conditions. A striking example of chromaffin to neuronal conversion is back-transplantation of embryonic adrenal cells from quail embryos into the migratory pathways of neural crest cells of chick hosts, which yields neurons in the sympathetic ganglia (Vogel and Weston, 1990). Even more remarkable, grafting of postnatal adrenal medullary tissue into the anterior chamber of the eye produces neurite outgrowth from the grafts (Olson, 1970). Also, treatment of dissociated chromaffin cells with NGF has a similar effect (Unsicker *et al.*, 1987). Moreover, injections of NGF into rat embryos results in a replacement of chromaffin cells by neurons (Aloe and Levi-Montalcini, 1979). NGF also promotes neurite outgrowth from the PC12 cell line, a tumor derived from rat adrenomedullary cells (Greene and Tischler, 1976). Where are the SIF cells localized in this transition? While converting into neurons, chromaffin cells transiently adopted SIF cell-like characteristics (Doupe *et al.*, 1985a,b). Taken together, the ability of chromaffin cells and SIF cells to convert into sympathetic neurons are excellent examples of phenotypic plasticity.

*5.5.2.3 The reverse conversion from a sympathetic neuron to a chromaffin cell or SIF cell is not attainable.* Although sympathetic neurons can switch neurotransmitter phenotype, they are unable to convert into either SIF or adrenomedullary cells, implying that commitment to a neuronal fate is a relatively irreversible process. Therefore, the chromaffin phenotype may represent a more primitive or even a constitutive pathway of differentiation. This idea is consistent with recent results showing that formation of chromaffin cells (though not adrenaline biosynthesis) is practically independent of glucocorticoid signalling (Section 5.4.4).

The immortalized MAH cell line, obtained from rat adrenal glands (Section 5.4.3.2) was considered to represent a restricted sympathoadrenal progenitor. However, while these cells are able to convert into neurons under appropriate conditions, treatment with glucocorticoids fails to induce them to express PNMT (Birren and Anderson, 1990). These observations suggest several possibilities. First, that this potentiality was abolished by immortalization. Second, that the MAH cell derives from a population of progenitors with noradrenergic properties present in the adrenal gland environment that would lack the ability to develop into PNMT and adrenaline-producing chromaffin cells. This possibility is strengthened by the fact that, in normal adrenal glands of rat and mice, about 20% of the chromaffin cells never acquire PNMT expression (see, e.g., Verhofstad *et al.*, 1989). If MAH cells indeed belong to this subpopulation, then one can assume that more than one type of adrenal precursors do exist and account for the phenotypic heterogeneity observed within the adrenal medulla.

## 5.6 Regulation of the survival of sympathetic and parasympathetic neurons

### 5.6.1 Differential roles of neurotrophins

Whereas recent evidence has established a requirement for NT-3, FGF, and CNTF/LIF during the early stages of sympathetic neuron development, it has been known for a long time that differentiated sympathetic neurons in culture respond to NGF. The first unambiguous demonstration for a physiological role of endogenous NGF arose from experiments in which anti-NGF antibodies were injected into early postnatal mice where they caused the virtual elimination of the paravertebral sympathetic chains (reviewed by Levi-Montalcini, 1987; Rohrer et al., 1988b). Conversely, injection of pharmacological doses of NGF into avian embryos around the period of programmed death of sympathetic neurons was shown to rescue a significant amount of cells, suggesting that this molecule has a survival-promoting activity in vivo, and that the amounts of NGF are limiting (reviewed in Oppenheim et al., 1992a). Further substantiation for a role of NGF as a target-derived factor in the sympathetic nervous system is the demonstration that both NGF mRNA and protein are expressed in the targets of sympathetic neurons, and that their levels are correlated with the density of sympathetic innervation (Korsching and Thoenen, 1983a; Heumann et al., 1984; Shelton and Reichardt, 1984). In addition, interruption of retrograde transport of NGF has the same effect as antibody treatment, suggesting that the physiologically relevant factor has to be transported from the target tissues to the neuronal soma (Hendry et al., 1974; Johnson et al., 1978). Full support for the role of NGF in survival of sympathetic neurons was obtained in mice carrying deletions in the NGF (Crowley et al., 1994) or the trkA genes (Smeyne et al., 1994), whose phenotypes closely resemble that of animals deprived of NGF (Table 4.3). In both types of mutants, the number of pycnotic cells is increased in the perinatal period corresponding to the peak of programmed cell death, a process that accounts for the virtual absence of sympathetic chains and of the SCG apparent at somewhat later stages.

Is the response to NGF uniquely transduced by the trkA receptor? As already discussed, NGF also binds to the p75 low-affinity receptor which acts in some systems by modulating NGF binding to trkA (Barker and Shooter, 1994). To investigate the possible involvement of p75, mutant mice carrying a deletion in the p75 gene were created. Analysis of the embryos homozygous for the mutation revealed no apparent change in neuronal numbers, suggesting either that p75 is not required for neurotrophin-mediated survival or that it is involved in other aspects of neurotrophin signaling. To test this idea, SCG from postnatal day 3 mice devoid of the p75 gene were cultured in the presence of NGF. A 2–3-fold decreased sensitivity to the exogenous factor was apparent in the mutants when compared to wild-type mice, suggesting that p75 gene modulates neurotrophin sensitivity in SCG neurons (Lee et al., 1994).

Like NGF, NT-3 promotes the survival of postmitotic embryonic and neo-natal sympathetic neuronal populations grown in culture, but higher concen-trations of NT-3 are required compared with NGF (Dechant *et al.*, 1993; Lee *et al.*, 1994). In agreement with the studies performed *in vitro*, only half of the normal number of sympathetic neurons are present postnatally in the SCG of NT-3$^{-/-}$ mice compared to wild-types (Table 4.3). Since NT-3 is known to affect the survival of sympathetic neuroblasts (see Section 5.4.3.1), the amount of cell loss detected postnatally may result from its early effects. During the period of programmed cell death, NT-3 was suggested instead to participate in neuritic branching in specific targets (Elshamy *et al.*, 1996). In contrast to NGF and NT-3, neither BDNF nor NT-4/5 promote the survival of cultured sympa-thetic neurons. Accordingly, no differences in neuronal number or size were detected in sympathetic ganglia of BDNF$^{-/-}$ or trkB$^{-/-}$ mice (Tables 4.2 and 4.3).

### 5.6.2  *Ciliary neurons and CNTF*

Ciliary neurons are derived from the mesencephalic neural crest (Le Douarin and Teillet, 1974; Landmesser and Pilar, 1974). In the chick embryo, the pro-liferative phase of ciliary ganglion progenitors is completed by E5, and target innervation by the ciliary neurons occurs by E6. Between E8 and E12, about 50% of the ciliary neurons degenerate. Removal of the eye before target inner-vation results in over 90% neuronal death. Conversely, transplantation of an additional eye was shown to reduce significantly the extent of neuronal death (Landmesser and Pilar, 1974; Narayanan and Narayanan, 1978b). In addition, interruption of one of the three nerve branches from the ciliary ganglion to the eye results in increased survival of neurons projecting in the remaining two branches (Pilar *et al.*, 1980). Altogether, these results are consistent with a role for the target in survival of ciliary neurons.

Cultured ciliary neurons served as a model system to screen for survival activities. Ciliary neurons were found to survive when cocultured with chick skeletal muscle cells (Nishi and Berg, 1977), rat eye tissue (Ebendal *et al.*, 1980), and sciatic nerve fragments (Richardson and Ebendal, 1982). The eye was found to be the richest source of survival activity, which was given the name of ciliary neurotrophic factor (CNTF). CNTF was partially purified by Varon and colleagues (Barbin *et al.*, 1984), and the cDNAs for rabbit, rat and human CNTF were cloned a few years later (reviewed in Sendtner *et al.*, 1994). CNTF was found to rescue many neuronal types from cell death *in vitro* in addition to ciliary neurons, including sensory, sympathetic, and motor neurons (Table 4.2). To assess whether CNTF is active *in vivo* during programmed cell death, the factor was administered to chick embryos. Whereas a significant rescue of spinal motor neurons was obtained, none of the other putative neuronal targets was affected by CNTF treatment (Oppenheim *et al.*, 1991). Surprisingly, mice bearing null mutations in the CNTF gene did not reveal any apparent abnormality during embryonic life, but atrophy and loss of motor neurons could be observed after birth (Masu *et al.*, 1993). These results suggest that, unlike the situation *in vitro*, CNTF does not play any crucial role in survival of the peripheral neurons *in vivo*. Alternatively, the lack of CNTF is fully compensated by other factors.

## 5.7   The enteric nervous system

The enteric nervous system is composed of neurons and glia aggregated into ganglia connected by nerve plexuses that innervate the bowel along its entire length. It contains a vast number of neurons and is endowed with an extremely complex network of neurons that act under the regulation of central inputs, but are also able to mediate reflex activity in their absence. These features place the enteric nervous system as a separate entity of the ANS that is distinct from the sympathetic and parasympathetic divisions.

The embryonic source of the enteric neurons and glia from the neural crest, the axial levels of origin, and the migratory routes followed by the enteric progenitor cells as well as the relationship with other components of the ANS have been presented earlier in this chapter. In this section, we will discuss the progress made toward elucidating the molecular basis of cell differentiation during normal development, and its implications in the understanding of the etiology of the megacolon phenotype, known as Hirschsprung's disease in humans.

### 5.7.1   Factors affecting the differentiation of enteric neurons: neurotrophin-3

NT-3, a neurotrophin with multiple activities on proliferation of multipotent neural crest cells and differentiation of sensory and sympathetic progenitors (reviewed in Kalcheim, 1996), was also found to influence the development of neural crest-derived cells in the gut. A system of enriched neural crest-derived cells was prepared from E14 fetal rat gut fragments by immunoselection with the NC1/HNK-1 antibody. Treatment of these cells with NT-3 was shown to stimulate a concentration-dependent increase in the proportion of both neurons and glial cells developing in the cultures (Chalazonitis *et al.*, 1994). Although NT-3 treatment did not affect the proliferation of these cells, the question of whether NT-3 acts on progenitor cell differentiation or survival remains open. In addition, NT-3 was found to stimulate neurite outgrowth of the immunoselected rat cells, as well as of cells that constitute the Remak ganglion in avian embryos (Ernfors *et al.*, 1990).

These *in vitro* results may reflect a paracrine interaction between NT-3 and trkC, as NT-3 mRNA can be found in the gut mesenchyme (Kahane and Kalcheim, unpublished data), while trkC mRNA is present along the whole gut in neural crest-derived cells (Chalazonitis *et al.*, 1994). Furthermore, NT-3 was found to be the unique neurotrophin with activity in this system, since neither NGF, BDNF, nor NT-4/5 had any significant effect on neuronal numbers. It is relevant in this context to stress that BDNF transcripts as well as BDNF-immunoreactive protein were found to be expressed in the rostral portion of the foregut mesenchyme in chick embryos (Fig. 5.7; Brill, Kahane and Kalcheim, unpublished data), suggesting that BDNF may exert a restricted effect on enteric progenitors that colonize the esophageal anlage. This possibility remains to be tested.

Surprisingly, analysis of mice bearing a deletion in the NT-3 gene revealed no apparent defect in gut innervation (Fariñas *et al.*, 1994). Although ruling out a generalized effect on the development of the enteric innervation, these data leave open the possibility of impaired development of specific cell types within a system characterized by such a rich variety of neuronal phenotypes. Likewise, it would be important to determine whether *in vitro* NT-3 acts as a general growth factor for all types of enteric neurons and glia, or only affects specific cellular subsets.

### 5.7.2 The molecular basis of the megacolon phenotype: the lethal spotted and piebald lethal mouse mutants as models for aneural hindgut

Aganglionic megacolon is a congenital disease of mammals which can be either sporadic or familial in humans and is characterized by the absence of enteric ganglion cells. In man, it is known as Hirschsprung's disease and affects 1 in 5000 live births. It is characterized by the absence of enteric ganglia in the rectum and in a variable length of the distal colon. In some cases the histopathological abnormalities also involve the proximal colon and even the distal small intestine (Passarge, 1973; Meier-Ruge, 1974). Total gut aganglionosis associated with sensorineural deafness and pigmentary defects has also been described (Hofstra *et al.*, 1996). Absence of ganglia in the bowel could be the result of a primary defect in the neural crest cells that give rise to these ganglia, either in their migration to the hindgut, or in their survival following an initially normal colonization of the bowel. Alternatively, a local abnormality in the splanchnopleural mesenchyme that encloses the gut epithelium and that serves the crest cells as substrate for migration could as well impair normal migration of the crest progenitors and colonization of the different segments of the gut. It is, of course, possible that both mechanisms may be operative, either simultaneously or under different sets of circumstances.

Animal models revealing a phenotype of aneural megacolon have been extensively exploited to investigate the etiology of this disease and to discriminate between the above possibilities. For example, *lethal spotted* mice (*ls/ls*) carry an autosomal recessive mutation associated with colonic aganglionosis (Lane, 1966). Rothman *et al.* (1993a), constructed interspecies chimeras in which fragments of hindgut derived from either normal or *ls/ls* mice were grafted between the neural tube and somites of quail embryos. Whereas the host quail cells successfully migrated into the grafted bowel that derived from

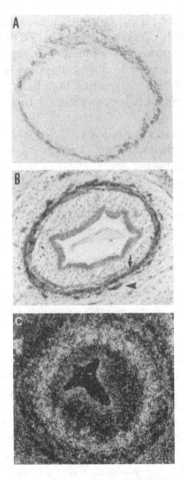

**Figure 5.7** Expression of BDNF mRNA and protein in the smooth muscle of the developing avian foregut. Transverse sections through a 7-day-old quail embryo at the level of the foregut, showing: (A) the expression of BDNF immunoreactivity in the enteric smooth muscle; (B) the expression of BDNF immunoreactivity (arrow) combined with HNK-1 immunolabeling of enteric neural crest-derived cells (arrowhead); (C) the corresponding localization of BDNF mRNA to the splanchnopleural-derived gut muscle. (From Brill, Kahane, and Kalcheim, unpublished).

normal mice, they failed to enter the bowel fragments derived from the *ls/ls* mutants. In agreement with the *in situ* results, *in vitro* studies showed that primary cultures of hindgut from *ls/ls* embryos would not support colonization and differentiation by wild-type or mutant neuroblasts, in contrast to similar cultures of wild-type hindgut that would support colonization by either type of neuroblasts (Jacobs-Cohen *et al.*, 1987). These results suggested that the aganglionic megacolon phenotype in *ls/ls* mutants is due to a primary defect in the mesenchyme. This conclusion was additionally supported by the observation that crest-derived cells with an *ls/ls* genotype give rise to enteric neurons, even in the terminal colon of *ls/ls* C3H chimeric mice, as long as 5% of the non-neuronal cells were of C3H origin.

Further support for a possible defect in the environment of the embryonic gut was provided in a study by Kapur *et al.* (1992, 1995). These authors generated transgenic mice bearing a DBH promoter that in the gut drives the expression of the reporter gene product β-galactosidase to a subpopulation of enteric neurons that express serotonin. Breeding these transgenic animals with mice bearing the *lethal spotted* allele enabled the establishment of a line of mice homozygous at both the *ls* and the transgene loci. The pattern of trans-gene expression in *ls/ls* mice was identical to wild-types and heterozygotes until E12.5. At that stage, cells expressing the transgene extended into the proximal colon of wild-type embryos. In striking contrast, expression of the transgene in the *ls/ls* embryos ended abruptly at the ileocecal valve, and the subsequent colonization of the proximal and midcolon was retarded and irregular. Moreover, the terminal portion of the gut was never colonized. These findings imply that the aneural colon phenotype arises as a consequence of an impair-ment in *de novo* gut colonization rather than as a consequence of cell death following an initial normal colonization of the gut. Since the initial stages of migration of crest cells appear, nevertheless, normal in the mutant embryos, Kapur and colleagues proposed that impaired neuroblast migration to the distal portion of the gut results from a local defect in the mesenchyme sur-rounding the gut epithelium rather than from a primary defect in the migrating cells.

What molecules are responsible for impaired hindgut colonization in the *ls/ls* mutants? Gershon and colleagues have documented an overabundance of extracellular matrix (ECM) components, in particular laminin, in the agangli-onic guts of *ls/ls* mice (Tennyson *et al.*, 1986; Payette *et al.*, 1988; Rothman *et al.*, 1996). Similar increases have been documented in humans with Hirschsprung's disease (Parikh *et al.*, 1992). Neural crest-derived cells in the gut were shown to express a 110-kDa laminin-binding protein (Pomeranz *et al.*, 1991b; Tennyson *et al.*, 1991), and to respond to laminin by enhanced neuronal differentiation *in vitro* (Pomeranz *et al.*, 1993; Chalazonitis *et al.*, 1997). Based on these results, it was proposed that a regional overproduction of laminin in the hindgut of *ls/ls* mutants would signal neural crest progenitors to stop their migration prior to complete colonization of the gut and to differentiate pre-maturely into neurons. The homeobox-containing gene *Hox1.4* was also impli-cated in the development of the megacolon phenotype. Mice in which a transgene bearing *Hox1.4* was overexpressed revealed an inherited abnormal megacolon phenotype (Wolgemuth *et al.*, 1989). This phenotype was correlated with an increased level of expression of *Hox1.4* transcripts in the gut mesen-chyme of the transgenic animals when compared to wild-type embryos. Though these data indicate as well that defects in the mesenchyme may lead to lack of enteric innervation, it is unknown at present whether abnormal levels and activity of this gene exist in the *ls/ls* mutants.

In addition to the *lethal spotted* mutation, another autosomal recessive mutation was found that produces aganglionosis in mice and was named *piebald lethal (sl)* (Lane and Liu, 1984). A striking similarity was reported to exist between these two types of mutations in both the dynamics and the pattern of gut colonization (compare for details Kapur *et al.,* 1992 and 1995).

### 5.7.3    The endothelin-3 peptide and the endothelin-B receptor are encoded respectively by the genes mutated in the lethal spotted *and* piebald lethal *mice*

A breakthrough in our understanding of the molecular basis of these mutations came recently, when it was found that the *lethal spotted* and the *piebald lethal* loci encode for the endothelin-3 (EDN3) peptide, and for its receptor (endothelin receptor B, EDNRB) (Greenstein-Baynash *et al.,* 1994; Hosoda *et al.,* 1994), respectively. The *lethal spotted* allele is a mutation that prevents conversion of the inactive precursor of EDN3 to its active form, and thereby leads to undetectable levels of functional peptide in the mutants. The *piebald lethal* mutation represents a complete deletion in the EDNRB gene (for discussion on EDN3 and EDNRB see Chapter 6).

To further understand the exact role mediated by interactions between EDN3 and its receptor, it is necessary to determine which cell types synthesize the two molecules. If the primary defect in the *ls/ls* mutants is indeed in the mesenchyme and not in the neural crest cells as proposed by grafting and *in vitro* data, then EDN3 would be expected to be expressed in the mesenchyme and the receptor in those neural crest-derived cells affected in the mutants. As far as the receptor is concerned, Nataf *et al.* (1996) have shown that most avian neural crest cells express receptor transcripts before and at the onset of their migration at all axial levels. In the gut, EDNRB mRNA signal was restricted to the neural crest cells as they progressively colonize the entire bowel, and no message was detected in the gut mesenchyme or endoderm. These results suggest that, in the avian gut, ENDRB acts in a cell autonomous manner. In contrast to avians, Inagaki *et al.* (1991) reported the expression of both EDN1 (but not EDN3) and EDNRB to enteric neuroblasts, indicative of autocrine- or paracrine-type interactions between the peptide and its receptor in neural crest-derived cells. In addition, EDNRB was also found to be expressed by smooth muscle cells from postnatal guinea pig cecum, suggesting that these molecules could also mediate neuroblast–mesenchymal interactions (Okabe *et al.,* 1995) were the receptor also to be found in the embryonic rodent mesenchyme at the time of neural crest migration. More recent data confirmed this view as EDN3 was found to be expressed in the gut mesenchyme of mice and its receptor in both neural crest cells along the axis and in the mesenchyme of the gut, regardless of the presence or absence of neural crest cells (Gershon, personal communication). These results suggest that, in mammals, the cell interactions mediated by these two molecules may be more complicated than originally assumed.

Based on the distinct expression patterns observed at least for the receptor between avians and rodents, one can assume the existence of species differences in the activity of the EDN3/EDNRB complex. Recent results report a dual activity for EDN3 on cultured mouse cells. On the one hand, this peptide was found to regulate the development of the smooth muscle of the bowel, assessed by the expression of actin. By doing so, the levels of laminin were found to decrease. On the other hand, EDN3 was found to directly inhibit the *in vitro* differentiation of neural crest cells into neurons. It would be interesting to examine whether EDN3 might act as a mitogen for murine neural crest cells, as it was found to be for the avian counterparts (Lahav *et al.*, 1996), thereby inhibiting neuronal differentiation. Based on the previously proposed role for laminin in precocious neuronal differentiation in the *ls/ls* mutants, these results can be reconciliated in a model in which the lack of EDN3 stimulates early neuronal differentiation prior to complete colonization of the gut, either by a direct effect on the crest progenitors or by an indirect effect on the overproduction of laminin which in turn causes early neurogenesis in inappropriate locations. Precocious differentiation would then lead to a lack of migration into the hindgut (Gershon, personal communication). Further experiments need to be done to discriminate between these possibilities. The exact cellular functions of EDN3 and EDNRB in normal development, and their possible interactions with other active factors, including laminin, *Hox1.4*, genes that regulate pigment development (i.e., c-*kit* and the *Steel factor*, etc.), GDNF, and its receptors (see below), remain to be clarified.

### 5.7.4 The ret receptor and glial cell line-derived neurotrophic factor (GDNF)

*5.7.4.1 The ret gene product is associated with Hirschsprung's disease and affects development of the enteric innervation.* The *ret* proto-oncogene encodes a member of the receptor tyrosine kinase superfamily. This proto-oncogene was originally cloned because of the transforming activity in *in vitro* and *in vivo* of forms of the gene in which the extracellular domain was missing, thus leading to a constitutive activation of the kinase domain of the molecule (Takahashi *et al.*, 1985; Santoro *et al.*, 1990). Most importantly, somatic mutations of *ret* have been associated with a large proportion of thyroid papillary carcinomas, while germline mutations at the *ret* locus were identified in patients with multiple endocrine neoplasias (MEN) of types 2A and 2B. The 2A type is characterized by the development of medullary thyroid carcinomas and pheochromocytomas, and type 2B also presents neuromas of the gastrointestinal tract. Moreover, a variety of mutations in the coding region of the *ret* gene were identified in patients with congenital Hirschsprung's disease (reviewed by Schuchardt *et al.*, 1995a). These findings clearly document that the c-ret gene product is responsible for a series of diseases that involve neural crest derivatives, and also suggest the possibility that the *ret* gene is involved in their normal development.

To study the role of *ret* in normal development, the mouse, human, and chicken genes were cloned. Sequence analysis revealed an evolutionary conservation of the molecule among species. For example, while the cytoplasmic domains of human and chicken c-*ret* are 91% similar, the extracellular domains are more divergent (68% homology). The human, mouse, and chicken c-*ret* all encode two protein isoforms which differ only in the c-terminal located in the cytoplasmic domain, but their possible differential patterns of expression and function have not yet been elucidated. At the mRNA level, *ret* is expressed at specific sites of the nervous and excretory systems (see Pachnis *et al.*, 1993; Robertson and Mason, 1995; Schuchardt *et al.*, 1995b). Concerning the enteric nervous system, *ret* mRNA is localized in the vagal neural tube region prior to the onset of neural crest migration. It is striking to see that in the chick the c-*ret*-positive neural fold cells correspond to the precise region of the neural axis from which most of the enteric ganglia are derived (i.e., somites 1–7; Le Douarin and Teillet, 1973). Later, the *ret* message is expressed in the migratory neural crest cells on their way to colonize the foregut and the SCG (see Robertson and Mason, 1995; Durbec *et al.*, 1996a), and at later stages in neural crest-derived neurons along the entire gut. Consistent with these expression patterns, mice homozygous for a mutation in the *ret* gene showed severe deficiencies in the enteric nervous system (mainly in the innervation of the midgut and hindgut), and renal agenesis or dysgenesis (Schuchardt *et al.*, 1994; Durbec *et al.*, 1996a; see also Table 5.1).

*5.7.4.2   GDNF is required for normal innervation of the enteric nervous system.* GDNF was initially identified in culture supernatants of a glial cell line by virtue of its ability to support *in vitro* the survival of midbrain dopaminergic neurons (Lin *et al.*, 1993). This discovery led to its subsequent cloning, expression, and initial functional characterization (Lin *et al.*, 1993, 1994). Further studies have shown that GDNF is also a potent survival factor for embryonic spinal motor, cranial sensory, sympathetic, and hindbrain noradrenergic neurons (see review by Unsicker, 1996, and references therein). To test for its physiological relevance *in vivo*, the expression pattern of GDNF was studied by *in situ* hybridization (Hellmich *et al.*, 1996; Suvanto *et al.*, 1996). Of relevance to enteric nervous system development was the localization of GDNF transcripts to the mesenchyme surrounding the gut endoderm and later to the muscle layers of the intestine. Demonstration of a role for GDNF was provided by three independent research groups which deleted the gene coding for GDNF and found, among various phenotypes, a lack of enteric innervation in the mutant mice (Fig. 5.8; see Moore *et al.*, 1996; Pichel *et al.*, 1996; Sanchez *et al.*, 1996). Although an initial colonization of the esophagus and stomach occurred in the mutants, these cells disappeared later by the time of birth (Sanchez *et al.*, 1996), consistent with a role of GDNF in cell survival. In contrast, no innervation of more distal regions of the gut could be detected at any stage of embryogenesis, suggestive of a primary defect in gut colonization (Table 5.1). More recent *in vitro* studies on crest-derived cells immunoselected from fetal rat guts revealed both mitogenic and differentiative

Table 5.1. *Comparison between the expression patterns and functions of GDNF and Ret*[a]

| | GDNF | Ret |
|---|---|---|
| Molecular identity | Secreted factor; belongs to the TGF-$\beta$ superfamily; binds to GDNFRα and to c-Ret | Encodes a tyrosine kinase receptor; binds GDNF (or a complex between GDNF and GDNFRα) |
| Sites of expression | Ventral midbrain, striatum, spinal cord, cerebellum, Schwann cells, skeletal muscle, skin, testes, heart, eye Gut mesenchyme (and muscle layers) throughout the length of the GI tract Metanephric mesenchyme | Autonomic precursors from migratory stages; sensory progenitors of cranial and dorsal root ganglia; ventral neural tube and later in motor neurons; neural retina Vagal neural tube prior to the onset of neural crest migration; neural crest cells at vagal levels of the axis and enteric ganglia |
| Phenotypes of knockout mice | Decreased neuronal numbers in petrosal nodose, DRG, SCG | Intermediate mesoderm and later the ureteric bud SCG forms normally by E10.5 but disappears by E12.5; SG and adrenal medulla are normal |
| | Lack of enteric innervation | Lack of enteric innervation. Esophagus and stomach only partially affected |
| | Renal agenesis or dysgenesis | Renal agenesis or dysgenesis |
| Possible roles during development | Proliferation, migration, differentiation, survival | Proliferation, migration, differentiation, survival |

*Note:*

[a] References to the literature are given in the text.

*Abbreviations:* DRG, dorsal root ganglion; GI, gastrointestinal tract; SCG, superior cervical ganglion.

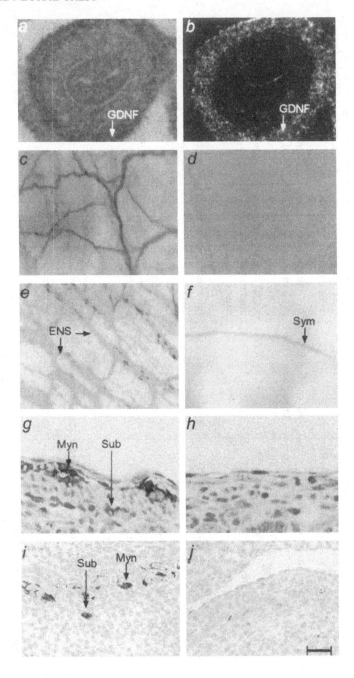

**Figure 5.8** The enteric nervous system in wild-type and GDNF$^{-/-}$ mice. (a, b) Expression of GDNF mRNA in E13.5 mouse gut as detected by *in situ* hybridization (arrows). (c–j) Whole-mounts of small intestine from P0 wild-type (c, e, g, i) and GDNF null mutants (d, f, h, j) stained with antibodies to neuron-specific enolase (c, d) showing enteric (ENS) neurons in the wild-type and only sympathetic (Sym) afferents in the

effects of GDNF at successive stages of ontogeny (Chalazonitis *et al.*, 1997; Gershon, 1997). Taken together, the severe aganglionosis observed along both the midgut and hindgut, and the results of *in vitro* work, are consistent with multiple effects of GDNF on enteric progenitors. Moreover, based on the relative severity of the GDNF/c-ret mutations when compared to the EDN3 mutation, one can postulate that, at least in mammals, GDNF acts earlier in development on a less-restricted population of target crest cells (see Gershon, 1997; see also Fig. 5.9).

*5.7.4.3  GDNF is a ligand for the* ret *tyrosine kinase receptor.* The similarity between the phenotypes observed in the mutant embryos lacking either *ret* or *GDNF* gene activity and the complementary patterns of expression of *ret* to the neural crest cells and their derivatives, and of GDNF to the mesenchyme adjacent to the *ret*-expressing cells, led to the hypothesis that the ret locus may encode a functional receptor for GDNF. Using biochemical, cellular, and molecular approaches, several independent studies have confirmed that a functional interaction exists between GDNF and the ret receptor (Durbec *et al.*, 1996b; Treanor *et al.*, 1996; Trupp *et al.*, 1996).

In addition, Treanor *et al.* (1996) have also shown that high-affinity binding of GDNF to ret and subsequent phosphorylation of tyrosine residues, requires an interaction between GDNF and a novel receptor protein that was subsequently cloned and called GFR-α. The formation of a physical complex between the three molecules was demonstrated in a transfection system. Consistent with a possible role as a binding protein, GFR-α mRNA was shown to be expressed in GDNF-responsive tissues. Establishment of a line of mice lacking GFR-α gene function should provide further evidence for its function in embryonic development.

### 5.7.5  Conclusions

In summary, the availability and strength of genetic models and molecular techniques have enabled the identification of several ligands and receptors that mediate the development of the enteric innervation, and may be responsible for aneural bowel phenotypes. Together with efforts aimed at finding new molecules with potential activity in the enteric system, the next stage in this research should concentrate, firstly, on clarifying the exact roles of these molecules, secondly, on studying the interactions between these families of ligands and receptors to determine whether these represent parallel mechanisms, or,

**Figure 5.8** (*cont.*)
mutant, neurofilament proteins (e, f) and peripherin (g, h). Small intestine of E13.5 wild-type (i) and GDNF$^{-/-}$ (j) embryos stained with peripherin antibodies. Myenteric (Myn) and submocosal (Sub) neurons, shown by arrows in the wild-type, are absent in the mutant mice. Bar = a,b: 65 μm; c–j: 5 μm. (Reprinted, in modified form, with permission, from Moore *et al.*, Copyright (1996) Macmillan Magazines Limited.)

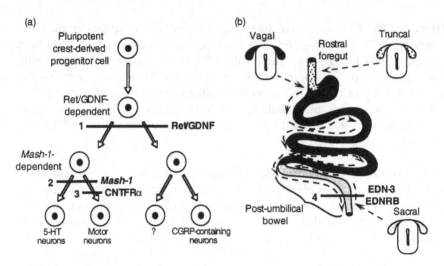

**Figure 5.9** The origin of gut innervation and the factors affecting the development of murine enteric cells. (a) Schematic view of the factors affecting the development of enteric lineages and the location of their putative effects along the neural crest developmental hierarchy. Scheme based mostly on results of targeted mutations of the respective genes. (b) Levels of colonization of the embryonic gut by neural crest cells from vagal, truncal, and sacral areas of the neuraxis. See text for details. (Reproduced, with permission of Current Biology Ltd, in modified form, from Gershon, 1997.)

alternatively, whether they all integrate into a sequential cascade whose hierarchy should then be established (see Gershon, 1997; Fig. 5.9). Thirdly, it will be critical to establish the validity of these results to the human disease by determining whether some or all the molecules in question are involved in the etiology of Hirschsprung's disease.

## 5.8 Developmental relationships between the neural crest and endocrine cells

### 5.8.1 Historical overview

As mentioned above, the origin of the adrenal medulla from the neural crest has long been recognized (see Strudel, 1953).

In the 1970s, a great deal of attention was devoted to the study of the production of polypeptide hormones by many cells either contained in well-recognized endocrine glands such as the pancreas or adenohypophysis, or dispersed in various epithelia. In particular, the gut, to which only digestive functions had been previously assigned, became fully recognized as being also an endocrine organ. Moreover, a number of these secreted polypeptides were found in nerve cells of both the CNS and PNS where they act as neuromediators or neuromodulators (see reviews by Guillemin, 1978a,b). This led to

Table 5.2. *Cytochemical characteristics of polypeptide hormone-secreting cells of the APUD series*

1. Fluorogenic amine content (catecholamine, 5-HT or other): (a) primary; (b) secondary uptake
2. Amine precursor uptake (5-HTP, DOPA)
3. Amino acid decarboxylase (AADC)
4. High side-chain carboxyl or carboxyamide (masked metachromasia)
5. High non-specific esterase or cholinesterase (or both)
6. High $\alpha$-GPD ($\alpha$-glycerophosphate menadione reductase)
7. Specific immunofluorescence

*Abbreviations*: 5-HT, 5-hydroxytryptamine; 5-HTP, 5-hydroxytryptophan; DOPA, 3,4-dihydroxyphenylalanine; $\alpha$-GPD, $\alpha$-glycerophosphate dehydrogenase.
*Source*: Pearse (1969).

the idea that peptide-secreting cells should be considered as being developmentally related to the neural primordium in the same way as are the adrenal medulla and the neurohypophysis.

Historically, the work of Feyrter is worth mentioning. In 1938, he described a system of clear cells (*helle Zellen*) dispersed in the gut epithelium and in various other parts of the body. He considered that these cells constituted a diffuse endocrine organ, acted on their immediate neighbors, and were therefore paracrine, rather than endocrine, in nature. As for the origin of these cells, he thought that they arose by a process called "endophytie" from the enterocytes of the gastrointestinal tract. This initial conception of a diffuse endocrine system was later reinvestigated and developed by Pearse from 1966 onward. On the basis of various cytochemical, biochemical, functional, and pathological observations, Pearse was led to emphasize the fact that a number of polypeptide hormone-producing cells, located mainly in the gut and its appendages but also found elsewhere, possess a common set of cytochemical and ultrastructural characteristics (Pearse, 1968, 1969). Pearse grouped these cells in the so-called APUD series, an acronym derived from the initial letters of their three more constant and important cytochemical properties: <u>A</u>mine content and/or <u>A</u>mine <u>P</u>recursor <u>U</u>ptake and <u>D</u>ecarboxylation (Table 5.2).

For Pearse, these features indicated closely related metabolic mechanisms and common synthetic storage and secretion properties. On that basis, he proposed that the APUD cells share a common embryological precursor which arises from the neural crest. The list of cells belonging to the APUD series, limited to 14 cell types in 1969 (including pituitary corticotrophs and melanotrophs, pancreatic islet cells, calcitonin-producing cells, carotid body type I cells, adrenal medulla, and various endocrine cells of the gut epithelium), increased to 40 less than 10 years later, by which time Pearse had included the parathyroid and a number of cells in the gut, lung, and skin that had been shown to produce peptides or neuropeptides (Pearse, 1976). The essence of Pearse's theory was that cells of the APUD series constituted a "diffuse neuroendocrine system" (DNES), which he viewed as a third branch of the nervous system, "acting with the second, autonomic division in the control of the function of all the intestinal organs" (Pearse, 1980).

In all these cells, according to Pearse (1969), the most characteristic properties can be evidenced by a simple cytochemical test. The L-isomer of one or the other of the two principal amino acid precursors of the fluorogenic monoamines (3,4-dihydroxyphenylalanine (DOPA) for catecholamines or 5-hydroxytryptophan (5-HTP) for serotonin) administered intravenously is taken up and decarboxylated by APUD cells. The amine is immobilized by freeze-drying and converted *in situ* to a fluorescent isoquinoline derivative by treatment with formaldehyde vapors at 70°C (formaldehyde-induced fluorescence (FIF) technique: Falck, 1962; Falck *et al.*, 1962). Radioactive amino acid precursors can also be used; the radioactivity appears concentrated in the cells responsible for the uptake (Pearse, 1969) (Table 5.2).

Some ultrastructural characters are shared by the APUD cells (Pearse, 1968), such as low levels of rough (granular) endoplasmic reticulum, high levels of smooth endoplasmic reticulum in the form of vesicles, high content of free ribosomes and membrane-bound secretion vesicles with osmophilic contents (average diameter 100–200 nm).

As far as the nature and variety of polypeptides secreted by cells of the diffuse endocrine organ, our knowledge has progressed considerably since the 1960s, when efficient purification procedures were used to isolate and sequence several polypeptide hormones from the digestive tract. The first of these were the mammalian gastrins isolated by Gregory and Tracy (1964) and secretin by Mutt *et al.* (1970).

Following Pearse's observation that endocrine and nerve cells share several characteristics, a number of other molecular markers were found to be present in both cell types. This is the case for synaptophysin, an integral membrane glycoprotein localized in presynaptic vesicles of neurons and in similar vesicles of the adrenal medulla, in pancreatic islet cells, as well as in a variety of epithelial tumors (including islet cells), and neuroendocrine carcinomas of the gastrointestinal and bronchial tracts, and medullary carcinomas of the thyroid (Wiedenmann *et al.*, 1986). This is also the case for receptors for tetanus toxin and for the antibody A2B5, both of which bind to specific gangliosides of nerve and glial cells (astrocytes), and pancreatic islet cells. Moreover, TH, the key enzyme of catecholamine synthesis, is transiently produced by islet cells but not by acinar exocrine cells of the pancreas (Teitelman and Lee, 1987; Alpert *et al.*, 1988). Finally, in culture, but not *in vivo*, islet cells extend long neurofilament-containing processes (Teitelman, 1990). Another connection between the nervous system and endocrine cells secreting polypeptide hormones is that the insulin promoter drives transgene expression in the neural tube as well as in the pancreas (Alpert *et al.*, 1988; Douhet *et al.*, 1993).

One of the most interesting aspects of the APUD cell concept is that it provided a clue to explain the relationships between a number of endocrine disorders and syndromes. The syndromes that best fit the APUD cell concept are the medullary thyroid carcinoma, associated with pheochromocytoma, also called familial chromaffinomatosis (Ljungberg *et al.*, 1967), Zollinger–Ellison syndrome (Zollinger and Ellison, 1955), Cushing's syndrome, the carcinoid syndrome, and some forms of the so-called multiple endocrine tumors.

The APUD cell concept has found a wide acceptance due to its far-reaching implications in fields as various as neurobiology, endocrinology, embryology, and pathology. It was, however, based on very circumstantial evidence, and a critical appraisal of the arguments upon which it rested as well as experimental verification are needed.

### 5.8.2   Critical appraisal of the concept of a diffuse neuroendocrine system

The presence of common molecular features were presented by Pearse and others as evidence for the existence of a common neurectodermal ancestor from which emerge divergent cell types. It is a common assessment among developmental biologists that, when cells differentiate from a common progenitor, certain metabolic features are conserved and these features can be regarded as "markers" for a family of cell types. In fact, this assumption should not be considered as an established fact since many observations show that there are no obligatory ancestral relationships between cells expressing the same structural genes. One example discussed at length in this book is provided by the mutation at the *W* locus, which in mice affects primordial germ cells, melanocytes, and hemopoietic cells, which do not share a common lineage. In fact, after careful examination, many of the molecular markers on which kinship was proposed for cells of the DNES display a distribution wider than initially stressed. The most characteristic common property of APUD cells, as stressed by Pearse, is the uptake and decarboxylation of amino acid precursors of fluorogenic monoamines and the presence of AADC (amino acid decarboxylase). In the embryo, however, this reaction is non-specific, since AADC is expressed, not only in neural and endocrine cells, but also transiently in tissues of mesodermal origin, such as notochord and muscles, and in pancreatic exocrine cells.

Another proposed marker for APUD cells is a cholinesterase. In chick embryos, a specific acetylcholinesterase (AChE) is found in developing pancreatic endocrine cells, in some unidentified cells in the gut groove (Drews *et al.*, 1967), in the basal plate of the neural tube, and in the neural crest (Cochard and Coltey, 1983). At later stages, however, AChE does not remain confined to neural and neurendocrine cells; thus its status as a molecular marker for the DNES is doubtful. Neuron-specific enolase (NSE), an enzyme present in nerve cells (Marangos *et al.*, 1978), is also found in some cells of the diffuse endocrine system, including enterochromaffin cells and the pancreatic islet cells (Schmechel *et al.*, 1978; Polak and Marangos, 1984). However, NSE has been detected in other cell types, such as megakaryocytes and platelets (Marangos *et al.*, 1980). Moreover, the whole early pancreatic rudiment expresses AADC (Teitelman and Lee, 1987).

Thus, although in certain instances the presence of common molecular markers is suggestive that different cell types share a common lineage, definitive evidence can come only through the use of cell tracing techniques.

### 5.8.3 Neurectodermal origin of the calcitonin-producing cells

One of the first applications of the quail–chick chimera system to follow the migration of neural crest cells was to test Pearse's hypothesis of the neural crest origin of cells of the APUD series.

No doubt remained about the ectodermal derivation of the adrenal medulla, nor about the pituitary APUD cells, since the hypophyseal placode (which later becomes Rathke's pouch) is closely related to the neural crest; it derives from the anterior neural fold (reviewed in Le Douarin et al., 1997). The first test case for the APUD concept was the cell type reported by Pearse (1968) as responsible for the production of the calcium-regulating hormone, calcitonin, discovered by Copp et al. (1962). The source of calcitonin in the thyroid gland was identified as the parafollicular cell, which was shown later also to contain somatostatin and 5-HT, and sometimes CGRP. In addition, those cells were shown to have the APUD characteristics (Pearse, 1966a,b); by fluorescent amine tracing, the mammalian parafollicular cells were demonstrated to arise, as in other vertebrates, from the ultimobranchial (UB) bodies (Fig. 5.10) (Pearse and Carvalheira, 1967). The calcitonin-producing cells are also distributed in other glandular tissues of embryonic pharyngeal derivation such as parathyroids, thymus, and along the large vessels of the neck.

In birds, the calcitonin-producing cells are present in the UB body, which develops from the last pair of branchial pouches in the most caudal region of the pharynx. In chick and quail, the UB rudiment becomes separated from the pharynx as a small epithelial vesicle at E5. The mesenchyme surrounding it is of neural crest origin, as attested by its quail nature in chimeras where the chick neural tube was replaced by its quail counterpart at the hindbrain level. From E8 onward the basement membrane lining the epithelial vesicle disrupts leading mesenchymal and epithelial cells to mingle. Considerable cell proliferation takes place; the initial lumen of the epithelial vesicle disappears at E10, a time point where most cells of the UB bodies contain dense secretory granules whose diameter varies from 60 to 150 nm. In some cells, larger granules (200–270 nm) are also present. At E15, cells with large secretory granules prevail, and the cords of secretory cells are lined by epithelial-type cells containing glycogen and devoid of dense granules. Examination of quail-to-chick neural chimeras have made it possible to assign an endodermal origin (from the branchial pouch) to the epithelial cells, a neural crest origin to the secretory cells possessing dense secretory granules and exhibiting calcitonin immunoreactivity, and a mesodermal origin to the endothelial cells lining the blood capillaries and sinusoids of the gland (Le Douarin and Le Lièvre, 1971, 1976; Le Douarin et al., 1974; Polak et al., 1974).

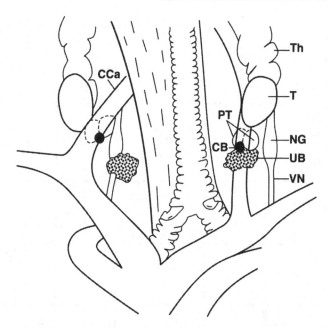

**Figure 5.10** Relationships of avian carotid bodies and ultimobranchial glands. CB, carotid body; CCa, common carotid artery; NG, nodose ganglion; PT, parathyroids; T, thyroid; Th, thymus; UB, ultimobranchial body; VN, vagus nerve.

In mammals, the UB bodies join the thyroid gland rudiment when the latter migrates caudally during neck morphogenesis (Pearse and Cavalheira, 1967; Stoekel and Porte, 1970). UB cells then invade the thyroid and become distributed mainly between the follicles, where they are termed parafollicular or clear cells, but they are sometimes also inserted into the follicular epithelium itself.

Cells with APUD characteristics have been described in the early mouse embryo in the vicinity of the fourth pharyngeal pouch and have been claimed to be C-cell precursors on their way to the UB endoderm (Pearse and Polak, 1971b). This claim, however, was made without proof of their neural crest origin or ultimate fate. Fontaine (1979) reinvestigated this question. Since the neural transplant technique used in avian embryo is not applicable to mammals, another experimental approach was devised. The pharyngeal region of 18–45-somite mouse embryos was dissected, yielding the thyroid rudiment isolated (Table 5.3, series A), or associated with the last pharyngeal pouches (Table 5.3, series B–D). It was then transplanted for about 14 days onto the chorioallantoic membrane of chick embryos which was used as a culture "medium." In series B, the endoderm and mesenchyme of the thyroid and branchial pouches were included in the explants, while in series C the mesenchyme was removed at the level of the branchial pouches; finally, in series D the mesenchyme was present, but the endoderm of the last pouch was removed.

Table 5.3. *Different types of explants and the presence of C cells in the thyroid gland after 14 days in graft on the chorioallantoic membrane of 6-day-old chick embryos*[a]

| Series | Type of explant | Stages of operation | | | Results: C-Cell differentiation + | Results: C-Cell differentiation − |
|---|---|---|---|---|---|---|
| A | | 9* | 7 | 9 | | 25 (100%) |
| B | | 6 | 10 | 7 | 23 (100%) | |
| C | | 17 | | 7 | 4 (24%) | 13 (76%) |
| | | | | 39 | 39 (100%) | |
| D | | 48 | 11 | 8 | 41 (86%) | 7 (14%) |
| | | | | | 5 (49%) | 6 (51%) |
| | | | | | | 8 (100%) |

Endoderm ☐    Mesenchyme ▨    BP, branchial pouch    * Numerals indicate number of cases throughout

BP 1st, 2nd, 3rd, 4th

Calcitonin-producing cells, identified either by their ability to take up L-DOPA and to convert it into dopamine, or by their content of dense secretory granules in electron micrographs, were never present in thyroid rudiments explanted alone. They were present, however, in every case when the complete UB primordium was included in the graft. Series C and D indicated, in addition, that the precursor C cells were mainly distributed in the mesenchymal component of the branchial arch until 28ss. Thereafter, and during a short period of time corresponding to 28–30ss, the presumptive C cells invaded the endoderm of the last pouch, where they remained until the UB body became confluent with the thyroid. Since the branchial arch mesenchyme has been shown to originate mostly from the neural crest in mammals as well as in birds (Johnston, 1966), one can assume from these experiments that the final localization of C cells in the thyroid gland involves a multistep migration, first from the neural primordium to the branchial arch mesenchyme, then from the latter to the endoderm of the UB rudiment, and finally from the UB body itself to the developing thyroid.

Interestingly, Barasch *et al.* (1987) have shown that cultured C cells from adult sheep thyroid respond to the $\beta$ subunit of NGF by extending neurites and switching expression from calcitonin to calcitonin gene-related peptide (CGRP), a peptide also found in a number of neuronal cell types in both the CNS and PNS. C cells therefore share with adrenomedullary cells a common origin from the neural crest, and a common response to NGF. In coculture with aganglionic chick gut, the sheep parafollicular cells invade the bowel, find their way between the muscle layers, and form a rudimentary myenteric plexus.

### 5.8.4  Origin of the carotid body

The carotid body (also called carotid glomus), the main function of which is chemoreception, is a paired glandular structure located close to the carotid artery and parathyroid glands on both sides and near the UB body on the left side. The diameter of the carotid body at hatching is 200 μm in the quail and 250 μm in the chick.

Investigations on the origin of the carotid body, based mainly on the classical techniques of descriptive embryology, have produced controversial results. Some authors (Rabl, 1922; Boyd, 1937) regarded it as a mesodermal derivative, while others (De Winiwarter, 1939; Celestino Da Costa, 1955) considered it to arise from the spinal or sympathetic ganglia or from ectobranchial placodes (Murillo-Ferrol, 1967). According to Rogers (1965), "the first visible sign of the developing carotid body is a primary condensation of cells of unknown origin, on the third aortic arch."

Ultrastructural observations of the carotid body make it possible to distinguish two main types of cells: the chemoreceptor cells (type I) more or less completely surrounded by supporting cells (type II) (Biscoe, 1971). Type I cells are characterized by secretory dense granules which, in the chick (Stoeckel and

Porte, 1969) and quail (Pearse *et al.*, 1973), measure about 50 nm in diameter and are separated by a clear halo from their limiting membrane. From an early developmental stage (8 days in chick and 6 days in quail embryos; Fontaine, 1973), type I cells are fluorescent if subjected to the FIF procedure. The fluorescence is yellowish in the chicken, and greenish in the quail. Emission spectra from the normal chicken carotid body cells are characteristic of 5-HT, with a peak at 540 nm. In the quail, the emission maximum was found at 480 nm, indicating the presence of catecholamines, which were shown to be mainly dopamine with some noradrenaline (Pearse *et al.*, 1973).

When a quail rhombencephalon was grafted into a 6–10-somite chick, the carotid body of the host was almost entirely made up of quail cells which emitted the same bright green fluorescence as the glomic cells of the normal quail after treatment by Falk's technique (Le Douarin *et al.*, 1972). Microspectrofluorometric analysis of the biogenic amine content of the carotid body type I cells of the chimeras revealed the presence of dopamine and the absence of 5-HT, as in the normal quail carotid body (Pearse *et al.*, 1973). Groups of cells similar to type I cells are present also in the wall of the large arteries arising from the heart. These also appear to derive from the neural crest in these experimental embryos as does the musculoconnective wall of these arteries (Le Lièvre and Le Douarin, 1975).

### 5.8.5 Embryological analysis of the origin of the pancreatic islet cells and of the APUD cells of the gut epithelium

Before the quail–chick marker system was applied to the problem of the embryonic origin of the APUD cells, Andrew (1963) presented convincing evidence that the differentiation of enterochromaffin cells of the gut epithelium was independent of an emigration of extrinsically derived cells. By explanting chick blastoderms deprived of the neural primordium (including the neural crest) onto the chorio–allantoic membrane, Andrew demonstrated that the gut which differentiated in the explants was devoid of innervation, but not of enterochromaffin cells. The same type of experiment was performed later on rat embryos and yielded similar results regarding several cell types of the APUD series, most notably the pancreatic islet cells (Pictet *et al.*, 1976). Moreover, aganglionic gut of the lethal spotting (*ls/ls*) mice also contains enterochromaffin cells (Gershon, personal communication).

The quail–chick chimera system, in which definite regions of the neural primordium of the chick embryo were substituted by their quail counterpart before the onset of neural crest cell emigration, was systematically applied to all the levels of the neuraxis (Le Douarin and Teillet, 1973). This allowed the contribution of the neural crest to the various cell types constituting the gut and its appendages to be established. As already mentioned, the gut was found to receive crest cells from two well-defined levels of the neuraxis: the vagal, and (to a lesser extent) lumbosacral levels. These cells build up the intrinsic enteric innervation but do not reach the gut epithelium where the endocrine and paracrine cells of the APUD series differentiate from the host's progenitors.

Regarding the pancreatic islet cells, Andrew (1976a) used a similar experimental approach to compare the results of the xenografts of neural epithelium with those grafts involving neural tubes of the same species (quail or chick according to the host) labeled by [$^3$H]TdR. She reached similar conclusions: regardless of the technique applied, no cells from the neural crest contributed to the pancreatic islets. However, some neural crest cells may colonize the pancreas, where they aggregate in small groups distinct from the endocrine islets. Fontaine et al. (1977) studied the type of differentiation expressed by these cells and observed that the clusters of quail cells were always separated from both exocrine and endocrine stuctures. The cells were subjected to the FIF technique after L-DOPA injection, and their affinity for lead hematoxylin was investigated. The cells originating from the neural crest, identified by the nuclear marker, did not exhibit the cytochemical properties of pancreatic endocrine islets. In fact, the crest cells that had migrated into the pancreas differentiated into parasympathetic ganglia. These pancreatic ganglia were shown to form by secondary migration from the gut in rat and mouse (Kirchgessner et al., 1992).

Antisera directed against glucagon, insulin, and somatostatin, marking A, B, and D cells respectively, were applied to the chimeric pancreas. As in the previous assays, the cells identified as being endocrine did not carry the quail marker (Fontaine-Pérus et al., 1980).

The possibility that these cells may be derived from the neurectoderm at a stage preceding the onset of neural crest formation was then tested experimentally (Fontaine and Le Douarin, 1977). The endomesoderm of chick embryos has been associated with the ectodermal germ layer of quail blastoderms at various stages, including the formation of the primitive streak, the head process and the neural plate, ranging from 12 to 24 hours of incubation. The recombined embryos were either cultured in vitro or on the chorioallantoic membrane. The intestinal structure which developed in the explants was analyzed for chimerism using various cytochemical techniques: FIF technique after L-DOPA injection, lead hematoxylin, and silver staining to indicate argentaffinity and argyrophily, all combined with the Feulgen–Rossenbeck reaction. In all the cases the enteric ganglia originated from the quail ectoderm, but the enterochromaffin cells, as well as the APUD cells, which developed normally in the epithelium, were always of the chick type. It therefore seems likely that no migration of cells from the ectoderm into the endoderm occurred before the formation of the neural crest. Thus a neurectodermal origin for the endocrine or paracrine cells of gastrointestinal tract epithelium had to be excluded.

More recent lineage studies have involved the production of transgenic mice, or of mice carrying a targeted mutation of genes expressed in the developing pancreas. In the first case, a promoter for one of the four principal endocrine hormones is used to drive a modifying gene or a toxin. These hormones are: insulin produced by $\beta$ (or B) cells which make up the majority of cells of the islets; glucagon produced by the $\alpha$ (or A or $A_2$) cells; somatostatin secreted by $\delta$ (or D or $A_1$) cells; and pancreatic polypeptide (PP) secreted by F cells. PP islets are mainly derived from the ventral pancreatic buds in the mouse. A proportion of the adult islet cells make peptide YY in addition to their principal product (Ali-Rachedi et al., 1984). Alpert et al. (1988) used the rat insulin promoter to drive SV40T antigen, the transforming effect of which is essentially efficient after birth. In the embryo, most of the endocrine cells expressed T antigen while post-natally only the $\beta$ cells do so. Similarly, Upchurch et al. (1994) utilized the peptide YY promoter to drive expression of T antigen, and the latter was found to be expressed by cells producing one of each of the four principal endocrine hormones. This supports the idea that an early endocrine progenitor exists in which several or all the hormones are produced.

Experiments were also performed using the diphtheria toxin A chain. This toxin is a potent enzyme causing adenoribosylation of elongation factor 2, hence blockage of protein synthesis, and cell death. When the elastase promoter was used, the result was the absence of an exocrine pancreas and a reduced complement of ductal tissue, and islets (Palmiter et al., 1987).

Recently, transcription factors of the Pax family have been found to play a key role in islet cell development. On the basis of targeted mutations by homologous recombination, Pax4 and Pax6 were shown to be involved in determining the fate of endocrine islet cells during development. According to St-Onge et al. (1997) and Sosa-Pineda et al. (1997), once the endodermal cells have acquired the potential to form the pancreas, Pax gene expression defines the lineage of the different endocrine cells. Thus Pax4 mutants lack $\beta$ cells but have a larger than normal population of $\alpha$ cells. Deletion of Pax6 eliminates the $\alpha$ cell lineage, or diverts $\alpha$ cell to the $\beta$ cell phenotype. Initially, cells expressing both Pax4 and Pax6 differentiate into $\beta$, $\gamma$, and $\delta$ cells, and cells expressing only Pax6 differentiate into $\alpha$ cells.

On the other hand, there are transcription factors which are expressed in the early pancreatic rudiment and condition the development of both the endocrine and exocrine cell lineages. This is the case of the homeobox gene IPF1 (otherwise known as IDX1, STF-1, or PDX: Ohlsson et al., 1993; Miller et al., 1995; Guz et al., 1995). When this gene is removed, the embryos lack a pancreas (Jonsson et al., 1994). In the adult this gene continues to be expressed in the $\beta$ cells only.

An additional reason to attribute an endodermal origin to enteroendocrine cells is the demonstration that they produce $\alpha$-fetoprotein, a reliable marker of endodermal derivatives. Further proof on the endodermal origin of the gastric endocrine cells was provided in an elegant series of experiments performed by Thompson et al. (1990). They used XX–XY tetraparental chimeric mice to show that the Y sex chromosome, as evidenced by in situ hybridization, was present or absent in the gut epithelium in groups of cells comprising endocrine and other epithelial cells. This strongly suggests that all these cell types are derived from a common endodermal stem cell.

*5.8.6  Concluding remarks*

We have, in this book, devoted a certain amount of space to the story of the "APUD cell hypothesis," which is now essentially only of historical interest for the following reasons. First, it was an ingenious and unifying view, which was put forward at the time of the discovery that many secreted peptides are commonly produced by endocrine–paracrine cells, and by neurons where they act as neuromediators or neuromodulators. The idea that these cells could have a common embryological origin was particularly attractive, since it led to the proposal that mutations in one of their progenitors could be responsible for tumors appearing simultaneously in several developmentally related cell types. The second reason is that Pearse's hypothesis has stimulated a large amount of research and speculation. For example, in the early 1970s, it appeared attractive to use the recently devised quail–chick chimera system to test Pearse's hypothesis, because it could provide the long-term lineage marker required for such a study. Le Douarin and coworkers, together with other groups, were able to confirm Pearse's hypothesis of a neural crest origin of the calcitonin producing cells, and of the carotid body type I cells, but showed that it was not tenable for the endocrine cells of the gut and its appendages. The clear-cut embryological demonstrations just described were confirmed by a number of other approaches as discussed above.

Andrew, one of the most active proponents of the endodermal origin of the pancreatic islet cells and whose work has largely contributed to clarifying this question, concluded in 1976 that "the factor sought in the genesis of the APUD cell, and in the tumors (the so-called APUDomas), may be the biochemical, rather than the embryological, relationship(s) of the progenitor cell types." Modern research amply proved that she was right. The case of the c-Ret/ GDNF receptor–ligand system beautifully illustrated her view: in the MEN type 2 syndrome, several neural crest derivatives are affected and can be related to definite types of mutations of the *ret* gene. However, Ret and GDNF are as crucial for the development of the kidney as they are for that of certain neural crest derivatives (see Section 5.7.4), showing once more that the sharing of gene activities is not restricted to lineally related cell types.

# 6

# The Neural Crest: Source of the Pigment Cells

The bright colors and the striking variety of patterns that characterize the skin and coat of many vertebrate species have long been a subject of interest and wonder. It seems that the first scientific studies of animal coloration started in the middle of the nineteenth century about the color changes of the African chameleon (see Bagnara and Hadley, 1973). Pertinent questions were raised concerning the physiological mechanisms underlying animal pigmentation, while chemists and physicists became interested in the nature of pigments. It was understood that some colorations resulted from the structural state of the biological material. Thus, it became apparent that the spectacular iridescent properties of certain insects and birds are not due to the presence of blue or green pigments but rather are the result of physical phenomena. Iridescent colors are produced when biological structures, such as fibrils and lamellae, are orderly arranged to generate light diffraction. Another phenomenon responsible for "structural coloration" results from differential scattering of light, which provides the basis for much of the blue colors seen in eyes, feathers, and skins of certain vertebrates.

By analyzing extracts of colored tissues, chemists discovered that lipid-soluble yellow, orange or red pigments of animals are carotenoids, while guanine deposits are responsible for the whitish reflecting surfaces of fishes and frogs. Most importantly, melanins, flavins, pteridines, and porphyrins were recognized as essential compounds in animal pigmentation.

The discovery that many of the diverse animal pigments are contained in specialized cells, designated chromatophores, stimulated research in this area. Rapid changes in the intensity of pigmentation in certain animals was first attributed to the capacity of chromatophores to change their form and to assume either an "expanded" or a "contracted" state. In the latter case, the amount of the pigmented skin surface is reduced, thus making the color lighter. With more sophisticated techniques of investigation, it became obvious that these changes resulted from either dispersion or aggregation of pigment granules inside the cell rather than in actual changes in cell shape.

A considerable impetus to the study of pigment cell development and differentiation followed the discovery that amphibian chromatophores originate from the neural crest (Du Shane, 1935). In fact, all the pigment cells of the body except those belonging to the CNS (i.e., the pigmented retina, and certain CNS neurons as those of the substantia nigra) are derived from the neural crest. These cells constitute a quantitatively and physiologically important derivative of the crest, owing to their function in protection against UV irradiation, the capacity they confer to certain species to undergo adaptive color changes, and the role that they play in sexual behavior. Moreover, melanocytes are the site of malignant transformation, leading to highly invasive melanomas which constitute an important problem in human pathology.

The physiology and pathology of pigment cells, as well as the chemical nature of the compounds which are responsible for their coloration, has been and still is the subject of intensive investigations. Of particular interest is the question of the mechanisms that control the pattern of skin pigmentation. Pigment patterns in vertebrates can be classified generally into two groups: dorsal–ventral pigment patterns, in which the ventrum is generally less pigmented than the back, and various pigment patterns such as spots, stripes, or mottlings. Many reports have been published on their developmental biology (see reviews by Bagnara and Hadley, 1973; Bagnara, 1987). The molecular mechanisms underlying pigment pattern formation are, however, still poorly understood. It was proposed in the past that the pigment cell precursors were already determined while in the neural crest, so that the pattern they generated was preordained before they started to migrate (see Volpe, 1964, and Bagnara, 1987, for a review). In fact, it appeared later on that the fate of neural crest derivatives, including that of chromatoblasts, was strongly influenced by the environment (Bagnara, 1972, 1987; Bagnara et al., 1979b). Thus, pattern formation appeared to be an interplay between the neural crest-derived cells and environmental influences, among which growth and survival factors, and components of the extracellular matrix play an important role (Bagnara, 1982; Ohsugi and Ide, 1983; Perris and Löfberg, 1986; Tucker and Erickson, 1986b; Fukuzawa and Ide, 1988; Richardson et al., 1990; Erickson, 1993a; Frost-Mason et al., 1994, and references therein). This chapter will focus on the segregation of the pigment cell lineage during neural crest ontogeny. The various types of pigment cells recorded in vertebrates and the nature of the pigment they produce will be considered first. Then the problem of the transition from the neural crest to the melanocytic lineage as revealed in the embryo in vivo and in vitro through the use of various markers will be addressed. Migration pathways, homing of melanoblasts to their sites of differentiation, determination of the neural crest cells to the melanocytic differentiation pathway will be considered. It will appear that novel data resulting from the identification of several genes whose mutations produce anomalies in coat color pigmentation in the mouse open new avenues in the identification of growth factors involved in the development of pigment cells, and on some aspects of pattern formation.

## 6.1 The various types of pigment cells

Pigment cells, also called chromatophores, are branched, migratory cells found in both epidermal and dermal layers, and also, particularly in lower vertebrates, in internal organs: perineural and perivascular tissues, as well as in the celomic wall.

Coloration is conferred to the skin by chromatophores, and the various color changes that many animals (essentially poikilotherm vertebrates) undergo in response to their environment result from movement of pigment granules within these cells.

The various types of pigment cells found in vertebrates will be described, together with the nature of the pigment that they contain.

### 6.1.1 Melanophores, melanocytes, and melanins

*6.1.1.1 Different types of melanin-producing cells.* According to the nomenclature established at the Third Conference on the Biology of Normal and Atypical Pigment Cell Growth (Gordon, 1953) the term "melanophore" is essentially used to designate the melanin-containing pigment cells of lower vertebrates. The term "melanocyte" is more often used for the mature melanin-producing and melanin-containing cells of higher vertebrates. Melanophores are generally stellate cells responsible for color changes in reptiles, amphibians, and fishes. Pigment granules may be either dispersed throughout the cell or concentrated in its center, depending upon the physiological state of the animal or the color of the background (Fig. 6.1).

Both melanocytes and melanophores have the same type of progenitor, the melanoblast, which, although itself unpigmented, has the potentiality to produce melanin. These cells must be distinguished from macrophages which contain phagocytized melanin and are sometimes wrongly called melanophores.

Among vertebrates, there exist two types of melanogenic pigment cells which differ markedly by their location and their differential response to hormones.

A. DERMAL MELANOPHORES. These are involved in rapid color changes, and are prevalent in poikilotherms. They may be very large cells several hundred micrometers in diameter. They are generally more abundant in dorsal than in ventral skin, which partly explains why the ventral surface of most lower vertebrates is light-colored. Melanophores quite similar to those found in the dermis are also distributed in the internal organs, and on nerve and blood

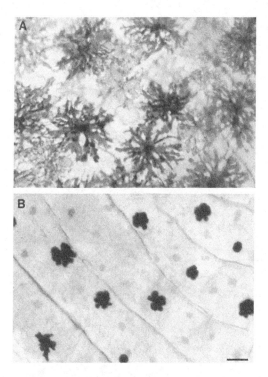

**Figure 6.1** Dispersed (A) and aggregated (B) melanophores in scales of the goldfish (*Carassius auratus*). Bar = 294 μm. (Kindly provided by J. Bagnara; reproduced, with permission, from Bagnara and Hadley, 1973.)

vessels. Color changes due to movements of pigment granules within dermal melanophores have been recently shown to result from a direct response of the cells to light. This response is mediated by melanopsin, and opsin discovered in *Xenopus laevis* by Provencio *et al.* (1998), and whose deduced amino-acid sequence shares greatest homology with cephalopod opsins.

B. EPIDERMAL MELANOPHORES. These are elongated "spindle-shaped" cells with long dendritic processes which under hormonal stimulation may become highly branched. In cold-blooded vertebrates they are distributed rather uniformly above the germinative layer but, as are their dermal counterparts, they are less abundant in the ventral than in the dorsal skin area. Dendrites from epidermal melanophores are responsible for deposition of melanin granules in other epidermal cells by their particular cytocrine activity. In birds and mammals coloration of the skin, feathers, and hair is due to pigment transfer from melanocytes into epithelial cells, the keratinocytes (Fig. 6.2). This deposition of pigment is a relatively slow process, and explains why epidermal melanophores are not involved in the rapid color changes in poikilotherms, although prolonged stimulation increases cytocrine activity with consequent darkening of

**Figure 6.2** Section through the bill of a male sparrow showing epidermal melanocytes with cytocrine activity resulting in pigment transfer to keratinocytes. (Reproduced, with permission, from Bagnara and Hadley, 1973.)

the animal. Since each melanocyte delivers its melanosomes to a given pool of epithelial cells, the notion of "epidermal melanin unit" was put forward by Fitzpatrick and Breathnach (1963) for mammals (see also Hadley and Quevedo, 1966, for amphibians). Environmental (e.g., UV light stimulation) and hormonal stimuli, acting on epidermal melanocytes, determine the number of epidermal cells which are recipients of cytocrine melanin donation from the melanocytes.

*6.1.1.2 Melanins and melanogenesis.* Pigments contained in melanophores and melanocytes are, like catecholamines, derived from phenylalanine and tyrosine. Generally black or brown in color, they are called eumelanins; the yellow and orange melanins are termed pheomelanins.

The distribution of these two basic types of tyrosine-derived pigments determines the color pattern in higher vertebrates. They differ by their biochemical composition and ultrastructural appearance in intracellular organelles (see Prota, 1992). Pheomelanosomes are spherical, lack internal structure, and contain a relatively soluble cysteine-rich material; eumelanosomes are elliptical, and contain a highly organized matrix with an insoluble cysteine-poor material. In humans, the highest amounts of pheomelanin are found in "fire red" hair. In black, grey, and blond hair the pigment is mostly composed of eumelanin. In mice, pelage pigmentation due to pheomelanin in animals homozygous for the *recessive yellow* mutation is colored rather like that of blond-haired humans.

The general metabolic scheme of melanogenesis is represented in Fig. 6.3.

Tyrosinase is the rate-limiting enzyme involved in synthesis of both melanins, and, when low levels of tyrosinase are present in the murine melanocyte, the majority of tyrosine is converted, via 3,4-dihydroxyphenyl-alanine (DOPA) and DOPAquinone, into cysteinylDOPA, which leads exclusively to pheomelanins (for reviews see Hearing and Tsukamoto, 1991, and Furumura *et al.*, 1996). At higher tyrosinase levels, excess DOPA and DOPAquinone may be diverted along a separate pathway leading specifically to synthesis of eumelanins from derivatives of DOPAchrome. Tyrosinase-related protein-2 (TRP2), also called DOPAchrome tautomerase (Dct or DT), converts DOPAchrome to 5,6-dihydroxyindole (DHI) instead of to the carboxylated derivative termed DHI-2-carboxylic acid (DHICA) (Tsukamoto *et al.*, 1992). DHICA has been shown to be a relatively stable intermediate which is present in coated vesicles in transit to melanosomes (see Kobayashi *et al.*, 1994b, and references therein).

*6.1.1.3  Mutations affecting the melanin synthetic pathway.* Many different loci have been identified in the mouse that regulate melanin production at different levels of its biosynthetic pathway. Several of those genes have been cloned. Thus *silver* (Kwon *et al.*, 1987, 1991; Kobayashi *et al.*, 1994a; Zhou *et al.*, 1994; Lee *et al.*, 1996; Chakraborty *et al.*, 1996) and *pink-eyed dilution* (Gardner *et al.*, 1992; Brilliant, 1992; Rinchik *et al.*, 1993) encode proteins involved in the structure or function of melanosomes. Others such as *albino* (*c*) and *slaty* (*slt*) encode enzymes of the melanogenic pathway: tyrosinase for the *albino* locus and TRP2 for the *slaty* gene. Tyrosinase is a trifunctional and rate-limiting enzyme for melanogenesis (see Fig. 6.3) and mutations in the *albino* locus prevent melanin production (Halaban *et al.*, 1988; King *et al.*, 1995) and alter the viability of the melanocyte which remains unpigmented.

The *brown* locus (*b*) was chronologically the first member of the tyrosinase gene family to be cloned (Shibahara *et al.*, 1986; Jackson, 1988). It encodes an enzyme TRP1 which functions as a DHICA oxidase in melanin biosynthesis, an essential step to the further metabolism of DHICA to a high molecular weight pigmented biopolymer (Kobayashi *et al.*, 1994b). Mutations of TRP1

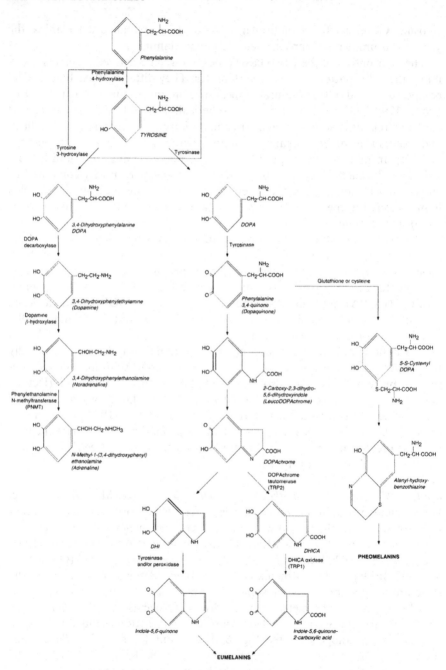

**Figure 6.3** Melanogenic pathway. The series of chemical reactions involved in the production of eumelanins and pheomelanins from tyrosine are indicated with the regulatory enzymes involved. (Partly from Kobayashi *et al.*, 1994b.)

result in formation of a light brown melanin of the hair (e.g., brown, $b/b$, or brown cordovan, $b^c/b^c$).

### 6.1.2  Iridophores

Iridophores are pigment cells that often appear iridescent and contain organelles orientated in such a way as to reflect the light. Localization of iridophores seems to be primarily restricted to the integument and, more precisely, to the dermis. Pigments of iridophores are contained in flat platelets usually arranged in highly orientated stacks. They are basically made of purines (primarily guanine), with also hypoxanthine, adenine, and uric acid entering into pigment composition. Iridophores are common in poikilotherms, but rare in homeotherms, with the possible exception of their presence in the irises of some birds (Bagnara and Ferris, 1971).

### 6.1.3  Xanthophores and erythrophores

Xanthophores and erythrophores are colored bright yellow, orange or red. They are often involved in color patterns in vertebrates but not as importantly as melanophores in color changes. Carotenoids are major pigments of the xanthophores and erythrophores of fishes, amphibians, and reptiles. This explains the old term "lipophores" used to designate these cells, which is related to the fat-soluble nature of carotenoids. Pteridines (concentrated in organelles called "pterinosomes") also play a major role in pigmentation of lower vertebrates, some species of which actually utilize pteridines almost exclusively as yellow and red pigments (Obika, 1963; Obika and Bagnara, 1964).

## 6.2  Development of pigment cells

### 6.2.1  A common precursor for all types of pigment cells

The various types of pigment cells are characterized by different organelles, the structure of which has been extensively revealed by electron microscopy. The melanosomes of both dermal and epidermal melanophores are electron-opaque ellipsoidal structures containing black-brown insoluble eumelanins. Melanosomes found in epidermal melanophores in the integument of some birds and mammals, including human red hair, may contain pheomelanin, a lighter-colored sulfur-bearing melanin. The pigmentary organelles of xanthophores and erythrophores, called pterinosomes, are spherical and contain concentric lamellae. The precise localization of the pteridines found in these organelles is not known. In the iridophores, the reflecting platelets are filled with purine crystals which usually disappear during staining and sectioning leaving empty spaces in their place. Their purine constituents, notably guanine, hypoxanthine, adenine, and uric acid, are not true pigments, but are involved in imparting structural colors. Although the various fully differentiated pigmentary organelles are different from one another, the early steps of their formation are similar since they are derived from the endoplasmic reticulum (Bagnara *et al.*, 1979a).

According to Bagnara *et al.* (1979b), the various types of pigment cells originate from a common neural crest precursor whose commitment towards a definite type of chromatophore is not established before its localization in a specific area of the body. External cues from the environment determine the final phenotype. This hypothesis is based on the existence of mosaic pigment cells containing more than one type of pigment. Many examples have been reported by Bagnara and his group, including the presence of melanosomes, reflecting platelets, and pterinosomes in crythrophores of the central red stripe area of the snake *Thamnophis proximus*. Organellogenesis in dermal chromatophores of the leaf frog *Agalychnis dacnicolor* during metamorphosis also offers a striking example of mosaicism. Melanophores in adults of this species possess large melanosomes made up of a eumelanin core surrounded by a fibrous mass (Bagnara *et al.*, 1973). This fibrous material is a pteridine pigment, pterorhodin (Misuraca *et al.*, 1977). The adult melanosome forms at metamorphosis from the transformation of the larval dermal melanosome by deposition of pterorhodin fibers on its surface (Bagnara *et al.*, 1978). In experiments involving premature metamorphosis induced by thyroxine–cholesterol pellets implanted under the skin in *Agalychnis*, electron microscopy revealed the presence of mosaic pigment cells containing reflecting platelets, pterinosomes, and carotenoid vesicles.

Although such an extreme case of mosaicism was never observed in normal chromatophores of this species at any stage of development, it reveals, along with the other examples mentioned, the close relationships between the various organelles found in different cell types. These observations led Bagnara *et al.* (1979b) to propose a model for pigment cell differentiation, according to which, in a common progenitor cell, a primordial organelle originating from the endoplasmic reticulum is able to differentiate into any of the known pigmentary organelles.

### 6.2.2 Neural crest origin of pigment cells

Extirpation and explantation of the neural crest have revealed that it is the sole source of all pigment cells of the body, except those which differentiate in the retina and are therefore derived from the optic cup.

The neural crest origin of pigment cells was first established by Du Shane (1934, 1935, 1936, 1938) in experiments with amphibians in which he removed the trunk crest and found that not only were the dorsal fin, spinal ganglia, and Rohon-Béard cells lacking in the operated region, but pigment cells were also absent. On the other hand, grafting pieces of neural folds to the ventral side of an embryo or explanting them *in vitro* resulted in the appearance of melanophores.

Shortly before Du Shane's work, Mangold (1929) and Holtfreter (1929, 1933) had brought forward some other evidence of a neural crest origin of melanophores: grafted or explanted presumptive neural plate gave rise not only to nervous tissues but also to pigment cells, whilst grafted presumptive epidermis developed without pigment.

Later, other workers (Raven, 1936; Twitty, 1936; Twitty and Bodenstein, 1939) confirmed these results through grafting experiments between different species. In the host, pigment cells were found to be of the donor type, and implantation of neural crest from frog to urodele showed that, also in anurans, not only the melanophores but also the other types of pigment cells have the same origin (Bytinsky-Salz, 1938; Baltzer, 1941).

In birds, the first experimental evidence that melanocytes are of neural crest origin is due to Dorris (1936, 1938, 1939), who found that explants of neural folds of the chick produced amoeboid, branched pigment cells, whereas pieces of the whole embryo lacking the neural fold gave no pigment. In addition, she showed that pieces of neural crest grafted into limb buds of 3-day-old embryos produced coloration of extensive areas of the epidermis in the host. A further proof of the migration of melanocyte precursors was given by Eastlick (1939), who transplanted limb buds from a pigmented strain of chick into the body wall of an unpigmented host. Pigmentation of the transplanted limb depended on whether or not neural crest cells were included in the graft. By cutting the limb bud at various distances from the neural primordium at different stages of development, Eastlick could follow the position of the advancing front of the migrating promelanocytes from the neural crest.

The source of the pigment cells has been traced back to earlier stages of development by grafting small areas of the blastoderm at the primitive streak and head-fold stages. The area capable of giving rise to pigment cells has been found to correspond to the whole presumptive neural plate at the earliest stages and to become restricted later on to the neural crest alone (Rawles, 1940a).

Equivalence of peritoneal and skin melanoblasts was demonstrated by Rawles (1945) in an experiment where a White Leghorn chick limb bud was grafted into the celomic cavity of Barred Rock embryos. Melanoblasts from the peritoneal wall invaded the graft and differentiated into epidermal melanocytes, although their normal fate would not have led them to contribute to the external color pattern.

Proof that melanocytes also originate from the crest in mammals was acquired later. An interesting experiment is due to Rawles (1947, 1948), who relied on the fact that mouse embryonic tissues transplanted into the celomic cavity of chick embryos are viable and develop for several days. Explants that included the neural crest were taken from black mouse embryos and transplanted into the celom of unpigmented White Leghorn chicks. During the following days, melanoblasts grew out of the crest material and invaded the mouse epidermis, whereas in control explants, which did not include the crest, no pigmentation appeared. Rawles was able to show that mouse embryos up to

23ss still possessed melanoblasts in their neural tubes anterior to the level of somite 6. However, neural tubes from 25ss embryos lost their pigment-forming capacity at the anterior levels.

In fish, removal of brain pieces in the groundling (*Misgurnus fossilis*) led to a total absence of pigment cells or to a reduction of their number. In the lamprey, Newth (1956) demonstrated clearly that the neural crest is the source of melanophores. As the embryonic zebrafish has become an important model for studies of early events in vertebrate development, many data have been gathered on neural crest cell migration in this species (see Eisen and Weston, 1993, for a review).

## 6.3   Migration, homing, and differentiation of presumptive pigment cells

Precursors of pigment cells move relatively long distances before reaching the sites of the embryo to which they home and differentiate. These sites include essentially the skin, but also internal organs such as the peritoneal epithelium in certain species, and the inner ear, where melanocytes play an important functional role in hearing. The production of melanocytes is not uniform in all parts of the neural crest. Niu (1947) stated that fragments of cranial neural crest transplanted *in vitro* or to pigment-free areas of the flank gave rise to fewer pigment cells than their truncal counterparts. The higher capacity of trunk than cephalic crest to produce pigment cells has also been found in heterotopic transplantation experiments of quail neural crest into White Leghorn chick embryos (Smith *et al.*, 1977).

The migration pathways of melanoblasts have for long been unknown, as the cells at this stage have not yet formed pigments and cannot be distinguished from other mesenchymal cells. Cell markers were to solve this question. Most of the knowledge we have about the dispersion of melanocytic precursors has been acquired by descriptive and experimental studies carried out in the avian embryo, which is particularly suitable for investigating the early stages of histogenesis and organogenesis. Results obtained in the avian embryos have recently been extended to the mouse through the use of DiI (Serbedzija *et al.*, 1990, 1992). Moreover, the discovery of several specific molecular markers of melanoblasts in mammals has contributed considerably to our present understanding of the successive steps of melanocytic development.

### 6.3.1   *Migration of the pigment cell precursors in the avian embryo*

During the 1930s and 1940s, pioneer studies established the timing of the invasion of the skin by neural crest cells along the dorsoventral axis of the chick embryo. This was achieved by examining the ability of different embryonic regions to produce pigments when isolated in ectopic positions at different developmental stages (Watterson, 1938; Eastlick, 1939; Willier and Rawles, 1940; Ris, 1941). Thus, wing buds begin to be colonized by pigment cell precursors by embryonic day 3.5 (E3.5) and leg buds at E.4.5, just after the lateral regions of the somites have been reached by these cells at the corresponding transverse levels. Weston (1963) was the first to use a cell marker to identify the migrating cells in the chick embryo. As indicated earlier, he implanted pieces of the neural primordium from [$^3$H]TdR-labeled chick embryos into unlabeled hosts and found labeled cells in the ectoderm of the host soon after grafting, and few or no cells in the infraepidermal mesenchyme. Weston concluded that the melanoblasts undergo their primary migration in the ectoderm. These results could not be confirmed in studies based on the use of the quail–chick marker system (Teillet and Le Douarin, 1970; Teillet, 1971a,b). From serial observations made at regular intervals following isotopic and isochronic grafts of quail neural primordium into White Leghorn chick embryos, it was shown that, in fact, prospective melanocytes migrated essentially (or totally) through the mesenchyme and not within the ectoderm. Only rarely were quail cells found localized in the ectoderm during the early stages of neural crest cell dispersion. Melanoblasts that seed the skin in the truncal area are derived from the stream of neural crest cells that migrate between the dermomyotome and the ectoderm. The pigment cells seeding the internal organs and the peritoneal wall follow the same dorsoventral migration route as the precursor cells of the peripheral nervous system. The most active migration in the skin takes place before the sixth day of incubation. Seeding of the epidermis by melanoblasts can be followed in chimeric embryos. It occurs massively at the end of E5 and during E6. However, certain cells remain much later in the dermal mesenchyme where they can be encountered even as late as at E11. Similar results were obtained more recently by Hulley et al. (1991). It should be noted that penetration of melanoblasts into the skin occurs first in dorsal regions and progresses ventrally, as do both the colonization process and the formation of the feather buds.

Melanocyte differentiation, evidenced by pigment deposition, takes place from E9 in chick. It is preceded by a period where epidermal melanoblasts of the developing feather buds divide actively. Nuclear and cytoplasmic volumes obviously increase dramatically soon before pigmentation appears. Only melanoblasts localized in the feather germ epidermis differentiate. Neural crest cells that have seeded the epidermis between the feather buds or have remained in the dermis do not acquire pigment. It is likely that they will function later as stem cells and contribute to the pigmentation of the further generations of feathers emerging after birth and during adulthood. The reason for the fact that melanocytic differentiation takes place only in the feather germ will be considered later in this chapter.

Grafting fragments of quail neural primordia into White Leghorn chick embryos results in the formation of a transverse stripe of quail-like pigmented feathers, whose position varies according to the level of the graft. However, the anterior and posterior limits of the neural tube implant and of the pigmented area do not coincide. The latter always projects beyond the former in the way

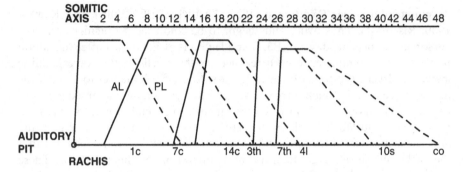

**Figure 6.4** Extension of the migration of the skin melanocytes after graft of fragments of the quail neural primordium into a chick embryo at various levels of the somitic axis. AL and PL, anterior and posterior limits of pigmentation. The vertebrae are taken as references to determine the transverse limits of migration of the melanocytes of the graft into the host skins. c, th, l, s, co, cervical, thoracic, lumbar, sacral, coccygial vertebrae. (From Teillet, 1971a.)

represented in Fig. 6.4. Moreover, when only a hemineural tube is grafted, the pigmented transverse stripe extends to both sides of the host embryo. Thus, from each particular point of the neural crest, the population of melanoblasts extends radially in all directions within the plane of the skin (see Fig. 1.8), even if the resulting dispersion of melanoblasts is predominantly dorsoventral and occurs in a direction perpendicular to the embryonic axis.

The problem of the cues responsible for generating the pigmentation pattern was addressed in experiments involving the graft of quail neural crest cells into non-pigmented White Leghorn chick embryos.

On the basis of previous grafting experiments by Rawles (1948), it was thought that the pigment pattern of chimeras is determined by the genotype of the donor melanocyte. This led Wolpert (1981) to suggest that melanoblasts from the donor bird can read the positional value of the ectoderm of the feather papillae of the recipient. In quail–chick combinations, neither positional information nor a prepattern that would exist in the feathers and could determine the pigment pattern can account for the results observed. In fact, migration of melanoblasts seems to lead them to invade all skin areas, even those which in the quail, or in quail–chick chimeras, remain unpigmented. Local cues, of still unknown nature, further influence positively or negatively the differentiation of melanocytes. In experiments where a supernumerary zone of polarizing activity was grafted to the anterior margin of the wing bud, a duplication not only of the skeletal elements but also of the pattern of feather papillae and of the feather pigment pattern was induced. This suggests that the same mechanism might be responsible for specifying these three characters (Richardson *et al.*, 1989, 1990).

The invasiveness of melanoblasts and melanocytes and their capacity to cross blood vessel walls and basement membranes have been demonstrated in several types of experiments. For example, the work performed by Weiss and Andres (1952) is worth mentioning. Heterogeneous cell suspensions from a colored strain of chicken were injected intravascularly into blood vessels of White Leghorn embryos. This resulted in the appearance of colored patches in the plumage of the host, indicating that the injected pigment cells crossed the wall of the blood vessels, migrated, and homed to the skin. Similar results were obtained by injecting a variety of cell types from quail embryos into the blood vessels of the chick (Cudennec, 1977). The quail–chick marker system allowed the localization of injected cells to be recognized in the host and showed that quail melanoblasts were able to reach the chick epidermis and to differentiate there into pigment cells, while the other injected cell types made little colonies in internal organs apparently without any kind of specificity. The main factor which seemed to control the position of all cells except melanocytes in the host was the diameter of the blood vessels: the smaller the diameter, the higher the frequency of quail cell clusters. Interestingly, matching between host and donor tissues influenced the size of the quail cell colonies. Thus, injected quail liver cells developed larger colonies when they homed to the liver or to the yolk sac than elsewhere (Cudennec, 1977).

Our knowledge of the precise timing of migration of neural crest cells along the dorsolateral pathway has benefited from more recent studies carried out on intact embryos and using either the HNK1/NC1 Mab or the vital dye, DiI.

Despite the fact that the HNK1/NC1 epitope disappears rapidly in these cells, particularly in the quail and to a lesser extent in the chick embryo, it was used by Erickson et al. (1992) in studies on the latter. As expected (see Chapter 3), the cells which are to take the dorsolateral pathway exit uniformly (and not segmentally) from the neural tube and remain "stuck" for a while (about 24 hours) close to the dorsolateral side of the neural tube in the so-called "staging area" (Wehrle-Haller and Weston, 1995). From about E3 to E3.5 according to the transverse level considered, the subectodermal pathway seems to "open up" since it is rapidly invaded by the neural crest cells. When the cells have reached the level of the lateral edge of the dermomyotome they start to lose their HNK1 epitope and cannot be further visualized by this method.

The problem is then raised of the nature of the molecular changes that make the dorsolateral pathway available for migration. If the dermatome is ablated prior to neural crest cell emigration, penetration of migratory cells in the subectodermal area takes place prematurely, suggesting that the state of development of the dermatome is critical in this process (Erickson et al., 1992). Molecules to which an inhibitory role in the migration process has been attributed (see Chapter 2), such as chondroitin 6-sulfate proteoglycan, identified by their ability to bind peanut agglutinin (PNA), are present in this area before crest cell invasion. They fail to appear in the subectodermal pathway in the dermatome-deprived areas (Erickson, 1993b).

A chemoattractive function for neural crest cells has also been attributed to the dermatome, particularly at the time its cells are in the dispersing phase (Tosney, 1992). This observation may provide an explanation for the homing of melanocytes to the skin when they are injected into the bloodstream as mentioned before (Weiss and Andres, 1952; Cudennec, 1977).

Some ECM (extracellular matrix) components may also support migration of neural crest cells in the dorsolateral pathway. Molecules like laminin (Rogers et al., 1986; Duband and Thiery, 1987), fibronectin (Newgreen and Thiery, 1980; Thiery et al., 1982), collagens type I, III, and IV (Perris et al., 1991a,b), and tenascin (Tan et al., 1987; Mackie et al., 1988) that favor neural crest cell adhesion and migration in culture (e.g., Erickson and Turley, 1983; Rovasio et al., 1983; Newgreen, 1984; Tucker and Erickson, 1984; Halfter et al., 1989; Perris et al., 1989; Lallier and Bronner-Fraser, 1991) have been found between the ectoderm and the dermatome.

The fact that the dorsolateral pathway is invaded by late-emigrating neural crest cells which have to wait until the pathway acquires the molecular characteristics appropriate for migration has long been noticed (Weston, 1963; Tosney, 1978; Serbedzija et al., 1989) and was fully confirmed by more recent observations (see Erickson, 1993b). Thus, by injecting the lipophilic vital dye DiI into the lumen of the neural tube of chick embryos at different stages of the migration process, Serbedzija et al. (1989) showed that injection of the dye at HH12 at the trunk level resulted in the labeling of all types of neural crest derivatives. In contrast, after injection at HH19 or HH21, respectively, only melanoblasts and DRG or melanoblasts alone were carrying the marker. Thus, it appears that the late-migrating neural crest cells generate only melanocytes. Whether this corresponds to an early commitment of this lineage has been answered by the following experiment: neural crest cells from E6 quail embryos differentiate into melanocytes, but also into glial and nerve cells (Nataf and Le Douarin, unpublished; Richardson and Sieber-Blum, 1993). Therefore, cells endowed with developmental potentialities pertaining to various lineages take the dorsolateral pathway.

### 6.3.2 Migration of melanocytic precursors in non-avian vertebrate species

The recent interest in zebrafish, in which mutants can be generated and analyzed on a large scale, together with the fact that the study of embryonic development can also be done at the cellular level by intracellular injection of the vital dye LRD (lysinated rhodamine dextran; see Chapter 1), has permitted the gathering of information about pigment cell development. In fish, neurulation proceeds by formation of an initially compact rod of neuroepithelial cells, the neural keel. Compared to avian embryos, zebrafish have many fewer crest cells at a given axial level (Raible et al., 1992). As in avian embryos, trunk crest cells of zebrafish migrate on a ventral pathway between the neural tube and the somite, and on a lateral pathway between ectodermal epithelium and the somite-derived dermomyotome. Neural crest cells also take the lateral pathway later than the medial one. Direct observation of crest cells in live embryos revealed that they seem to probe the entrance of both pathways before selecting which one to follow. Thus, some cells still enter the dorsoventral direction even after migration has begun on the lateral pathway (Raible et al., 1992). This means that the temporal segregation of the two pathways described in the avian embryo (see Erickson et al., 1992) is not as strict in zebrafish. As in avian and mammalian embryos (Weston, 1991), zebrafish neural crest cells that migrate along the two different pathways have different fates (Raible et al., 1992). Cells migrating along the dorsoventral pathway yield sensory and sympathetic ganglia as well as pigment cells; those taking the mediolateral pathway give rise to pigment cells only. Whether these cell fate differences correspond to an early commitment of neural crest cells cannot be inferred from these observations. Lineage analysis of individual trunk crest cells suggest that the early migrating crest cells produce multiple phenotypes in their progeny. These progenitors then generate restricted precursors able to yield only a single type of derivative (Raible and Eisen, 1994).

The establishment of certain aspects of the pigment pattern in the zebrafish embryo could be related to definite gene activities. Thus, analysis of defects in pigment pattern development caused by sparse (spa), rose (ros), and leopard (leo) single and double mutant combinations suggests that $spa^+$ and $ros^+$ functions are required for development of separate populations of pigment cells in the adult. The role of leo is to control assembly of melanocytes into stripes. Thus, a first migration wave of melanoblasts takes place between 2 and 3 weeks of zebrafish development and is dependent for its dispersion and differentiation in the spa gene. These cells become further arranged into stripes under the influence of the leo gene. From 3 weeks of development, a distinct ros-dependent population of melanocytes differentiates in the stripe. Both early and late differentiating melanocytes then cooperate for the formation of the silver stripes, ensuring expression of melanocyte and iridophore stripes (Johnson et al., 1995).

Recently the large-scale mutation screen for embryonic/early larval mutations in zebrafish has led to the defining of 94 genes affecting all aspects of larval pigmentation: specification, patterning, proliferation, survival and differentiation. Several genes are strong candidates for controling pigment cell specification and proliferation. As examples, the genes salz (sal) and pfeffer (pfe) seem to be involved in xanthophore specification, while shady (shd) and colourless (cls) may be necessary for the specification of iridophores and melanophores, respectively (Kelsh et al., 1996).

Further studies revealed that 17 genes are specifically required for the development of xanthophores. Apart from sal and pfe, mentioned above, that are required for xanthophore formation and migration, others control pigment synthesis (edison, yobo, yocca, and brie) or pigment translocation (esrom, tilsit, and tofu) (Odenthal et al., 1996).

In amphibians, Löfberg et al. (1980) have studied the migration of neural crest cells in the axolotl and seen that it is into the dorsolateral pathway that the cells enter first as they leave the neural primordium. In the naturally occurring white mutant (Dalton, 1949; Keller et al., 1982), neural crest cells take this pathway after a significant delay compared to wild-type embryos. The fact that, in the wild-type, the ECM is rich in chondroitin 6-sulfate proteoglycans while these compounds are missing in the mutants was proposed to account for this delay (Löfberg *et al.*, 1989; Perris *et al.*, 1990; Olsson *et al.*, 1996a,b). Further studies of the white mutants of axolotl involved the construction of chimeras between embryos belonging to the white (d/d) and the albino (a/a) strains and showed that white melanoblasts can migrate into the Albino skin where they differentiate into melanocytes (Houillon and Bagnara, 1996). The interpretation favored by these authors is that the skin of the white mutant does not provide a favorable "milieu" for melanocyte differentiation because it either contains a melanocyte inhibitory factor (MIF) or does not provide these cells with the appropriate stimulating factor (MSF) (see Houillon and Bagnara, 1996, for discussion and references).

In *Xenopus*, cells of the melanocytic lineage are abundant not only in the dorsolateral pathway but also in the dorsoventral pathway (Sadaghiani and Thiebaud, 1987; Krotoski *et al.*, 1988; Epperlein and Löfberg, 1990; Collazo *et al.*, 1993) (Fig. 6.5).

In the mouse, dispersion of presumptive melanoblasts takes place between 8.5 and 12.5 dpc. This was deduced from experiments in which the capacity of isolated fragments of embryonic skin to differentiate into melanocytes in culture was tested. After 12.5 dpc, the precursors proliferate massively in the dermis and then invade the epidermis where they become incorporated into the hair follicles and differentiate into pigment cells (Rawles, 1940b, 1947; Mayer, 1965, 1973). The presence of two, dorsoventral and mediolateral, migration pathways for the trunk neural crest cells, as is the case in birds, was also evidenced in the mouse by DiI injection inside the lumen of the neural tube (Serbedzija *et al.*, 1990). Cells were seen taking the dorsolateral pathway following dye injection from the onset of (E8.5) and throughout (up to E10.5) the crest cell emigration process. Therefore, in contrast to the observations made in birds, the mouse crest cells do not seem to remain in the "staging area" before invading the subectodermal pathway of migration.

The critical stage at which the melanoblasts enter the epidermis was determined by Mayer (1973). Ectoderm–mesoderm recombination experiments were carried out in which pieces of dorsolateral skin from *albino* (*c/c*) and *pigmented (black)* (*C/C*) mouse embryos were separated into their two initial tissue components. Ectoderm and mesoderm were recombined and grown in the chick embryo celom for a sufficient period to allow melanin formation. Recombined skin from 11-day-old embryos formed pigment only when the mesodermal component was from the black strain. "Black ectoderm–albino mesoderm" associations failed to produce pigment in all cases. Combinations made with tissues taken from 12- and 13-day embryos produced pigmented hair in both types of association. This shows that ectoderm had already begun to be invaded by melanoblasts which were present in both components of the skin. In recombinations of skin from 14-day embryos, the "albino ectoderm–black mesoderm" grafts had only white hair. The reverse associations ("black ectoderm–albino mesoderm") possessed melanocytes in the hair as well as in the dermis, and in the chick celom around the graft. The latter

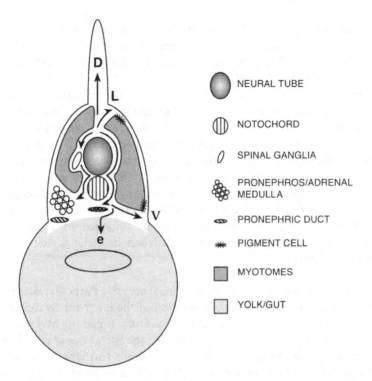

| | |
|---|---|
| ⬤ | NEURAL TUBE |
| ⬤ | NOTOCHORD |
| ⬤ | SPINAL GANGLIA |
| ⬤ | PRONEPHROS/ADRENAL MEDULLA |
| ⬤ | PRONEPHRIC DUCT |
| ✳ | PIGMENT CELL |
| ⬛ | MYOTOMES |
| ⬜ | YOLK/GUT |

**Figure 6.5** Summary of the derivatives to which the neural crest contributes and the paths of migration taken by neural crest cells in *Xenopus laevis* embryo. D, dorsal pathway; L, lateral pathway; V, ventral pathway; e, enteric region. Pigment cells migrate along pathways L and V. (Reproduced, with permission of Company of Biologists Ltd, from Collazo *et al.*, 1993.)

localization is seen in all types of combinations. These results can be interpreted by assuming that the presumptive amelanotic melanocytes of the albino strain already present in the epidermis at 14 days of gestation prevent melanoblasts of the wild-strain from entering it. Thus, it seems that the skin can receive a definite number of presumptive pigment cells, beyond which the seeding process is blocked. This suggests that the epidermis of the developing hair follicle exhibits a constant ratio of melanoblasts to epidermal cells and that this ratio is established at 14 days in the mouse, a notion that can be related to that of the "epidermal melanin unit" mentioned earlier.

That more than one pigment cell precursor can colonize a hair follicle has been demonstrated in tetraparental mice[1] obtained by combining pigmented and non-pigmented strains. These mice displayed hair of the parental types

---

[1]Tetraparental mice result from aggregation of two cleaving mouse embryos. Embryos aggregate, form a single blastocyst in culture and, after transplantation to the uterus of a foster-mother, develop into fetuses of normal size and morphology (Tarkowski, 1961, 1963, 1964; Mintz, 1962a,b, 1964). They are also called aggregation chimeras (Tarkowski, 1961) or allophenic mice (Mintz, 1967).

(i.e., either fully pigmented or totally devoid of pigment) and in addition possessed many chimeric hairs with adjacent septa differing widely in the amount of pigment they contained (McLaren and Bowman, 1969; Mintz and Silvers, 1970; Cattanach et al., 1972).

An interesting experimental system to study the migratory capacities of melanocytic precursors in the mouse was designed by Jaenisch (1985) and also used later for rat embryos (Kajiwara et al., 1988). Cultured mouse neural crest cells were microinjected in utero into mouse embryos at 9 dpc. The cells are mostly deposited in the amniotic cavity and enter the embryo via the anterior and posterior neuropores. The donor cells can migrate over considerable distances to the skin where they participate in hair pigmentation (Jaenicsh, 1985; Huszar et al., 1991) and also to certain internal organs like the inner ear. This was shown in cases where the recipient belonged to a mutant strain ($W^v/W^v$) which exhibits a large or total deficit in melanocytes (see below) (Cable et al., 1994).

*6.3.2.1 Colonization of the inner ear by melanocytes.* Particular attention has been paid in recent years to the colonization of the inner ear by melanoblasts, and to the role played by these cells in auditory function. Melanocytes are abundant within the cochlea, especially in the stria vascularis. The stria, which lines the lateral wall of the cochlea, contains two types of melanocytes, called light and dark intermediate cells (Cable and Steel, 1991). The strial epithelium produces the potassium-rich endolymph in the cochlea which has a resting or endocochlear potential (EP) of about 100 mV in the mouse (Cable and Steel, 1991).

As described below, *Dominant White Spotting* mutations (*W*) affect (to various degrees according to the allele considered) the gene encoding the tyrosine kinase receptor c-*kit* which is involved in a signaling pathway essential for melanocyte survival. Drastic impairment, more or less complete, of this pathway leads to the formation of white skin areas where melanocytes are missing, and to various degrees of deafness. In mice affected with the *Viable Dominant Spotting* ($W^v/W^v$) mutation (see below), inner ear melanocytes are missing and the EP is close to zero. Such mice have elevated thresholds for evoked cochlear activity. These functional impairments have been related to the absence of melanocytes to which a role was assigned in generating the endocochlear potential (Deol, 1970a,b; Schrott and Spoendlin, 1987; Steel et al., 1987; Steel and Barkway, 1989; Cable et al., 1994).

White spotting patterns associated with congenital deafness are common in many mammal species, including man (e.g., Deol, 1970a; Searle, 1968a; Steel, 1991). One example is the Dalmatian dog, which has a spotted coat with a high incidence of deafness associated with deficiencies in pigmentation of the stria of the inner ear (Cable et al., 1994). On the other hand, human piebaldism, caused by a mutated *KIT* gene (Giebel and Spritz, 1991) in the homozygote state, was reported to involve deafness (Hultén et al., 1987).

### 6.3.3 Clonal model of melanoblast development deduced from the pigment pattern of tetraparental mice

Tetraparental mouse chimeras have proved to be of great interest in the analysis of cell lineages, particularly in cases where suitable cell markers such as pigments were available. The first analysis of the coat color patterns of mouse chimeras was done by Mintz (1967). In a number of albino-colored or black $(B/B)$–brown $(b/b)$ strain combinations, she observed a high incidence of patterns with transverse bands of the two colors on the head, body, and tail. On the borders of the bands, adjacent colors were mingled to various degrees, and in some cases mingling extended throughout the band. At the mid-dorsal line there was a sharp discontinuity, and the pattern of the two sides of the animal seemed to be established independently.

According to Mintz, the entire coat pigmentary system is determined early in embryonic life. She postulated that, from the neural crest, 17 pairs of cells are set aside as clonal initiators of melanoblasts which then proliferate and migrate laterally, forming transverse bands. At clonal boundaries, cells of adjacent clones are admixed. Adjacent descendants of identical precursors give rise to clones that cannot be distinguished. Therefore, the number of initial clones had to be inferred from the maximum number of bands found in a single allophenic mouse. This has been evaluated at 17 melanoblast clones on each side, including the tail (i.e., 34 clonal initiator cells for the whole body). Such a disposition corresponds, according to Mintz, to an "archetypal or standard pattern" around which variations are attributed to what she terms "developmental noise." This may be due to several factors such as cell mingling (McLaren and Bowman, 1969), cell death, and differential rates of proliferation (Mintz, 1971). From her initial observations, Mintz thought that in the allophenic mice the two genetically different clones (e.g., black and white) occupy "alternating rather than random positions in the chains" of melanocyte precursors (Mintz, 1967). In fact, a non-random arrangement seemed implausible, and a much larger study led her to conclude there was a random distribution of the initial melanoblasts of both types. Then, the "standard pattern" is only established when clones of different colors happen to alternate, making individual clone identification possible (Mintz, 1970, 1971).

Support for Mintz's model has been provied by the pigmentation patterns of mice heterozygous for X-linked color genes (Cattanach et al., 1972).

Another experimental approach to study this problem was to generate single genotype mosaic animals by labeling single cells in the embryo in utero. This was acheived by microinjection of replication defective retroviral vectors (Compere et al., 1989a,b) into neurulating mouse embryos. A retroviral vector (MLV-HuTyr) bearing a human tyrosinase cDNA under the transcriptional control of MLV LTR (murine leukemia virus long terminally repeated) was microinjected into midgestation albino embryos. The retrovirus rescues the defect of the white melanocytes of the recipient albino strain mouse and clonal progenies of these infected cells were pigmented. The pigmented bands generated by this approach were very similar to the "standard" stripes observed in aggregation chimeras. These results thus support the notion that the unit bands seen in aggregation chimeras are clonal in origin and confirm Mintz's (1967) contention that coat pigmentation is generated by a few melanoblast precursors (Huszar *et al.*, 1991).

## 6.4   The differentiation of pigment cell precursors

As for the other derivatives of the neural crest, the problem is raised as to when the pigment cell lineage becomes segregated during ontogeny. This question will be treated in depth in Chapter 7, but some specific points related to pigment cells will be alluded to here. A significant amount of the work aimed at elucidating the problem of the commitment of neural crest cells toward the melanocytic differentiation pathway, as well as the requirement of melanoblasts for differentiation, was done in culture. The most important breakthrough in the latter respect was, however, provided by the molecular cloning of genes whose mutations affect pigmentation in the mouse. This novel avenue of research will be discussed in Sections 6.5 and 6.6 (see also Section 6.1.1.3). One of the requirements for deciphering the sequential molecular events leading from the undifferentiated neural crest cell to the fully differentiated melanocyte is to discover the molecular markers that characterize the successive steps of melanocyte development.

### 6.4.1   Molecular markers of melanocytic precursor cells

Apart from the molecular markers which are shared by other neural crest cells (see Chapter 1), some are specific for pigment cell lineage.

A number of these markers have been produced in the last decade either through the use of Mab (monoclonal antibody) technology or after the cloning of genes expressed in these cells.

The human marker of prepigment cells recognized by the Mab HMB-45 (Gown *et al.*, 1986) detects melanoblasts in the embryo as early as at 40–50 days of gestation. The antigenic epitope it binds is carried by the glycoprotein pmel17 (Kwon *et al.*, 1991), which turned out to be recognized also by two other reagents obtained independently by Adema *et al.* (1993). *pmel17* expression disappears during late gestation but reappears in melanomas.

In mouse embryos, melanoblasts can be detected as early as at 10 dpc by an antibody directed against the product of the *slaty* locus, the enzyme tyrosinase-related protein-2 (TRP2), which belongs to the melanin synthesis catalytic cascade (Jackson *et al.*, 1992; Steel *et al.*, 1992; Pavan and Tilghman, 1994). This enzyme is also designated DOPAchrome tautomerase (Dct) (see Fig. 6.3).

In quail and chick embryos several markers are also available to follow the early steps of melanocyte differentiation. MEBL-1, discovered by means of immunological methods by Kitamura *et al.* (1992), is present in chick melanoblasts from E3 when crest cells begin to emigrate from the neural tube along the dorsolateral pathway.

Nataf *et al.* (1993) obtained several Mabs which recognize antigenic determinants carried by cells of the melanocytic lineage of chick and quail by immunizing a mouse with a crude preparation of membranes of quail neural crest cells cultured in conditions (i.e., with high concentration of chick embryo extract) which favor melanocyte differentiation. Mel1 and Mel2 Mabs detect antigens present in melanosomes from E8 onward in the quail, after pigmentation has begun. More interesting, owing to its more precocious expression which takes place prior to the appearance of pigments, is the antigen MelEM (melanoblast early marker) (Fig. 2.3), which turned out to be an $\alpha$ subunit of the enzyme glutathione *S*-transferase (Nataf *et al.*, 1995). This particular form of the enzyme is specific to neural crest-derived melanocytes since it is not present in the pigmented retina. Moreover, it is not present in any other derivative of the neural crest, but is abundant in liver cells.

### 6.4.2 In vitro *studies of melanocyte differentiation from neural crest progenitors*

The first attempts to culture neural crest cells were those of Cohen and Königsberg (1975). These authors took advantage of the tendency of neural crest cells to spread around the explanted neural primordium to devise a simple technique for the isolation of neural crest cells.

Quail neural tubes isolated at E2 yield cells which exit from the neural folds and migrate around the explant on the substrate provided by the Petri dish. If the neural tube itself is removed after 1 or 2 days, the remaining cells are predominantly of crest origin with some possible contaminatior of cells of neural tube origin. After a period of growth the cells can be replated at low density to establish clones.

Three types of clones were obtained in these conditions: in some of them all cells were pigmented, others were entirely unpigmented and some contained a mixture of both types (Cohen and Königsberg, 1975; Sieber-Blum and Cohen, 1980). The plating efficiency of pigmented clones was two or three times higher that of unpigmented or mixed colonies. In the last two types, histofluorescence for catecholamines (CA) revealed a small proportion of clones containing CA cells. Interestingly, the number of clones with CA fluorescence depended on the nature of the culture substrate. Thus, ECM (extracellular matrix) produced by somites or skin fibroblasts enhanced the percentage of CA-positive clones from 0.9% to 6.9% (Sieber-Blum and Cohen, 1980). Certain CA-positive colonies also had a pigmented component, showing that a common precursor derived from the neural crest is capable of giving rise to both melanocytes and CA-producing cells. By transplanting mixed clones, which had first developed *in vitro,* into the lumen of trunk somites, Bronner-Fraser and Cohen (1980) obtained the survival of both CA-containing and pigmented cell populations. The quail nuclear marker allowed their identification in the host tissues.

Clonal analysis of quail neural crest cells *in vitro* was also pursued by seeding single cephalic neural crest cells from 10–13ss quail embryos directly (i.e., without a primary step of "mass" culture), under the control of a microscope, on to a feeder layer of growth-inhibited 3T3 cells (Baroffio *et al.*, 1988, 1991; Dupin *et al.*, 1990; Dupin and Le Douarin, 1995). These experiments showed that, at the time of their migration, the neural crest cells are highly heterogeneous with respect to their capacity to proliferate and to differentiate. Most crest cells are pluripotent, whereas a few are already committed toward a definite phenotype as attested by the fact that they yield colonies containing only a single phenotype (Fig. 6.6). The culture conditions on the 3T3 fibroblastic feeder layer turned out not to be favorable to melanocytic differentiation since the numbers of pigmented cells in the clones were generally modest. However, it appeared clearly that, during their migratory phase, the several types of neural crest progenitors could be evidenced by the phenotypes recorded in the colonies they produced: the capacity to produce melanocytes was found in a variety of progenitors to be associated with that of producing glial cells and/or no TH-containing neurons, and/or TH-containing cells, and/or cartilage (i.e., mesectoderm). These potentialities can be found in bipotent or pluripotent precursors as indicated in Fig. 6.6. If quail trunk neural crest is cultured directly on plastic in non-clonal ("mass" culture) conditions (Dupin and Le Douarin, 1993, 1995), TH-positive cells and melanocytes begin to differentiate after 4 days of culture in permissive conditions [i.e., in the presence of 10% of chick embryo extract (CEE) + 10% fetal calf serum (FCS) in Dulbecco's modified Eagle's medium (DMEM)], whereas with 2% of CEE, pigment cells appear in smaller numbers and virtually no TH-positive cells developed in the culture. Retinoic acid (RA) at a concentration of 100 nM increased by 44-fold the number of TH-positive cells and by 2.4-fold that of pigment cells in a 2% CEE-containing medium. Moreover, while the effect of RA on TH activity has to take place from the onset of the culture to be effective, it seems to act on pigment cell differentiation at later stages: from day 3 to day 6 of culture.

The strong stimulation of CEE on melanogenesis and catecholaminergic cell differentiation was also observed by Howard and Bronner-Fraser (1985), and Howard and Gershon (1993). Whether RA is one of the factors present in CEE which is responsible for the stimulation of adrenergic and melanogenic differentiation is a possibility. Other factors like endothelin-3 (see below) have recently been shown to be potent inducers of melanogenesis in cultured quail neural crest (Lahav *et al.*, 1996).

The important related question eventually concerns the role of RA in NC cell development *in vivo*. Is RA (or related retinoids) available to crest cells at the correct differentiation sites and developmental times and, if so, does RA actually influence the normal differentiation of crest-derived cells? Several lines of evidence argue for the presence of retinoids in the skin. The fetal ectoderm in mouse coexpresses alcohol dehydrogenase 1, the enzyme responsible for conversion of retinol into RA (Vonesh *et al.*, 1994), and CRBP1 which functions as

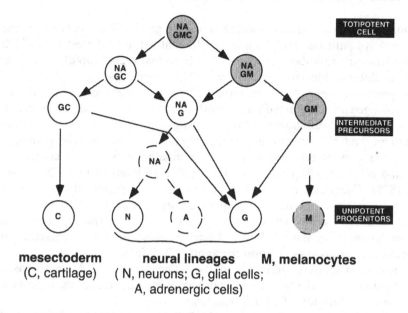

**mesectoderm**          **neural lineages**   **M, melanocytes**
(C, cartilage)      ( N, neurons; G, glial cells;
                     A, adrenergic cells)

**Figure 6.6** Diagram illustrating the different types of progenitors from quail cephalic neural crest cells as revealed by the phenotypic analysis of the colonies generated by cloned cells. The cell types recorded in the clones were cartilage (C), neuron (N), glial cell (G), adrenergic cell (A), and melanocyte (M). The putative progenitors have been classified according to the number of these distinct cell types in their progeny; the combinations of phenotypes corresponding to the progenitors actually recorded in the experiments are shown in full circles. The results are thus consistent with the generation of unipotent progenitors from a "totipotent" neural crest stem-like cell through several intermediate oligopotent precursors. Filiations between precursors are only hypothetical. The precursors endowed with melanogenic potencies are shaded. (Reproduced, with permission of Current Biology Ltd, in modified form, from Le Douarin *et al.*, 1994.)

a substrate carrier in local RA synthesis (Gustafson *et al.*, 1993), suggesting that melanoblasts very likely receive ectodermally derived RA.

Clonal analysis by the limit dilution method provided further evidence that pluripotent precursors endowed with the capacity to yield sensory and autonomic neuroblasts together with melanocytes are present in the quail trunk neural crest (Sieber-Blum, 1989c; Ito and Sieber-Blum, 1991).

Pigment cell precursors were found to differentiate in cultures of various types of ganglia such as DRG from E5 chick embryos (Cowell and Weston, 1970; Nichols and Weston, 1977; Nichols *et al.*, 1977). These melanocytes are likely to be derived from remaining pluripotent precursors.

In fact, Sieber-Blum *et al.* (1993) demonstrated by their cloning technique that pluripotent progenitor crest cells still exist in several sites of the embryo at advanced stages of migration. Thus, some progenitors that had colonized the skin could, in clonal cultures, give rise to pigment cells associated with sensory and adrenergic neurons. Also of interest in the present context is the fact that neuron/pigment cell progenitors were found at stage 29 of Zacchei (1961) in the DRG and sympathetic ganglion anlagen. In contrast, neural crest cells which migrate from the rhombencephalon to the branchial arches rapidly lose their ability to differentiate into pigment cells. This means that survival of pluripotent precursors depends on the target site where neural crest cells migrate.

A prolonged association with the neural tube that can be obtained in certain culture conditions was shown to favor the engagement of the just emigrating pluripotent crest cells toward the melanocytic differentiation pathway. In conditions that prevent cell dispersal (i.e., a non-adhesive substrate containing agar) and promote cell proliferation (i.e., with high levels of FCS (20%) and CEE (8%), Glimelius and Weston (1981) obtained practically one single phenotype in crest cultures from quail embryos. Forty-eight hours after plating, crest cells formed clusters which remained at the surface of the neural tube, each containing several thousand cells. Most of these cells differentiated into melanocytes upon cultivation for another 3–5 days.

If such clusters were subjected to the influence of TGF$\beta$, there was a dramatic decrease in the number of melanocytes arising in culture, regardless of the onset or duration of TGF$\beta$ treatment.

In contrast, some neural crest-derived cells from quail DRG and peripheral nerves (Schwann cells) treated in culture with 12-$O$-tetradecanoylphorbol 13-acetate (TPA) are able to differentiate into melanocytes (Ciment et al., 1986; Kanno et al., 1987; Stocker et al., 1991). Similarly FGF2-induced pigmentation in about 20% of such cultures. TPA and FGF2 can act synergistically on pigmentation of glial cell cultures, an effect which is antagonized by TGF$\beta$1 (Stocker et al., 1991) and was later shown to be mediated via an autocrine pathway (Sherman et al., 1993; also see discussion in Chapter 7).

## 6.5 Spotting mutants in mouse reveal the nature of growth and survival factors for melanocytic precursors

Mutations affecting the development of integumental melanocytes while having no effect on eye pigmentation are generally referred to as "spotting mutants." The loci responsible for these mutations act on melanocytic precursors early in their developmental history and their dysfunction results in either local or global loss of skin pigmentation. According to Silvers (1979), these loci include *Lethal spotting, Belted, Dominant spotting, Patch, Rump-white, Steel, Flexed tailed, Splotch, Varitint-waddler, Mottled, Microphthalmia, Fleck,* and *Belly spot and tail.* Some have been characterized to some extent at the molecular level while for others no molecular data are available (Chabot et al., 1988; Geissler et al., 1988; Copeland et al., 1990; Huang et al., 1990; Zsebo et al., 1990a,b; Epstein et al., 1991; Stephenson et al., 1991; Hodgkinson et al., 1993; Greenstein-Baynash et al., 1994; Hosoda et al., 1994; Levinson et al., 1994; Nagle et al., 1994). Some of the mutants, which are particularly relevant for understanding the development of pigment cells, will be discussed in this chapter.

### 6.5.1 Dominant spotting (W) *and* Steel (Sl) *mutants*

Among the many mutations known in the mouse that interfere with the normal sequence of pigment cell development (see Deol, 1973), *Dominant spotting* (*W*), identified by Little (1915), and *Steel* (*Sl*), first described by Sarvella and Russell (1956), are of particular interest because of the association of the pigment defect with macrocytic anemia and deficiency of primordial germ cells. It has been shown that the failure in melanocyte development and the anemia in the *W* mutants result from a defect affecting the melanoblast and the hemopoietic cells themselves. In the case of *Sl*, in contrast, it is the tissue environment into which both presumptive pigment and blood cells have to home in order to differentiate that is responsible for producing the abnormalities (see Russell, 1979, and Silvers, 1979).

Twenty-seven independent alleles of *W* (located on chromosome 5) and at least 50 of the *Sl* (on chromosome 10) loci have been recognized (see Lyon and Searle, 1990).

Pigmentation and inner ear colonization by melanocytes are affected in animals carrying alleles at the *W* locus. The characteristic spotting pattern of *W* heterozygotes varies in the extent and locations of the white areas, which in all cases are limited by clear-cut boundaries between pigmented and non-pigmented regions (Geissler *et al.*, 1981). In contrast, in *Sl* heterozygotes the coat color shows a slight dilution of the pigment with, however, unpigmented feet and tail tip (Sarvella and Russell, 1956). Severely affected, but viable, mutants of either strain (e.g., *W/W* and *Sl/Sl^d*) are not distinguishable with completely white coat and black eyes in which the retina is pigmented, while the choroid and the skin are totally devoid of pigment cells (Bennett, 1956; Market and Silvers, 1956).

The fact that in *W* mutants the defects observed reside in the three affected cell types (melanoblasts, primordial germ cells, and hemopoietic stem cells), whereas in *Sl* mutants it is the microenvironment in which these cells differentiate which is affected, suggested that the genes of these loci respectively encode a receptor and its ligand (Russell, 1979).

*6.5.1.1  W encodes a tyrosine kinase receptor.* The discovery that *W* encodes a tyrosine kinase receptor was made possible thanks to the isolation of the v-*kit* oncogene as the transforming gene in a new strain of feline sarcoma virus (Besmer *et al.*, 1986). The human c-*kit* proto-oncogene was thereafter cloned and characterized by sequence homology. The deduced amino-acid sequence of c-*kit* reveals a transmembrane receptor molecule structurally related to the receptors for CSF-1 and PDGF, with a tyrosine kinase cytoplasmic domain and an extracellular region containing five immunoglobulin (Ig)-like domains (Yarden *et al.*, 1987; Qiu *et al.*, 1988).

It was then found that c-*kit* is the product of *W* (Chabot *et al.*, 1988; Geissler *et al.*, 1988). The distinct *W* phenotypes, related to all the examined *W* alleles, result from a deficiency in c-*kit*-associated kinase activity due to various types of mutations. The latter can be due either to the disruption of the gene by genomic rearrangements or missense mutations that replace one amino acid in the c-*kit* protein product (Reith *et al.*, 1990; Tan *et al.*, 1990). Mutations that confer reduced levels of an apparently normal protein produce wild phenotypes in the heterozygotes. In contrast, mutations that impair the kinase activity of c-*kit* give more strongly dominant heterozygous phenotypes. This suggested that coexpression in the same cell of normal and defective c-*kit* proteins alters signal transduction by the normal protein. This can be accounted for by the fact that the c-*kit* receptor functions in a homodimeric form and that heterodimers of a wild-type and a functionally inactive receptor are unable to transduce the signal.

Further studies revealed that the c-*kit* gene is also present in all the other vertebrate species where it was looked for (e.g., rat, chick, quail, etc.; Tsujimura *et al.*, 1991; Sasaki *et al.*, 1993; Lecoin *et al.*, 1995).

*6.5.1.2 Identification of the* Sl *gene.* The gene of the *Sl* locus was cloned simultaneously by several groups using different approaches based on the various biological activities of the protein it encodes (Anderson *et al.*, 1990; Huang *et al.*, 1990; Martin *et al.*, 1990; Williams *et al.*, 1990; Zsebo *et al.*, 1990a). The murine gene is located on chromosome 10, where the *Sl* locus has been assigned by genetic analysis.

The protein encoded by the *Sl* gene has been identified as a transmembrane molecule and as a secreted factor which consists of the first 164 or 165 amino acids of the extracellular domain encoded by the cDNA (Flanagan and Leder, 1990; Nocka *et al.*, 1990a; Williams *et al.*, 1990; Zsebo *et al.*, 1990a,b). Alternative splicing and proteolytic cleavage of the transmembrane protein have been found to be responsible for generating the soluble factor (Flanagan *et al.*, 1991; Huang *et al.*, 1992; Majumdar *et al.*, 1994). Several names have been coined to designate the secreted factor: stem cell factor (SCF), Steel factor (SF), kit ligand or mast cell growth factor (MCGF) (see Williams *et al.*, 1992, for a review).

Mice that lack the membrane-bound form of SF are severely affected by the mutation (Flanagan *et al.*, 1991; Brannan *et al.*, 1992). This suggests that short-range cell-to-cell signaling mediated by c-kit/SF is critical for the development of the cell types (i.e., primordial germ cells, hemopoietic stem cells and mela-noblasts) affected by the *Sl* or *W* mutations.

*6.5.1.3 Expression pattern of* Sl *and* W *genes.* As expected, *Sl* and *W* genes show a complementary expression pattern during development with the cells affected by the mutations expressing c-*kit*, whereas the tissues forming the environment in which they differentiate express the *Sl* gene which encodes the kit ligand.

In the mouse, the first few cells expressing c-*kit* mRNA are detected dorsal to the somites within the migratory pathway of neural crest cells at E10, approximately 24–36 hours after crest cells emerge from the neural tube. As development proceeds, c-*kit*-positive cells appear in more and more ventral locations of the dermis, and starting from E13.5 some are already observed in the epidermis. Labeled cells become more numerous in these two sites at E15.5 and E19.5. At these stages, however, presumptive melanoblasts which are still devoid of pigment could not be distinguished from mast cells, known to express c-*kit* and to be located in the dermis. At birth, these labeled cells were frequent in the dermis, epidermis, and hair follicles (Manova and Bachvarova, 1991).

*Sl* expression is first detected in the somitic compartment that is the precursor of dermis at a time preceding the influx of pigment cells. At E13 when melanoblast colonization is recently completed in the forelimb, high levels of *Sl* transcripts are found in mesenchymal cells underlying the epidermis. The same pattern is observed at E15.5 when melanoblasts are in the process of proliferating and differentiating (Keshet *et al.*, 1991). A role for SF in melanocyte differentiation was further suggested by the finding that in transgenic mice *Sl* regulatory sequences conferred expression of a *LacZ* reporter in the dermal papillae of hair follicles (Yoshida *et al.*, 1996a).

In the avian embryo, the pattern of expression of c-*kit* in the neural crest cells could be precisely defined through the joint use of *in situ* hybridization using the c-*kit* probe and the quail–chick chimera system to selectively labeling neural crest cells as they migrate and home to their various locations (Lecoin *et al.*, 1995).

It was first found that the neural crest cells migrating along the dorsoventral pathway do not express the c-*kit* gene. In the chick embryo the first c-*kit*-expressing cells are found at E4 under the superficial ectoderm in the dorsal mesenchyme lateral to the neural tube. They progressively invade more lateral areas between ectoderm and somite. By E5, the number of c-*kit*-positive cells have increased in subectodermal position while some have already migrated into the ectoderm itself. They reach the lateral limit of the dermomyotome at the end of E5. From E5 on, some cells in the dorsal area of the dorsal root ganglion (DRG) were c-*kit* positive. This is in agreement with the *in vitro* studies of Carnahan *et al.* (1994) who found that the Steel factor has a survival effect on mediodorsal DRG neurons. The mesenchyme of the branchial arches was also strongly labeled by the c-*kit* probe.

In chimeric embryos in which a quail neural tube was grafted into a chick at the cervicotruncal level, quail neural crest cells located underneath the ectoderm in the mediodorsal pathway of migration at E3 did not express c-*kit*. In contrast, most quail cells located in this position at E4 were c-*kit*-positive, thus confirming the observations made in the intact chick. Lecoin *et al.* (1995) also studied two pigmentation mutants: White Leghorn, a cell-associated mutant that involves melanocyte programmed cell death from E13 onward (Jimbow *et al.*, 1974), and the Silky Fowl, a strain of chickens which, in addition to the silky appearance of their feathers which retain the down structure in the adult, also shows hyperpigmentation of the internal organs (Kuklenski, 1915). Most tissues, including meninges, gonads, nerves, blood vessels, muscles, body wall, and periosteum, contain numerous melanocytes which pigment them from midincubation time onward. In contrast, the feathers are pigmented only transiently in the Silky Fowl embryo and remain white throughout life. The dermis and body wall are so heavily pigmented that the comb and wattles have a black color. Eastlick and Wortham (1946) showed that the internal pigmented cells are true neural crest-derived melanocytes and not melanin-storing macrophages such as those encountered in certain forms of melanosis. Moreover, the pigment defect was shown to be due to an environmental factor and not to a cell-autonomous defect of melanocytes (Hallet and Ferrand, 1984).

The aim of this investigation was to see if the abnormalities in the pigmentation observed in these two mutants might be related to aberrant expression of c-kit and/or Sl genes. Silky Fowl chick embryos were used as hosts for engrafting quail neural tubes at E2. Up to E5, the number and distribution of c-kit-positive cells was the same in Silky and wild-type chick embryos. From E5 onward the number of c-kit-positive cells had increased significantly, principally in chimeras (normal neural tube-grafted in Silky embryos); some of them had entered the ectoderm while others were still present in the underlying mesenchyme. In these chimeras, many c-kit-positive cells found around the dorsal aorta at E5 and E6 had quail-type nuclei, indicating that these c-kit-positive cells did not belong to the hemopoietic lineage but were melanoblasts of neural crest origin which had invaded that region of the Silky embryo at that stage.

Sl transcripts were first found at E3, in the notochord, the dorsal mesentery, the lateral plate mesoderm, the yolk sac, and amniotic mesoderm, but not in the embryonic ectoderm where they appear from E4 only (Fig. 6.7).

The complementarity of expression of c-kit and Sl is particularly striking during development of down feathers. Down feathers start to develop in the chick at E8; Sl mRNA is abundant in the epidermis and absent in the dermis. During the following days, as the feather germ elongates, Sl mRNA becomes more abundant at the apex of the feather germ. In chimeric embryos, from E12, Sl expression becomes restricted to the apical part of the feather filaments.

In chimeric embryos the relationships between the melanoblasts (of quail origin) and the chick dermis and epidermis could be followed throughout feather development. In the dorsal pterylae, melanoblasts which have started to invade massively the epidermis from E5 onward are detectable in the whole surface of the epidermis in the E11 feather bud (Fig. 6.8). Those that are located at the sites where Sl mRNA is the most abundant are already pigmented, while those which are at the basis of the feather germs are not. When barbs and barbules start to develop (see Fig. 6.9) melanocytes are aligned at the basis of the barb ridges extending pigmented processes towards the periphery of the feather filaments as described by Watterson (1942). Their cell bodies were strongly c-kit positive. While at E11, the Sl gene was still uniformly expressed by most epidermal cells, at E13 its expression was restricted to the outermost cells of the ridges (Fig. 6.9). One can propose that a concentration gradient of

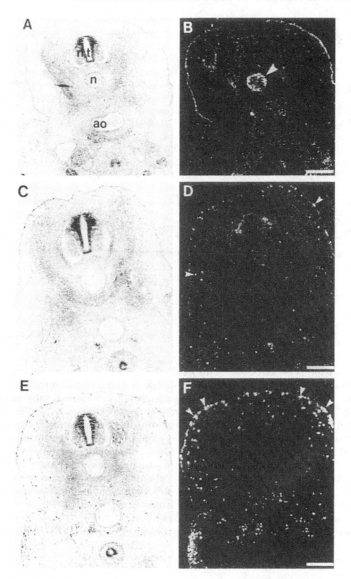

**Figure 6.7** *Sl* and c-*kit* expression in early chick embryos (HH24). (A, B) *In situ* hybridization with *Sl* probe on a normal chick. *Sl* expression is detached in the ectoderm and in the notochord (arrowhead). The signal is more intense in lateral than in dorsal ectoderm. (A) Bright-field; (B) dark-field. n, notochord; nt, neural tube; ao, aorta. (C–F) Comparison of c-*kit* expression pattern between wild-type embryos (C, D) and embryos of the Silky Fowl strain which is hyperpigmented (E, F). In both types of embryos, the c-*kit*-positive cells (arrowheads) are distributed in the ectoderm and the underlying mesenchyme (arrowheads). They are much more numerous in the Silky Fowl embryo where many c-*kit*-positive cells (arrowheads) are distributed in the ectoderm and the underlying mesenchyme (arrowheads). They are much more numerous in the Silky Fowl embryo where many c-*kit*-positive cells are also distributed around the dorsal aorta. (C, E) Bright-field; (D, F) dark-field. Bar = 24 μm. (Reproduced, with permission, from Lecoin *et al.*, 1995.)

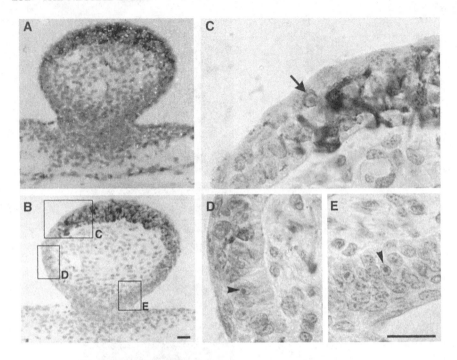

**Figure 6.8** *Sl* expression associated with Feulgen–Rossenbeck staining showing quail cells of neural crest origin in a quail–chick chimera. *Sl* expression in a feather bud of a quail–chick chimera at the beginning of melanogenesis. A quail neural tube was grafted in a 15ss embryo isotopically and isochronically at level of somites 8–14. Fixation was performed at E11 (HH 37). (A) *In situ* hybridization with a radioactive *Sl* riboprobe on a feather bud of the dorsal pteryla. *Sl* is mainly expressed in the epidermis of the top of the bud and is no longer expressed in the rest of the skin epidermis. (B) Adjacent sections were subjected to Feulgen–Rossenbeck staining. The framed areas are enlarged in C, D, E. At the top of the feather bud, quail cells are pigmented (arrow in C), whereas quail cells at the basis of the feather bud are still undifferentiated (arrowheads). Bar = 17 μm. (Reproduced, with permission, from Lecoin *et al.*, 1995.)

*For a colored version of this figure, see www.cambridge.org/9780521122252.*

Steel factor (SF) is thus created along which the differentiating melanocytes extend their processes. At E17–18, c-*kit* and *Sl* expression were extinguished in the feather filaments while epidermal cells and melanocytes were dying.

Comparison of c-*kit* and *Sl* expression in normal chicken (JA57 strain) and mutant White Leghorn and Silky revealed that, in both strains, expression of *Sl* is similar to that observed in the wild type. Cells positive for c-*kit* were present but in smaller number than in other breeds in the E9–11 skin. They had disappeared at E14 in the White Leghorn and were, in contrast, very numerous in Silky embryos particularly in internal organs where they are not encountered in the wild-type chicken (Fig. 6.5). As a conclusion of this comparative study of wild-type and Silky mutant chick embryos, it appears that an abnormal

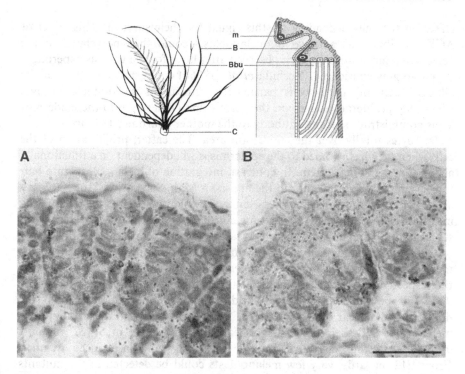

**Figure 6.9** Top: Schematic drawing of a down feather modified from Watterson (1942). The melanocytes (m) are located at the inner side of the barb ridge (B) and deliver their pigment granules to the outermost barbule cells (Bbu). C, calamus. Bottom: Complementary expression pattern of *c-kit* and *Sl* in the developing feather. *In situ* hybridization with *c-kit* and *Sl* radioactive riboprobes on E13 feathers sectioned transversally. (A) Melanocytes expressing the *c-kit* gene are located at the basis of the barb ridge, whereas *Sl* is expressed by the barbule cells at the periphery of the feather (B). It is therefore conceivable that the Steel factor attracts the melanocyte processes towards the outermost aspect of the barb while they deliver pigment to the barbule cells. Thus, the c-kit/Steel system is likely to play a role in the transfer of the melanosomes from the melanocyte to the keratinocyte. Bar = 26 μm. (Reproduced, with permission, from Lecoin *et al.*, 1995.)

*For a colored version of this figure, see www.cambridge.org/9780521122252.*

production of SF is not responsible for the increase and spreading of melanocytes in the Silky Fowl strain.

*6.5.1.4 Functional studies on the Sl/c-kit signaling pathway in the melanocytic lineage.* The role of c-*kit* during melanocyte development was first investigated *in vivo* in the mouse by the administration of an anti-c-*kit* blocking antibody (ACK2 Mab) to pregnant mice at different stages of pregnancy (Nishikawa *et al.*, 1991) and by observing the distribution of ACK2-positive cells in the skin of embryos at different stages (Yoshida *et al.*, 1996a). Moreover, the blocking antibody supresses the c-kit/SF signaling process and eventually has a lethal effect on the cells depending on this signal for their survival. This effect of ACK2 antibody on survival of melanoblasts could then be related to the defects in pigmentation pattern observed after birth for the various experimental paradigms considered. Yoshida *et al.* (1996b) could demonstrate that the c-kit/SF signaling pathway is particularly critical for melanoblast survival when they proliferate actively in the dermis upon entering into the epidermis. Thus administration of the antibody to the mother at around 13.0 dpc depletes melanocytes totally over the entire skin area. Thereafter, proliferation of the cells when they have homed to the epidermis is still dependent on a functionally active c-*kit* signaling system. In contrast, integration of melanoblasts into hair follicles makes them resistant to the anti-c-*kit* Mab. Such a resistant stage was also identified before melanoblasts enter the epidermis. Thus, differentiating melanoblasts go through alternating c-*kit*-dependent and c-*kit*-independent stages and SF functions as a survival factor for proliferating melanoblasts.

Comparative studies involving normal mice and $Sl^d/Sl^d$ mutants in which membrane-bound SF is missing (Steel *et al.*, 1992) were essential in delineating the respective role of the soluble and the membrane-bound forms of SF. The *TRP-2* probe was used in *in situ* hybridization to identify melanoblasts in the developing inner ear. In normal mice, *TRP-2* was found to detect migratory melanocytes and their precursors as early as E10, while in $Sl^d/Sl^d$ mutants *TRP-2* expressing cells were found at E11 only and in reduced number. From E12 onwards, very few melanoblasts could be detected in the mutants and none after birth (Steel *et al.*, 1992). Wehrle-Haller and Weston (1995) have shown that the soluble ligand for c-*kit* induces dispersion of the earliest melanoblasts into the dorsolateral pathway to the skin, while its membrane-bound form is necessary for supporting migration within this pathway.

The period when cells of the melanocytic lineage require SF was further investigated by Morrison-Graham and Weston (1993) using *in vitro* cultures of murine neural crest cells. It turned out that survival of neural crest-derived melanoblasts requires SF for a critical period which begins only after the second day of cell dispersal *in vitro* and lasts for about 4 days, ending when pigment cells differentiate. It was also demonstrated that, although differentiation does not seem to require SF, this process may be enhanced by extended exposure to high levels of SF *in vitro* (Morrison-Graham and Weston, 1993). These results are in agreement with those previously showing that SF is required for the maintenance of murine melanocytes (Murphy *et al.*, 1992). The fact that the *W* mutation impairs the survival of melanocytes was elegantly confirmed by Huszar *et al.* (1991). These authors showed that microinjection of cultured C57Bl6 neural crest cells into *W* embryos resulted in chimeras which displayed extensive pigmentation throughout, often exeeding 50% of the coat. In contrast, similar injection to Balb/c albino mice in which unpigmented melanocytes are present in the skin, yielded only very limited donor cell pigment contribution. This stresses the fact already mentioned above that only a definite number of melanoblasts can colonize the epidermis. Experiments carried out *in vitro* with quail neural crest cells also showed that chicken recombinant SF has a rather modest effect on neural crest cell proliferation but promotes the survival and differentiation rate of melanocytic precursors (Lahav *et al.*, 1994). Therefore, all the results available clearly indicate that the c-kit/SF signaling pathway is required for the survival and differentiation of melanocytic precursors in the embryonic skin after neural crest cell migration has started.

### 6.5.2   Piebald lethal (s$^l$) *and* lethal-spotted (ls) *mutants*

*Piebald (s* locus, chromosome 14) is a recessive mouse mutation that has long been known. Homozygotes have dark eyes and show irregular white spotting of the coat, especially on the belly, sides and back. The white areas of the coat are completely devoid of melanocytes. There is a reduction in the number of melanocytes in the choroid layer of the eye. In addition, homozygotes may develop megacolon which is always associated with lack of ganglion cells in the distal portion of the colon. Mice with a more severe mutation at the *piebald* locus, *piebald lethal* (s$^l$), have dark eyes and an almost completely white coat with pigmented hair restricted to small areas on the head and base of the tail. All s$^l$/s$^l$ homozygotes develop megacolon. They usually die at about 2 weeks of age, but some live a year or more and may breed (Lyon and Searle, 1990). By explantation of embryonic tissues, consisting in the association of neural tubes and skins from wild-type and *s/s* mice, Mayer (1965) concluded that the primary cause of spotting in *piebald* lies in the neural crest.

Another recessive mutation with a phenotype similar to that of s$^l$ mice is *lethal spotting* (*ls* locus on chromosome 2). *Lethal spotting* mice usually die in the third week of life; however, some survive and are fertile. Homozygotes have considerable white spotting and megacolon. As in *piebald lethal,* the most distal portion of the bowel is aganglionic in *ls/ls* mice. The absence of the intrinsic reflexes of the enteric nervous system in the terminal gut causes dilatation of the normally innervated bowel proximal to the aganglionic region, thus forming a megacolon. The colons of these mice resemble those of human patients with Hirschsprung's disease, in which aganglionosis also occurs (Gershon, 1995, for a review). By tissue recombination experiments, Mayer and Maltby (1964) concluded that the *ls* mutation exerts its effect on pigmentation by reducing the number of melanoblasts. Since melanoblasts and intrinsic ganglion cells are both neural crest derivatives, it is possible that *ls* acts by causing defective migration or function of these cells or by reducing their number. The possibility that the *ls* gene acts by causing an intrinsic abnormality of the terminal 2 mm of the gut has also been proposed (Rothman and Gershon, 1984; Kapur *et al.*, 1993; Rothman *et al.*, 1993a). The discovery of the genes involved in these mutations was a determinant factor in increasing the knowledge about the mechanisms involved in these abnormalities.

*6.5.2.1   The* s$^l$ *and* ls *loci encode a receptor and its ligand.* The breakthrough in the study of the developmental abnormalities characteristic of the s$^l$ and *ls* mutants came with the demonstration by Yanagisawa and his colleages that the wild-type genes allelic to these mutations are encoding a G-coupled hepta-helical receptor: endothelin receptor type B for endothelins, EDNRB for $s^l$, and its ligand endothelin-3 (EDN3) for *ls*. Moreover, related defects referred to piebaldism associated with megacolon (or Waardenburg type 2 Hirschsprung phenotype also called Shah–Waardenburg syndrome) in humans carry muta-tions at corresponding loci (Puffenberger *et al.*, 1994; Edery *et al.*, 1996; Hofstra *et al.*, 1996).

The endothelins-1, 2, and 3 (EDN1, EDN2, and EDN3) are a family of 21-amino-acid peptides that activate one or both of the two heptahelical, G protein-coupled endothelin receptors A, and B (EDNRA, and EDNRB). EDNRA exhibits different affinities for endothelin peptides with the highest affinity for EDN1 and very low or no affinity for EDN3 (Yanagisawa, 1994). EDNRB accepts all three peptides equally (Sakurai *et al.*, 1990) (Fig. 6.10).

Endothelins are each produced from large polypeptide precursors which are first cleaved by furine protease(s) to yield biologically inactive intermediates called big endothelins. Big endothelins which contain 38–41 amino acids are further processed at a tryptophan residue (Trp-21 site) by endothelin-converting enzymes (ECE) (Xu *et al.*, 1994). This second cleavage produces the active 21-residue mature forms of endothelins (Fig. 6.10).

Targeted mutations of EDN3 and EDNRB genes in the mouse reproduced the phenotypes of *ls* and $s^l$ mutants, respectively (Greenstein-Baynash *et al.*, 1994; Hosoda *et al.*, 1994). It thus appears that both the ligand EDN3 and its receptor EDNRB are necessary for the development of neural crest-derived melanocytes and nerve cells in the terminal bowel. Moreover, the striking similarity of the phenotypes of mice deficient in either EDN3 or EDNRB suggests that the absence of EDN3 is sufficient to reproduce EDNRB null phenotype and that EDN1 and EDN2 do not play a major role in the devel-opment of these two cell lineages.

*6.5.2.2 Functional studies on the EDN3/EDNRB signaling pathway in neural crest cells.* In vitro culture assays and the construction of tetraparental mice were applied to the problem of the mechanism of action of EDN3 on pigment cells and enteric ganglion cell development. The well-established *in vitro* culture systems of quail neural crest cells was used by Lahav *et al.* (1996). Addition of EDN3 to a basic culture medium was found to promote the increase in cell number per culture by a factor of 75 at day 5 of culture at a concentration of 100 nM. The increase in cell proliferation is detectable from day 1 of cultiva-tion and accompanied by a delay in cellular differentiation as evaluated by the percentage of pigmented melanocytes and of cells expressing the premelano-cytic MelEM marker. In the first 3 days of culture, EDN3 inhibited the appear-ance of MelEM-labeled cells in a dose-dependent manner. By day 16, the proportion of MelEM-labeled cells (incubated with 100 nM EDN3) reached 90%, while only 5% of the cells in control cultures were labeled (Fig. 6.11). The same effect of EDN3 on melanogenesis was also revealed by evaluating the proportion of pigmented cells grown for different periods of time in culture (Fig. 6.11). From day 10 to day 16 the culture containing EDN3 became

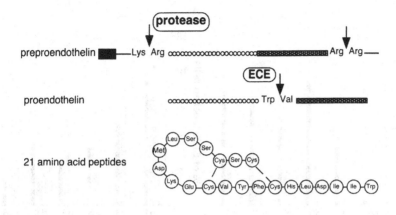

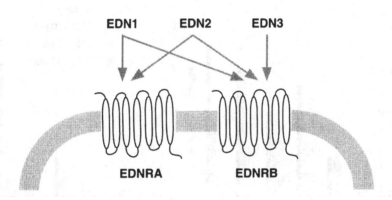

**Figure 6.10** Endothelins and endothelin receptors. Endothelins are a family of three peptides (EDN1, EDN2, EDN3) that are cleaved from large precursors by the endothelin converting enzyme (ECE). The receptors (EDNR) have been identified in mammals: EDNRA which binds preferentially EDN1, and EDNRB which binds any of the three peptides with the same affinity. These receptors belong to the G protein-coupled seven transmembrane domain protein family. (Reproduced, in part, from Yanagisawa *et al.*, 1988, with permission.)

progressively pigmented. Finally, fully differentiated melanocytes aggregated forming a reproducible pattern with a network of pigmented cells surrounding islets of unpigmented but mostly MelEM-positive cells (Fig. 6.12). Interestingly, the presence of EDN3 did not prevent the differentiation of other cell types such as neurons and glia in the culture. Similar results were obtained for mouse crest cells in culture (Reid *et al.*, 1996).

The expression pattern of the EDN3 receptor was first investigated in the avian embryo. The quail homolog of the mouse EDNRB was cloned by Nataf *et al.* (1996) and its expression was found to take place in the neural fold as the neural tube begins to close. Neural crest cells express this receptor as they take the dorsoventral migration pathway at E2 and E3 in both quail and chick.

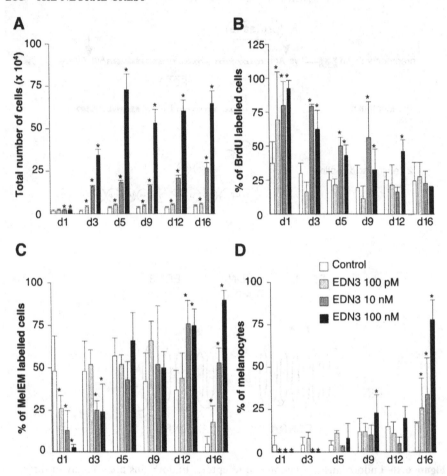

**Figure 6.11** Quantification of EDN3-induced changes at different time points in cultures of quail neural crest cells. Cultures in enriched medium with different EDN3 concentrations were analyzed during a culture period of 16 days. (A) Total cell numbers. (B) Cell proliferation was determined by the proportion of cells incorporating BrdU. (C) Proportion of premelanocytes labeled by the ME1EM MAb. (D) Proportion of melanocytes. Significant differences are indicated by the presence of an asterisk at the top of the column. (Reproduced, with permission, from Lahav *et al.*, 1996. Copyright (1996) National Academy of Sciences, U.S.A.)

Neural crest-derived structures such as DRG, sympathetic ganglia (SG) and plexuses, adrenal glands, and the enteric ganglia continue to express the EDNRB receptor at least up to midincubation time, the latest stage observed (Fig. 6.13). The cells which take the dorsolateral pathway, the fate of which is to give rise to the skin melanocytes, express EDNRB as long as they remain in close contact with the neural tube in the so-called "staging area" (Loring and Erickson, 1987; Erickson *et al.*, 1992). In contrast, when they invade the sub-ectodermal pathway the neural crest cells, recognizable by the fact that they express c-*kit* or are labeled by the MelEM Mab or carry the quail nuclear

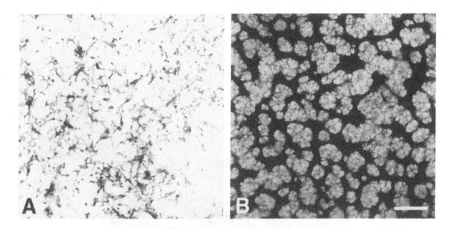

**Figure 6.12** Effect of EDN3 (100 nM) on neural crest cells cultured for 11 days in medium supplemented with serum and chick embryo extract. (A) control; (B) EDN3-treated culture. Note the striking pattern of pigmented and unpigmented cells in B. Bar = 800 μm. (Reproduced, with permission, from Lahav *et al.*, 1996. Copyright (1996) National Academy of Sciences, U.S.A.)

marker in quail-to-chick neural chimeras, do not express the EDNRB gene. A large majority of quail neural crest cells cultivated *in vitro* express EDNRB as long as they are not engaged in the melanocytic differentiation pathway. None of the cells carrying pigment granules were found to contain detectable amounts of EDNRB mRNA (Nataf *et al.*, 1996). This observation was inconsistent with the fact that EDN3 not only stimulates the proliferation of neural crest cells, but also strongly enhances their differentiation into melanocytes (Lahav *et al.*, 1996). This paradox was solved by the discovery by Lecoin *et al.* (1998b) that, in birds, premelanocytes express a different type of EDNR, referred to as EDNRB2, which is strongly induced in neural crest cells cultured in conditions that favor melanocytic differentiation (e.g., addition of EDN3 at 100 nM concentration or high content (10%) of chick embryo extract to the culture medium). Interestingly, EDNRB2 is expressed exclusively by the neural crest cells which take the dorsolateral skin migration pathway. Activation of the EDNRB2 gene takes place at about the time when the neural crest cells start to express c-*kit* (Fig. 6.14).

The question as to whether EDNRB2 is also present in mammals has not yet received a response.

### 6.5.3  Respective roles of c-kit/SF and EDNRB-B2/EDN3 signaling systems on pigmentation

The available data reviewed above indicate that EDN3 acts on neural crest cell expansion as they are still in the neural fold, when they migrate along the dorsoventral pathway, and as they invade the large field constituted by the

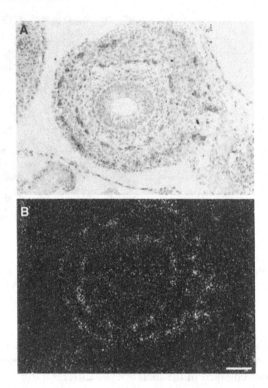

**Figure 6.13** EDNRB expression in the enteric nervous system of the quail embryo. HNK1 immunocytochemistry was combined with radioactive *in situ* hybridization on transverse sections of embryonic gut (duodenum) at E6. The bright-field view (A) shows the HNK1-positive plexuses; the epipolarization view (B) shows that the HNK1-positive cells express EDNRB whereas the gut wall mesenchyme and the endoderm do not. Bar = 60 μm. (Reproduced, with permission from Nataf *et al.*, 1996. Copyright (1996) National Academy of Sciences, U.S.A.)

digestive tract. It is clear that a deficit in neural crest cell proliferation, related to the deficiency in the EDN3/EDNRB signaling pathway, leads to the incapacity of the precursors of the enteric nervous system to colonize the entire gut; hence the aganglionic/megacolon syndrome is observed in both $s^l$ and $l^s$ mice.

The skin, like the gut, also constitutes a large field to be invaded by neural crest cells. The important role of EDN3 in neural crest cell proliferation can also be invoked to explain spotting or total absence of pigmentation seen in the mice deficient for the EDN3/EDNRB pathway.

As far as the c-*kit*/*sl* system is concerned, *in vivo* and *in vitro* studies point to the essential role of this pathway in ensuring survival of premelanocytes.

Interestingly, the spotting patterns exhibited in the two groups of mutants ($W$ and $S^l$ on the one hand and $s^l$ and $l^s$ on the other) differ: in the former, the pigmented areas are scattered or diluted (with a mixture of pigmented and non-pigmented hair), while, in the latter, spotted areas have well defined margins.

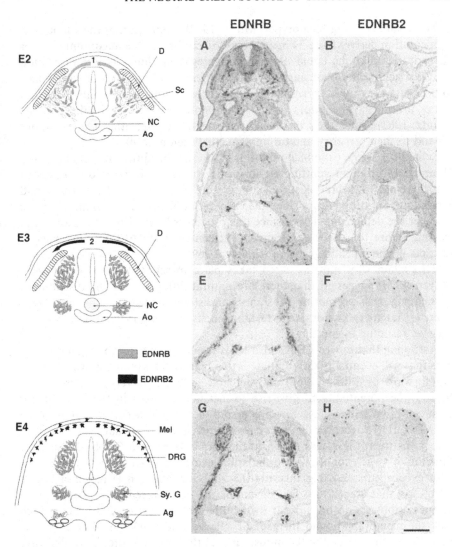

**Figure 6.14** Expression pattern of the two EDNRB subtypes identified in birds. Digoxygenin-labeled EDNRB (A, C, E, G) (reproduced, with permission, from Nataf *et al.*, 1996. Copyright (1996) National Academy of Sciences, U.S.A.) and EDNRB2 (B, D, F, H) (from Lecoin *et al.*, 1988). Probes have been hybridized on cryostat sections of the same quail embryo. The results are summarized on the left. At E2, neural crest cells that migrate along the dorsoventral pathway (1) express EDNRB. There is no EDNRB labeling in cells taking the dorsolateral pathway (E, G). Neural crest cells migrating along this pathway (2) start to express EDNRB2 at E3 (F) and become more numerous at E4 (H). D, dermomyotome; Sc, sclerotome; NC, notochord; Ao, aorta; Mel, melanoblast; DRG, dorsal root ganglia; Sy.G, sympathetic ganglia; Ag, adrenal gland. Bar = 145 μm.

To explain spotting phenotypes, Mintz (1967) proposed that the whole coat pigmentation derives from 34 melanoblasts according to the maximum number of territories that she could observe in aggregation chimeras (see discussion above). Each melanoblast migrates and proliferates along the dorsolateral axis to populate a strip-like territory. Spots occur when the descendants of certain clones die, their progenitors being "preprogrammed" to die in the mutants (Mintz, 1967). Since their death happens relatively late, other clones cannot expand to their territory and an unpigmented region occurs.

Schaible (1969) examined spotting patterns in different mutants and observed 14 different spotting territories in the coat. He therefore suggested that there are 14 melanoblasts, each migrating to the centre of a territory and expanding. Spotting occurs when the early melanoblasts die or cannot expand to cover the depigmented area since this process is limited in time. Clones can expand before the tissue environment differentiates up to the point when it restricts pigment cell migration (Schaible, 1969).

To explain why spotting often takes place in selected areas of the skin, Mayer (1977) suggested that there are natural differences in the distribution of melanogenesis-promoting substances in distinct tissue environments. Thus, in areas that are less favorable to melanogenesis, deficient melanoblasts will not survive, resulting in spotting.

Assuming that *in vivo* EDN3 also leads to a large expansion of neural crest cells that will eventually differentiate into melanocytes would confirm the view put forward by Mintz (1967) and Schaible (1969) that very few precursors bearing the potentiality to give rise to melanocytes exist in the early neural crest cell population. These few precursors thus would need the action of the EDNRB/EDN3 receptor–ligand system to proliferate. Our data show that, in avians, between the stages at which melanoblasts require EDNRB and c-kit function, a large expansion in population size occurs. This conclusion agrees with experiments showing that murine EDNRB function is required before E10.5. At that stage, very few melanoblasts are detected in wild-type mice, and melanoblasts do not appear in the affected skin areas of $Sl^d$ mice which lack the membrane-bound form of SF (Pavan and Tilghman, 1994). By the time c-*kit* is first needed in the mouse (E11.5) the dermis is already seeded with numerous melanoblasts (Wehrle-Haller and Weston, 1995).

Taken together, spotting in EDNRB or EDN3 mutant mice might occur according to the explanation provided by Schaible since the precursors are probably affected before they expand. In contrast, in mice deficient for c-*kit* or SF, the mechanisms that lead to spotting resemble more those described by Mintz where melanoblasts that are destined to die (either because they carry the mutation or because they are not provided with the SF ligand) first expand, but cannot survive when in the skin. These differences might result in the differences in the nature of spots. In EDNRB or EDN3 mutant mice, the spots have sharp margins since pigment cell precursors were eliminated early, resulting in the absence of all their descendants in a certain area of the skin. In contrast, in c-*kit*- or SF-deficient mice, melanoblasts are affected relatively late. At this stage, many melanoblasts are found in the dermis. Partial elimination of melanoblasts would thus result in small distances among pigmented and unpigmented hair. This corresponds of course to mutants in which penetrance of the mutation is not complete. In the latter case, the phenotype would be entirely unpigmented.

The spotting pattern in mice bearing mutations in both endothelin and exit/steel receptor–ligand system shares a similar character: in both cases it is the trunk region that is usually depigmented while the head and hindlimb regions often retain pigmentation. Interestingly, it was recently demonstrated that melanoblasts show an uneven distribution along the murine neuraxis. In early developmental stages, many more melanocytic precursors are found in the head and tail as compared to the trunk, consistent with the fact that those areas are often pigmented in spotted mice (Pavan and Tilghman, 1994; Wehrle-Haller and Weston, 1995).

The different steps of melanoblast development and the result in terms of color pattern have therefore been fully explored in these mouse mutants. However, it has to be kept in mind that these processes might differ somehow among species.

### 6.5.4  Dependence of melanocyte development on the Mitf basic helix–loop–helix–zipper transcription factor

The murine *microphtalmia* gene (*mi*, now designated *Mitf* for microphtalmia-associated transcription factor) is a multi-allelic, classic coat-color gene that has been recently cloned from a transgenic insertional mutation at the *microphtalmia* locus in the mouse and human (Hodgkinson *et al.*, 1993; Hughes *et al.*, 1993; Hemesath *et al.*, 1994; Tassabehji *et al.*, 1994; Steingrimsson *et al.*, 1994). It encodes a member of the large bHLH-Zip family of transcription factors which is exemplified by the proto-oncogene *Myc* or by proteins such as Mash-1 and NeuroD, known as potent regulators of cell fate (Jan and Jan, 1993). The three conserved motifs that constitute this family are: the basic region, which recognizes a canonical CANNTG DNA-binding sequence, the HLH motif and the Zip motif. The HLH and Zip motifs participate in protein dimerization, a prerequisite for DNA binding. In other parts of the proteins there are sequences that are specific to individual members of the bHLH-Zip family. The *Mitf* gene has undergone multiple independent mutations in the mouse so that at least 21 different *Mitf* alleles have been identified (Mouse Genome Database, 1997).

Most alleles of this gene ablate or severely reduce pigmentation and produce white belly spots when homozygous in mice. In addition to a role in pigmentation, *Mitf* mutants have a number of pleiotropic phenotypes, such as small eyes, reduced eye pigmentation, retinal degeneration, early onset deafness, reduced mast and natural killer (NK) cell numbers, and osteopetrosis. Moreover, the human homologous gene (*MITF*) has been shown to be mutated in two families with Waardenburg's syndrome type II (WS2). Patients heterozygous for inherited WS2 show varying degrees of deafness and abnormal patchy hair, skin, and eye pigmentation.

In mice homozygous for *Mitf*, the coat is white due to the absence of melanocytes. Thus, they differ from albino mice which carry a null mutation in the gene encoding for the enzyme tyrosinase (*c*), in which melanocytes are structurally normal except that they cannot synthesize melanins.

Studies carried out *in vivo* and *in vitro* by Opdecamp *et al.* (1997) have shown that neural crest cells located in the dorsolateral migration pathway express first *Mitf* together with a low level of the c-kit receptor. At that stage they are designated "melanoblast precursors." Such cells soon after express TRP2 and become "early melanoblasts." Finally, "late melanoblasts" are characterized by the following phenotype: they are $Mitf^+$, $TRP2^+$, and exhibit a high level of c-*kit*.

In embryos homozygous for a *Mitf* allele encoding a non-functional Mitf protein, $Mitf^+$, and c-$kit^+$ cells were scarce and never acquired the TRP2 marker. Later, the number of c-$kit^+$ cells was severely reduced. Wild-type neural crest cells in culture yielded rapidly $Mitf,^+$ c-$kit^+$ cells. With time c-*kit* expression increased and typical melanocytes appeared in the culture. Addition of SF and EDN3 considerably increased the number of cells of the melanocytic lineage as already described in the quail by Lahav *et al.* (1996) and by Reid *et al.* (1996) for the mouse. Cultures from *Mitf* mutant embryos contained initially a number of $Mitf,^+$ c-$kit^+$ cells comparable to those of control cultures. However, in the mutant cultures c-*kit* expression did not increase with time and the cultures did not respond to SF and EDN3. Thereafter, *Mift* expression was lost. This suggests that the product of *Mitf* plays a crucial role in the transition between "melanoblast precursor" and "early melanoblast" stages.

The fact that *Trp1*, *Trp2*, and also *tyrosinase* (*c*) genes might be targets for *Mitf* action is compatible with results of studies carried out on their promoter region. The pigmentation genes encoding tyrosinase (*c*) and tyrosinase-related proteins 1 and 2 (*Trp1*, and *Trp2*) both from human and from mouse have an 11-bp consensus sequence in the 5′ promoter region, the M box, which contains a CATGTG bHLH factor-binding site. Cotransfection assays show that MITF can activate both the *c* (Bentley *et al.*, 1994; Yasumoto *et al.*, 1997) and the *Trp1* (Yavuzer *et al.*, 1995) promoters.

In conclusion, *Mift* appears to be a critical factor in melanogenesis.

### 6.6 Some aspects of the hormonal control of pigmentation

It is well established that pigmentation is regulated by several hormonal influences (see Bagnara and Hadley, 1973). Only recent data concerning this large physiological question will be discussed in some depth in this chapter.

The mechanism responsible for pigment pattern formation which is largely under genetic control is still little understood. In amphibians, manifestation of adult pigmentation is regulated by thyroid hormones which, however, do not determine the pattern of pigmentation itself but merely permit its expression. Prolactin and other steroids also act as potent regulators of pigmentation. Thus, prolactin has long been known to play a major role in bringing about the various changes occurring in second metamorphosis in newts (Grant, 1961), and sex hormones are responsible for the onset of adult bright pigmentation in amphibians and other vertebrates as well. They are also involved in the changes in pigmentation related to sexual activity.

The major hormone regulating pigmentation is melanocyte-stimulating hormone (MSH). A significant contribution to its role in pigmentation in lower vertebrates has come essentially from work carried out in amphibians.

### 6.6.1 Control of pigmentation by melanocyte-stimulating hormone

It has long been known that pigmentation is to a certain extent under hormonal control. One example of such an action is provided by the seasonal changes in fur color observed in arctic mammals, another by skin hyperpigmentation consecutive to insufficiency of the adrenal gland in humans. The latter example is consecutive to increased production by the pituitary gland of the adrenocorticotrophic hormone (ACTH) derived by post-translational processing from a single polypeptide, proopiomelanocortin (POMC). Two other peptides derived from POMC are $\alpha$- and $\gamma$-melanocyte-stimulating hormones ($\alpha$- and $\gamma$-MSH, also designated melanocortin) (reviewed in Bertagna, 1994) which have a strong effect on pigmentation.

*6.6.1.1 Regulation of pigment pattern in amphibians.* One of the first indications that hormone(s) of pituitary origin influence the development of pigment cells was provided by the fact that hypophysectomized *Xenopus* embryos yield larvae that possess many fewer melanophores than do their intact siblings. Moreover, such larvae have iridophores in areas of the integument where only melanophores normally occur. It was concluded that MSH stimulates melanocyte proliferation while inhibiting the differentiation of iridophores (Bagnara, 1957). Similarly, Pehlemann (1967) showed that high levels of circulating MSH in *Xenopus* larvae lead to the proliferation of melanophores and the darkening of the skin. Moreover, implants of pars intermedia in the normally unpigmented lower jaw of *Xenopus* larvae induce the differentiation of melanophores at the vicinity of the graft (Bagnara and Fernandez, 1993). Thus, it is clear that MSH exerts a positive effect on proliferation and differentiation of melanocyte precursors present in the skin during larval and adult stages in amphibians.

The problem as to whether MSH affects early melanophore development was addressed directly by subjecting whole *Xenopus* and *Pleurodeles* embryos to MSH or cAMP during the first days of embryonic development. It was found that the number of melanophores in treated and untreated embryos was the same (Wahn *et al.*, 1976; Turner *et al.*, 1977). It was also shown that

MSH is not involved in the establishment of basic pigmentation patterns (e.g., spot patterns in *Rana pipiens*; Hanaoka, 1967).

Although amphibian pigment patterns have long been studied on a number of different species, *Xenopus laevis* has recently become a widely used model system for such investigations. As in many vertebrates, adults of this species have a dorsoventral pigment pattern with a lighter pigmentation in the ventrum than dorsally. In *Xenopus laevis* the back skin is darkly mottled by regions where melanocytes are present in high density, and a white ventrum devoid of melanocytes. In fact, developing melanocytes actually exist in the ventral skin of the larva where they can be evidenced by their dendritic morphology and dopa content, but they fail to differentiate (Ohsugi and Ide, 1983; Fukuzawa and Ide, 1986). The ventral skin produces a melanization inhibiting factor (MIF) which was proposed to be responsible for the dorsoventral pattern (Fukuzawa and Ide, 1988). This view was substantiated by the fact that MIF inhibits both outgrowth and melanization of neural crest cells cultured *in vitro*.

In a continuing investigation, Fukuzawa and Bagnara (1989) demonstrated that MIF can override the stimulatory effects of MSH and of a melanizing factor present in the serum of frogs. They used a neural crest culture assay system similar to that initiated for quail neural crest cultures by Cohen and Königsberg (1975). Based on this assay the partial purification of MIF could be performed from extracts of ventral skin of the Mexican leopard frog (*Rana forreri*). The partially purified MIF was used to generate an anti-MIF Mab with which it was revealed that MIF is localized in specific regions of the ventral integuments of adult frogs (Samaraweera *et al.*, 1991). Another interesting observation was that a factor present in the ventral skin (likely to be MIF) stimulates the differentiation of iridophores (Bagnara and Fukuzawa, 1990). Recently Lopez-Contreras *et al.* (1996) showed that MIF has a biological activity on mammalian melanocytes since it blocks the α-MSH effect on a mouse malignant melanoma cell line seemingly by interacting with the MSH receptor. It is therefore likely that MIF might be related to the Agouti protein (see below). This view is supported by the fact that the mammalian agouti protein inhibits the dispersion of melanophores in the *Xenopus* neural crest cell culture assay (Ollman and Barsh, 1996).

Factors which increase the melanization process have also been demonstrated in the dorsal skin, such as an intrinsic melanization-stimulating activity present in leopard frogs (Mangano *et al.*, 1992) (also found in mammalian skin, see below). A similar factor extracted from catfish skin has been designated MSF (melanization-stimulating factor) (Johnson *et al.*, 1992; Zuasti *et al.*, 1992). Apparently MSF activity is especially concentrated in the dorsally located black spots of the skin of *Rana pipiens* (Mangano *et al.*, 1992). A similar activity was found in the pigmented skin of the teleost *Sparus auratus* (Zuasti *et al.*, 1993).

While comparative studies on putative intrinsic pigmentary factors present in the skin of various vertebrates is interesting, of more immediate significance are recent discoveries concerning the basis for pigment pattern determination in mammals. Thus, important progress in our understanding of the mode of action of MSH in pigment pattern was brought about by the identification of certain genes affecting coat color in the mouse.

### 6.6.2 Coat color mutants in mice show that skin and hair pigmentation is regulated by endocrine and paracrine mechanisms

In mice, nearly 100 genes that affect coat color have been identified. Many coat color variants have been selected in Europe and Asia by mouse fanciers who prize animals with unusual or striking color variations in their pelage for their beauty and singularity. Such is the case for color variants such as brown, silver, pink-eyed dilution, and yellow (see Silver, 1995). The various colors depend on the distribution of the two basic pigments produced by melanocytes, eumelanin and pheomelanin. Fancy mice were among the initial tools used early in this century to create inbred strains whose development laid down the grounds for much of mammalian genetics. In spite of the long history of the classical genetics of mouse coat color, it is only in the recent years that a certain number of color genes have been isolated and the proteins they encode characterized.

The relative amounts of eumelanin and pheomelanin in the melanocyte are controled primarily by two loci in mammals: *extension* and *agouti*. The *extension* locus, known in many mammalian species (Searle, 1968a,b), increases brown/black pigment when dominant and when recessive blocks eumelanin synthesis, thereby extending the incidence of red/yellow pigment. The cloning of the genes of the *extension* and *agouti* loci has provided an explanation for their interaction and has revealed that the *extension* encoded gene is involved in an hormonal control of pigmentation. In contrast, the protein product of *agouti* acts in a paracrine manner. Previous experiments had established that the *extension* locus acts on the melanocytes residing within the hair follicle, in a cell autonomous manner, to determine whether pheomelanin or eumelanin is synthesized (Geschwind, 1966; Geschwind et al., 1972; Lamoreux and Mayer, 1975), whereas the *agouti* locus acts within the cells of the hair follicle surrounding the melanocytes to control eumelanin synthesis both spatially and temporally (Silvers and Russel, 1955).

It has been found in the recent years that MSH activity on pigmentation is mediated in humans by the melanocortin-1 receptor (MC1R), a seven transmembrane G protein-coupled receptor expressed by melanocytes. Activation of MC1R produces elevated levels of intracellular cAMP (Mountjoy et al., 1992), and triggers the pathway leading to melanin synthesis. Classically, POMC-derived peptides are considered as pituitary hormones. However, they have been shown to be produced also in the mammalian brain and in a variety of peripheral tissues, including the gastrointestinal tract, the reproductive organs, and in cells of the immune system (see Wintzen and Gilchrest, 1996, for a review). More recently, POMC-derived peptides have been found in mammalian epidermal cells, including human keratinocytes (Slominski et al., 1992; Farooqui et al., 1993), and even in melanocytes. POMC transcripts were detected first in human and rodent melanoma cell lines and more recently also in UV-stimulated human melanocytes. All these cells were found to contain α-MSH peptide and ACTH immunoreactivity (reviewed by Wintzen and Gilchrest, 1996).

A great deal of interest in the MSH receptor MC1R followed the discovery that a loss-of-function mutation in the mouse of the the *Mc1r* gene is responsible for the long-known mutation *recessive yellow* (Robbins *et al.*, 1993). Formerly designated *e*, *recessive yellow* is one of several alleles at the *extension* locus. Recessive alleles diminish the amount of black pigment while extending that of yellow pigment as in the *nonextension of black* (*e*) rabbits, mice, and dogs (Robbins *et al.*, 1993), or in the red fox (Adalsteinsson *et al.*, 1987).

Mutations of the *MC1R* gene have been detected in red-haired human subjects as well as in mice with yellow hair. Eight of the nine human *MC1R* variants lie within the N-terminal domain of the protein. It is likely that the mutations prevent normal receptor signaling, but it is not clear whether the alterations in MC1R proteins affect membrane insertion, ligand binding, or interaction with intracellular effectors. As in the case of other G protein-coupled receptors, allelic variation of the *Mc1r* can result either in an increase or in a decrease in the signaling efficiency. Thus, alterations of the mouse *Mc1r* have been found which lead to hyperactive or constitutively active receptors responsible for graded degrees of hyperpigmentation such as in mice carrying the *sombre, sombre3-J*, and *tobacco* alleles. *Dominant extension* locus alleles are thought to be responsible for some examples of uniform brown or black pigmentation, such as those found in the rabbit, or in the melanic forms of cats, and in the black leopard, *Panthera pardus* (Robbins *et al.*, 1993).

### 6.6.3  Paracrine control of pigmentation

The incidence of the "fire red" hair phenotype in human with a normal *MC1R* gene indicated that other factors interfere with eumelanin synthesis. One answer to this problem was provided by the discovery of the Agouti protein, a novel paracrine signaling molecule that inhibits the effect of melanocortin signaling. Thus, the most obvious color pattern in rodents which varies according to position along the dorsoventral axis results from regional expression of the *agouti* gene (reviewed by Silvers, 1979). The accumulated evidence indicates that the constitutive expression of the *agouti* gene results in yellow hairs, lack of expression results in black hairs, modulation of expression results in the switch from black to yellow to black pigment as seen in the wild-type *agouti* whose phenotype is a single subapical band of phaeomelanin on an eumelanin hair. In animals which carry the wild-type *white bellied agouti* ($A^w$) allele, nearly all the dorsal hairs are banded, but ventral hairs are entirely yellow, or cream-colored.

In fact, the *agouti* gene controls the deposition of yellow and black pigment in developing hairs. Dominant alleles, like *lethal yellow* ($A^y$), result in the exclusive production of yellow pigment, and inhibition of eumelanin. While the homozygous ($A^y/A^y$) condition is lethal, heterozygotes ($A^y/a^x$) exhibit pleiotropic effects including yellow coat color, insulin-resistant diabetes, a tendency to develop tumors, obesity, and increased somatic growth (for review see Herberg and Coleman, 1977). $A^y$ was the first recognized obesity gene (Danforth, 1927; for review see Herberg and Coleman, 1977), and the oldest known embryonic lethal mutation (Cuénot, 1908; Castle and Little, 1910).

About 19 distinct phenotypic classes of *agouti* mutations have been recorded, and alleles of several classes have arisen many times (Dickie, 1969; Cattanach *et al.*, 1987; Siracusa, 1991). Besides dominant alleles, such as *lethal yellow* ($A^y$), that are associated with excessive production of pheomelanin, some recessive alleles are characterized by the production of eumelanin but not pheomelanin: *non-agouti* (*a*), *lethal non-agouti* ($a^\alpha$), *non-agouti lethal* ($a^l$), and *extreme non-agouti* ($a^e$). In a wild-type background, $a^\alpha/a$ and $a/a$ mice are mostly black, and $a^l/a^e$ or $a^e/a^e$ mice are completely black (Papaioannou and Mardon, 1983; Lyon *et al.*, 1985).

Transplantation experiments in which melanocytes from one mouse strain are allowed to migrate into the developing hair follicles of another have shown that the banded pattern is determined by the *agouti* genotype of the skin, and not of the melanocytes (Silvers and Russel, 1955; Silvers, 1958a,b).

Cloning of the *agouti* gene (Bultman *et al.*, 1992; Miller *et al.*, 1993) has provided the clue to the understanding of the agouti signaling pathway, and the pleiotropic effects caused by $A^y$. The *agouti* gene codes for a 131-amino-acid protein of $\approx$ 16 kDa, which can be divided into three domains, as shown in Fig. 6.15 (Willard *et al.*, 1995).

The *agouti* gene is expressed in hair follicles, and increased levels of expression coincide with the synthesis of pheomelanin, meaning that the banding pattern observed in black hairs of the homozygous $A$ or $A^w$ mice corresponds to a transient upregulation of *agouti* gene transcription during the growth of the hair follicle, and its permanent activation in hair follicles of the ventrum of white belly animals (Bultman *et al.*, 1992; Miller *et al.*, 1993). Thus, the transcript is highly expressed in the yellow belly of *black* and *tan* ($a^t$) mice, but cannot be detected in their black backs (Bultman *et al.*, 1992). The transcript was not detected in other tissues from wild-type ($A$, $A^w$) mice, nor was it in the skin from neonatal mice homozygous for the classical *non-agouti* (*a*) mutation with mostly black coats (Bultman *et al.*, 1992; Miller *et al.*, 1993).

The genomic region encoding the *agouti* gene initially revealed four exons. The first one (exon D) resides 18 kb upstream from the coding exons 2, 3, and 4 (Bultman *et al.*, 1992; Miller *et al.*, 1993). Further analysis revealed that separate regulatory elements control the alternative use of two sets of untranslated 5' *agouti* exons: one set is expressed only in the ventrum, the other set is hair-cycle-specific, and is expressed only during hair growth (Bultman *et al.*, 1994; Vrieling *et al.*, 1994) (Fig. 6.16).

The mode of action of the agouti protein was shown to be related to an interaction with the melanocortin 1 receptor (MC1R). It was observed that the recessive *Mc1r* alleles in mice mimic the pigmentary effect of dominant *agouti* alleles. This observation and others (reviewed in Siracusa, 1994) provided the genetic background to the work of Lu *et al.* (1994).

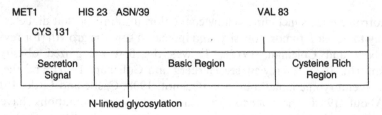

**Figure 6.15** Depiction of the primary structure of murine agouti. The agouti sequence includes a secretion signal, a predominantly basic region loosely defined from His23 to Lys82 containing an *N*-linked glycosylation site, and a C-terminal cysteine-rich region 2. (Reprinted, with permission, from Willard *et al.*, 1995. Copyright (1995) American Chemical Society.)

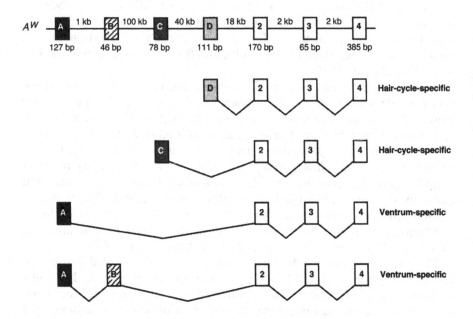

**Figure 6.16** Genomic organization and multiple forms of wild-type $A^w$ transcripts. The structure of the *agouti* gene is shown at the top from proximal (left) to distal (right). *Agouti* exons are shown as large shaded boxes; numbers below the boxes indicate the length of exons. Exons A–D are noncoding, whereas exons 2–4 contain protein-coding sequences. Numbers between the exons indicate the length of introns. Two different forms with unique 5′ untranslated exons are expressed during hair growth (labeled hair-cycle specific); two other alternatively spliced forms are expressed continuously in the belly (labeled ventrum-specific). (Reproduced, with permission, from Siracusa, 1994.)

As described above, production of eumelanin by the hair-bulb melanocytes depends upon the binding of $\alpha$-MSH to its receptor which is the product of the *extension* (*E*) locus on mouse chromosome 8. Lu *et al.* (1994) showed that the Agouti protein influences or alters this process by acting as an antagonist that prevents $\alpha$-MSH binding to the product of *Mc1r*.

Mice carrying any one of the hundred or more *agouti* mutations can range in color from yellow to black. Molecular analysis of the genomic region mutated at the *A* locus has provided an explanation for the high mutation rate of the locus. The *a* locus contains an 11-kb insertion in the intron between exons D and 2 (Fig. 6.17).

This insertion includes a 5.5-kb VL30 element, a retroviral-like structure, interrupted by a 5.5-kb fragment of genomic origin. This 5.5-kb fragment is composed of two 526-bp direct repeats flanking the remaining 4.5 kb. Mutations *a* to *a*$^{l}$ and *a* to *A*$^{w}$ are due to homologous recombination and excision of sequences between either the two 526-bp direct repeats or between the long terminal repeats (LTRs) or the VL30 element, respectively. The different sizes of the insertion must interfere with normal transcription and/or splicing since they are related with the expression pattern of the *agouti* gene and with the phenotype of the various mutants (reviewed in Siracusa, 1994).

Several *agouti* mutations result in recessive embryonic lethality. The *A*$^{y}$ and *lethal light-bellied non-agouti* (*a*$^{x}$) mutations define two complementation groups at the *agouti* locus (Siracusa, 1991). *A*$^{y}$/*a*$^{x}$ mice are viable and completely yellow. The *A*$^{y}$ mutation involves a chromosomal rearrangement that results in the production of a chimeric mRNA expressed in nearly every tissue of the body. In its 5$'$ portion, this chimeric RNA contains 5$'$ sequences that do not belong to the *agouti* gene, while its 3$'$ portion retains the protein-coding potential of the non-mutant allele (Miller *et al.*, 1993). The first exon of the abnormal 1.1-kb *A*$^{y}$ transcript is the first untranslated exon of an ubiquitously expressed gene *Raly*, which resides 120–170 kb upstream of *a*. The deletion removes all coding and 3$'$ sequences of *Raly*, and leaves *agouti* exons in place. Thus, in *A*$^{y}$ mutants, the stronger *Raly* promoter drives expression of the *agouti* gene ubiquitously (Michaud *et al.*, 1994a; Siracusa, 1994). Absence of *Raly* seems to be the cause of *A*$^{y}$/*A*$^{y}$ embryonic lethality since treatment of embryos with *Raly* antisense mRNA causes embryonic death before blastocyte formation (Duhl *et al.*, 1994).

Phenotypic studies and molecular analysis have shown that the *agouti* gene is conserved in most orders of mammals (Searle, 1968a; Bultman *et al.*, 1992). The function of the gene in humans has not been related to banded hair pigmentation patterns which are not found in human hair. A role for *agouti* in hair coloration in humans cannot be ruled out, but most likely the gene plays another physiological role (Lu *et al.*, 1994).

The fact that *agouti* may influence various biological functions when ectopically expressed (e.g., in *A*$^{y}$ mutants; see Siracusa, 1994) led to the assumption that there are receptors for the Agouti protein in cells where this gene is not normally expressed. Thus agouti is probably a member of a family of related ligands, and its abnormal presence in certain cells alters or mimics the normal

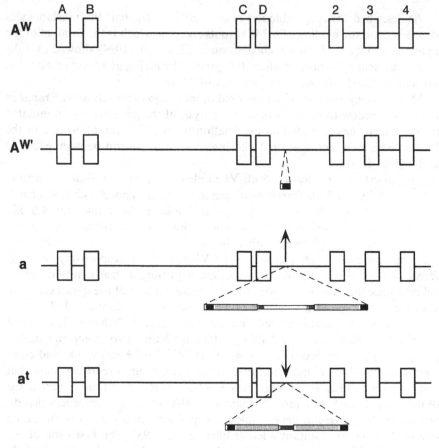

**Figure 6.17** Genomic structure of the *a*, *a^t^* and *A^w^* alleles of the *agouti* gene reveals a mechanism for the high reversion rate of *a* to *a^t^* and *A^w^*. The wild-type *A^w^* allele is shown at the top, with exons represented as large boxes. Distances between exons and exon sizes are not drawn to scale. The dashed lines indicate the sites of inserted sequences in the *A^w′^*, *a* and *a^t^* alleles. The *a* mutation is caused by insertion of an 11-kb fragment just distal to exon D. The insertion contains a complete 5.5-kb VL30 element (shown in light dotted line) interrupted by a 5.5-kb fragment of genomic origin (dark dotted boxes bordering a white bar). Of particular significance are two 526-bp direct terminal repeats (dark dotted boxes) flanking the remaining 4.5 kb of inserted sequences (white bar). The 526-bp repeats show strong homology to a region located between two T-cell receptor γ variable region gene segments, whose function is unknown. Mutations from *a* to *a^t^* appear to involve homologous recombination between the two 526-bp direct repeats, thus decreasing the 5.5-kb part of the insertion to 526 bp between the two portions of the VL30 element (single dark dotted box). Similarly, reverse mutations from *a* to *A^w^* involve recombination between the long terminal repeats (light dotted boxes) of the VL30 element, leaving behind a single long terminal repeat. The prime symbol distinguishes the true wild-type *A^w^* chromosome from the *A^w′^* chromosome, which contains only one VL30 long terminal repeat. (See Siracusa, 1994, for details and references; reproduced, with permission, from Siracusa, 1994.)

functions of its related family members. It may alternatively antagonize other members of the melanocortin receptor family produced in tissues other than the epidermis (Lu *et al.*, 1994). These assumptions have been substantiated by recent results which have shown that such receptors (MC3R and MC4R) exist in the brain and that a recombinant agouti-related protein (ARGP) is a potent selective antagonist of melanocortin, the natural ligand of these receptors. Moreover, ubiquitous expression of *ARGP* cDNA in transgenic mice causes obesity without altering pigmentation (Ollmann *et al.*, 1997). Furthermore, *in vitro*, recombinant agouti protein inhibits the binding of radiolabeled melanocortins to cells that express not only MC1R but also related receptors MC2R or MC3R. In addition, recent gene targeting studies indicate that animals lacking MC4R have metabolic defects and abnormalities in body weight regulation similar to those caused by ectopic expression of agouti, in $A^y$ mice, suggesting that agouti-induced obesity is caused by MC4R antagonism (Huszar *et al.*, 1997).

# 7

# Cell Lineage Segregation During Neural Crest Ontogeny

## 7.1 Introduction

The large variety of neural and non-neural derivatives that arise from the neural crest raises the fundamental question of how cell fates become specified during its ontogeny. In other words, what is the relative influence of intrinsic cell commitment compared to cell–cell interactions in promoting differentiation of distinct cell types.

Neural crest cells have long been implicitly considered as forming a population of homogeneous pluripotent cells that become specified to distinct lineages only after the arrest of migration and homing to their final destinations. This view would imply that local signals at the sites of homing play an instructive role in determining the precise cell types that develop. This notion has challenged investigators in the field of neural crest development to search for the existence of a stem cell from which all derivatives would arise.

The extreme opposite view of the ontogeny of the neural crest is to assume that this structure arises as, or rapidly becomes, a sum of heterogeneous cell subsets already committed to differentiate along different phenotypes. This would mean that cell fate is lineage-determined. Consequently, progressive cell divisions would specify the fate of successive daughter cells by giving rise to limited sets of crest cells. Then, upon migration, different subsets of committed cells would be expected to arrive at distinct homing sites. In this case, the interactions between crest cells and the microenvironment would be of a permissive rather than of an instructive type, allowing for the selection of those progenitors able to survive and differentiate in the conditions offered by the microenvironment. Such a view presents similarities with one of the models proposed to account for cell lineage segregation in the hemopoietic system, in which environmental factors are likely to affect the survival or proliferation of lineage-committed precursors that have been generated in a stochastic manner (Fairbairn et al., 1993; Ogawa et al., 1993; Davis and Littman, 1994).

From research performed over the past 15 years, it has become increasingly evident that, already at the beginning of emigration from the neural tube, the neural crest is formed by heterogeneous populations of cells, some pluripotent and others already restricted to different degrees in their developmental potentials, including precursors committed to one particular fate. These results suggest that environmental signals encountered during migration and homing are likely to operate both by instructive and permissive mechanisms on target cells with varying degrees of developmental restriction.

How was this early diversification among crest progenitors unraveled? First, performing isotopic and isochronic grafts of neural primordia prior to the onset of neural crest migration between quail and chick embryos allowed the fate map of the neural crest to be established. Subsequently, heterotopic transplantations revealed that neural crest cells develop, in general, according to phenotypes normally encountered in the new sites to which they home, thus revealing a high degree of plasticity at the population level. Furthermore, the development of efficient culture techniques for neural crest cells has allowed the analysis of the effects of various morphogens and growth factors on their differentiation to be undertaken. Taken together, the above techniques have revealed the paramount influence of the environment in the development of neural crest progenitors. Although pluripotentiality is undoubtedly a property of the neural crest cell population at many axial levels, the developmental potentiality of individual cells or of their progeny could not be established in heterotopic grafts or in mass cultures. More recent approaches have attempted to resolve this issue by examining the developmental potentials of individual crest cells by clonal analysis in vitro and their state of specification by in vivo lineage tracing.

In vitro techniques enabled neural crest and crest-derived cells to be challenged with culture conditions which virtually allowed all the potentialities of individual precursors to be expressed. Neural crest cells were thus cultured either at clonal density or as populations in which single cells were marked. Statistical analysis of the resulting phenotypes shed light on the developmental capacities of the founder cells. On the other hand, it was possible to follow in vivo the actual fate of individual progenitors by injecting or infecting them with non-diffusable lineage tracers. Thus, the in vitro and in vivo clonal approaches are complementary. A third strategy was to find molecules whose expression within the crest population is heterogeneous, with the assumption that this molecular heterogeneity underlies differences in population composition.

## 7.2 Heterogeneity among neural crest cells is revealed by differential expression of a variety of molecules

### 7.2.1 Monoclonal antibodies have led to subpopulations being distinguished in the early migratory neural crest

Since cell diversification is the result of the expression of different sets of genes, efforts were made to demonstrate gene products as early as possible within the migrating and differentiating neural crest cell population. With the advent of the Mab technology, several laboratories embarked on programs aimed at identifying proteins which are specific to particular neural crest-derived phenotypes, and which can lead, back in development, to the identification of the founder population. With such a purpose, the immunogen chosen was already differentiated cells. Barald (1988) was the first to raise a Mab against ciliary ganglion cells which also recognized small subsets of mesencephalic neural crest cells. Marusich et al. (1986) reported an antibody directed against dorsal root ganglion (DRG) cells which also recognized subsets of trunk neural crest cells. Likewise, the E/C8 Mab raised against chicken DRG cells (Ciment and Weston, 1982), and the A2B5 marker produced against neuronal plasma membranes (Eisenbarth et al., 1979), have been shown to mark early sub-populations of neural crest cells (Ciment and Weston, 1982; Girdlestone and Weston, 1985). In addition, the GLN1 marker is expressed by satellite and Schwann cells as well as by subsets of sensory and autonomic neurons. Earlier in development, it is carried by a subpopulation of migrating neural crest cells (Barbu et al., 1986).

Another very interesting marker is the SSEA-1 Mab. It was raised against F9 teratocarcinoma cells, and was shown to react with a cell-membrane carbohydrate antigen on blastomeres of eight-cell stage mouse embryos (Solter and Knowles, 1978). Notably, SSEA-1 was shown to be expressed by a sub-population of primary sensory neurons in the rat where it was suggested to play a role in the specific innervation of the dorsal horn by binding to complementary surface molecules on the target neurons (Regan et al., 1986; Jessell et al., 1990). Subsequently, Sieber-Blum (1989a) has shown that this antigen is also carried specifically by quail sensory neuroblasts in spinal ganglia at all axial levels, and she has further explored the possibility that SSEA-1 is expressed at earlier stages by quail neural crest cells. Two populations of neural crest cells that develop into sensory neuroblasts were observed to express SSEA-1 in culture: a non-migrating/early differentiating and an early migrating/late differentiating subset of precursors. Based on these observations, it was proposed that SSEA-1 may be a specific marker for the sensory neuron lineage. Implicit in this conclusion is that sensory neurons are established as a separate lineage during early ontogeny. Although enough experimental evidence points to the fact that this is certainly the case for a subset of early specified progenitors (Le Douarin, 1986; Ziller et al., 1987; Sieber-Blum, 1989b), it is not likely to be true for all sensory neurons (see discussion on sensory and sympathetic lineages below).

In the search for markers of neuronal lineages, Marusich and Weston (1992) have assessed the ability of a human autoantibody, anti-Hu, to identify crest-derived cells with neurogenic potential. Anti-Hu antibodies are found in sera of patients that exhibit both small cell lung carcinoma and subacute sensory neuropathy. This antibody was shown to recognize 34–40-kDa proteins that are mainly expressed in the nuclei of neurons of both the PNS and CNS (Graus et al., 1986). Earlier in development, it stains neurons in peripheral ganglia, and back to the progenitors, it marks a subpopulation of proliferating neural crest cells with non-neuronal morphology that has the ability to develop into neurons (Marusich et al., 1994). From these observations it was inferred that the Hu proteins identify early neurogenic precursors within the neural crest. This view was substantiated by the finding that Hu proteins share homology with the elav gene of Drosophila, a neuronal-specific RNA-binding protein that plays a role in maintenance of the fly nervous system (Yao et al., 1993).

Attempts were also made to find molecular markers using still undifferentiated neural crest cells to immunize the mice. This approach proved in most cases unsuccessful, perhaps because the routine methods of immunization were not adapted to the small numbers of cells available for injection. Using a direct intrasplenic injection of a neural crest cell suspension, Heath et al. (1992) raised Mabs against premigratory neural crest cells, and isolated several clones. These clones recognized a heterogeneity among early migrating crest cells in culture (15 hours), but they behaved in a totally homogeneous manner (either all cells stained or none of them stained with the different markers tested) on the premigratory population that served as immunogen. These observations led the authors to suggest that the premigratory crest is formed by homogeneous populations of cells that begin diversifying following the onset of migration.

Phenotypic heterogeneity defined by antigen expression does not necessarily imply differences in the state of commitment or in the developmental potential of progenitor cells. To demonstrate directly that the expression of a differential trait by subsets of neural crest cells is relevant to the selection of a subsequent developmental fate, the progenitors expressing a specific marker should be positively or negatively selected. Showing that the fate of the selected progenitors is different from that of their siblings that do not express the marker would provide evidence that this molecular heterogeneity has also a functional meaning. Such a strategy can, however, be applied only in the case of cell-surface markers, in which case the progenitors can be isolated by fluorescence-activated cell sorting and further analyzed (Maxwell et al., 1988), or alternatively, negatively selected by complement-mediated lysis (Vogel and Weston, 1988). Positive immunoselection can also be employed to enrich selected populations of neural crest-derived cells from the cells in their close vicinity. For example, Gershon et al. (1993) have incubated suspensions of embryonic gut either with the NC-1 antibody or with an antibody to the 110-kDa laminin-binding protein, both expressed on the surface of crest-derived cells in the bowel. Immunopositive cells were then isolated using a secondary antibody coupled to magnetic beads. Their reactivity to identified growth factors could then be assessed without any further influence of the non-crest cells present in the gut microenvironment.

### 7.2.2 The presence of various transcription factors allows recognition of different subpopulations of neural crest cells

About 10 years ago, we began witnessing what was soon to become a major contribution in the field of molecular biology to our understanding of neural crest development. A number of transcription factors have been described with interesting expression patterns in neural crest subsets and/or in specific neural crest derivatives. These data suggested the notion that they are intermediate signaling molecules in early processes such as lineage specification, or in later events of cell differentiation and/or survival. The possible functions of these factors are being challenged by combined cell culture and molecular genetic approaches.

An important step in research on transcription factors was to isolate vertebrate homologs of *Drosophila* genes that in the fly control early stages of neuronal or glial determination, with the idea that these genes would be conserved both structurally and functionally. Many genes were cloned using this strategy, such as MASH-1, NeuroD and Neurogenin (see discussion on neuronal lineages). Other sets of vertebrate genes that were cloned based on this homology belong to the engrailed family. Engrailed proteins contain a homeodomain region that binds to DNA, and they also localize to the cell nuclei where they act as transcription factors (Joyner *et al.*, 1985; Joyner and Martin, 1987). In the grasshopper, the *engrailed* gene product was shown to control, in the median neuroblast progenitor, the switch responsible for the generation of neurons versus glia (Condron *et al.*, 1994; Condron and Zinn, 1995). In vertebrates, Engrailed proteins are expressed by some cultured mesencephalic neural crest cells (Gardner *et al.*, 1988). *In situ* localization revealed that engrailed-positive cells are found in the neural folds prior to neural tube closure, and in the pathway of migration of neural crest cells at cephalic regions (Darnell *et al.*, 1992; Gardner and Barald, 1992). Their significance to neural crest development was examined in mouse embryos in which the *engrailed* genes were inhibited by antisense targeting, leading, among various defects, to a hypoplasia of neural crest-derived areas such as the face, the first and second pharyngeal arches, and the heart (Augustine *et al.*, 1995; Sadler *et al.*, 1995). Thus, the *engrailed* gene provides an example of an invertebrate nuclear factor whose function in vertebrates was apparently not conserved. It will be interesting to determine whether the function of other genes controlling the binary switch between neuronal versus glial fates in *Drosophila*, such as *glial cells missing* (Jones *et al.*, 1995; Hosoya *et al.*, 1995), remain conserved throughout evolution. Various other genes encoding such transcription factors have been discussed in Chapter 3 since they play important roles in patterning neural crest derivatives in the head and face. Such is the case for genes of the *HOM-C* and of the *Pax* gene family.

## 7.3 Studying developmental potencies of neural crest cell populations

### 7.3.1 The behavior of cell populations in heterotopic grafts

Results of heterotopic transplantations of different regions of the neural crest provided the notion that the neural crest cell population at many axial levels is pluripotential at least as far as the PNS derivatives of the crest are concerned, with only few exceptions. Fragments of neural primordia (neural tube containing premigratory neural crest) were taken from quail embryos and grafted into chick hosts, or vice versa. In most cases, the neural crest cells that individualized from the implants at ectopic sites migrated to homing sites characteristic of the new axial level rather than to their origin. In these anlage, the labeled cells differentiated according to a fate appropriate to their new environment (Fig. 7.1).

Noden (1975, 1978a,b) performed grafts of forebrain neural crest in the mesencephalic/metencephalic region. Whereas forebrain-level crest cells do not differentiate normally into neural derivatives, they give rise to ciliary and trigeminal ganglia in the new location. Moreover, when the vagal neural crest of a chick embryo was replaced by the thoracic crest from a quail, which does not give rise to enteric innervation in normal development, the grafted cells were able to colonize the gut and differentiate there into neurons that display appropriate cholinergic, peptidergic, and serotonergic phenotypes (Le Douarin and Teillet, 1974; Le Douarin et al., 1975; Fontaine-Pérus et al., 1982; Rothman et al., 1986). Furthermore, the cephalic and vagal neural crest, when transplanted in the adrenomedullary level of the axis (somites 18–24) gave rise to adrenergic cells in the host sympathetic ganglia, and to adrenal glands (Le Douarin and Teillet, 1974).

Thus, heterotopic transplantations revealed that premigratory neural crest cells are highly plastic. This raised the question as to whether populations of neural crest-derived cells retain a similar plasticity. One such example will be discussed below concerning neural crest-derived cells taken from the gut following colonization. Rothman et al. (1990) back-grafted pieces of E4 quail bowel containing recently migrated neural crest cells into the adrenomedullary level of young E2 chick hosts. Donor cells remigrated in the host embryos and colonized sensory ganglia and roots, peripheral nerves, sympathetic ganglia, and the adrenal medulla. Nevertheless, no melanocytes were detected. Gut-derived crest also failed to colonize the host gut when placed at an axial level that does not normally lead the migratory cells to the bowel. By contrast, if E4 foregut fragments were placed into the vagal or sacral levels of the axis of host embryos, successful colonization of the host's bowel was achieved, in addition to the development of adrenergic cells in sympathetic ganglia at sacral levels, but not of melanocytes at any of the levels tested (Rothman et al., 1993b). Several conclusions were drawn from these results. First, neural crest cells of the gut are able to remigrate when placed in an appropriate environment. Second, the targets colonized by grafted cells are determined by the

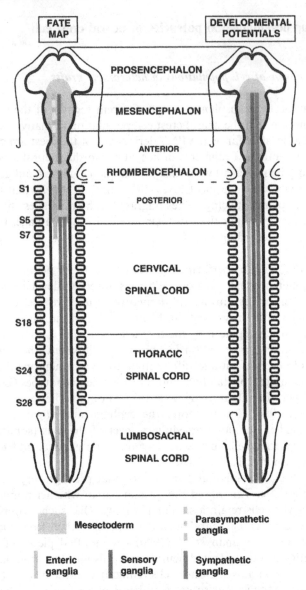

FATE MAP

DEVELOPMENTAL POTENTIALS

PROSENCEPHALON

MESENCEPHALON

ANTERIOR

RHOMBENCEPHALON

POSTERIOR

S1
S5
S7

CERVICAL

SPINAL CORD

S18

THORACIC

SPINAL CORD

S24

S28

LUMBOSACRAL

SPINAL CORD

Mesectoderm

Parasympathetic ganglia

Enteric ganglia

Sensory ganglia

Sympathetic ganglia

**Figure 7.1** Fate map and development potentials of the neural crest along the axis. Left: Fate map of the presumptive territories along the neural crest yielding the mesectoderm, the sensory, sympathetic and parasympathetic ganglia in normal development. Right: Development potentials for the same cell types as shown in the fate map are indicated. Results are based on isotopic and heterotopic grafting of neural primordia between quail and chick embryos. See text for details. (Reprinted, with permission, from Le Douarin, 1986. Copyright (1986) American Association for the Advancement of Science.)

*For a colored version of this figure, see www.cambridge.org/9780521122252.*

routes of migration characteristic of the axial level of the host embryo. Third, neural crest-derived cells of the gut may be largely multipotent inasmuch as they have the ability to develop into derivatives characteristic of the axial level of the host. Yet, some developmental restriction already exists, because in all types of transplants the grafted cells failed to develop into melanocytes. Similar restrictions evidenced by other types of back-transplantation experiments are described in Section 7.4.1. Fourth, given the high plasticity of the grafted cells, one must assume that among the crest-derived cells that colonize the gut, there are still multipotent progenitors upon which the microenvironment of the new arrest sites induce an appropriate pattern of phenotypic differentiation.

The plasticity revealed in the above experiments does not necessarily imply a complete equivalence in developmental potencies of neural crest cells along the entire axis. For example, unlike the cephalic crest, cells from the trunk are unable to form cartilage, even when transplanted in the head (Chibon, 1967; Nakamura and Ayer-Le-Lièvre, 1982). An additional restriction of thoracic-level crest cells is their inability to develop into mesenchyme that participates in septation of the truncus arteriosus of the forming heart (Kirby, 1989). These findings are consistent with the existence of progenitor cells in the early neural crest endowed with various states of commitment.

### 7.3.2 A time-dependent topographical segregation of peripheral progenitor cells: possible significance for understanding neural crest plasticity

The segregation of subpopulations of progenitors from the multipotent neural crest population may take place before, during, or after migration. To test whether all populations of crest cells emigrating from the neural tube have equivalent capacities to colonize different target sites, the fates of early- versus late-migrating neural crest cells were examined in various species. Labeling of neural crest cells throughout the axis at progressively later stages of development with DiI, with the quail marker, or by position relative to the tube, revealed that the different crest derivatives are colonized in a ventrodorsal order, with the last emigrating cells localized to the dorsolateral pathway (Weston and Butler, 1966; Serbedzija et al., 1989; Raible and Eisen, 1996; Baker et al., 1997). For instance, in the mesencephalic region, early-migrating crest cells were found to populate both dorsal (melanocytes, ciliary neurons, Schwann cells along the oculomotor and trigeminal nerves, dorsal dermis, etc.) and ventral (skeleton of the jaws) derivatives. By contrast, late-migrating crest cells gave rise only to dorsally localized derivatives (Baker et al., 1997).

This sequential order of topographical colonization was subsequently interpreted to reflect, at least in the trunk of avian and zebrafish embryos, a progressive restriction in the developmental potential of the migrating crest cell populations, because early-migrating cells developed into melanocytes, neurons, and adrenergic cells, whereas late migrating cells lost the ability to give rise to neurons (Raible and Eisen, 1996), or to adrenergic derivatives

(Artinger and Bronner-Fraser, 1992) upon retrotransplantation into young hosts. In contrast to the behavior of the truncal crest, when isochronic and heterochronic grafts of mesencephalic crest cells were performed between quail and chick embryos, it was found that both early- and late-migrating cells were able to form melanocytes, neurons, glia, cartilage, and bone (although late crest cells made comparatively much less of the two last tissue types). These results suggest that, at the population level, both early- and late-migrating mesencephalic crest cells are plastic, even though their derivatives become topographically segregated along the ventrodorsal direction.

In spite of the fact that late-migrating crest cells do not normally contribute to the cartilage of the ventral jaws, when the late crest cells were grafted into a late-stage host whose neural crest had previously been ablated, they succeeded in migrating ventrally toward the jaw (Baker *et al.*, 1997). Similarly, in zebra-fish, late-migrating cells retained their ability to give rise to neurons if the early cells were ablated (Raible and Eisen, 1996). These findings suggest that inter-actions between crest cells found in sequential migratory waves may play a role in fate specification. Alternatively, since there is a temporal order of coloniza-tion of the different targets by corresponding cells in the various migratory waves, it is likely that interactions between crest cells and the environment of the homing sites determine their fate. Thus, in the absence of one component, the normal sequence is moved to the next generation of cells so as to ensure that the normal pattern of target colonization is kept.

Surprisingly, Sharma *et al.* (1995) have reported that some cells located in the spinal cord of stage 25–26 chick embryos, well after neural crest cell emi-gration is complete, are able to migrate afield from the CNS and populate selected peripheral derivatives. These progenitor cells were proposed to orig-inate near the dorsal root entry zone and to migrate out through the dorsal roots to colonize the DRG where they differentiate into neurons and glia. The presence of melanocytes was also documented, although the pathways leading to the colonization of the skin were not studied. Provided the procedures employed exclusively label neuroepithelial cells with no contamination by neural crest-derived cells, such as satellite or Schwann cells present along the dorsal roots, two important issues deserve attention. The first is related to the concept that the CNS is directly at the origin of some "classical" neural crest derivatives at a time that follows the transient lifespan of the neural crest. If correct, this concept is certainly relevant to our understanding of lineage restrictions in the nervous system. Yet, the relative significance of these findings is affected by quantitative considerations, such as the proportion of cells fur-nished to the various structures via this late migratory process. As for the DRG, cell counts revealed a very modest contribution, while the number of melanocytes was not determined. The second issue of relevance concerns the type of derivatives issued from this late migration. In agreement with the data discussed above, these spinal cord cells gave rise only to dorsal structures. However, contrary to the results discussed above, Sharma *et al.* (1995) have shown that the late wave of progenitors also has a neurogenic fate.

### 7.3.3 Single-cell analysis of the neural crest

The various *in vivo* and *in vitro* methods described above for studying the development of neural crest cells revealed the significant plasticity of this cell population along the neuraxis. At the same time, they uncovered an intrinsic heterogeneity within this population. These results raised critical questions as to the timing and mechanisms of segregation of different lineages. For example, do the multipotent progenitors directly generate differentiated phenotypes or do they indeed undergo a progressive restriction in their developmental capacities? What is the proportion of multipotent compared to more restricted types of neural crest progenitors at any stage in development in different species? What are the genes involved in lineage specification and do specific environmental signals act on different sublineages? What are the lineage relationships among cells in different ganglia? Clearly, it is not possible to infer accurately the potential of cells from population studies. To this end, it became necessary to devise techniques that make possible the phenotypic analysis of the progeny of individual neural crest cells.

Analysis of single cells can be performed either by *in vivo* lineage tracing or by *in vitro* cloning. *In vivo* lineage tracing permits the *fate* of individual progenitor cells to be assessed under apparently normal conditions. This means that the progenitors and their descendents whose fate is being followed are subjected to the same stringent selection that operates in the intact embryo, and that is mediated by signals distributed in a region-specific manner. Thus, this approach does not enable evaluation of the entire range of developmental potentials of neural crest cells revealing its actual state of commitment. This issue can be analyzed instead by *in vitro* cloning of individual progenitors challenged with environments more permissive than those encountered in the embryo. *In vitro* cloning techniques seek to establish optimal conditions for growth and proliferation of single cells, aimed at uncovering the largest possible repertoire of sublineages inherent to a given progenitor. A complementary approach to those described above has been recently implemented in the neural crest field. It consists of growing neural crest cells in mass cultures; at defined times after seeding, a single cell is labeled randomly with a lineage tracer in each culture dish and its derivatives are monitored. The advantages and limitations of each of these approaches are discussed below.

### 7.3.4 In vitro *cloning of neural crest cells*

The *in vitro* analysis of the progeny deriving from single founder neural crest cells provides a substantial means of assessing the developmental potentialities of individual precursor cells. Random sampling of single progenitors was achieved either by direct seeding of single cells picked up from a neural crest suspension (Baroffio *et al.*, 1988; Dupin *et al.*, 1990; Dupin and Le Douarin, 1995), or by culturing cell populations at low-density conditions, sometimes followed by serial subcloning of the primary clones (Cohen and Königsberg, 1975; Sieber-Blum and Cohen, 1980; Stemple and Anderson, 1992).

As the state of commitment of the founder cells is inferred from the number of resulting phenotypes, an essential requirement of the method is the availability of appropriate markers that allow the sublineages developing in the cultures to be defined. Therefore, those cell types for which appropriate identification tags are lacking cannot be unequivocally distinguished, and their phenotypes remain undefined. It is worth mentioning that some cells which remain unstained might also represent undifferentiated progenitors. Notably, in order to identify in a single clone as many phenotypic traits as possible, it is necessary to combine several stainings on single cultures.

Another important issue that affects the reliability of *in vitro* cloning procedures is the percentage of single cells seeded that give rise to clones (cloning efficiency). To determine the relative proportion of progenitors with given properties that exist at a given axial level of the crest and at a precise time point, it is crucial that the clones be a good representation of the overall population from which the founder cells are randomly taken. A very low cloning efficiency would suggest that the growth conditions employed exert a strong selection on particular cells over others, perhaps favoring the development of those progenitors displaying stronger adhesiveness to the substrate, higher proliferative capacity, or other less obvious properties. On the other hand, many neural crest cells might require cell–cell contacts to survive and further develop. Cloning on extracellular matrix substrates may not be sufficient for such progenitors. One way to improve cloning efficiency would be to seed individual cells on a cellular substrate. Baroffio *et al.* (1988) cultured single crest cells on a feeder layer of Swiss 3T3 fibroblasts whose growth was arrested by treatment with mitomycin (Fig. 7.2). This substrate together with the presence of a complex and rich medium greatly favored attachment, proliferation, and differentiation into the major crest-derived lineages.

Another method for clonal analysis consists of intracellularly labeling individual neural crest cells in primary cultures with a lineage dye, and growing the cells under enriched medium that permits the differentiation of multiple phenotypes (Henion and Weston, 1997). This technique is likely to minimize the selection of certain progenitors over others since, while emerging as outgrowths from neural tubes, neural crest cells interact with each other and with the tube, and are known to give rise to many phenotypes. Consequently, random labeling of single progenitors in this type of culture is likely to reflect accurately the proportion of progenitors with different properties within the population.

### 7.3.5 Heterogeneity of neural crest cells is revealed by in vitro clonal analysis

Single neural crest cells were obtained from trunk levels of quail embryos by the limited dilution method (Cohen and Königsberg, 1975; Sieber-Blum and

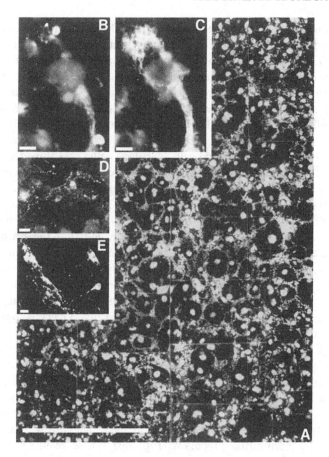

**Figure 7.2** Clonal analysis of neural crest cells grown on growth-arrested 3T3 fibroblasts. (A) Hoechst DNA staining of a 10-day clone that contained approximately 23 000 neural crest cells (small nuclei) grown on 3T3 cells (large nuclei). Several types of neuronal cells are seen, containing TH (B), SP (D), or VIP (E), as well as numerous SMP glial cells (C). Scals bars = (A) 1 mm; (B–E) 20 μm. (From Le Douarin and Dupin, 1993.)

Cohen, 1980; Sieber-Blum, 1989b). Analysis of their progeny revealed the presence of three types of clones: (1) clones formed exclusively by pigmented cells; (2) unpigmented clones that contain neurons expressing SSEA-1 and tyrosine hydroxylase or dopamine β-hydroxylase immunoreactivities, indicative of the presence of sensory neurons and autonomic cells, respectively; and (3) mixed clones which are formed by cells that have the capacity to generate neuronal as well as pigmented progenies. These results suggested the existence of cells with various degrees of restriction: tripotent (melanogenic, sensory, and autonomic), bipotent (sensory, and autonomic), and monopotent (melanogenic) progenitors that coexist in the trunk neural crest population (reviewed by Sieber-Blum, 1990).

This approach was further extended by Le Douarin and coworkers. In a series of studies, the potentialities of the migrating population of neural crest cells from the cephalic level of the neuraxis was analyzed. Large clones containing between one and four different cell types were found (Fig. 7.2). Substance P-positive neurons were found to coexist with adrenergic cells within the same clone either in distinct cells or colocalizing to individual cells. Moreover, non-neuronal cells expressing the HNK-1 marker were usually found associated with neurons in mixed clones. A common precursor for different neuronal types as well as for non-neuronal cells (cartilage cells) was thus demonstrated (see below for details). Furthermore, founder cells whose progeny became melanocytes also gave rise to adrenergic cells, and/or to cells expressing substance P (Baroffio et al., 1988). These studies, however, failed to show the development of clones containing only melanocytes, a clonal population previously reported by Sieber-Blum. It should be noted that the proportion of these unipotent melanogenic precursors is reduced in rhombencephalic as compared to trunk neural crest (9% versus 62%, respectively; see Ito and Sieber-Blum, 1991, and Ito et al., 1993, for review). However, comparison between these data and the results of Dupin and Le Douarin (1995) for trunk levels of the neural crest are consistent with the assumption that 3T3 cells inhibit differentiation and/or survival of melanogenic monopotent progenitors.

The large amount of clones analyzed in all the above studies enabled quantification of the distribution of potentialities of the different progenitors. It was found that about 80% of the clones were derived from multipotent progenitor cells with a progeny composed at least of 2–4 different cell types in various combinations. Furthermore, about 20% of the clones were composed of a single cell type, either neurons, glia, or cartilage, suggesting that these are derived from a committed precursor (Fig. 7.3). These results indicate that, although a majority of progenitors within the neural crest are pluripotent, a considerable proportion of cells derives from early-specified precursors. Cloning studies by Ito and Sieber-Blum (1991) provided consistent results concerning the various potentialities of rhombencephalic-level neural crest cells.

The last notion received further support in a recent line of experiments performed by Henion and Weston (1997). These authors labeled single thoracic neural crest cells growing in mass cultures with a lineage tracer. Individual crest cells were labeled immediately after leaving the neural tube and at different time intervals prior to overt differentiation. Under these conditions, almost half of the initial crest population (44.5% labeled between 1 and 6 hours after leaving the tube) was found to generate a single cell type such as neurons, glia, or melanocytes, suggesting that the early-emigrating neural crest generated in culture already contains a significant proportion of fate-restricted precursors (Table 7.1). Analysis of the potentialities of older crest cells revealed that 52.2% and 77% of the progenitors were fate-restricted at 13–16 and 30–36 hours after explantation, respectively (Table 7.1). Furthermore, distinct neurogenic and melanogenic precursors were present in the outgrowth population very soon after explantation, but melanogenic cells dispersed from the tube

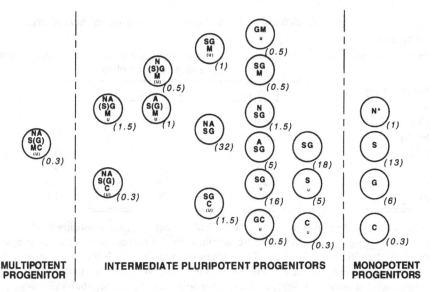

| MULTIPOTENT PROGENITOR | INTERMEDIATE PLURIPOTENT PROGENITORS | MONOPOTENT PROGENITORS |

**Figure 7.3** Diagram illustrating all the types of progenitors and their frequency as revealed by the phenotypic analysis of neural crest clones. A total of 533 clones was obtained from cephalic crest and analyzed under identical medium conditions. Each circle corresponds to one type of progenitor, defined and classified according to the combinations of cell types found in the clone obtained from that progenitor. N, NF + cells; A, TH + cells; S, SMP⁻ glial cells; G, SMP–/HNK-1 + non-glial cells; M, melanocytes; C, cartilage, u, phenotypically undefined cells that were negative for all markers. (G), (S), (u), presence not determined; N*, neuron-like cells carrying the HNK-1 epitope. The numbers in parentheses are the percentages of each type of progenitor. (See Baroffio *et al.*, 1991, for details; reproduced with permission of Company of Biologists Ltd.)

relatively late compared to neurogenic cells. Taken together, these results show that segregation of cell lineages is a progressive, and non-random process, and, more specifically, that melanogenic and neurogenic lineages are likely to segregate early from a common progenitor.

A major finding of the *in vitro* clonal approach was the characterization of a progenitor cell that is able to give rise to the majority of neural crest-derived lineages such as neurons, glia, melanocytes, and cartilage. This cell appeared at the extremely low frequency of one out of 305 clones in cultured cephalic avian crest (Baroffio *et al.*, 1991). Because of its high multipotency it was suggested that this founder cell is a stem cell, in analogy to the stem cells of the hemopoietic system (Anderson, 1989) that are both capable of self-renewal as well as able to give rise to all types of cells (totipotency) (Lajtha, 1979). The self-renewal capacity of these highly multipotent progenitors in the avian neural crest still remains to be demonstrated. More recently, Stemple and Anderson (1992) have reported on the isolation of a mammalian neural crest stem cell based on its capacity for self-renewal through at least six to ten generations. Still, this progenitor cell gives rise to a rather limited set of derivatives such as

Table 7.1. *Temporally defined precursor-type composition of neural crest cell populations*

| | Percentage of total clones | | |
| --- | --- | --- | --- |
| Precursor type | 1–6 hours | 13–16 hours | 30–36 hours |
| N | 11.1 | 9.3 | 24.1 |
| NG | 44.4 | 20.4 | 3.7 |
| NGM | 1.9 | 0 | 0 |
| NM | 0 | 0 | 0 |
| G | 31.5 | 24.1 | 48.1 |
| GM | 9.2 | 7.4 | 9.3 |
| M | 1.9 | 38.8 | 14.8 |

*Note*: Labeled clonal precursors were classified (precursor type) according to the phenotype of cells they generated as determined by the expression of cell-type specific markers after 96 hours of culture. These phenotypes identify both the precursor and the resulting clonal descendants, as follows: N, neuronal; NG, neuron–glial; NGM, neuron–glial–melanocyte; NM, neuron–melanocyte; G, glial; GM, glial–melanocyte; M = melanocyte. The clonal descendants of 54 individual precursor cells labeled at each time-point were analyzed (total of 162 clones). The data are presented as the percentage of all labeled precursors represented by each precursor type at the indicated time.
*Source*: Reproduced, with permission of Company of Biologists Ltd, from Henion and Weston (1997).

neurons, Schwann cells, and smooth muscle cells (Shah *et al.*, 1996). Thus, formal demonstration for its totipotency is still lacking. Perhaps the strongest analogy that can be drawn at present between the neural crest and the hemopoietic system is that in both systems the segregation of lineages from initially multipotent precursors appears to involve a progressive restriction of developmental potencies. Clones expressing multiple lineages in variable combinations could be identified both in the cephalic crest and in cultures of single hemopoietic precursors (Suda *et al.*, 1983). Statistical analysis of these findings for the neural crest have suggested that the intermediate pluripotent progenitors are created by stochastic events rather than in a preexisting order (Baroffio and Blot, 1992; Henion and Weston, 1997).

The progressive restriction in developmental capacities of neural crest cells is clearly reflected as a function of embryonic age. It was shown that the capacity for proliferation and the range of differentiation abilities of cloned DRG cells is inferior to that of their progenitor crest cells (Duff *et al.*, 1991; Sextier-Sainte-Claire Deville *et al.*, 1992). Likewise, neural crest-derived cells from E4–12 quail embryonic gut revealed that the clonal efficiency and the phenotypic diversity obtained within individual clones were highest between E4 and E6 but decreased sharply by E8 (Sextier-Sainte-Claire Deville *et al.*, 1994). In spite of reducing their developmental capacities, some multipotent progenitors still prevail in young avian sensory and sympathetic ganglia (Duff *et al.*, 1991;

Le Douarin *et al.*, 1991) and in other structures containing neural crest-derived cells such as the skin, and the visceral arches (Ito and Sieber-Blum, 1993; Richardson and Sieber-Blum, 1993).

In summary, *in vitro* cloning of avian (Cohen and Königsberg, 1975; Sieber-Blum and Cohen, 1980; Baroffio *et al.*, 1988, 1991; Sieber-Blum, 1989b; Dupin *et al.*, 1990; Henion and Weston, 1997), mammalian (Lo *et al.*, 1991a; Stemple and Anderson, 1992; Ito *et al.*, 1993), and amphibian (Akira and Ide, 1987) neural crest cells has revealed that, in a particular developmental stage and axial level, pluripotent neural crest cells coexist with fully committed progenitors and with cells displaying intermediate developmental potentials. Although there is full agreement as to the heterogeneity of the crest population, the different cloning methods employed led to discrepancies concerning the proportion of multipotent compared to restricted progenitors within the population. Whereas results of clonal seeding favored the notion that the early crest cells are largely pluripotent, single cell labeling in mass cultures led to the view that many cells are already fate-restricted early during the ontogeny of the neural crest (see previous section for possible explanations). Clearly, clarification of this issue is required in order to further approach the question of the mechanisms whereby these progenitors respond to environmental cues.

### 7.3.6 *Single-cell lineage analysis* in vivo

As previously mentioned, *in vivo* lineage tracing permits the fate of individual progenitor cells to be assessed under apparently normal conditions because the labeled cells remain subjected to the selective environments of the embryo. Two methods are now available to label individual progenitors for lineage analysis: the first is iontophoretic microinjection of a vital dye, such as lysinated rhodamine dextran (LRD). LRD is a fluorescently tagged dextran whose high molecular weight prevents its diffusion out of the injected cell. Thus, the label is transferred exclusively by cell division. The second technique consists of infecting single cells with a recombinant retrovirus bearing the *Escherichia coli* β-galactosidase *(lacZ)* gene. This gene is activated only upon integration into the host cell DNA, and is further propagated upon cell division. Activity of the *lacZ* gene product can be monitored on tissue sections by the addition of the appropriate substrate. The appearance of a blue color indicates enzymatic activity within the cells bearing the construct.

It is worth mentioning briefly the strengths and weaknesses of each method. Frank and Sanes (1991) pointed to the following differences: (1) Dye injection permits targeting to more or less defined progenitors whereas viral infection is a random event. (2) The probability of hitting more than a single progenitor per event is much higher in the case of viral infection (even if using highly diluted virions) than after direct intracellular microinjection during which membrane potential is recorded; this issue is very critical in the case of neural crest cells that actively disperse in different embryonic sites, thus obscuring the possible clonal identity of the progeny. (3) The provirus is a heritable, non-diluting label, whereas injected tracers are diluted as cells divide. (4) The technique of viral infection enables cells to be reached that would be otherwise difficult to attain or impale with a microelectrode.

In spite of the differences between the techniques, similar results were obtained regarding labeling of premigratory and migrating neural crest cells (Bronner-Fraser and Fraser, 1988, 1989; Fraser and Bronner-Fraser, 1991; Frank and Sanes, 1991). It was found that a majority of trunk neural crest precursors in avian embryos gave rise to multiple types of progenitors (Figs 7.4 and 7.5). These included combinations of cells in at least two of the following sites: DRG, sympathetic ganglia, adrenal medulla, ventral roots (Bronner-Fraser and Fraser, 1988, 1989; Fraser and Bronner-Fraser, 1991). Moreover, if localized to DRG only, clones were usually composed of neurons and non-neuronal cells (Figs 7.4 and 7.5), or neurons in the lateroventral and dorso-medial areas of the ganglia known to innervate different targets (Frank and Sanes, 1991; Fraser and Bronner-Fraser, 1991). The lineage of individual neural crest cells was also traced in *Xenopus* embryos using LRD microinjection (Collazo *et al.*, 1993). As in birds, the majority of clones had progeny in multiple derivatives including spinal ganglion cells, pigment cells, enteric cells, fin cells, and/or neural tube cells in all combinations. Taken together, these results suggest that most premigratory and migrating crest cells are multipotent. In addition, a minority of clones in all cases contained only a single type of derivative. Although such an outcome could be accounted for by death of some progenitors that homed to inappropriate sites, this result conforms with the notion that a certain proportion of neural crest cells is restricted in potential, as suggested by data from *in vitro* cloning.

In contrast to bird and *Xenopus* embryos, the zebrafish neural crest appears to be lineage-restricted to a large extent, generating progenitors that produce single types of derivatives. For example, 82% of the clones arising from individual LRD-injected cells were composed of single derivatives, whereas 11.6% contained combinations of two cell types, and only 6.5% of the clones were composed of three cell types (Raible and Eisen, 1994). Moreover, cells that migrate through different pathways appear to be restricted to different extents, probably as a function of time from the onset of migration (Raible *et al.*, 1992; Eisen and Weston, 1993; Raible and Eisen, 1994, 1996): i.e., the earliest cells migrating on the medial pathway between neural tube and somites generated multiple derivatives, while progenitors migrating later on the lateral pathway generated only pigment cells. This situation is similar to that found in avian embryos in which crest cells were labeled at progressively older stages and their fate followed both *in vivo* and *in vitro* (Weston and Butler, 1966; Serbedzija *et al.*, 1989).

Aside from the significant contribution of *in vivo* lineage tracing to our understanding of the fate of single neural crest cells, this methodology suffers from two main weaknesses. The first is that the phenotypes of the progeny of injected or infected cells were essentially diagnosed by their localization in the embryo and by their morphology (Fig. 7.6). Few markers were used to discriminate within a ganglion between putative neurons (neurofilament immuno-

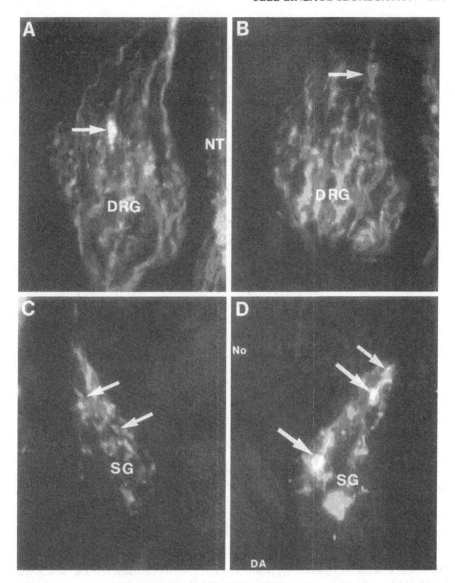

**Figure 7.4** Lineage analysis of trunk neural crest. Neurofilament expression in lysinated rhodamine dextran (LRD)-labeled descendants. LRD labeling is shown in red, and staining with an antibody against neurofilament (NF) protein is shown in green. (A–C) Images from an embryo that contained LRD-labeled cells in the DRG, sympathetic ganglion (SG), and ventral root (VR, not shown). (A) An LRD-labeled cell (arrow) in the DRG has bright NF staining in its axon. Orange color indicates double NF/LRD staining. (B) Another cell in the DRG is NF– and has the appearance of a support cell. (C) The SG of the same embryo showing two NF– cells. (D) The SG of another embryo contains numerous cells (arrows) with large cell bodies and NF+ axons. (See Fraser and Bronner-Fraser, 1991, for details; reproduced with permission of Company of Biologists Ltd.)

*For a colored version of this figure, see www.cambridge.org/9780521122252.*

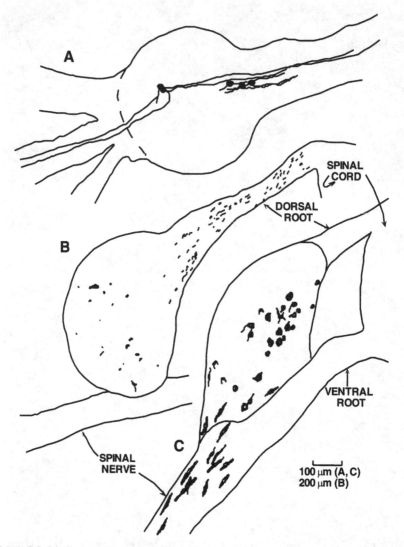

**Figure 7.5** Lineage analysis of the neural crest with a recombinant retrovirus. Camera lucida reconstructions of three clones, illustrating the varieties encountered. In A, the clone contained five neurons only. In B, the clone contained 28 non-neuronal cells in the ganglion and 107 Schwann cells in the dorsal root, but no neurons. In C, the clone contained 18 neurons, three Schwann cells, and eight ganglionic non-neuronal cells in the DRG, plus 12 Schwann cells in the spinal neve. (Reproduced, with permission of Company of Biologists Ltd, from Frank and Sanes, 1991.)

reactivity) and satellite cells or between distinct neuronal subtypes (Fig. 7.4). Consequently, the presence of labeled cells in both DRG and sympathetic ganglia might not necessarily imply that the founder cell was multipotent. The second limitation of this technique resides in the relatively small number of embryos injected for each stage of the neural crest. Clearly enough, the

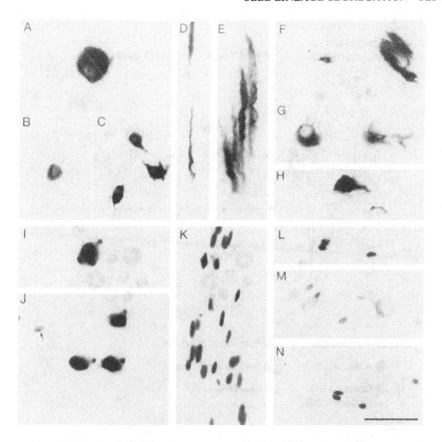

**Figure 7.6** Lineage analysis of the neural crest with a recombinant retrovirus. Morphology of neural crest derivatives labeled with cytoplasmic (LZ10; A–H) or nuclear (LZ12; I–N) viral strains. (A–C, I, J) neurons; (D, E, K) Schwann cells; (F–H, L–N) ganglionic non-neuronal cells. A non-neuronal nucleus is labeled in J (arrow), and neurons neighbor satellite cells in G and H. Bar = 50 μm. (Reproduced, with permission of Company of Biologists Ltd, from Frank and Sanes, 1991.)

relative complexity of these experiments makes it extremely difficult to gather large samples of injected embryos required for a statistical analysis of the resulting phenotypes.

## 7.4   The segregation of specific neural crest-derived lineages

### 7.4.1   *Sensory and autonomic neuronal lineages*

As described earlier in the chapter, back-transplantation of embryonic quail peripheral ganglia into the chick neural crest migration pathway subjected the cells to microenvironments where various potentialities, not expressed during

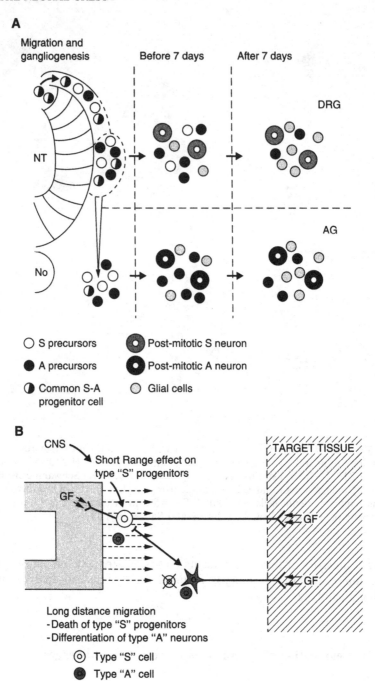

**A**

Migration and gangliogenesis    Before 7 days    After 7 days

NT

No

DRG

AG

○ S precursors

● A precursors

◐ Common S-A progenitor cell

◉ Post-mitotic S neuron

◉ Post-mitotic A neuron

○ Glial cells

**B**

CNS

Short Range effect on type "S" progenitors

GF

TARGET TISSUE

GF

GF

Long distance migration
- Death of type "S" progenitors
- Differentiation of type "A" neurons

◉ Type "S" cell

● Type "A" cell

normal development, could be uncovered (reviewed by Le Douarin, 1986). The localization of the grafted cells varied with the type of ganglion and its age. Quail cells populating the host DRG and differentiating there into sensory neurons and glia were found only after grafting of DRG. In contrast, Schwann cells and autonomic derivatives (sympathetic neurons, chromaffin cells, and in some cases enteric ganglia) were obtained after both DRG and autonomic ganglion grafts. For instance, autonomic ganglia such as the ciliary, Remak and sympathetic ganglia never gave rise to sensory neurons in the host DRG. Moreover, the capacity of quail DRG cells to populate the host sensory ganglia was restricted to the time when sensory neuroblasts were still mitotically active. When all sensory neurons had withdrawn from the cell cycle at about day 7 of embryonic development in the quail, DRG grafts gave rise exclusively to ganglion cells of the autonomic type (Schweizer *et al.*, 1983). These results suggested that postmitotic neurons die under conditions of back-transplantation, and that the plastic behavior revealed in the grafts resides within the non-neuronal population. This notion was confirmed experimentally by using the vagal ganglia in which the non-neuronal cell subset of the ganglion can be selectively labeled since it is of neural crest origin, while the neurons, which derive from a placode, are of the chick host type. If such a chimeric ganglion is back-grafted into a E2 chick neural crest migration pathway, only the non-neuronal cells survive and are able to differentiate into autonomic (sympathetic and enteric) neurons when subjected to the inducive conditions of the younger host (Ayer-Le Lièvre and Le Douarin, 1982).

Based on the above results, Le Douarin (1986) proposed a model of dual sensory–autonomic segregation which implies the following (Fig. 7.7): (1) the existence of distinct sensory and autonomic neuronal progenitors in the migrating neural crest in addition to common precursors with at least bipotent cap-

**Figure 7.7** (A) Hypothesis accounting for cell lineage segregation during PNS ontogeny. Two types of precursors – sensory (S) and autonomic (A) – arise from a common progenitor during neural crest cell individualization and/or migration and in the early steps of gangliogenesis. During gangliogenesis, the S progenitors can survive only in ganglia developing in close proximity to the neural tube, from which they benefit from the effect(s) of growth factor(s). They rapidly become exhausted in their proliferation capacities and, by 7 days, they have reached the postmitotic state. In the non-neuronal population of these sensory ganglia, A progenitor cells survive while glial cells differentiate. In the autonomic ganglia (AG) developing at a distance from the CNS, the S progenitors disappear rapidly while the A precursor cells survive at least until hatching. NT, neural tube; No, notochord. (B) Diagram showing the different survival requirements of the S and A progenitors at the time of gangliogenesis. S progenitors need a growth factor (GF) from the CNS to survive and differentiate. It is only later upon arrival of its fibers to peripheral targets that they find neurotrophins and other factors for survival. A progenitors can survive and differentiate far apart from the CNS. In sensory ganglia they survive for long periods in a latent form, but do not express differentiated traits either in the chick or in quail. S precursors do not survive if migrating into ganglia localized at a distance from the CNS. (Reprinted, with permission, from Le Douarin, 1986. Copyright (1986) American Association for the Advancement of Science.)

abilities; (2) a transient lifespan of the sensory progenitors until neurons are born, in contrast to a prolonged subsistence of autonomic progenitors, at least until hatching; (3) differential requirements for survival and differentiation based on the fact that sensory progenitors can only survive in sensory ganglia close to the CNS primordium, while autonomic precursors remain alive in all types of ganglia. We now evaluate the validity of this model in light of evidence accumulated over the past 10 years.

### 7.4.2 Existence of distinct sensory and autonomic neuronal progenitors in the migrating neural crest

The differentiation of various cell types in culture depends upon the composition of the culture medium. Thus, different media should allow the differentiation of distinct cell types. Ziller *et al.* (1983, 1987) have explanted mesencephalic neural crest from migratory stages in a serum-free, defined medium. Under these conditions, a subset of neural crest cells rapidly leaves the cell cycle and differentiates into neurons with sensory properties. Adrenergic cells do not differentiate under these conditions. In contrast, if the medium is supplemented with serum and chick embryo extract, the first neuronal category dies, and, 5–7 days later, adrenergic cells develop from a different population of cycling precursors. These experiments are consistent with the notion that some sensory and autonomic progenitors are segregated within the migratory neural crest and that they have distinct growth requirements. Moreover, the rapid differentiation of the sensory-like neurons indicates that they may derive from an early committed progenitor upon which the environment acts in a selective way, whereas the appearance of autonomic cells, based on the time and complex medium requirements, might arise as a result of instructive signals. Furthermore, Sieber-Blum (1989b) found a similar group of progenitors that developed in close proximity to the neural tube, appeared to be postmitotic at the time of explantation, and differentiated into sensory neurons within 48 hours.

Results of *in vivo* clonal analysis of the neural crest enabled a more direct assessment of the lineage relations of sensory neurons. Most clones were found to populate DRG, sympathetic ganglia, and peripheral nerves, suggesting that they derive from a multipotent founder cell (see Figs. 7.4 and 7.5). Moreover, certain clones were shown to colonize exclusively the DRG or the sympathetic ganglia and contained either a mixture of neurons and glia-like cells, or sensory neurons of the dorsomedial and ventrolateral type; others consisted of apparently homogeneous types of cells (either neurons or glia), suggesting that they derive from monopotent committed precursors (Bronner-Fraser and Fraser, 1988, 1989; Frank and Sanes, 1991; Fraser and Bronner-Fraser, 1991). These results are therefore consistent with the notion that some progenitors become precociously segregated into sensory as compared to sympathetic lineages, a process that is likely to take place during the migration of neural crest cells.

Nevertheless, they indicate that a majority of neurons in sensory and sympathetic ganglia arise from multipotent precursors.

*In vitro* cloning experiments by Le Douarin and colleagues (Baroffio *et al.*, 1988) and Sieber-Blum (1989a,b) revealed sensory neurons that develop in mixed clones with other cell types. However, this method failed to demonstrate in a systematic manner the existence of monopotent sensory progenitors in clonal conditions, although clones containing few neurofilament-positive cells were detected as well (Baroffio *et al.*, 1991). Thus, the subset of neural crest cells that is committed to develop along the sensory lineage may be more sensitive to growth in clonal conditions when compared to mitotically active progenitors. It is therefore not possible to evaluate from *in vitro* cloning what is the relative proportion of sensory neurons that derives from either type of progenitor.

### 7.4.3   Existence of autonomic progenitors in all peripheral ganglia

The observation that back-transplanted DRG gave rise to adrenergic neurons in autonomic ganglia implied that progenitors with the ability to differentiate into autonomic neurons are present among the non-neuronal population of sensory ganglia. Consistent with these findings, transient catecholaminergic cells were observed in sensory ganglia as well as in the enteric nervous system of rodent embryos. Whereas such cells could not be demonstrated in avian embryos *in vivo*, evidence for their presence in a latent form was provided by culturing DRG and gut tissue, a procedure that allowed their expansion and phenotypic differentiation (for further discussion, see Section 7.6 and Chapter 5).

### 7.4.4   Differential environmental requirements of sensory and autonomic neurons

The dual segregation model proposed by Le Douarin (1986) (see Fig. 7.7B) inferred the existence of short-range signals emanating from the early CNS. These signals would affect the development of sensory neurons, in contrast to that of sympathetic neurons which leave the CNS soon after the onset of neural crest migration and are, therefore, not exposed to a CNS input. Microsurgical manipulations, which consisted in depriving the early forming DRG of the neural tube, interfered with DRG development in a selective manner. Subsequently, BDNF was shown to mediate this effect in a specific way, while not affecting development of sympathetic progenitors at any stage (see Chapter 4 for a detailed discussion). Ongoing research on the molecular requirements of peripheral ganglion cells led to the further identification of additional neurogenic factors whose activities are specific either to sensory or to autonomic neuroblasts, as well as some factors acting on both cell types (see sections on early development of sensory and sympathetic neurons in Chapters 4 and 5).

Several transcription factors were found to act as distinctive signals for sensory and autonomic lineages. The achaete–scute complex of *Drosophila* contains four genes whose function is required for neuroblast generation (Ghysen and Dambly-Chaudière, 1988). These genes encode nuclear regulatory proteins of the basic helix–loop–helix (bHLH) class (Villares and Cabrera, 1987; Alonso and Cabrera, 1988). Johnson *et al.* (1990) isolated a mammalian *achaete–scute* homologous gene (*MASH-1*), which was shown to be expressed in autonomic but not sensory ganglia of the mammalian PNS (Johnson *et al.*, 1990; Lo *et al.*, 1991b; Guillemot and Joyner, 1993). Analysis of homozygous embryos with a deletion in this gene and complementary cell culture experiments revealed that the MASH-1 product is involved in promoting the differentiation of a committed sympathetic neuronal progenitor, but it does not affect sensory neurogenesis (Guillemot *et al.*, 1993; Sommer *et al.*, 1995). This was followed by the isolation and characterization of NeuroD and Neurogenin, two transcription factors expressed in sensory, but not autonomic, ganglia. While overexpression of both factors in *Xenopus* embryos can induce ectopic neurogenesis, the timing of expression of *NeuroD in vivo* suggests that it is more likely to play a role in neuronal differentiation. In contrast, *neurogenin* is expressed very early and also regulates *NeuroD* expression, raising the possibility that *neurogenin* is a neuronal determination gene (Lee *et al.*, 1995; Ma *et al.*, 1996b). Analysis of the appropriate knockout embryos will help clarify its function in mammalian development.

Specific members of another family of transcription factors, the class IV POU domain factor *Brn-3.0*, and the highly related factors *Brn-3.1*, and *Brn-3.2*, were recently shown to be differentially expressed in populations of neural crest cells, and later to become restricted to sensory neurons within the PNS. Analysis of null mutations of *Brn-3.0* revealed that it is required for the migration and survival of subsets of sensory neurons expressing the neurotrophin receptors TrkA, TrkB, TrkC and p75 at different times of ontogeny, and that deletion of the gene affects the expression of specific neurotrophin genes such as BDNF (McEvilly *et al.*, 1996; Xiang *et al.*, 1996). These results suggest that Brn-3.0 is indeed necessary for modulating aspects of neurotrophin signaling in sensory ganglia. The restricted biological effects and localization patterns of several such transcription factors, suggests that these molecules may confer an additional level of spatial and temporal specificity to that conveyed by neurotrophic factors.

## 7.5 Glial lineages in neural crest ontogeny

### 7.5.1 Origin and diversity of glial cells

The neural crest-derived peripheral glia is composed of three main categories of cells: the satellite cells of the sensory and autonomic ganglia, the Schwann cells lining the peripheral nerves, and the glia of the enteric nervous system (reviewed by Le Douarin *et al.,* 1991). There is now general agreement that all the above types of peripheral glia originate from the neural crest (Le Douarin and Teillet, 1973; Ayer-Le Lièvre and Le Douarin, 1982; D'Amico-Martel and Noden, 1983; Teillet *et al.,* 1987; Carpenter and Hollyday, 1992). Yet, it has been claimed that the ventral neural tube itself contributes some Schwann cells that localize in the proximal part of the peripheral nerve *in vivo* (Rickmann *et al.,* 1985; Lunn *et al.,* 1987; but not confirmed in quail–chick chimeras by Teillet, unpublished results), or migrate away from ventral neural tube explants *in vitro* (Loring and Erickson, 1987).

The axially restricted origin and distribution of neuronal derivatives is also followed by glial progenitors. For instance, enteric glia, like their neuronal counterparts, originate from the vagal and sacral levels of the neural crest. In addition, Schwann cell precursors that arise from a length equivalent to one somite populate the dorsal and ventral roots and the peripheral nerves of two consecutive segments, much like neural crest cells that colonize the DRG (Teillet *et al.,* 1987; Carpenter and Hollyday, 1992).

### 7.5.2 *Early markers of glial development*

Understanding the cellular and molecular events involved in the generation of glial sublineages from the neural crest depends upon the availability of specific markers with an early expression pattern. The best characterized subtype of glial cells belongs to the Schwann cell sublineage. Their development can be considered in two parts: first, the generation of differentiated Schwann cells from the neural crest; and second, their maturation into myelin- and non-myelin-forming cells that ensheath the peripheral nerves.

A set of antigens has been identified that characterizes the second phase of Schwann cell development. S100 and myelin-associated glycoprotein become expressed, for instance, by the time of myelinization (Owens and Bunge, 1989; Jessen *et al.,* 1994). Additional markers of Schwann cell maturation have also been described (Mirsky *et al.,* 1980; Trapp *et al.,* 1984; Jessen *et al.,* 1985, 1990; Martini and Schachner, 1986; Eccleston *et al.,* 1987; Curtis *et al.,* 1992; Rudel and Rohrer, 1992; Snipes *et al.,* 1992). Few markers are available, however, that stain early neural crest-derived Schwann cells. Although the HNK-1 marker stains young glial cells and their progenitors in avian and rat embryos, it is not specific to this lineage. So is the low-affinity NGF receptor p75 which is expressed on the surface of many neural crest cells in the mouse, including glial precursors.

Attempts to identify novel markers specific to early glial cells, led to the characterization of a Mab, termed Schwann cell myelin protein (SMP), that recognizes a surface glycoprotein of the immunoglobulin superfamily that carries the HNK-1 epitope and is present in quail myelinating and non-myelinating Schwann cells from E6, but in embryo sections it is never encountered on enteric glia or satellite cells of the peripheral ganglia (Dulac *et al.,* 1988, 1992).

This localized expression led the authors to propose that SMP is a specific marker for the Schwann cell sublineage *in vivo*. Subsequent experiments revealed that dissociation of sensory or enteric ganglia and growth in culture upregulate the expression of SMP in the satellite cells, implying that, in the embryo, the expression of the SMP antigen is normally inhibited in the enteric plexuses and in all types of peripheral ganglia (Dulac and Le Douarin, 1991; Cameron-Curry *et al.*, 1993). Thus, it appears that SMP is a constitutive protein of all types of peripheral glia. Consistent with this view is the appearance of SMP in glial cells developing in clonal cultures of neural crest cells in the absence of neurons (Dupin *et al.*, 1990). Yet, the expression of the SMP phenotype in glial cells of the DRG can be regulated by ganglionic neurons, and by neuron-derived factors (Pruginin-Bluger *et al.*, 1997; see below). Thus, the SMP gene is constitutively linked to the development of the glial lineage and its expression is modulated by specific environmental signals as part of a multi-step glial differentiation program.

To study the earliest events in glial segregation, however, an earlier marker is required. SMP is not present in neural crest subsets *in vivo*, and it develops in crest cultures about a week after initial seeding. In search for such molecules, a Mab, 1E8, was generated that recognizes myelinating and non-myelinating Schwann cells in mature chickens. During embryogenesis, it recognizes a subpopulation of neural crest cells following initial emigration from the neural tube; subsequently, it stains a subset of cells that migrates in the ventral pathway between neural tube and somites and in the sclerotome. These cells were shown to represent a subpopulation of HNK-1-positive cells, an observation that confirmed their neural crest identity. Upon homing, 1E8-positive cells were found to be present in the dorsal and ventral roots and along the growing peripheral nerves. Biochemical characterization of the 1E8 determinant revealed that it is a sugar carried by proteins comigrating with P0, the major glycoprotein of myelin. Thus, it appears that, in avians, P0 (or a closely related protein) is an early marker for the Schwann cell lineage (Bhattacharyya *et al.*, 1991). Depletion experiments or immunopositive selection of 1E8-positive cells should critically test whether the cells expressing this marker throughout crest ontogeny are indeed lineally related.

### 7.5.3 The lineage relations of glial cells

*7.5.3.1 Clones containing only glia or combinations of neurons and glia.* With the discovery of the SMP marker for glial cells, it became possible to study the lineage relationships between the SMP-expressing derivatives and other cell types. Dupin *et al.* (1990) provided evidence for the coexistence of a pluripotent precursor for glia, neurons, and melanocytes, as well as of a more restricted common precursor for neurons and glia, an even more limited bipotential precursor cell for SMP-positive and SMP-negative satellite cells, and, finally, a committed progenitor which develops exclusively into SMP-positive progeny. These results show that glial cells expressing SMP can derive from progenitors with different degrees of specification that coexist within the neural crest at migratory stages (see Figs 7.2 and 7.3). In view of the finding discussed in the previous section that the SMP marker recognizes in culture all types of satellite cells, the question remains open as to the exact identity of the SMP-immuno-reactive derivatives that arise from committed compared to multipotent crest progenitors.

A similar picture was revealed when the fate of single neural crest cells was followed in avian embryos with a recombinant retrovirus (Frank and Sanes, 1991). Whereas Schwann cells were easily identified by their position along nerves and roots, the discrimination of satellite cells was more difficult and was done essentially by morphological features (irregular shape of the cyto-plasm, smaller nucleus than that of neurons), and also by their positions rela-tive to ganglionic neurons (Fig. 7.6). In 45% of the clones, a mixture of glia and neurons was obtained, 43% of the clones contained only neurons, and 12% only glia in different combinations, suggesting that the heterogeneity of neural crest potentials is also reflected in their actual fates.

*7.5.3.2  The bipotent glia–melanocyte progenitor.* Several years ago, the obser-vation was made that peripheral nerves and DRG from young quail embryos (E4–6) can give rise to melanocytes under permissive *in vitro* conditions while similar explants from older embryos are unable to do so (Nichols and Weston, 1977). This finding raised the possibility that Schwann cell precursors, common to these two structures, have the transient capacity to differentiate into mel-anocytes, suggesting the existence of a restricted intermediate precursor in the neural crest with dual Schwann cell–melanocyte potentials. Subsequent experi-ments have shown that development of pigment cells from older nerves was also possible provided the cells were treated with the tumor-promoting phorbol ester drug 12-*O*-tetradecanoylphorbol 13-acetate (TPA) (Ciment *et al.*, 1986; Kanno *et al.*, 1987). In addition, the activity of TPA could be specifically mimicked by FGF2 (Stocker *et al.*, 1991), which was later shown to act via an autocrine pathway (Sherman *et al.*, 1993). As a whole, these results were interpreted to reflect a reversal of a developmental restriction to melanogenesis in avian Schwann cells.

Full confirmation for the existence of such a dual progenitor cell required a common ancestor for the two phenotypes be directly demonstrated. *In vitro* cloning of mesencephalic-level neural crest cells revealed that, among several combinations of cell types obtained in the analyzed clones, about 2% of the clones contained melanocytes together with non-neuronal cells (which were either positive or negative for HNK-1 and SMP) (Dupin *et al.*, 1990). Similar results were obtained *in vivo*, where some individual neural crest cells were found to give rise both to descendants localized in the ventral roots or the spinal nerves, and to presumptive melanocytes under the ectoderm (Bronner-Fraser and Fraser, 1988).

## 7.6 Factors affecting development of neuronal and glial lineages from the neural crest

### 7.6.1 Determination versus differentiation

The results of clonal analysis of neural crest cells have shown that many of these progenitors are pluripotent during migration, while others have already undergone restrictions in their developmental potentials. These findings imply that distinct neural crest cells reach their sites of differentiation with hetero-geneous states of commitment. As a consequence, the role of the microenviron-ment at the homing sites is to induce a specific fate in multipotent progenitors, as well as to stimulate the expression of a restricted developmental repertoire (differentiative effect) in fully specified progenitors. These options raise funda-mental questions as to the mechanisms of phenotypic segregation and final differentiation. The first is whether fate specification of a multipotent progeni-tor is a cell autonomous process, or is it modulated by environmental cues. The literature dealing with development of invertebrate embryos suggests that some lineages are determined by cell-autonomous mechanisms while others appear to be specified by cell–cell interactions because their development is related to the relative position adopted by the cells in the embryo (reviewed for *Caenorhabditis elegans* in Hill and Sternberg, 1993). Both mechanisms might also apply for the neural crest. For instance, the specification of the earliest cells that is likely to take place at premigratory stages and during early migration might be governed by cell-autonomous mechanisms (stochastic or deterministic). These mechanisms generate developmentally distinct sub-populations that respond differentially to environmental cues along the migratory paths and at the homing sites (see Chapter 2 for detailed discussion). Furthermore, those progenitors that arrive in a multipotent state to the various homing sites might become specified by local interactions. Consistent with this notion, the population studies previously described have stressed that intercel-lular interactions determined by position are of paramount importance in neural crest development.

The best approximation, however, to possibly discriminate between induc-tive and selective mechanisms of cell specification is to study the effects of environmental factors on homogeneous (or at least initially homogeneous) populations. The clonal progeny of single primary or immortalized cells was chosen for such studies (Lo *et al.*, 1991a; Sieber-Blum, 1991; Stemple and Anderson, 1992). One approach consisted of generating a rat clonal cell line upon immortalization of neural crest cells with a v-*myc*-containing retrovirus (Lo *et al.*, 1991a). Initially, these multipotent cells express the low-affinity NGF receptor (NGFR) and give rise, at high frequency, to elongated glial-like cells, a process which is accelerated by serum withdrawal and prevented by TGF$\beta$. At low frequency, these cells irreversibly generate progeny lacking the NGFR, NGFR(–). After exposure to FGF and dexamethasone some cells will express tyrosine hydroxylase and neurofilament proteins. Has FGF an instructive effect on these cells? The data suggest that the population that lacks the

NGFR is more restricted in its developmental potential when compared to the NGFR(+) founder cell. However, since the loss of the low-affinity NGF receptor occurs in the absence of added FGF, one must assume that the conversion of an NGF(+) to an NGF(−) cell is factor-independent. Thus, most likely, the presence of FGF and dexamethasone stimulates the development of adrenergic traits (expansion and/or differentiation) in a subset of progenitor cells that underwent an FGF-independent restriction process.

Along the same line, Sieber-Blum (1991) has undertaken a study to test the possible role of BDNF on commitment of pluripotent neural crest cells in clonal cultures. This work was inspired by previous results showing that BDNF is required both in the embryo and in culture for the survival and/or differentiation of neural crest cells that give rise to sensory neurons (Kalcheim et al., 1987; Kalcheim and Gendreau, 1988). Sieber-Blum has shown that treatment of cloned neural crest cells with this neurotrophin caused an increase of up to 21-fold in the number of sensory neuronal precursors per colony, and an equivalent decrease in the number of undifferentiated precursors. Moreover, these changes were not accompanied by a corresponding increase in total cell number, suggesting that, under these conditions, BDNF is not a survival factor for the neural crest precursors. Notably, analysis of resultant phenotypes was performed between 1 and 3 weeks after initial clonal seeding. Consequently, the possibility cannot be ruled out that, instead of instructing a pluripotent precursor to become a specified sensory neuroblast, BDNF, albeit present in the cultures from the beginning, starts acting at a later stage on daughter cells that had in fact become restricted in a cell autonomous fashion. In such a case, BDNF would presumably stimulate their differentiation into sensory neurons (expression of the SSEA-1 antigen, characteristic neuronal morphology, etc.), as previously proposed.

The above studies reveal that the question of the influence of a growth factor at the level of founder multipotent crest progenitors remains an open issue. This is because, even if a given factor is present in the cultures from the single cell stage, its actual influence on the development of certain phenotypes is analyzed following several rounds of cell division. Under these circumstances, it is not possible to discriminate whether the growth factor has acted directly on the multipotent cell, on its identical progeny (instructive mechanism), or, alternatively, on one of its more restricted descendents (possible trophic mechanism). The ability to fit a model of environmental determination of cell commitment to experimental data depends upon the ability to demonstrate that the multipotent cell is the actual cell in the lineage that responds to the extrinsic signal.

A step forward in this direction was made in the experiments of Shah et al. (1994), in which it was shown that primary rat neural crest cells grown in clonal density are able to differentiate into glia and neurons expressing MASH-1. In contrast, in the presence of glial growth factor-2 (GGF2), a Schwann cell mitogen of neuronal origin (Lemke and Brockes, 1984), no neurons developed. Sequential analysis of the development of such clones revealed that neurons failed to develop at any time tested; consequently, no expression of the neuronal marker MASH-1 was detected. These results were interpreted to mean that GGF2 acts directly on the multipotent progenitors to supress the neuronal fate. While this conclusion derives from the data, the lack of quantification of the relative proportions of glial cells precludes establishing whether GGF2 has, in addition, an instructive effect on the commitment of multipotent crest cells to become glia, or, alternatively, whether the glial fate is a default pathway under the conditions employed. Further work is also necessary to assess whether GGF2 is the endogenous factor responsible for the reported effects.

Using the same system, Shah *et al.* (1996) have shown that BMP2 has an opposite effect to that of GGF2. In the presence of BMP2, glial cells do not develop. In contrast, neurons are apparent earlier than in controls, and MASH-1 is upregulated. The early timing of this effect is highly suggestive of an activity of BMP2 on the initial neural crest cells or their immediate progeny. Curiously, whereas two-thirds of the BMP2-treated clones contained only neurons, about one-third also contained cells that stained for markers of smooth muscle phenotype, a cell type induced to develop in this type of culture by TGFβ1. Again these results leave open the possibility that the factors assayed affect partially restricted crest cells rather than stem cells with equal developmental potentials.

### 7.6.2  *Phenotypic diversification in a homogeneous environment*

A second question related to phenotypic diversification concerns the mechanisms whereby multipotent progenitors segregate into distinct phenotypes within the apparently homogeneous milieu of a nascent ganglion. Clonal analysis *in vitro* has clearly shown that Schwann cells derive from multipotent as well as from restricted progenitors that coexist in the mesencephalic neural crest at the time of migration (Dupin *et al.*, 1990). Likewise, sensory neurons can differentiate from multipotent precursors (Baroffio *et al.*, 1988; Bronner-Fraser and Fraser, 1988; Sieber-Blum, 1989b), and also from specified progenitor cells (Ziller *et al.*, 1987; Baroffio *et al.*, 1988; Bronner-Fraser and Fraser, 1988; Sieber-Blum, 1989b). Whether all sublineages of sensory neurons can differentiate from either precursor type remains to be established. Thus, the local environment of the organizing ganglia, or the proximity to the CNS, or to a peripheral nerve may readily stimulate differentiation of the arriving committed precursors. These same milieux could also exert a negative selection (maintenance in a latent mode, or death) on cells whose fate is inappropriate to their homing sites (see Le Douarin, 1986). On the other hand, the fate of multipotent cells that colonize a given embryonic anlage may be determined by contact-mediated signalling with already differentiated cells, and, also in this case, the cues might be either of a stimulatory or inhibitory nature.

An example of positive cell–cell interactions is the stimulation of glial cell development by ganglionic neurons. The differentiation of neurons in the ganglia precedes that of satellite cells (McMillan-Carr and Simpson, 1978), suggesting a sequential order by which glial differentiation would depend upon signals from the recently formed neurons. Holton and Weston (1982a,b) have shown that, in sensory ganglia, specific contact with neurons is required by developing glial cells for the synthesis and accumulation of a glial marker, the S100 protein. Recently, Pruginin-Bluger *et al.* (1997) have shown that coculturing neurons with developing glial cells of E7–8 quail DRG stimulates within a day a significant increase in the proportion of non-neuronal cells that express the glial marker SMP. Moreover, this paracrine effect was found to be mediated by BDNF, as treatment of the cocultures with a fusion protein that contains the extracellular domain of TrkB and thus sequesters BDNF in a specific way completely abolished the stimulation. These results suggested that certain interactions between neurons and developing glia in DRG are mediated by BDNF, a neurotrophin synthesized by the developing sensory neurons in the avian ganglia, but not by the satellite cells. Therefore, one mechanism that accounts for the differentiation of distinct cell types within the ganglionic milieu is the occurrence of sequential paracrine interactions between already differentiated cells and responsive progenitors. It remains to be tested whether, in addition to stimulating glial differentiation, neuronal cells also exert a negative effect on glial progenitors to prevent them from becoming neurons.

Examples of local inhibition within the ganglionic environment are provided, first, by the previously discussed inhibition of SMP expression, and second, by the dormant autonomic progenitors that populate the avian DRG but never differentiate along the adrenergic sublineage within the ganglia (see also Section 7.4.1). These latent cells are, however, able to express catecholaminergic traits in mass cultures of dissociated sensory ganglia (Xue *et al.*, 1985, 1987; see also Sextier-Sainte-Claire Deville *et al.*, 1992). The inhibition of expression of the autonomic phenotype in the DRG milieu may reflect the absence of appropriate signals that are required for autonomic development. Alternatively, it may be explained by lateral inhibition exerted by one cell type on its adjacent neighbors, in analogy to the well-demonstrated lateral-inhibition process discovered in *Drosophila* neurogenesis, which is mediated by the *Delta* and *Notch* genes (Artavanis-Tsakonas, 1988; Greenspan, 1990).

The existence of such a mechanism in vertebrate embryos was also demonstrated. Chitnis *et al.* (1995) and Henrique *et al.* (1995) have shown that the integral membrane protein Delta is expressed by postmitotic cells with neurogenic potential in the neural tube of both avian and *Xenopus* embryos. These cells signal via Delta to adjacent cells expressing the Notch receptor that are located in the luminal portion of the neuroepithelium and prevent them from becoming neurons prematurely. Consequently, absence of Delta results in the formation of an extra set of neurons that does not exist under normal conditions. Delta and Notch might also be involved in lateral inhibition in neural crest-derived cells, based on the expression of Notch transcripts in early sensory ganglia of mouse embryos (Weinmaster *et al.*, 1991).

# Concluding Remarks and
# Perspectives

In the 15-year interval separating the first edition of *The Neural Crest* from the present book, our understanding of the ontogeny of this structure has progressed considerably. New derivatives of the neural crest have not been discovered but truly novel perspectives have illuminated the field. Considering the list of neural crest derivatives summarized in Table 8.1, one can only be overwhelmed by the impressive diversification of phenotypes arising from this discrete and transient embryonic structure and by its paramount participation to the vertebrate organism. By means of the PNS, the neural crest provides the body with an efficient communication system with the outside world, and within the organism itself. By means of pigment cells, the neural crest provides protection from UV light and a means of undergoing adaptive color changes. Furthermore, the neural crest has accommodated the large increase in brain size which took place during vertebrate evolution by providing it with the skull. Thus, it appears to be a highly adaptive structure and, as it is absent from their chordate ancestors, the neural crest can be considered as a "spectacular invention" of vertebrates as pointed out by Gans in 1987.

One is also incited to seek for common denominators between the highly diversified neural crest derivatives. Such a line of thinking is stimulated by the existence of congenital anomalies in which multisystem neural crest defects (neuronal and pigmentary, neuronal and craniofacial, neuronal and endocrine, etc.) are found to be associated. Molecules mutated in specific neurocristopathies have begun to be elucidated, and provide precious tools to tackle these links. Yet, the mode of action of these gene products either on common intermediate progenitors or on already diversified cell types, and their relationships with other active factors, are only examples of open questions that await to be solved. One of the most striking developmental features of the neural

Table 8.1. *The derivatives of the neural crest*

## Peripheral nervous system

### I. SENSORY GANGLIA

*Sensory neurons*

*Spinal ganglia*
Trigeminal nerve (V) root ganglion: *trigeminal ganglion* (partly)
*Facial nerve* (VII) *root ganglion*
Glossopharyngeal nerve (IX) root ganglion: *Superior ganglion*
Vagal nerve (X) root ganglion: *jugular ganglion*
Cells of Rohon-Béard (amphibians)

*Satellite cells*
In all sensory ganglia including geniculate (nerve VII); otic (nerve VIII); petrosum (nerve IX); nodose (nerve X)

### II. AUTONOMIC NERVOUS SYSTEM

Sympathetic ganglia and plexuses: neurons and satellite cells
Parasympathetic ganglia and plexuses: neurons and satellite cells

### III. SCHWANN CELLS OF THE PERIPHERAL NERVES

Endocrine and paraendocrine cells
Carotid body type I cells (and type II, satellite-type cells)
Calcitonin-producing (C cells)
Adrenal medulla

## Pigment cells

### *Mesectodermal derivatives*

*Cephalic crest*
SKELETON
Nasal and orbitary
Palate and maxillary
Trabeculae (partly)
Sphenoid complex (small contribution)
Cranial vault (squamosal and small contribution to frontal)
Otic capsule (partly)
Visceral skeleton

CONNECTIVE TISSUE
*Dermis, smooth muscles* and *adipose tissue* of the *skin* in the face and ventral part of the neck
Ciliary muscles
Small contribution of striated muscle cells of the face and neck
*Connective* and *muscular* tissues of the *wall* of the *large arteries* derived from the *aortic arches* (except endothelial cells)
*Tooth papillae* (except endothelium of blood vessels)
*Cornea*: corneal "endothelium" and stromal fibroblasts
*Meninges*: in prosencephalon and partly in mesencephalon
*Connective component* of the pituitary, lacrymal, salivary, thyroid, parathyroid glands, and thymus

### *Trunk crest*
Dorsal fin mesenchyme (lower vertebrates)

crest is the migratory behavior of its component cells. The discovery that neural crest cell guidance involves negative as well as positive cues has opened new avenues in the search for mechanisms that provide the spatial patterning of crest derivatives. Direct evidence for active chemoattraction is still missing, and much remains to be done to understand how so many molecules act in a coordinated fashion to pattern the migration of the neural crest.

The crucial question of whether crest cell migration is conditioned by a precommitment toward a given fate is still open. Cloning experiments have outlined the largely multipotent status of neural crest cells, while also stressing the existence of early-emerging developmental restrictions. Thus, *a priori,* no unique answer is likely to explain the behavior of all migrating cells. In the future, researchers will be faced with the uneasy task of relating the state of specification of individual progenitors with their ability to respond differentially to regional cues.

In the near future, molecular genetics is likely to provide invaluable information. In particular, the production of conditional mutants will allow the assessment of gene function to be tested at restricted times and embryonic sites. In avian embryos the lack of genetic techniques is gradually being overcome by virus-mediated gene misexpression. It is expected that the increasing power of this technique, combined with quail–chick chimerism to restrict the spreading of certain viral subtypes, will become a most useful tool in future research.

In the first edition of this book, the chapter devoted to the mesenchymal derivatives of the neural crest reviewed experimental data essentially gathered in the avian embryo during the 1970s. These data showed, not only that the neural crest was at the origin of the facial skeleton in higher vertebrates, as already reported for amphibians in Horstadius's monograph, but they also showed that the contribution of the neural crest was even more diversified than previously expected. The availability of a stable marker established without ambiguity that all the connective tissue of the head associated with mesodermally derived blood vessels, with myocytes as well as dermis, were also of neural crest origin. The paramount quantitative and qualitative contribution of ectodermal derivatives to the vertebrate head was thus recognized.

Another major event in the field was the recognition of the segmental structure of the brain primordium, which has implications in the development not only of the CNS but also of its neural crest derivatives. The advent of molecular genetics and the unveiling of multiple regulatory pathways are now providing new tools to unravel the developmental mechanisms involved in head and facial morphogenesis, as well as in patterning the highly diversified derivatives of the neural crest.

# References

Abo, T. and Balch, C.M. (1981). A differentiation antigen of human NK and K cells identified by a monoclonal antibody (HNK-1). *J. Immunol.* **127**, 1024–1029.

Acampora, D., Mazan, S., Lallemand, Y., Avantaggiato, V., Maury, M., Simeone, A. and Brulet, P. (1995). Forebrain and midbrain regions are deleted in Otx2(–/–) mutants due to a defective anterior neuroectoderm specification during gastrulation. *Development* **121**, 3279–3290.

Acheson, A., Conover, J.C., Fandl, J.P., Dechiara, T.M., Russell, M., Thadani, A., Squinto, S.P., Yancopoulos, G.D. and Lindsay, R.M. (1995). A BDNF autocrine loop in adult sensory neurons prevents cell death. *Nature* **374**, 450–453.

Adalsteinsson, S., Hersteinsson, P. and Gunnarsson, E. (1987). Fox colors in relation to colors in mice and sheep. *J. Hered.* **78**, 235–237.

Adams, A.E. (1924). An experimental study of the development of the mouth in the amphibian embryo. *J. Exp. Zool.* **40**, 311–379.

Adema, G.J., de Boer, A.J., van't Hullenaar, R., Denijn, M., Ruiter, D.J., Vogel, A.M. and Figdor, C.G. (1993). Melanocyte lineage-specific antigens recognized by monoclonal antibodies NKI-beteb, HMB-50, and HMB-45 are encoded by a single cDNA. *Am. J. Pathol.* **143**, 1579–1585.

Aebischer, P., Schluep, M., Deglon, N., Joseph, J.M., Hirt, L., Heyd, B., Goddard, M., Hammang, J.P., Zurn, A.D., Kato, A.C., Regli, F. and Baetge, E.E. (1996). Intrathecal delivery of CNTF using encapsulated genetically modified xenogeneic cells in amyotrophic lateral sclerosis patients. *Nat. Med.* **2**, 696–699.

Affolter, M., Schier, A. and Gehring, W.J. (1990). Homeodomain proteins and the regulation of gene expression. *Curr. Opin. Cell Biol.* **2**, 485–495.

Airaksinen, M.S., Koltzenburg, M., Lewin, G.R., Masu, Y., Helbig, C., Wolf, E., Brem, G., Toyka, K.V., Thoenen, H. and Meyer, M. (1996). Specific subtypes of cutaneous mechanoreceptors require neurotrophin-3 following peripheral target innervation. *Neuron* **16**, 287–295.

Akam, M. (1987). The molecular basis for metameric pattern in the *Drosophila* embryo. *Development* **101**, 1–22.

Akam, M. (1989). *Hox* and *HOM*: homologous gene clusters in insects and vertebrates. *Cell* **57**, 347–349.

Akira, E. and Ide, H. (1987). Differentiation of neural crest cells of *Xenopus laevis* in clonal culture. *Pigment Cell Res.* **1**, 28–36.

Akitaya, T. and Bronner-Fraser, M. (1992). Expression of cell adhesion molecules during initiation and cessation of neural crest cell migration. *Dev. Dyn.* **194**, 12–20.

Albano, R.M., Arkell, R., Beddington, R.S. and Smith, J.C. (1994). Expression of inhibin subunits and follistatin during postimplantation mouse development: decidual expression of activin and expression of follistatin in primitive streak, somites and hindbrain. *Development* **120**, 803–813.

Albers, B. (1987). Competence as the main factor determining the size of the neural plate. *Dev. Growth Differ.* **29**, 535–545.

Alexandre, D., Clarke, J.D.W., Oxtoby, E., Yan, Y.L., Jowett, T. and Holder, N. (1996). Ectopic expression of *Hoxa-1* in the zebrafish alters the fate of the mandibular arch neural crest and phenocopies a retinoic acid-induced phenotype. *Development* **122**, 735–746.

Ali-Rachedi, A., Varndell, I.M., Adrian, T.E., Gapp, D.A., Van Noorden, S., Bloom, S.R. and Polak, J.M. (1984). Peptide YY (PYY) immunoreactivity is co-stored with glucagon-related immunoreactants in endocrine cells of the gut and pancreas. *Histochemistry* **80**, 487–491.

Allan, I.J. and Newgreen, D.F. (1977). Catecholamine accumulation in neural crest cells and the primary sympathetic chain. *Am. J. Anat.* **149**, 413–421.

Allan, I.J. and Newgreen, D.F. (1980). The origin and differentiation of enteric neurons of the intestine of the fowl embryo. *Am. J. Anat.* **157**, 137–154.

Allin, E.F. (1975). Evolution of the mammalian middle ear. *J. Morphol.* **47**, 403–437.

Allsopp, T.E., Wyatt, S., Paterson, H.F. and Davies, A.M. (1993). The proto-oncogene *bcl-*2 can selectively rescue neurotrophic factor-dependent neurons from apoptosis. *Cell* **73**, 295–307.

Aloe, L. and Levi-Montalcini, R. (1979). Nerve growth factor-induced transformation of immature chromaffin cells *in vivo* into sympathetic neurons: effect of antiserum to nerve growth factor. *Proc. Natl. Acad. Sci. USA* **76**, 1246–1250.

Alonso, M.C. and Cabrera, C.V. (1988). The *achaete-scute* gene complex of *Drosophila melanogaster* comprises four homologous genes. *EMBO J.* **7**, 2585–2591.

Alpert, S., Hanahan, D. and Teitelman, G. (1988). Hybrid insulin genes reveal a developmental lineage for pancreatic endocrine cells and imply a relationship with neurons. *Cell* **53**, 295–308.

Andermarcher, E., Surani, M.A. and Gherardi, E. (1996). Co-expression of the *HGF/SF* and *c-met* genes during early mouse embryogenesis precedes reciprocal expression in adjacent tissues during organogenesis. *Dev. Genet.* **18**, 254–266.

Anderson, C.B. and Meier, S. (1981). The influence of the metameric pattern in the mesoderm on migration of cranial neural crest cells in the chick embryo. *Dev. Biol.* **85**, 385–402.

Anderson, D.J. (1989). The neural crest cell lineage problem: neuropoiesis? *Neuron* **3**, 1–12.

Anderson, D.J. (1993a). Cell fate determination in the peripheral nervous system: the sympathoadrenal progenitor. *J. Neurobiol.* **24**, 185–198.

Anderson, D.J. (1993b). Molecular control of cell fate in the neural crest: the sympathoadrenal lineage. *Annu. Rev. Neurosci.* **16**, 129–158.

Anderson, D.J. and Axel, R. (1986). A bipotential neuroendocrine precursor whose choice of cell fate is determined by NGF and glucocorticoids. *Cell* **47**, 1079–1090.

Anderson, D.J. and Michelson, A. (1989). Role of glucocorticoids in the chromaffin–neuron developmental decision. *Int. J. Dev. Neurosci.* **12**, 83–94.

Anderson, D.J., Carnahan, J.F., Michelson, A. and Patterson, P.H. (1991). Antibody markers identify a common progenitor to sympathetic neurons and chromaffin cells *in vivo* and reveal the timing of commitment to neuronal differentiation in the sympathoadrenal lineage. *J. Neurosci.* **11**, 3507–3519.

Anderson, D.M., Lyman, S.D., Baird, A., Wignall, J.M., Eisenman, J., Rauch, C., March, C.J., Boswell, H.S., Gimpel, S.D., Cosman, D. and Williams, D.E. (1990). Molecular cloning of mast cell growth factor, a hematopoietin that is active in both membrane bound and soluble forms. *Cell* **63**, 235–243.

Andrew, A. (1963). A study of the developmental relationship between enterochromaffin cells and the neural crest. *J. Embryol. Exp. Morphol.* **11**, 307–324.

Andrew, A. (1964). The origin of intramural ganglia. I. The early arrival of precursor cells in the presumptive gut of chick embryos. *J. Anat.* **98**, 421–428.

Andrew, A. (1969). The origin of intramural ganglia. II. The trunk neural crest as a source of enteric ganglion cells. *J. Anat.* **105**, 89–101.

Andrew, A. (1970). The origin of intramural ganglia. III. The "vagal" source of enteric ganglion cells. *J. Anat.* **107**, 327–336.

Andrew, A. (1971). The origin of intramural ganglia. IV. The origin of enteric ganglia: a critical review and discussion of the present state of the problem. *J. Anat.* **108**, 169–184.

Andrew, A. (1976a). An experimental investigation into the possible neural crest origin of pancreatic APUD (islet) cells. *J. Embryol. Exp. Morphol.* **35**, 577–593.

Andrew, A. (1976b). APUD cells, Apudomas and the neural crest: a critical review. *S. Afr. Med. J.* **50**, 890–898.

Andrews, P.W. (1984). Retinoic acid induces neuronal differentiation of a cloned human embryonal carcinoma cell line *in vitro*. *Dev. Biol.* **103**, 285–293.

Andrews, P.W. (1988). Human teratocarcinomas. *Biochim. Biophys. Acta* **948**, 17–36.

Andrews, P.W., Gonczol, E., Plotkin, S.A., Dignazio, M. and Oosterhuis, J.W. (1986). Differentiation of TERA-2 human embryonal carcinoma cells into neurons and HCMV permissive cells. Induction by agents other than retinoic acid. *Differentiation* **31**, 119–126.

Ang, S.L. and Rossant, J. (1994). HNF-3 beta is essential for node and notochord formation in mouse development. *Cell* **78**, 561–574.

Ang, S.L., Conlon, R.A., Jin, O. and Rossant, J. (1994). Positive and negative signals from mesoderm regulate the expression of mouse *Otx2* in ectoderm explants. *Development* **120**, 2979–2989.

Arai, H., Hori, S., Aramori, I., Ohkubo, H. and Nakanishi, S. (1990). Cloning and expression of a cDNA encoding an endothelin receptor. *Nature* **348**, 730–732.

Artavanis-Tsakonas, S. (1988). The molecular biology of the Notch locus and the fine tuning of differentiation in Drosophila. *Trends Genet.* **4**, 95–100.

Artinger, K.B. and Bronner-Fraser, M. (1992). Partial restriction in the developmental potential of late emigrating avian neural crest cells. *Dev. Biol.* **149**, 149–157.

Artinger, K.B., Fraser, S. and Bronner-Fraser, M. (1995). Dorsal and ventral cell types can arise from common neural tube progenitors. *Dev. Biol.* **172**, 591–601.

Auerbach, R. (1960). Morphogenetic interactions in the development of the mouse thymus gland. *Dev. Biol.* **2**, 271–284.

Augustine, K.A., Liu, E.T. and Sadler, T.W. (1995). Antisense inhibition of *engrailed* genes in mouse embryos reveals roles for these genes in craniofacial and neural tube development. *Teratology* **51**, 300–310.

Averof, M. and Akam, M. (1993). *HOM/Hox* genes of *Artemia*: implications for the origin of insect and crustacean body plans. *Curr. Biol.* **3**, 73–78.

Ayer-Le Lièvre, C.S. and Le Douarin, N.M. (1982). The early development of cranial sensory ganglia and the potentialities of their component cells studied in quail–chick chimeras. *Dev. Biol.* **94**, 291–310.

Baetge, G. and Gershon, M.D. (1989). Transient catecholaminergic (TC) cells in the vagus nerves and bowel of fetal mice: relationship to the development of enteric neurons. *Dev. Biol.* **132**, 189–211.

Baetge, G., Pintar, J.E. and Gershon, M.D. (1990). Transiently catecholaminergic (TC) cells in the bowel of the fetal rat: precursors of noncatecholaminergic enteric neurons. *Dev. Biol.* **141**, 353–380.

Bagnara, J.T. (1957). Hypophysectomy and tail darkening reaction in *Xenopus. Proc. Soc. Exp. Biol. Med.* **94**, 573–575.

Bagnara, J.T. (1972). Interrelationships of melanophores, iridophores, and xanthophores. In *Pigmentation: its Genesis and Control*, ed. W. Riley, pp. 171–180. Appleton-Century-Crofts, New York.

Bagnara, J.T. (1982). Development of the spot pattern in the leopard frog. *J. Exp. Zool.* **224**, 283–287.

Bagnara, J.T. (1987). The neural crest as a source of stem cells. In *Developmental and Evolutionary Aspects of the Neural Crest*, ed. P. Maderson, pp. 57–87. Wiley & Sons.

Bagnara, J.T. and Fernandez, P.J. (1993). Hormonal influences on the development of amphibian pigmentation patterns. *Zool. Sci.* **10**, 733–748.

Bagnara, J.T. and Ferris, W.R. (1971). Interrelationship of chromatophores. In *Biology of the Normal and Abnormal Melanocyte*, eds. T. Kawamura, T.B. Fitzpatrick and M. Seiji, pp. 57–76. University of Tokyo Press, Tokyo.

Bagnara, J.T. and Fukuzawa, T. (1990). Stimulation of cultured iridophores by amphibian ventral conditioned medium. *Pigment Cell Res.* **3**, 243–250.

Bagnara, J.T. and Hadley, M.E. (1973). *Chromatophores and Color Change*. Prentice Hall, Englewood Cliffs, NJ.

Bagnara, J.T., Taylor, J.D. and Prota, G. (1973). Color changes, unusual melanosomes, and a new pigment from leaf frogs. *Science* **182**, 1034–1035.

Bagnara, J.T., Ferris, W., Turner W.A., Jr, and Taylor, J.D. (1978). Melanophore differentiation in leaf frogs. *Dev. Biol.* **64**, 149–163.

Bagnara, J.T., Turner, W.A., Rothstein, J., Ferris, W. and Taylor, J.D. (1979a). Chromatophore organellogenesis. In *Pigment Cell*, Vol 4, ed. S. Klaus, pp. 13–27. Karger, Basel.

Bagnara, J.T., Matsumoto, J., Ferris, W., Frost, S.K., Turner Wa, Jr, Tchen, T.T. and Taylor, J.D. (1979b). Common origin of pigment cells. *Science* **203**, 410–415.

Baker, C.V.H., Bronner-Fraser, M., Le Douarin, N.M. and Teillet, M.-A. (1997). Early- and late-migrating cranial neural crest cell populations have equivalent developmental potential *in vivo*. *Development* **124**, 3077–3087.

Balaban, E. (1997). Changes in multiple brain regions underlie species differences in a complex, congenital behavior. *Proc. Natl. Acad. Sci. USA* **94**, 2001–2006.

Balaban, E., Teillet, M.-A. and Le Douarin, N. (1988). Application of the quail–chick chimera system to the study of brain development and behavior. *Science* **241**, 1339–1342.

Balkan, W., Colbert, M., Bock, C. and Linney, E. (1992). Transgenic indicator mice for studying activated retinoic acid receptors during development. *Proc. Natl. Acad. Sci. USA* **89**, 3347–3351.

Ballester, R., Marchuk, D., Boguski, M., Saulino, A., Letcher, R., Wigler, M. and Collins, F. (1990). The NF1 locus encodes a protein functionally related to mammalian GAP and yeast IRA proteins. *Cell* **63**, 851–859.

Baltzer, F. (1941). Untersuchungen an Chimären von Urodelen und Hyla. I. Die Pigmentierung chimärischer Molch und Axolotllarven mit Hyla- (Laubfrosch) Ganglienleiste. *Rev. Suisse Zool.* **48**, 413–482.

Bannerman, P.G. and Pleasure, D. (1993). Protein growth factor requirements of rat neural crest cells. *J. Neurosci. Res.* **36**, 46–57.

Barald, K.F. (1988). Monoclonal antibodies made to chick mesencephalic neural crest cells and to ciliary ganglion neurons identify a common antigen on the neurons and a neural crest subpopulation. *J. Neurosci. Res.* **21**, 107–118.

Barasch, J.M., MacKey, H., Tamir, H., Nunez, E.A. and Gershon, M.D. (1987). Induction of a neural phenotype in a serotonergic endocrine cell derived from the neural crest. *J. Neurosci.* **7**, 2874–2883.

Barbacid, M. (1994). The Trk family of neurotrophin receptors. *J. Neurobiol.* **25**, 1386–1403.

Barbacid, M. (1995). Structural and functional properties of the TRK family of neurotrophin receptors. *Ann. N.Y. Acad. Sci.* **766**, 443–458.

Barbin, G., Manthorpe, M. and Varon, S. (1984). Purification of the chick eye ciliary neuronotrophic factor. *J. Neurochem.* **43**, 1468–1478.

Barbu, M., Ziller, C., Rong, P.M. and Le Douarin, N.M. (1986). Heterogeneity in migrating neural crest cells revealed by a monoclonal antibody. *J. Neurosci.* **6**, 2215–2225.

Barbu, M., Pourquié, O., Vaigot, P., Gateau, G. and Smith, J. (1992). Phenotypic plasticity of avian embryonic sympathetic neurons grown in a chemically defined medium: direct evidence for noradrenergic and cholinergic properties in the same neurons. *J. Neurosci. Res.* **32**, 350–362.

Bard, J.B., Hay, E.D. and Meller, S.M. (1975). Formation of the endothelium of the avian cornea: a study of cell movement *in vivo*. *Dev. Biol.* **42**, 334–361.

Barde, Y.-A. (1989). Trophic factors and neuronal survival. *Neuron* **2**, 1525–1534.

Barde, Y.-A. (1994). Neurotrophic factors: an evolutionary perspective. *J. Neurobiol.* **25**, 1329–1333.

Barde, Y.-A, Lindsay, R.M., Monard, D. and Thoenen, H. (1978). New factor released by cultured glioma cells supporting survival and growth of sensory neurones. *Nature* **274**, 818–818.

Barde, Y.-A., Edgar, D. and Thoenen, H. (1980). Sensory neurons in culture: changing requirements for survival factors during embryonic development. *Proc. Natl. Acad. Sci. USA* **77**, 1199–1203.

Barghusen, H.R. and Hopson, J.A. (1979). The endoskeleton: the comparative anatomy of the skull and the visceral skeleton. In *Hyman's Comparative Vertebrate Anatomy*, ed. M.H. Wake, pp. 265–326. University of Chicago Press, Chicago.

Barker, P.A. and Shooter, E.M. (1994). Disruption of NGF binding to the low affinity neurotrophin receptor p75$^{LNTR}$ reduces NGF binding to TrkA on PC12 cells. *Neuron* **13**, 203–215.

Barlow, A.J. and Francis-West, P.H. (1997). Ectopic application of recombinant BMP-2 and BMP-4 can change patterning of developing chick facial primordia. *Development* **124**, 391–398.

Baroffio, A. and Blot, M. (1992). Statistical evidence for a random commitment of pluripotent cephalic neural crest cells. *J. Cell Sci.* **103**, 581–587.

Baroffio, A., Dupin, E. and Le Douarin, N.M. (1988). Clone-forming ability and differentiation potential of migratory neural crest cells. *Proc. Natl. Acad. Sci. USA* **85**, 5325–5329.

Baroffio, A., Dupin, E. and Le Douarin, N.M. (1991). Common precursors for neural and mesectodermal derivatives in the cephalic neural crest. *Development* **112**, 301–305.

Bashkin, P., Neufeld, G., Gitay-Goren, H. and Vlodavsky, I. (1992). Release of cell surface-associated basic fibroblast growth factor by glycosyl-phosphatidylinositol-specific phospholipase C. *J. Cell. Physiol.* **151**, 126–137.

Basler, K., Edlund, T., Jessell, T.M. and Yamada, T. (1993). Control of cell pattern in the neural tube: regulation of cell differentiation by *dorsalin-1*, a novel TGFbeta family member. *Cell* **73**, 687–702.

Bateson, W. (1894). *Materials for the Study of Variation*. Macmillan, London.

Batini, C., Teillet, M.-A., Naquet, R. and Le Douarin, N.M. (1996). Brain chimeras in birds: application to the study of a genetic form of reflex epilepsy. *Trends Neurosci.* **19**, 246–252.

Baynash, A.G., Hosoda, K., Giaid, A., Richardson, J.A., Emoto, N., Hammer, R.E. and Yanagisawa, M. (1994). Interaction of endothelin-3 with endothelin-B receptor is essential for development of epidermal melanocytes and enteric neurons. *Cell* **79**, 1277–1285.

Beard, J. (1888). Morphological studies. II. The development of the peripheral nervous system of vertebrates. I. Elasmobranchii and Aves. *Q. J. Microsc. Sci.* **29**, 153–234.

Beato, M. (1989). Gene regulation by steroid hormones. *Cell* **56**, 335–344.

Becker, N., Seitanidou, T., Murphy, P., Mattei, M.G., Topilko, P., Nieto, M.A., Wilkinson, D.G., Charnay, P. and Gilardi-Hebenstreit, P. (1994). Several receptor tyrosine kinase genes of the Eph family are segmentally expressed in the developing hindbrain. *Mech. Dev.* **47**, 3–17.

Bennett, D. (1956). Developmental analysis of a mutation with pleiotropic effect in the mouse. *J. Morphol.* **98**, 199–233.

Bennett, J.H., Hunt, P. and Thorogood, P. (1995). Bone morphogenetic protein-2 and -4 expression during murine orofacial development. *Arch. Oral Biol.* **40**, 847–854.

Bentley, N.J., Eisen, T. and Goding, C.R. (1994). Melanocyte-specific expression of the human tyrosinase promoter: activation by the microphthalmia gene product and role of the initiator. *Mol. Cell. Biol.* **14**, 7996–8006.

Bergemann, A.D., Cheng, H.J., Brambilla, R., Klein, R. and Flanagan, J.G. (1995). *ELF-2*, a new member of the Eph ligand family, is segmentally expressed in mouse embryos in the region of the hindbrain and newly forming somites. *Mol. Cell. Biol.* **15**, 4921–4929.

Berkemeier, L.R., Winslow, J.W., Kaplan, D.R., Nikolics, K., Goeddel, D.V. and Rosenthal, A. (1991). Neurotrophin-5: a novel neurotrophic factor that activates trk and trkB. *Neuron* **7**, 857–866.

Bernd, P. (1987). Neuron-like cells in long-term neural crest cultures are not targets of nerve growth factor. *Dev. Brain Res.* **33**, 31–38.

Bernd, P. (1988). Selective expression of high-affinity nerve growth factor receptors on tyrosine hydroxylase-containing neuron-like cells in neural crest cultures. *J. Neurosci.* **8**, 3549–3555.

Bernhard, W. (1968). Une méthode de coloration régressive à l'usage de la microscopie électronique. *C. R. Acad. Sci., Ser. III* **267**, 2170–2173.

Berreta, C. and Zurn, A.D. (1991). The neuropeptide VIP modulates the neurotransmitter phenotype of cultured chick sympathetic neurons. *Dev. Biol.* **148**, 87–94.

Bertagna, X. (1994). Proopiomelanocortin-derived peptides. *Endocrinol. Metab. Clin. North Am.* **23**, 467–485.

Besmer, P., Murphy, J.E., George, P.C., Qiu, F.H., Bergold, P.J., Lederman, L., Snyder, H.W. Jr., Brodeur, D., Zuckerman, E.E. and Hardy, W.D. (1986). A new acute transforming feline retrovirus and relationship of its oncogene v-*kit* with the protein kinase gene family. *Nature* **320**, 415–421.

Bhattacharyya, A., Frank, E., Ratner, N. and Brackenbury, R. (1991). $P_0$ is an early marker of the Schwann cell lineage in chickens. *Neuron* **7**, 831–844.

Biffo, S., Offenhäuser, N., Carter, B.D. and Barde, Y.-A. (1995). Selective binding and internalisation by truncated receptors restrict the availability of BDNF during development. *Development* **121**, 2461–2470.

Birchmeier, W. and Birchmeier, C. (1994). Mesenchymal–epithelial transitions. *BioEssays* **16**, 305–307.

Birgbauer, E. and Fraser, S.E. (1994). Violation of cell lineage restriction compartments in the chick hindbrain. *Development* **120**, 1347–1356.

Birgbauer, E., Sechrist, J., Bronner-Fraser, M. and Fraser, S. (1995). Rhombomeric origin and rostrocaudal reassortment of neural crest cells revealed by intravital microscopy. *Development* **121**, 935–945.

Birren, S.J. and Anderson, D.J. (1990). A v-*myc*-immortalized sympathoadrenal progenitor cell line in which neuronal differentiation is initiated by FGF but not NGF. *Neuron* **4**, 189–201.

Birren, S.J., Lo, L. and Anderson, D.J. (1993). Sympathetic neuroblasts undergo a developmental switch in trophic dependence. *Development* **119**, 597–610.

Biscoe, T.J. (1971). Carotid body: structure and function. *Physiol. Rev.* **51**, 437–495.

Blaugrund, E., Pham, T.D., Tennyson, V.M., Lo, L., Sommer, L., Anderson, D.J. and Gershon, M.D. (1996). Distinct subpopulations of enteric neuronal progenitors defined by time of development, sympathoadrenal lineage markers and *Mash-1*-dependence. *Development* **120**, 309–320.

Blum, M., Gaunt, S.J., Cho, K.W., Steinbeisser, H., Blumberg, B., Bittner, D. and De Robertis, E.M. (1992). Gastrulation in the mouse: the role of the homeobox gene goosecoid. *Cell* **69**, 1097–1106.

Blumberg, B., Wright, C.V., De Robertis, E.M. and Cho, K.W. (1991). Organizer-specific homeobox genes in *Xenopus laevis* embryos. *Science* **253**, 194–196.

Blume-Jensen, P., Claesson-Welsh, L., Siegbahn, A., Zsebo, K.M., Westermark, B. and Heldin, C.H. (1991). Activation of the human c-kit product by ligand-induced dimerization mediates circular actin reorganization and chemotaxis. *EMBO J.* **10**, 4121–4128.

Bockman, D.E. and Kirby, M.L. (1984). Dependence of thymus development on derivatives of the neural crest. *Science* **223**, 498–500.

Bodenstein, D. (1952). Studies on the development of the dorsal fin in Amphibians. *J. Exp. Zool.* **120**, 213–245.

Boncinelli, E., Simeone, A., Acampora, D. and Mavilio, F. (1991). *HOX* gene activation by retinoic acid. *Trends Genet.* **7**, 329–334.

Borasio, G.D., Markus, A., Wittinghofer, A., Barde, Y.-A. and Heumann, R. (1993). Involvement of ras p21 in neurotrophin-induced response of sensory, but not sympathetic neurons. *J. Cell Biol.* **121**, 665–672.

Bouillet, P., Oulad-Abdelghani, M., Vicaire, S., Garnier, J.M., Schuhbaur, B., Dollé, P. and Chambon, P. (1995). Efficient cloning of cDNAs of retinoic acid-responsive genes in P19 embryonal carcinoma cells and characterization of a novel mouse gene, *Stra1* (mouse *LERK-2/Eplg2*). *Dev. Biol.* **170**, 420–433.

Bouillet, P., Oulad-Abdelghani, M., Ward, S.J., Bronner, S., Chambon, P. and Dollé, P. (1996). A new mouse member of the *Wnt* gene family, *mWnt-8*, is expressed during early embryogenesis and is ectopically induced by retinoic acid. *Mech. Dev.* **58**, 141–152.

Bovenkamp, D.E. and Greer, P. (1997). Novel Eph-family receptor tyrosine kinase is widely expressed in the developing zebrafish nervous system. *Dev. Dyn.* **209**, 166–181.

Boyd, J.D. (1937). The development of the human carotid body. *Contrib. Embryol. Carnegie Inst.* **26**, 1–31.

Brachet, A. (1907). Recherches sur l'ontogenèse de la tête chez les amphibiens. *Arch. Biol.* **23**, 165–257.

Brändli, A.W. and Kirschner, M.W. (1995). Molecular cloning of tyrosine kinases in the early Xenopus embryo: identification of Eck-related genes expressed in cranial neural crest cells of the second (Hyoid) arch. *Dev. Dyn.* **203**, 119–140.

Brannan, C.I., Bedell, M.A., Resnick, J.L., Eppig, J.J., Handel, M.A., Williams, D.E., Lyman, S.D., Donovan, P.J., Jenkins, N.A. and Copeland, N.G. (1992). Developmental abnormalities in *Steel$^{17H}$* mice result from a splicing defect in the steel factor cytoplasmic tail. *Genes Dev.* **6**, 1833–1842.

Brannan, C.I., Perkins, A.S., Vogel, K.S., Ratner, N., Nordlund, M.L., Reid, S.W., Buchberg, A.M., Jenkins, N.A., Parada, L.F. and Copeland, N.G. (1994). Targeted disruption of the neurofibromatosis type-1 gene leads to developmental abnormalities of the heart and various neural crest-derived tissues. *Genes Dev.* **8**, 1019–1029.

Breier, G., Bucan, M., Francke, U., Colberg-Poley, A.M. and Gruss, P. (1986). Sequential expression of murine homeo box genes during F9 EC cell differentiation. *EMBO J.* **5**, 2209–2215.

Brill, G., Vaisman, N., Neufeld, G. and Kalcheim, C. (1992). BHK-21–derived cell lines that produce basic fibroblast growth factor, but not parental BHK-21 cells, initiate neuronal differentiation of neural crest progenitors. *Development* **115**, 1059–1069.

Brill, G., Kahane, N., Carmeli, C., Von Schack, D., Barde, Y.-A. and Kalcheim, C. (1995). Epithelial-mesenchymal conversion of dermatome progenitors requires neural tube-derived signals: characterization of the role of neurotrophin-3. *Development* **121**, 2583–2594.

Brilliant, M.H. (1992). The mouse pink-eyed dilution locus: a model for aspects of Prader-Willi syndrome, Angelman syndrome, and a form of hypomelanosis of Ito. *Mamm. Genome* **3**, 187–191.

Bronner, M.E. and Cohen, A.M. (1979). Migratory patterns of cloned neural crest melanocytes injected into host chicken embryos. *Proc. Natl. Acad. Sci. USA* **76**, 1843–1847.

Bronner-Fraser, M. (1986a). Analysis of the early stages of trunk neural crest migration in avian embryos using monoclonal antibody HNK-1. *Dev. Biol.* **115**, 44–55.

Bronner-Fraser, M. (1986b). An antibody to a receptor for fibronectin and laminin perturbs cranial neural crest development *in vivo*. *Dev. Biol.* **117**, 528–536.

Bronner-Fraser, M.E. and Cohen, A.M. (1980). The neural crest: what can it tell us about cell migration and determination? *Curr. Top. Dev. Biol.* **15** (Pt 1), 1–25.

Bronner-Fraser, M. and Fraser, S.E. (1988). Cell lineage analysis reveals multipotency of some avian neural crest cells. *Nature* **335**, 161–164.

Bronner-Fraser, M. and Fraser, S. (1989). Developmental potential of avian trunk neural crest cells *in situ*. *Neuron* **3**, 755–766.

Bronner-Fraser, M. and Lallier, T. (1988). A monoclonal antibody against a laminin-heparan sulfate proteoglycan complex perturbs cranial neural crest migration *in vivo. J. Cell Biol.* **106**, 1321–1329.

Bronner-Fraser, M., Sieber-Blum, M. and Cohen, A.M. (1980). Clonal analysis of the avian neural crest: migration and maturation of mixed neural crest clones injected into host chicken embryos. *J. Comp. Neurol.* **193**, 423–434.

Bronner-Fraser, M., Wolf, J.J. and Murray, B.A. (1992). Effects of antibodies against N-cadherin and N-CAM on the cranial neural crest and neural tube. *Dev. Biol.* **153**, 291–301.

Brümmendorf, T., Wolff, J.M., Frank, R. and Rathjen, F.G. (1989). Neural cell recognition molecule F11: homology with fibronectin type III and immunoglobulin type C domains. *Neuron* **2**, 1351–1361.

Buchberg, A.M., Bedigian, H.G., Jenkins, N.A. and Copeland, N.G. (1990). Evi-2, a common integration site involved in murine myeloid leukemogenesis. *Mol. Cell. Biol.* **10**, 4658–4666.

Buchman, V.L. and Davies, A.M. (1993). Different neurotrophins are expressed and act in a developmental sequence to promote the survival of embryonic sensory neurons. *Development* **118**, 989–1001.

Buj-Bello, A., Adu, J., Pinon, L.G.P., Horton, A., Thompson, J., Rosenthal, A., Chinchetru, M., Buchman, V.L. and Davies, A.M. (1997). Neurturin responsiveness requires a GPI-linked receptor and the Ret receptor tyrosine kinase. *Nature* **387**, 721–724.

Bulfone, A., Kim, H.J., Puelles, L., Porteus, M.H., Grippo, J.F. and Rubenstein, J.L.R. (1993). The mouse *Dlx-2* (*Tes-1*) gene is expressed in spatially restricted domains of the forebrain, face and limbs in midgestation mouse embryos. *Mech. Dev.* **40**, 129–140.

Bultman, S.J., Michaud, E.J. and Woychik, R.P. (1992). Molecular characterization of the mouse agouti locus. *Cell* **71**, 1195–1204.

Bultman, S.J., Klebig, M.L., Michaud, E.J., Sweet, H.O., Davisson, M.T. and Woychik, R.P. (1994). Molecular analysis of reverse mutations from nonagouti (*a*) to black-and-tan (*a'*) and white-bellied agouti (*A^w*) reveals alternative forms of agouti transcripts. *Genes Dev.* **8**, 481–490.

Burke, B.A., Johnson, D., Gilbert, E.F., Drut, R.M., Ludwig, J. and Wick, M.R. (1987). Thyrocalcitonin-containing cells in the Di George anomaly. *Human Pathol.* **18**, 355–360.

Burns, A.J. and Le Douarin, N.M. (1998). The sacral neural crest contributes neurons and glia to the post-umbilical gut: spatiotemporal analysis of the development of the enteric nervous system. *Development* **125**, 4335–4347.

Burns, A.J., Lomax, A.E.J., Torihashi, S., Sanders, K.M. and Ward S.M. (1996). Interstitial cells of Cajal mediate inhibitory neurotransmission in the stomach. *Proc. Natl. Acad. Sci. USA* **93**, 12008–12013.

Butler, P.M. (1939). Studies of the mammalian dentition. Differentiation of the post-canine dentition. *Proc. Zool. Soc. London* **109B**, 329–356.

Buxton, P., Hunt, P., Ferretti, P. and Thorogood, P. (1997). A role for midline closure in the reestablishment of dorsoventral pattern following dorsal hindbrain ablation. *Dev. Biol.* **183**, 150–165.

Bytinsky-Salz, H. (1938). Chromatophorenstudien. II. Struktur und Determination des adepidermalen Melanophorennetzes bei *Bombina. Arch. Exp. Zellforsch. Gewebezücht.* **22**, 132–170.

Cable, J. and Steel, K.P. (1991). Identification of two types of melanocyte within the *stria vascularis* of the mouse inner ear. *Pigment Cell Res.* **4**, 87–101.

Cable, J., Huszar, D., Jaenisch, R. and Steel, K.P. (1994). Effects of mutations at the *w* locus (c-kit) on inner ear pigmentation and function in the mouse. *Pigment Cell Res.* **7**, 17–32.

Cameron-Curry, P., Dulac, C. and Le Douarin, N.M. (1993). Negative regulation of Schwann cell myelin protein gene expression by the dorsal root ganglionic microenvironment. *Eur. J. Neurosci.* **5**, 594–604.

Campenot, R.B., Walji, A.H. and Draker, D.D. (1991). Effects of sphingosine, staurosporine, and phorbol ester on neurites of rat sympathetic neurons growing in compartmented cultures. *J. Neurosci.* **11**, 1126–1139.

Carnahan, J.F. and Patterson, P.H. (1991a). The generation of monoclonal antibodies that bind preferentially to adrenal chromaffin cells and the cells of embryonic sympathetic ganglia. *J. Neurosci.* **11**, 3493–3506.

Carnahan, J.F. and Patterson, P.H. (1991b). Isolation of the progenitor cells of the sympathoadrenal lineage from embryonic sympathetic ganglia with the SA monoclonal antibodies. *J. Neurosci.* **11**, 3520–3530.

Carnahan, J.F., Anderson, D.J. and Patterson, P.H. (1991). Evidence that enteric neurons may derive from the sympathoadrenal lineage. *Dev. Biol.* **148**, 553–561.

Carnahan, J.F., Patel, D.R. and Miller, J.A. (1994). Stem cell factor is a neurotrophic factor for neural crest-derived chick sensory neurons. *J. Neurosci.* **14**, 1433–1440.

Carpenter, E. (1937). The head pattern in *Amblystoma* studied by vital staining and transplantation methods. *J. Exp. Zool.* **75**, 103–129.

Carpenter, E.M. and Hollyday, M. (1992). The distribution of neural crest-derived Schwann cells from subsets of brachial spinal segments into the peripheral nerves innervating the chick forelimb. *Dev. Biol.* **150**, 160–170.

Carpenter, E.M., Goddard, J.M., Chisaka, O., Manley, N.R. and Capecchi, M.R. (1993). Loss of *Hox-A1 (Hox-1.6)* function results in the reorganization of the murine hindbrain. *Development* **118**, 1063–1075.

Carrasco, A.E., McGinnis, W., Gehring, W.J. and De Robertis, E.M. (1984). Cloning of an *X. laevis* gene expressed during early embryogenesis coding for a peptide region homologous to *Drosophila* homeotic genes. *Cell* **37**, 409–414.

Carroll, S.L., Silos-Santiago, I., Frese, S.E., Ruit, K.G., Mibrandt, J. and Snider, W.D. (1992). Dorsal root ganglion neurons expressing trk are selectively sensitive to NGF deprivation *in utero*. *Neuron* **9**, 779–788.

Carter, B.D., Kaltschmidt, C., Kaltschmidt, B., Offenhäuser, N., Böhm-Matthaei, R., Baeuerle, P.A. and Barde, Y.-A. (1996). Selective activation of NF-kappaB by nerve growth factor through the neurotrophin receptor p75. *Science* **272**, 543–545.

Castle, W.E. and Little, C.C. (1910). On a modified Mendelian ratio among *yellow* mice. *Science* **32**, 868–870.

Catala, M., Teillet, M.-A. and Le Douarin, N.M. (1995). Organization and development of the tail bud analyzed with the quail–chick chimaera system. *Mech. Dev.* **51**, 51–65.

Cattanach, B.M., Wolfe, H.G., and Lyon, M.F. (1972). A comparative study of the coats of chimaeric mice and those of heterozygotes for X-linked genes. *Genet. Res.* **19**, 213–228.

Cattanach, B.M., Lyon, M.F., Peters, J. and Searle, A.G. (1987). *agouti* locus mutations at Harwell. *Mouse News Lett.* **77**, 123–125.

Celestino Da Costa, A. (1955). La notion de métaneurogonie. *C. R. Assoc. Anat.* **86**, 644–653.

Chabot, B., Stephenson, D.A., Chapman, V.M., Besmer, P. and Bernstein, A. (1988). The proto-oncogene *c-kit* encoding a transmembrane tyrosine kinase receptor maps to the mouse *W* locus. *Nature* **335**, 88–89.

Chakraborty, A.K., Platt, J.T., Kim, K.K., Kwon, B.S., Bennett, D.C. and Pawelek, J.M. (1996). Polymerization of 5,6-dihydroxyindole-3-carboxylic acid to melanin by the pmel 17/silver locus protein. *Eur. J. Biochem.* **236**, 180–188.

Chalazonitis, A., Rothman, T.P., Chen, J.X., Lamballe, F., Barbacid, M. and Gershon, M.D. (1994). Neurotrophin-3 induces neural crest-derived cells from fetal rat gut to develop *in vitro* as neurons or glia. *J. Neurosci.* **14**, 6571–6584.

Chalazonitis, A., Tennyson, V.M., Kibbey, M.C., Rothman, T.P. and Gershon, M.D. (1997). The alpha1 subunit of laminin-1 promotes the development of neurons by interacting with LBP110 expressed by neural crest-derived cells immunoselected from the fetal mouse gut. *J. Neurobiol.* **33**, 118–138.

Chambon, P. (1994). The retinoid signaling pathway: molecular and genetic analyses. *Semin. Cell Biol.* **5**, 115–125.

Changeux, J.-P. and Danchin, A. (1976). Selective stabilization of developing synapses as a mechanism for the specification of neuronal networks. *Nature* **264**, 705–712.

Chao, M.V. (1992). Neurotrophin receptors: a window into neuronal differentiation. *Neuron* **9**, 583–593.

Chao, M.V. (1994). The p75 neurotrophin receptor. *J. Neurobiol.* **25**, 1373–1385.

Charite, J., de Graaff, W., Shen, S. and Deschamps, J. (1994). Ectopic expression of *Hoxb-8* causes duplication of the ZPA in the forelimb and homeotic transformation of axial structures. *Cell* **78**, 589–601.

Chavrier, P., Vesque, C., Galliot, B., Vigneron, M., Dollé, P., Duboule, D. and Charnay, P. (1990). The segment-specific gene *Krox-20* encodes a transcription factor with binding sites in the promoter region of the *Hox-1.4* gene. *EMBO J.* **9**, 1209–1218.

Chen, C.L., Ager, L., Gartland, L. and Cooper, M. (1986). Identification of a T3/T cell receptor complex in chickens. *J. Exp. Med.* **164**, 375–380.

Chen, C.L., Cihak, J., Losch, U. and Cooper, M.D. (1988). Differential expression of two T cell receptors, TcR1 and TcR2, on chicken lymphocytes. *Eur. J. Immunol.* **18**, 539–543.

Chen, Y.P., Huang, L. and Solursh, M. (1992). A concentration gradient of retinoids in the early *Xenopus laevis* embryo. *Dev. Biol.* **161**, 70–76.

Chen, Z.F. and Behringer, R.R. (1995). *Twist* is required in head mesenchyme for cranial neural tube morphogenesis. *Genes Dev.* **9**, 686–699.

Cheng, H.-J. and Flanagan, J.G. (1994). Identification and cloning of ELF-1, a developmentally expressed ligand for the Mek4 and Sek receptor thyrosine kinases. *Cell* **79**, 157–168.

Chevallier, A. and Kieny, M. (1982). On the role of connective tissue in the patterning of the chick limb musculature. *Roux Arch. Dev. Biol.* **191**, 277–280.

Chevallier, A., Kieny, M. and Mauger, A. (1977). Limb–somite relationship: origin of the limb musculature. *J. Embryol. Exp. Morphol.* **41**, 245–258.

Chiang, C., Litingtung, Y., Lee, E., Young, K.E., Corden, J.L., Westphal, H. and Beachy, P.A. (1996). Cyclopia and defective axial patterning in mice lacking *Sonic hedgehog* gene function. *Nature* **383**, 407–413.

Chibon, P. (1964). Analyse par la méthode de marquage nucléaire à la thymidine tritiée des dérivés de la crête neurale céphalique chez l'Urodèle *Pleurodeles waltlii*. *C. R. Acad. Sci., Ser. III* **259**, 3624–3627.

Chibon, P. (1966). Analyse expérimentale de la régionalisation et des capacités morphogénétiques de la crête neurale chez l'Amphibien Urodèle *Pleurodeles waltlii* Michah. *Mem. Soc. Fr. Zool.* **36**, 1–107.

Chibon, P. (1967). Nuclear labelling by tritiated thymidine of neural crest derivatives in the amphibian Urodele *Pleurodeles waltlii* Michah. *J. Embryol. Exp. Morphol.* **18**, 343–358.

Chibon, P. (1970). L'origine de l'organe adamantin des dents. Etude au moyen du marquage nucléaire de l'ectoderme stomodeal. *Ann. Embryol. Morphog.* **3**, 203–213.

Chisaka, O. and Capecchi, M.R. (1991). Regionally restricted developmental defects resulting from targeted disruption of the mouse homeobox gene *Hox-1.5*. *Nature* **350**, 473–479.

Chisaka, O., Musci, T.S. and Capecchi, M.R. (1992). Developmental defects of the ear, cranial nerves and hindbrain resulting from targeted disruption of the mouse homeobox gene *Hox-1.6*. *Nature* **355**, 516–520.

Chitnis, A., Henrique, D., Lewis, J., Ish Horowicz, D. and Kintner, C. (1995). Primary neurogenesis in *Xenopus* embryos regulated by a homologue of the Drosophila neurogenic gene *Delta*. *Nature* **375**, 761–766.

Chou, D.K.H., Ilyas, A.A., Evans, J.E., Quarles, R.H. and Jungalawa, F.B. (1985). Structure of a glycolipid reacting with monoclonal IgM in neuropathy and with HNK-1. *Biochem. Biophys. Res. Commun.* **128**, 383–388.

Chou, D.K.H., Ilyas, A.A., Evans, J.E., Costello, C., Quarles, R.H. and Jungalwala, F.B. (1986). Structure of sulfated glucuronyl glycolipids on the nervous system reacting with HNK-1 antibody and some IgM paraproteins in neuropathy. *J. Biol. Chem.* **261**, 11717–11725.

Christ, B., Jacob, H.J. and Jacob, M. (1977). Experimental analysis of the origin of the wing musculature in avian embryos. *Anat. Embryol. (Berl).* **150**, 171–186.

Christ, B., Poelmann, R.E., Mentink, M.M.T. and Gittenberger-De Groot, A.C. (1990). Vascular endothelial cells migrate centripetally within embryonic arteries. *Anat. Embryol.* **181**, 333–339.

Christ, B., Grim, M., Wilting, J., Von Kirschhofer, K. and Wachtler, F. (1991). Differentiation of endothelial cells in avian embryos does not depend on gastrulation. *Acta Histochem.* **91**, 193–199.

Ciani, F., Contestabile, A. and Villani, L. (1973). Ultrastructural differentiation *in vivo* and *in vitro* of chick embryo trigeminal ganglion. *Acta Embryol. Exp.* **3**, 341–354.

Cihak, J., Ziegler-Heitbrock, H.W., Trainer, H., Schranner, I., Merkenschlager, M. and Losch, U. (1988). Characterization and functional properties of a novel monoclonal antibody which identifies a T cell receptor in chickens. *Eur. J. Immunol.* **18**, 533–537.

Ciment, G. and Weston, J.A. (1982). Early appearance in neural crest and crest-derived cells of an antigenic determinant present in avian neurons. *Dev. Biol.* **93**, 355–367.

Ciment, G. and Weston, J.A. (1983). Enteric neurogenesis by neural crest-derived branchial arch mesenchymal cells. *Nature* **305**, 424–427.

Ciment, G., Glimelius, B., Nelson, D.M. and Weston, J.A. (1986). Reversal of a developmental restriction in neural crest-derived cells of avian embryos by a phorbol ester drug. *Dev. Biol.* **118**, 392–398.

Ciossek, T., Millauer, B. and Ullrich, A. (1995). Identification of alternatively spliced mRNAs encoding variants of MDK1, a novel receptor tyrosine kinase expressed in the murine nervous system. *Oncogene* **10**, 97–108.

Claude, P., Parada, I.M., Gordon, K.A., D'Amore, P.A. and Wagner, J.A. (1988). Acidic fibroblast growth factor stimulates adrenal chromaffin cells to proliferate and to extend neurites, but is not a long-term survival factor. *Neuron* 1, 783–790.

Clouthier, D.E., Hosoda, K., Richardson, J.A., Williams, S.C., Yanagisawa, H., Kuwaki, T., Kumada, M., Hammer, R.E. and Yanagisawa, M. (1998). Cranial and cardiac neural crest defects in endothelin-A receptor-deficient mice. *Development* 125, 813–824.

Cochard, P. and Coltey, P. (1983). Cholinergic traits in the neural crest: acetylcholinesterase in crest cells of the chick embryo. *Dev. Biol.* 98, 221–238.

Cochard, P., Goldstein, M. and Black, I.B. (1978). Ontogenetic appearance and disapearance of tyrosine hydroxylase and catecholamines in the rat embryo. *Proc. Natl. Acad. Sci. USA* 75, 2986–2990.

Cochard, P., Goldstein, M. and Black, I.B. (1979). Initial development of the noradrenergic phenotype in autonomic neuroblasts of the rat embryo *in vivo*. *Dev. Biol.* 71, 100–114.

Coelho, C.N.D., Sumoy, L., Rodgers, B.J., Davidson, D.R., Hill, R.E., Upholt, W.B. and Kosher, R.A. (1991). Expression of the chicken homeobox-containing gene *ghox-8* during embryonic chick limb development. *Mech. Dev.* 34, 143–154.

Coghill, G.E. (1916). Correlated anatomical and physiological studies of the growth of the nervous system of Amphibia. II. The efferent system of the head of *Amblystoma*. *J. Comp. Neurol.* 26, 247–340.

Cohen, A. (1974) DNA synthesis and cell division in differentiating avian adrenergic neuroblasts. In *Wenner-Gren Center International Symposium Series*, eds. K. Fuxe, L. Olson and Y. Zotterman, pp. 359–370. Pergamon Press, Oxford.

Cohen, A.M. and Königsberg, I.R. (1975). A clonal approach to the problem of neural crest determination. *Dev. Biol.* 46, 263–280.

Cohen, S.M. and Jürgens, G. (1989). Proximal-distal pattern formation in Drosophila: cell autonomous requirement for *Distal-less* gene activity in limb development. *EMBO J.* 8, 2045–2055.

Colberg-Poley, A.M., Voss, S.D., Chowdhury, K. and Gruss, P. (1985). Structural analysis of murine genes containing homeo box sequences and their expression in embryonal carcinoma cells. *Nature* 314, 713–718.

Cole, T.J., Blendy, J.A., Monaghan, A.P., Krieglstein, K., Schmid, W., Aguzzi, A., Fantuzzi, G., Hummler, E., Unsicker, K. and Schütz, G. (1995). Targeted disruption of the glucocorticoid receptor gene blocks adrenergic chromaffin cell development and severely retards lung maturation. *Genes Dev.* 9, 1608–1621.

Collazo, A., Bronner-Fraser, M. and Fraser, S.E. (1993). Vital dye labelling of *Xenopus laevis* trunk neural crest reveals multipotency and novel pathways of migration. *Development* 118, 363–376.

Compere, S., Baldacci, P., Sharpe, A. and Jaenisch, R. (1989a). Retroviral transduction of the human c-Ha-ras-1 oncogene into midgestation mouse embryos promotes rapid epithelial hyperplasia. *Mol. Cell. Biol.* 9, 6–14.

Compere, S., Baldacci, P., Sharpe, A., Thompson, T., Land, H. and Jaenisch, R. (1989b). The ras and myc oncogenes cooperate in tumor induction in many tissues when introduced into midgestation mouse embryos by retroviral vectors. *Proc. Natl. Acad. Sci. USA* 86, 2224–2228.

Condron, B.G. and Zinn, K. (1995). Activation of cAMP-dependent protein kinase triggers a glial-to-neuronal cell-fate switch in an insect neuroblast lineage. *Curr. Biol.* 5, 51–61.

Condron, B.G., Patel, N.H. and Zinn, K. (1994). *engrailed* controls glial/neuronal cell fate decisions at the midline of the central nervous system. *Neuron* **13**, 541–554.

Conlon, R.A. and Rossant, J. (1992). Exogenous retinoic acid rapidly induces anterior ectopic expression of murine *Hox-2* genes *in vivo*. *Development* **116**, 357–368.

Conover, J.C., Erickson, J.T., Katz, D.M., Bianchi, L.M., Poueymirou, W.T., McClain, J., Pan, L., Helgren, M., Ip, N.Y., Boland, P., Friedman, B., Wiegand, S., Vejsada, R., Kato, A.C., DeChiara, T.M. and Yancopoulos, G.D. (1995). Neuronal deficits, not involving motor neurons, in mice lacking BDNF and/or NT4. *Nature* **375**, 235–238.

Conrad, G.W. (1970). Collagen and mucopolysaccharide biosynthesis in the developing chick cornea. *Dev. Biol.* **21**, 293–317.

Conrad, G.W., Bee, J.A., Roche, S.M. and Teillet, M.-A. (1993). Fabrication of microscalpels by electrolysis of tungsten wire in a meniscus. *J. Neurosc. Meth.* **50**, 123–127.

Copeland, N.G., Gilbert, D.J., Cho, B.C., Donovan, P.J., Jenkins, N.A., Cosman, D., Anderson, D., Lyman, S.D. and Williams, D.E. (1990). Mast cell growth factor maps near the steel locus on mouse chromosome 10 and is deleted in a number of steel alleles. *Cell* **63**, 175–183.

Copp, D.H., Cameron, E.C., Cheney, E.A., Davidson, A.G. and Henze, K.G. (1962). Evidence for calcitonin. A new hormone from the parathyroid that lowers blood calcium. *Endocrinology* **70**, 638–649.

Cordes, S.P. and Barsh, G.S. (1994). The mouse segmentation gene *kr* encodes a novel basic domain leucine zipper transcription factor. *Cell* **79**, 1025–1034.

Costaridis, P., Horton, C., Zeitlinger, J., Holder, N. and Maden, M. (1996). Endogenous retinoids in the zebrafish embryo and adult. *Dev. Dyn.* **205**, 41–51.

Coulombe, J.N. and Bronner-Fraser, M. (1986). Cholinergic neurones acquire adrenergic neurotransmitters when transplanted into an embryo. *Nature* **324**, 569–572.

Coulombre, A.J. (1965). Problems in corneal morphogenesis. *Adv. Morphog.* **4**, 81–109.

Couly, G.F. and Le Douarin, N.M. (1985). Mapping of the early neural primordium in quail–chick chimeras. I. Developmental relationships between placodes, facial ectoderm, and prosencephalon. *Dev. Biol.* **110**, 423–439.

Couly, G.F. and Le Douarin, N.M. (1987). Mapping of the early neural primordium in quail–chick chimeras. II. The prosencephalic neural plate and neural folds: implications for the genesis of cephalic human congenital abnormalities. *Dev. Biol.* **120**, 198–214.

Couly, G.F. and Le Douarin, N.M. (1988). The fate map of the cephalic neural primordium at the presomitic to the 3-somite stage in the avian embryo. *Development* **103** (Suppl.), 101–113.

Couly, G. and Le Douarin, N.M. (1990). Head morphogenesis in embryonic avian chimeras: evidence for a segmental pattern in the ectoderm corresponding to the neuromeres. *Development* **108**, 543–558.

Couly, G.F., Coltey, P.M. and Le Douarin, N.M. (1992). The developmental fate of the cephalic mesoderm in quail–chick chimeras. *Development* **114**, 1–15.

Couly, G.F., Coltey, P.M. and Le Douarin, N.M. (1993). The triple origin of skull in higher vertebrates: a study in quail–chick chimeras. *Development* **117**, 409–429.

Couly, G.F., Coltey, P., Eichmann, A. and Le Douarin, N.M. (1995). The angiogenic potentials of the cephalic mesoderm and the origin of brain and head blood vessels. *Mech. Dev.* **53**, 97–112.

Couly, G., Grapin-Botton, A., Coltey, P. and Le Douarin, N.M. (1996). The regeneration of the cephalic neural crest, a problem revisited: the regenerating cells originate from the contralateral or from the anterior and posterior neural fold. *Development* **122**, 3393–3407.

Couly, G., Grapin-Botton, A., Coltey, P., Ruhin, B. and Le Douarin, N.M. (1998). Determination of the identity of the derivatives of the cephalic neural crest: incompatibility between Hox gene expression and lower jaw development. *Development* **125**, 3445–3459.

Cowell L.A. and Weston, J.A. (1970). An analysis of melanogenesis in cultured chick embryo spinal ganglia. *Dev. Biol.* **22**, 670–697.

Creedon, D.J., Tansey, M.G., Baloh, R.H., Osborne, P.A., Lampe, P.A., Fahrner, T.J., Heuckeroth, R.O., Milbrandt, J. and Johnson E.M., Jr. (1997). Neurturin shares receptors and signal transduction pathways with glial cell line-derived neurotrophic factor in sympathetic neurons. *Proc. Natl. Acad. Sci. USA* **94**, 7018–7023.

Crossin, K.L., Hoffman, S., Grumet, M., Thiery, J.P. and Edelman, G.M. (1986). Site-restricted expression of cytotactin during development of the chicken embryo. *J. Cell Biol.* **102**, 1917–1930.

Crossley, P.H. and Martin, G.R. (1995). The mouse *Fgf8* gene encodes a family of polypeptides and is expressed in regions that direct outgrowth and patterning in the developing embryo. *Development* **121**, 439–451.

Crowley, C., Spencer, S.D., Nishimura, M.C., Chen, K.S., Pitts-Meek, S., Armanini, M.P., Ling, L.H., McMahon, S.B., Shelton, D.L., Levinson, A.D. and Phillips, H.S. (1994). Mice lacking nerve growth factor display perinatal loss of sensory and sympathetic neurons yet develop basal forebrain cholinergic neurons. *Cell* **76**, 1001–1011.

Cserjesi, P., Lilly, B., Bryson, L., Wang, Y., Sassoon, D.A. and Olson, E.N. (1992). MHox: a mesodermally resticted homeodomain protein that binds an essential site in the muscle creatine kinase enhancer. *Development* **115**, 1087–1101.

Cudennec, C. (1977). Reconnaissances cellulaires au cours du développement: étude *in vivo* au moyen de chimères interspécifiques chez l'embryon d'oiseau. *Biol. Cell* **30**, 41–48.

Cuénot, L. (1908). Sur quelques anomalies apparentes des proportions mendéliennes. *Arch. Zool. Exp. Genet.* **9**, 7–15.

Curtis, R., Stewart, H.J.S., Hall, S.M., Wilkin, G.P., Mirsky, R. and Jessen, K.R. (1992). GAP-43 is expressed by nonmyelin-forming Schwann cells of the peripheral nervous system. *J. Cell Biol.* **116**, 1455–1464.

Curtis, R., Adryan, K.M., Zhu, Y., Harkness, P.J., Lindsay, R.M. and DiStefano, P.S. (1993). Retrograde axonal transport of ciliary neurotrophic factor is increased by peripheral nerve injury. *Nature* **365**, 253–255.

Dalton, H.C. (1949). Developmental analysis of genetic differences in pigmentation in the axolotl. *Proc. Natl. Acad. Sci. USA* **35**, 277–283.

D'Amico-Martel, A. and Noden, D.M. (1980). An autoradiographic analysis of the development of the chick trigeminal ganglion. *J. Embryol. Exp. Morphol.* **55**, 167–182.

D'Amico-Martel, A. and Noden, D.M. (1983). Contributions of placodal and neural crest cells to avian cranial peripheral ganglia. *Am. J. Anat.* **166**, 445–468.

Danforth, C.H. (1927). Hereditary adiposity in mice. *J. Hered.* **18**, 153–162.

Darnell, D.K., Schoenwolf, G.C. and Ordahl, C.P. (1992). Changes in dorsoventral but not rostrocaudal regionalization of the chick neural tube in the absence of cranial notochord, as revealed by expression of *engrailed-2*. *Dev. Dyn.* **193**, 389–396.

Davies, A.M. (1994). Intrinsic programmes of growth and survival in developing vertebrate neurons. *Trends Neurosci.* **17**, 195–199.

Davies, A.M. and Lindsay, R.M. (1985). The cranial sensory ganglia in culture: difference in the response of placode-derived and neural crest-derived neurons to nerve growth factor. *Dev. Biol.* **11**, 63–72.

Davies, A. and Lumsden, A. (1984). Relation of target encounter and neuronal death to nerve growth factor responsiveness in the developing mouse trigeminal ganglion. *J. Comp. Neurol.* **223**, 124–137.

Davies, A.M. and Wright, E.M. (1995). Neurotrophic factors: neurotrophin autocrine loops. *Curr. Biol.* **5**, 723–726.

Davies, A.M., Thoenen, H. and Barde, Y.-A. (1986). The response of chick sensory neurons to brain-derived neurotrophic factor. *J. Neurosci.* **6**, 1897–1904.

Davies, A.M., Lumsden, A.G. and Rohrer, H. (1987). Neural crest-derived proprioceptive neurons express nerve growth factor receptors but are not supported by nerve growth factor in culture. *Neuroscience* **20**, 37–46.

Davies, A.M., Horton, A., Burton, L.E., Schmelzer, C., Vandlen, R. and Rosenthal, A. (1993a). Neurotrophin-4/5 is a mammalian-specific survival factor for distinct populations of sensory neurons. *Neuroscience* **13**, 4961–4967.

Davies, A.M., Lee, K.-F. and Jaenisch, R. (1993b). p75-deficient trigeminal sensory neurons have an altered response to NGF but not to other neurotrophins. *Neuron* **11**, 565–574.

Davies, A.M., Minichiello, L. and Klein, R. (1995). Developmental changes in NT3 signalling via TrkA and TrkB in embryonic neurons. *EMBO J.* **14**, 4482–4489.

Davies, J.A., Cook, G.M., Stern, C.D. and Keynes, R.J. (1990). Isolation from chick somites of a glycoprotein fraction that causes collapse of dorsal root ganglion growth cones. *Neuron* **4**, 11–20.

Davis, C.B. and Littman, D.R. (1994). Thymocyte lineage commitment: is it instructed or stochastic? *Curr. Opin. Immunol.* **6**, 266–272.

Davis, J.B. and Stroobant, P. (1990). Platelet-derived growth factors and fibroblast growth factors are mitogens for rat Schwann cells. *J. Cell Biol.* **110**, 1353–1360.

Davis, S., Gale, N.W., Aldrich, T.H., Maisonpierre, P.C., Lhotak, V., Pawson, T., Goldfarb, M. and Yancopoulos, G.D. (1994). Ligands for EPH-related receptor tyrosine kinases that require membrane attachment or clustering for activity. *Science* **266**, 816–819.

De Beer, G.R. (1947). The differentiation of neural crest cells into visceral cartilages and odontoblasts in *Amblystoma* and a re-examination of the germ-layer theory. *Proc. R. Soc. Lond. Ser. B* **134**, 377–398.

De Beer, G. (1985). *The Development of the Vertebrate Skull.* University of Chicago Press, Chicago.

De Thé, H., Vivanco-Ruiz, M.D.M., Tiollais, P., Stunnenberg, H. and Dejean, A. (1990). Identification of a retinoic acid responsive element in the retinoic acid β-receptor gene. *Nature* **343**, 177–180.

De Winiwarter, H. (1939). Origine et développement du ganglion carotidien. Appendice: participation de l'hypoblaste à la constitution des ganglions crâniens. *Arch. Biol.* **50**, 67–94.

Debby-Brafman, A., Burstyn-Cohen, T., Klar, A. and Kalcheim, C. (1999). F-Spondin, expressed in somite regions avoided by neural crest cells, mediates inhibition of distinct somite domains to neural crest migration. *Neuron* **22**, 475–488.

Dechant, G., Biffo, S., Okazawa, H., Kolbeck, R., Pottgiesser, J. and Barde, Y.-A. (1993a). Expression and binding characteristics of the BDNF receptor chick *trk*B. *Development* **119**, 545–558.

Dechant, G., Rodríguez-Tébar, A., Kolbeck, R. and Barde, Y.-A. (1993b). Specific high-affinity receptors for neurotrophin-3 on sympathetic neurons. *J. Neurosci.* **13**, 2610–2616.

Delannet, M. and Duband, J.L. (1992). Transforming growth factor-beta control of cell–substratum adhesion during avian neural crest cell emigration *in vitro*. *Development* **116**, 275–287.

Deol, M.S. (1970a). The relationship between abnormalities of pigmentation and of the inner ear. *Proc. R. Soc. Lond. Ser. A* **175**, 201–217.

Deol, M.S. (1970b). The origin of the acoustic ganglion and effects of the gene dominant spoting (*W*ᵛ) in the mouse. *J. Embryol. Exp. Morphol.* **23**, 773–784.

Deol, M.S. (1973). The role of the tissue environment in the expression of spotting genes in the mouse. *J. Embryol. Exp. Morphol.* **30**, 483–489.

Detwiler, S.R. (1919). An experimental study of spinal nerve segmentation in amblystoma with reference to the plurisegmental contribution to the brachial plexus. *J. Exp. Zool.* **67**, 395–441.

Detwiler, S.R. (1937a). Application of vital dyes to the study of sheath cell origin. *Proc. Soc. Exp. Biol. Med.* **37**, 380–382.

Detwiler, S.R. (1937b). Observations upon the migration of neural crest cells, and upon the development of the spinal ganglia and vertebral arches in *Amblystoma*. *Am. J. Anat.* **61**, 63–94.

Deutsch, U., Dressler, G.R. and Gruss, P. (1988). *Pax 1*, a member of a paired box homologous murine gene family, is expressed in segmented structure during development. *Cell* **53**, 617–625.

Di George, A.M. (1968). Congenital absence of the thymus and its immunological consequences: concurrence with congenital hypoparathyroidism. In *Birth Defects: Immunologic Deficiency Diseases in Man*, eds. D. Bergsma and R.A. Good, p. 116. National Foundation March of Dimes, New York.

Diamond, J., Holmes, M. and Coughlin, M. (1992). Endogenous NGF and nerve impulses regulate the collateral sprouting of sensory axons in the skin of the adult rat. *J. Neurosci.* **12**, 1454–1466.

DiCicco-Bloom, E., Townes-Anderson, E. and Black, I.B. (1990). Neuroblast mitosis in dissociated culture: regulation and relationship to differentiation. *J. Cell Biol.* **110**, 2073–2086.

DiCicco-Bloom, E., Friedman, W.J. and Black, I.B. (1993). NT-3 stimulates sympathetic neuroblast proliferation by promoting precursor survival. *Neuron* **11**, 1101–1111.

Dickie, M.M. (1969). Mutations at the *agouti* locus in the mouse. *J. Hered.* **60**, 20–25.

Dickinson, M.E., Selleck, M.A.J., McMahon, A.P. and Bronner-Fraser, M. (1995). Dorsalization of the neural tube by the non-neural ectoderm. *Development* **121**, 2099–2106.

Dickson, M.C., Martin, J.S., Cousins, F.M., Kulkarni, A.B., Karlsson, S. and Akhurst, R.J. (1995). Defective hematopoiesis and vasculogenesis in transforming growth factor beta1 knockout mice. *Development* **121**, 1845–1854.

Dietrich, S. and Gruss, P. (1995). *undulated* phenotypes suggest a role of *Pax-1* for the development of vertebral and extravertebral structures. *Dev. Biol.* **167**, 529–548.

Dimberg, Y., Hedlund, K.O. and Ebendal, T. (1987). Effects of nerve growth factor on sensory neurons in the chick embryo: a stereological study. *Int. J. Dev. Neurosci.* **5**, 207–213.

Disse, J. (1897). Die erste Entwicklung des Riechnervenn. *Anat. Hefte, Abt. 1*, **9**, 255–300.

DiStefano, P.S., Friedman, B., Radziejewski, C., Alexander, C., Boland, P., Schick, C.M., Lindsay, R.M. and Wiegand, S.J. (1992). The neurotrophins, BDNF, NT-3, and NGF display distinct patterns of retrograde axonal transport in peripheral and central neurons. *Neuron* **8**, 983–993.

Dodd, J. and Jessell, T.M. (1985). Lactoseries carbohydrates specify subsets of dorsal root ganglion neurons projecting to the superficial dorsal horn of rat spinal cord. *J. Neurosci.* **5**, 3278–3294.

Dodson, J.W. and Hay, E.D. (1971). Secretion of collagenous stroma by isolated epithelium grown *in vitro*. *Exp. Cell Res.* **65**, 215–220.

Dollé, P., Izpisua-Belmonte, J.C., Boncinelli, E. and Duboule, D. (1991a). The *Hox-4.8* gene is localized at the 5' extremity of the *Hox-4* complex and is expressed in the most posterior parts of the body during development. *Mech. Dev.* **36**, 3–13.

Dollé, P., Izpisua-Belmonte, J.C., Brown, J.M., Tickle, C. and Duboule, D. (1991b). *HOX-4* genes and the morphogenesis of mammalian genitalia. *Genes Dev.* **5**, 1767–1767.

Dollé, P., Price, M. and Duboule, D. (1992). Expression of the murine *dlx-1* homeobox gene during facial, ocular and limb development. *Differentiation* **49**, 93–99.

Dollé, P., Lufkin, T., Krumlauf, R., Mark, M., Duboule, D. and Chambon, P. (1993). Local alterations of *Krox-20* and *Hox* gene expression in the hindbrain suggest lack of rhombomere 4 and rhombomere 5 in homozygote null *Hoxa-1* (*Hox-1.6*) mutant embryos. *Proc. Natl. Acad. Sci. USA* **90**, 7666–7670.

Dorris, F. (1936). Differentiation of pigment cells in tissue cultures of chick neural crest. *Proc. Soc. Exp. Biol. Med.* **34**, 448–449.

Dorris, F. (1938). The production of pigment *in vitro* by chick neural crest. *Roux Arch. Entwicklungsmech. Org.* **138**, 323–335.

Dorris, F. (1939). The production of pigments by chick neural crest in grafts to the 3-day limb bud. *J. Exp. Zool.* **80**, 315–345.

Douhet, P., Bucchini, D., Jami, J. and Calas, A. (1993). Demonstration by *in situ* hybridization and immunohistochemistry of human insulin gene expression cells in medial habenula of transgenic mice. *C. R. Acad. Sci., Ser. III.* **316**, 400–403.

Doupe, A.J., Landis, S.C. and Patterson, P.H. (1985a). Environmental influences in the development of neural crest derivatives: glucocorticoids, growth factors, and chromaffin cell plasticity. *J. Neurosci.* **5**, 2119–2142.

Doupe, A.J., Patterson, P.H. and Landis, S.C. (1985b). Small intensely fluorescent cells in culture: role of glucocorticoids and growth factors in their development and interconversions with other neural crest derivatives. *J. Neurosci.* **5**, 2143–2160.

Drews, U. (1975). Cholinesterase in embryonic development. *Prog. Histochem. Cytochem.* **7**, 1–52.

Drews, U., Kussather, E. and Usadel, K.H. (1967). Histochemischer Nachweis der Cholinesterase in der Frühentwicklung der Hühnerkeimscheibe. *Histochemie* **8**, 65–89.

Du Shane, G.P. (1934). The source of pigment cells in Amphibia. *Science* **80**, 620–621.

Du Shane, G.P. (1935). An experimental study of the origin of pigment cells in Amphibia. *J. Exp. Zool.* **72**, 1–31.

Du Shane, G.P. (1936). The Dopa reaction in Amphibia. *Proc. Soc. Exp. Biol. Med.* **33**, 593–595.

Du Shane, G.P. (1938). Neural fold derivatives in the Amphibia: pigment cells, spinal ganglia and Rohon-Béard cells. *J. Exp. Zool.* **78**, 485–503.

Duband, J.L. and Thiery, J.P. (1982). Appearance and distribution of fibronectin during chick embryo gastrulation and neurulation. *Dev. Biol.* **94**, 337–350.

Duband, J.L. and Thiery, J.P. (1987). Distribution of laminin and collagens during avian neural crest development. *Development* **101**, 461–478.

Duband, J.L., Rocher, S., Chen, W.T., Yamada, K.M. and Thiery, J.P. (1986a). Cell adhesion and migration in the early vertebrate embryo: location and possible role of the putative fibronectin receptor complex. *J. Cell Biol.* **102**, 160–178.

Duband, J.L., Rocher, S., Yamada, K.M. and Thiery, J.P. (1986b). Interactions of migrating neural crest cells with fibronectin. *Prog. Clin. Biol. Res.* **226**, 127–139.

Duband, J.L., Volberg, T., Sabanay, I., Thiery, J.P. and Geiger, B. (1988). Spatial and temporal distribution of the adherens-junction-associated adhesion molecule A-CAM during avian embryogenesis. *Development* **103**, 325–344.

Duband, J.L., Belkin, A.M., Syfrig, J., Thiery, J.P. and Koteliansky, V.E. (1992). Expression of alpha 1 integrin, a laminin–collagen receptor, during myogenesis and neurogenesis in the avian embryo. *Development* **116**, 585–600.

Duboule, D. (1991). Patterning in the vertebrate limb. *Curr. Opin. Genet. Dev.* **1**, 211–216.

Duboule, D. (1992). The vertebrate limb: a model system to study the *Hox/HOM* gene network during development and evolution. *BioEssays* **14**, 375–384.

Duboule, D. (1994). *Guidebook to the Homeobox.* Oxford University Press, Oxford.

Duboule, D. and Dollé, P. (1989). The structural and functional organization of the murine *HOX* gene family resembles that of Drosophila homeotic genes. *EMBO J.* **8**, 1497–505.

Duff, R.S., Langtimm, C.J., Richardson, M.K. and Sieber-Blum, M. (1991). *In vitro* clonal analysis of progenitor cell patterns in dorsal root and sympathetic ganglia of the quail embryo. *Dev. Biol.* **147**, 451–459.

Dufour, S., Duband, J.L., Kornblihtt, A.R. and Thiery, J.P. (1988). The role of fibronectins in embryonic cell migrations. *Trends Genet.* **4**, 198–203.

Duhl, D.M., Stevens, M.E., Vrieling, H., Saxon, P.J., Miller, M.W., Epstein, C.J. and Barsh, G.S. (1994). Pleiotropic effects of the mouse *lethal yellow* (*A^y*) mutation explained by deletion of a maternally expressed gene and the simultaneous production of *agouti* fusion RNAs. *Development* **120**, 1695–1708.

Dulac, C. and Le Douarin, N.M. (1991). Phenotypic plasticity of Schwann cells and enteric glial cells in response to the microenvironment. *Proc. Natl. Acad. Sci. USA* **88**, 6358–6362.

Dulac, C., Cameron-Curry, P., Ziller, C. and Le Douarin, N.M. (1988). A surface protein expressed by avian myelinating and nonmyelinating Schwann cells but not by satellite or enteric glial cells. *Neuron* **1**, 211–220.

Dulac, C., Tropak, M.B., Cameron-Curry, P., Rossier, J., Marshak, D.R., Roder, J. and Le Douarin, N.M. (1992). Molecular characterization of the Schwann cell myelin protein, SMP: structural similarities within the immunoglobulin superfamily. *Neuron* **8**, 323–334.

Dupin, E. (1984). Cell division in the ciliary ganglion of quail embryos *in situ* and after back-transplantation into the neural crest migration pathways of chick embryos. *Dev. Biol.* **105**, 288–299.

Dupin, E. and Le Douarin, N.M. (1993). Culture of avian neural crest cells. In *Essential Developmental Biology: A Practical Approach*, eds. C.D. Stern and P.W.H. Holland, pp. 153–166. Oxford University Press, New York.

Dupin, E. and Le Douarin, N.M. (1995). Retinoic acid promotes the differentiation of adrenergic cells and melanocytes in quail neural crest cultures. *Dev. Biol.* **168**, 529–548.

Dupin, E., Baroffio, A., Dulac, C., Cameron-Curry, P. and Le Douarin, N.M. (1990). Schwann-cell differentiation in clonal cultures of the neural crest, as evidenced by the anti-Schwann cell myelin protein monoclonal antibody. *Proc. Natl. Acad. Sci. USA* **87**, 1119–1123.

Dupin, E., Maus, M. and Fauquet, M. (1993). Regulation of the quail tyrosine hydroxylase gene in neural crest cells by cAMP and beta-adrenergic ligands. *Dev. Biol.* **159**, 75–86.

Durbec, P., Marcos-Gutierrez, C.V., Kilkenny, C., Grigoriou, M., Wartiowaara, K., Suvanto, P., Smith, D., Ponder, B., Costantini, F., Saarma, M., Sariola, H. and Pachnis, V. (1996a). GDNF signalling through the Ret receptor tyrosine kinase. *Nature* **381**, 789–793.

Durbec, P.L., Larsson-Blomberg, L.B., Schuchardt, A., Constantini, F. and Pachnis, V. (1996b). Common origin and developmental dependence on c-ret of subsets of enteric and sympathetic neuroblasts. *Development* **122**, 349–358.

Durston, A.J., Timmermans, J.P., Hage, W.J., Hendriks, H.F., de Vries, N.J., Heideveld, M. and Nieuwkoop, P.D. (1989). Retinoic acid causes an anteroposterior transformation in the developing central nervous system. *Nature* **340**, 140–144.

Duttlinger, R., Manova, K., Chu, T.Y., Gyssler, C., Zelenetz, A.D., Bachvarova, R.F. and Besmer, P. (1993). *W-sash* affects positive and negative elements controlling *c-kit* expression: ectopic *c-kit* expression at sites of kit-ligand expression affects melanogenesis. *Development* **118**, 705–717.

Eagleson, G.W., Ferreiro, B. and Harris, W.A. (1995). Fate of the anterior neural ridge and the morphogenesis of the *Xenopus* forebrain. *J. Neurobiol.* **28**, 146–158.

Eastlick, H.L. (1939). The point of origin of the melanophores in chick embryos as shown by means of limb bud transplants. *J. Exp. Zool.* **82**, 131–157.

Eastlick, H.L. and Wortham, R.A. (1946). An experimental study on the feather-pigmenting and subcutaneous melanophores in the silkie fowl. *J. Exp. Zool.* **103**, 233–258.

Ebendal, T. (1976). The relative roles of contact inhibition and contact guidance in orientation of axons extending on aligned collagen fibrils *in vitro*. *Exp. Cell Res.* **98**, 159–169.

Ebendal, T. (1977). Extracellular matrix fibrils and cell contacts in the chick embryo. Possible roles in orientation of cell migration and axon extension. *Cell Tissue Res.* **175**, 439–458.

Ebendal, T. and Hedlund, K.O. (1974). Histology of the chick embryo trigeminal ganglion and initial effects of its cultivation with and without nerve growth factor. *Zoon* **2**, 25–35.

Ebendal, T., Olson, L., Seiger, A. and Hedlund, K.O. (1980). Nerve growth factors in the rat iris. *Nature* **286**, 25–28.

Eccleston, P.A., Jessen, K.R. and Mirsky, R. (1987). Control of peripheral glial cell proliferation: a comparison of the division rates of enteric glia and Schwann cells and their response to mitogens. *Dev. Biol.* **124**, 409–417.

Echelard, Y., Vassileva, G. and McMahon, A.P. (1994). Cis-acting regulatory sequences governing *Wnt-1* expression in the developing mouse CNS. *Development* **120**, 2213–2224.

Eckenstein, F.P. (1994). Fibroblast growth factors in the nervous system. *J. Neurobiol.* **25**, 1467–1480.

Eckenstein, F.P., Esch, F., Holbert, T., Blacher, R.W. and Nishi, R. (1990). Purification and characterization of a trophic factor for embryonic peripheral neurons: comparison with fibroblast growth factors. *Neuron* **4**, 623–631.

Eckenstein, F., Woodward, W.R. and Nishi, R. (1991). Differential localization and possible functions of aFGF and bFGF in the central and peripheral nervous systems. *Ann. N.Y. Acad. Sci.* **638**, 348–360.

Edery, P., Attié, T., Amiel, J., Pelet, A., Eng, C., Hofstra, R.M.W., Martelli, H., Bidaud, C., Munnich, A. and Lyonnet, S. (1996). Mutation of the endothelin-3 gene in the Waardenburg–Hirschsprung disease (Shah–Waardenburg syndrome). *Nat. Genet.* **12**, 442–444.

Ehinger, B. and Falck, B. (1970). Uptake of some catecholamines and their precursors into neurons of the rat ciliary ganglion. *Acta Physiol. Scand.* **78**, 133–141.

Eichmann, A., Grapin-Botton, A., Kelly, L., Graf, T., Le Douarin, N.M. and Sieweke, M. (1997). The expression pattern of the *mafB/kr* gene in birds and mice reveals that the *Kreisler* phenotype does not represent a null mutant. *Mech. Dev.* **65**, 111–122.

Eisen, J.S. and Weston, J.A. (1993). Development of the neural crest in the Zebrafish. *Dev. Biol.* **159**, 50–59.

Eisenbarth, G.S., Walsh, F.S. and Nirenberg, M. (1979). Monoclonal antibody to a plasma membrane antigen of neurons. *Proc. Natl. Acad. Sci. USA* **76**, 4913–4917.

Ellis, H.M. and Horvitz, H.R. (1986). Genetic control of programmed cell death in the nematode *C. elegans*. *Cell* **44**, 817–829.

Ellis, J., Liu, Q.R., Breitman, M., Jenkins, N.A., Gilbert, D.J., Copeland, N.G., Tempest, H.V., Warren, S., Muir, E., Schilling, H., Fletcher, F.A., Ziegler, S.F. and Rogers, J.H. (1995). Embryo brain kinase: a novel gene of the *eph/elk* receptor tyrosine kinase family. *Mech. Dev.* **52**, 319–341.

Elshamy, W.M. and Ernfors, P. (1996). A local action of neurotrophin-3 prevents the death of proliferating sensory neuron precursor cells. *Neuron* **16**, 963–972.

Elshamy, W.M., Linnarsson, S., Lee, K.-F., Jaenisch, R. and Ernfors, P. (1996). Prenatal and postnatal requirements of NT-3 for sympathetic neuroblast survival and innervation of specific targets. *Development* **122**, 491–500.

Enemar, A., Falck, B. and Hakanson, R. (1965). Observations on the appearance of norepinephrine in the sympathetic nervous system of the chick embryo. *Dev. Biol.* **11**, 268–283.

Eph Nomenclature Committee (1997). Unified nomenclature for Eph family receptors and their ligands, the Ephrins. *Cell* **90**, 403–404.

Epperlein, H.H. and Löfberg, J. (1990). The development of the larval pigment patterns in *Triturus alpestris* and *Ambystoma mexicanum*. *Adv. Anat. Embryol. Cell Biol.* **118**, 1–101.

Epstein, D.J., Vekemans, M. and Gros, P. (1991). *Splotch (Sp2H)*, a mutation affecting development of the mouse neural tube, shows a deletion within the paired homeodomain of *Pax-3*. *Cell* **67**, 767–774.

Eranko, O. (1978). Small intensely fluorescent (SIF) cells and nervous transmission in sympathetic ganglia. *Annu. Rev. Pharmacol. Toxicol.* **18**, 417–430.

Erickson, C. (1987). Behavior of neural crest cells on embryonic basal laminae. *Dev. Biol.* **120**, 38–49.

Erickson, C.A. (1993a). From the crest to the periphery: control of pigment cell migration and lineage segregation. *Pigment Cell Res.* **6**, 336–347.

Erickson, C.A. (1993b). Morphogenesis of the avian trunk neural crest: use of morphological techniques in elucidating the process. *Microsc. Res. Tech.* **26**, 329–351.

Erickson, C.A. and Goins, T.L. (1995). Avian neural crest cells can migrate in the dorsolateral path only if they are specified as melanocytes. *Development* **121**, 915–924.

Erickson, C.A. and Perris, R. (1993). The role of cell–cell and cell matrix interactions in the morphogenesis of the neural crest. *Dev. Biol.* **159**, 60–74.

Erickson, C.A. and Turley, E.A. (1983). Substrata formed by combinations of extracellular matrix components alter neural crest cell mobility *in vitro. J. Cell Sci.* **61**, 299–323.

Erickson, C.A., Tosney, K.W. and Weston, J.A. (1980). Analysis of migratory behavior of neural crest and fibroblastic cells in embryonic tissues. *Dev. Biol.* **77**, 143–156.

Erickson, C.A., Loring, J.F. and Lester, S.M. (1989). Migratory pathways of HNK-1-immunoreactive neural crest cells in the rat embryo. *Dev. Biol.* **134**, 112–118.

Erickson, C.A., Duong, T.D. and Tosney, K.W. (1992). Descriptive and experimental analysis of the dispersion of neural crest cells along the dorsolateral path and their entry into ectoderm in the chick embryo. *Dev. Biol.* **151**, 251–272.

Ericson, J., Muhr, J., Placzek, M., Lints, T., Jessell, T.M. and Edlund, T. (1995). *Sonic hedgehog* induces the differentiation of ventral forebrain neurons: a common signal for ventral patterning within the neural tube. *Cell* **81**, 747–756.

Ernfors, P., Wetmore, C., Olson, L. and Persson, H. (1990). Identification of cells in rat brain and peripheral tissues expressing mRNA for members of the nerve growth factor family. *Neuron* **5**, 511–526.

Ernfors, P., Lee, K.-F. and Jaenisch, R. (1994a). Mice lacking brain-derived neurotrophic factor develop with sensory deficits. *Nature* **368**, 147–150.

Ernfors, P., Lee, K.-F., Kucera, J. and Jaenisch, R. (1994b). Lack of neurotrophin-3 leads to deficiencies in the peripheral nervous system and loss of limb proprioceptive afferents. *Cell* **77**, 503–512.

Ernsberger, U. and Rohrer, H. (1988). Neuronal precursor cells in chick dorsal root ganglia: differentiation and survival *in vitro. Dev. Biol.* **126**, 420–432.

Ernsberger, U., Sendtner, M. and Rohrer, H. (1989). Proliferation and differentiation of embryonic chick sympathetic neurons: effects of ciliary neurotrophic factors. *Neuron* **2**, 1275–1284.

Ernsberger, U., Patzke, H., Tissier-Seta, J.P., Reh, T., Goridis, C. and Rohrer, H. (1995). The expression of tyrosine hydroxylase and the transcription factors cPhox-2 and Cash-1: evidence for distinct inductive steps in the differentiation of chick sympathetic precursor cells. *Mech. Dev.* **52**, 125–136.

Fairbairn, L.J., Cowling, G.J., Reipert, B.M. and Dexter, T.M. (1993). Suppression of apoptosis allows differentiation and development of a multipotent hemopoietic cell line in the absence of added growth factors. *Cell* **74**, 823–832.

Fairman, C.L., Clagett-Dame, M., Lennon, V.A. and Epstein, M.L. (1995). Appearance of neurons in the developing chick gut. *Dev. Dyn.* **204**, 193–201.

Falck, B. (1962). Observations on the possibilities of the cellular localization of monoamines by a fluorescence method. *Acta Physiol. Scand.* **56** (Suppl. 197), 1–25.

Falck, B., Hillarp, N.A., Thieme, G. and Torp, A. (1962). Fluorescence of catecholamines and related compounds condensed with formaldehyde. *J. Histochem. Cytochem.* **10**, 348–354.

Fan, C.-M. and Tessier-Lavigne, M. (1994). Patterning of mammalian somites by surface ectoderm and notochord: evidence for sclerotome induction by a hedgehog homolog. *Cell* **79**, 1175–1186.

Fan, C.-M., Porter, J.A., Chiang, C., Chang, D.T., Beachy, P.A. and Tessier-Lavigne, M. (1995). Long-range sclerotome induction by *Sonic hedgehog*: direct role of the amino-terminal cleavage product and modulation by the cyclic AMP signaling pathway. *Cell* **81**, 457–465.

Fan, G. and Katz, D.M. (1993). Non-neuronal cells inhibit catecholaminergic differentiation of primary sensory neurons: role of leukemia inhibitory factor. *Development* **118**, 83–93.

Fariñas, I., Jones, K.R., Backus, C., Wang, X.-Y. and Reichardt, L.F. (1994). Severe sensory and sympathetic deficits in mice lacking neurotrophin-3. *Nature* **369**, 658–661.

Fariñas, I., Yoshida, C.K., Backus, C. and Reichardt, L.F. (1996). Lack of neurotrophin-3 results in death of spinal sensory neurons and premature differentiation of their precursors. *Neuron* **17**, 1065–1078.

Farooqui, J.Z., Medrano, E.E., Abdel-Malek, Z. and Nordlund, J. (1993). The expression of proopiomelanocortin and various POMC-derived peptides in mouse and human skin. *Ann. N. Y. Acad. Sci.* **680**, 508–510.

Federoff, H.J., Grabczyk, E. and Fishman, M.C. (1988). Dual regulation of GAP-43 gene expression by nerve growth factor and glucocorticoids. *J. Biol. Chem.* **263**, 19290–19295.

Feijen, A., Goumans, M.J. and Van Den Eijnden-Van Raaij, A.J. (1994). Expression of activin subunits, activin receptors and follistatin in postimplantation mouse embryos suggests specific developmental functions for different activins. *Development* **120**, 3621–3637.

Fernholm, M. (1971). On the development of the sympathetic chain and the adrenal medulla in the mouse. *Z. Anat. Entwicklungsgesch.* **133**, 305–317.

Feulgen, R. and Rossenbeck, H. (1924). Mikroskopisch-chemischer Nachweis einer Nukleinsäure von Typus der Thymonukleinsäure und die darauf beruhende elektive Färbung von Zellkernen in mikroskopischen Präparaten. *Hoppe-Seylers Z. Physiol. Chem.* **135**, 203–248.

Feyrter, F. (1938). *Uber diffüse endokrine epitheliale Organe.* Barth JA, Leipzig.

Figdor, M.C. and Stern, C.D. (1993). Segmental organization of embryonic diencephalon. *Nature* **363**, 630–634.

Finkelstein, R. and Perrimon, N. (1991). The molecular genetics of head development in *Drosophila melanogaster*. *Development* **112**, 899–912.

Fitzpatrick, T.B. and Breathnach, A.S. (1963). Das epidermale Melanin-Einheit System. *Derm. Wochenschr.* **147**, 481–489.

Fjose, A., Izpisua-Belmonte, J.C., Fromental-Ramain, C. and Duboule, D. (1994). Expression of the Zebrafish gene *Hlx-1* in the prechordal plate and during CNS development. *Development* **120**, 71–81.

Fjose, A., Weber, U. and Mlodzik, M. (1995). A novel vertebrate svp-related nuclear receptor is expressed as a step gradient in developing rhombomeres and is affected by retinoic acid. *Mech. Dev.* **52**, 233–246.

Flanagan, J.G. and Leder, P. (1990). The kit ligand: a cell surface molecule altered in *steel* mutant fibroblasts. *Cell* **63**, 185–194.

Flanagan, J.G., Chan, D.C. and Leder, P. (1991). Transmembrane form of the kit ligand growth factor is determined by alternative splicing and is missing in the *Sl^d* mutant. *Cell* **64**, 1025–135.

Flenniken, A.M., Gale, N.W., Yancopoulos, G.D. and Wilkinson, D.G. (1996). Distinct and overlapping expression patterns of ligands for Eph-related receptor tyrosine kinases during mouse embryogenesis. *Dev. Biol.* **179**, 382–401.

Fontaine, J. (1973). Development of the carotid body and the ultimobranchial body in birds. Monoamine content and L-dopa uptake capacity in glomic and calcitonin cells during embryonic development. *Arch. Anat. Microsc. Morphol. Exp.* **62**, 89–100.

Fontaine, J. (1979). Multistep migration of calcitonin cell precursors during ontogeny of the mouse pharynx. *Gen. Comp. Endocrinol.* **37**, 81–92.

Fontaine, J. and Le Douarin, N.M. (1977). Analysis of endoderm formation in the avian blastoderm by the use of quail–chick chimaeras. The problem of the neurectodermal origin of the cells of the APUD series. *J. Embryol. Exp. Morphol.* **41**, 209–222.

Fontaine, J., Le Lièvre, C. and Le Douarin, N.M. (1977). What is the developmental fate of the neural crest cells which migrate into the pancreas in the avian embryo? *Gen. Comp. Endocrinol.* **33**, 394–404.

Fontaine-Pérus, J., Le Lièvre, C. and Dubois, M.P. (1980). Do neural crest cells in the pancreas differentiate into somatostatin-containing cells? *Cell Tissue Res.* **213**, 293–299.

Fontaine-Pérus, J., Chanconie, M., Polak, J.M. and Le Douarin, N.M. (1981). Origin and development of VIP and substance P containing neurons in the embryonic avian gut. *Histochemistry* **71**, 313–323.

Fontaine-Pérus, J.C., Chanconie, M. and Le Douarin, N.M. (1982). Differentiation of peptidergic neurones in quail–chick chimaeric embryos. *Cell Diff.* **11**, 183–193.

Fontaine-Pérus, J., Chanconie, M. and Le Douarin, N.M. (1985). Embryonic origin of substance P containing neurons in cranial and spinal sensory ganglia of the avian embryo. *Dev. Biol.* **107**, 1–12.

Fontaine-Pérus, J., Chanconie, M. and Le Douarin, N.M. (1988). Developmental potentialities in the nonneuronal population of quail sensory ganglia. *Dev. Biol.* **128**, 359–375.

Forbes, M.E., Curtis, M.L. and Maxwell, G.D. (1993). Temporal restriction of neural crest development *in vitro*. Changes in the adrenergic differentiation of neural crest clusters in the presence and absence of an overlay of reconstituted basement membrane-like matrix. *Exp. Neurol.* **120**, 114–122.

Forehand, C.J., Ezerman, E.B., Rubin, E. and Glover, J.C. (1994). Segmental patterning of rat and chicken sympathetic preganglionic neurons: correlation between soma position and axon projection pathway. *J. Neurosci.* **14**, 231–241.

Francis-West, P.H., Tatla, T. and Brickell, P.M. (1994). Expression patterns of the bone morphogenetic protein genes Bmp-4 and Bmp-2 in the developing chick face suggest a role in outgrowth of the primordia. *Dev. Dyn.* **201**, 168–178.

Frank, E. and Sanes, J.R. (1991). Lineage of neurons and glia in chick dorsal root ganglia: Analysis *in vivo* with a recombinant retrovirus. *Development* **111**, 895–908.

Fraser, S.E. and Bronner-Fraser, M. (1991). Migrating neural crest cells in the trunk of the avian embryo are multipotent. *Development* **112**, 913–920.

Fraser, S., Keynes, R. and Lumsden, A. (1990). Segmentation in the chick embryo hindbrain is defined by cell lineage restrictions. *Nature* **344**, 431–435.

Fredette, B.J., Miller, J. and Ranscht, B. (1996). Inhibition of motor axon growth by T-cadherin substrata. *Development* **122**, 3163–3171.

Friedrich, G. and Soriano, P. (1991). Promoter traps in embryonic stem cells: a genetic screen to identify and mutate developmental genes in mice. *Genes Dev.* **5**, 1513–1523.

Frohman, M.A., Martin, G.R., Cordes, S.P., Halamek, L.P. and Barsh, G.S. (1993). Altered rhombomere-specific gene expression and hyoid bone differentiation in the mouse segmentation mutant, *kreisler (kr)*. *Development* **117**, 925–936.

Frost-Mason, S.K., Morrison, R. and Mason, K.A. (1994). Pigmentation. In *Biology of Amphibia*, eds. G. Heatwole and H. Berthalmus, Chapter 3. Surrey Beatty and Sons, Sydney.

Fukiishi, Y. and Morriss-Kay, G.M. (1992). Migration of cranial neural crest cells to the pharyngeal arches and heart in rat embryos. *Cell Tissue Res.* **268**, 1–8.

Fukuzawa, T. and Bagnara, J.T. (1989). Control of melanoblast differentiation in amphibia by alpha-melanocyte stimulating hormone, a serum melanization factor, and a melanization inhibiting factor. *Pigment Cell Res.* **2**, 171–181.

Fukuzawa, T. and Ide, H. (1986). Further studies on the melanophores of periodic albino mutant of *Xenopus laevis*. *J. Embryol. Exp. Morphol.* **91**, 65–78.

Fukuzawa, T. and Ide, H. (1988). A ventrally localized inhibitor of melanization in *Xenopus laevis* skin. *Dev. Biol.* **129**, 25–36.

Furness, J.B. and Costa, M. (1971). Morphology and distribution of intrinsic adrenergic neurons in the proximal colon of the guinea-pig. *Z. Zellforsch.* **120**, 346–363.

Furshpan, E.J., MacLeish, P.R., O'League, P.H. and Potter, D.D. (1976). Chemical transmission between rat sympathetic neurons and cardiac myocytes developing in microcultures: evidence for cholinergic, adrenergic and dual-function neurons. *Proc. Natl. Acad. Sci. USA* **73**, 4225–4229.

Furumura, M., Sakai, C., Abdel-Malek, Z., Barsh, G.S. and Hearing, V.J. (1996). The interaction of agouti signal protein and melanocyte stimulating hormone to regulate melanin formation in mammals. *Pigment Cell Res.* **9**, 191–203.

Gabella, G. (1976). *Structure of the Autonomic Nervous System.* Chapman and Hall, London.

Gabella, G. (1989). Development of smooth muscle: ultrastructural study in the chick embryo gizzard. *Anat. Embryol.* **180**, 213–226.

Gaese, F., Kolbeck, R. and Barde, Y.-A. (1994). Sensory ganglia require neurotrophin-3 early in development. *Development* **120**, 1613–1619.

Gagliardini, V., Fernandez, P.A., Lee, R.K.K., Drexler, H.C., Rotello, R.J., Fishman, M.C. and Yuan, J. (1994). Prevention of vertebrate neuronal death by the *crmA* gene. *Science* **263**, 826–828.

Gait, G.C. and Farbman, A. (1973). The chicken trigeminal ganglion. II. Fine structure of neurons during development. *J. Morphol.* **141**, 57–76.

Gale, N.W., Holland, S.J., Valenzuela, D.M., Flenniken, A., Pan, L., Ryan, T.E., Henkemeyer, M., Strebhardt, K., Hirai, H., Wilkinson, D.G., Pawson, T., Davis, S. and Yancopoulos, G.D. (1996). Eph receptors and ligands comprise two major specificity subclasses and are reciprocally compartmentalized during embryogenesis. *Neuron* **17**, 9–19.

Galileo, D.D., Gray, G.E., Owens, G.C., Majors, J. and Sanes, J.R. (1990). Neurons and glia arise from a common progenitor in chicken optic tectum: demonstration with two retroviruses and cell-type specific antibodies. *Proc. Natl. Acad. Sci. USA* **87**, 458–462.

Gallien L. and Durocher, M. (1957). Table chronologique du développement chez *Pleurodeles waltlii* Michah. *Bull. Biol. Fr. Belg.* **91**, 97–114.

Ganju, P., Shigemoto, K., Brennan, J., Entwistle, A. and Reith, A.D. (1994). The Eck receptor tyrosine kinase is implicated in pattern formation during gastrulation, hindbrain segmentation and limb development. *Oncogene* **9**, 1613–1624.

Gans, C. (1987). The neural crest: a spectacular invention. In *Developmental and Evolutionary Aspects of the Neural Crest*, ed. P.F.A. Maderson, pp. 361–379. John Wiley & Sons, New York.

Gans, C. and Northcutt, R.G. (1983). Neural crest and the origin of vertebrates. A new head. *Science* **220**, 268–274.

Garcia-Arraras, J.E., Chanconie, M. and Fontaine-Pérus, J. (1984). *In vivo* and *in vitro* development of somatostatin-like immunoreactivities in the peripheral nervous system of quail embryos. *J. Neurosci.* **4**, 1549–1558.

Garcia-Arraras, J.E., Chanconie, M., Ziller, C. and Fauquet, M. (1987). *In vivo* and *in vitro* expression of vasointestinal active polypeptide-like immunoreactivity by neural crest derivatives. *Dev. Brain Res.* **33**, 255–265.

Garcia-Arraras, J.E., Lugo-Chinchilla, A.M. and Chevere-Colon, I. (1992). The expression of Neuropeptide Y immunoreactivity in the avian sympathoadrenal system conforms with two models of coexpression development for neurons and chromaffin cells. *Development* **115**, 617–627.

Garcia-Bellido, A. (1975). Genetic control of wing disc development in *Drosophila*. *CIBA Found. Symp.* **29**, 161–182.

Garcia-Bellido A., Ripoll, P. and Morata, G. (1973). Developmental compartmentalisation of the wing disk of *Drosophila*. *Nature New Biol.* **245**, 251–253.

Gardner, C.A. and Barald, K.F. (1992). Expression patterns of *engrailed*-like proteins in the chick embryo. *Dev. Dyn.* **193**, 370–388.

Gardner, C.A., Darnell, D.K., Poole, S.J., Ordahl, C.P. and Barald, K.F. (1988). Expression of an *engrailed-like* gene during development of the early embryonic chick nervous system. *J. Neurosci. Res.* **21**, 426–437.

Gardner, J.M., Nakatsu, Y., Gondo, Y., Lee, S., Lyon, M.F., King, R.A. and Brilliant, M.H. (1992). The mouse pink-eyed dilution gene: association with human Präder–Willi and Angelman syndromes. *Science* **257**, 1121–1124.

Gariepy, C.E., Cass, D.T. and Yanagisawa, M. (1996). Null mutation of endothelin receptor type B gene in spotting lethal rats causes aganglionic megacolon and white coat color. *Proc. Natl. Acad. Sci. USA* **93**, 867–872.

Gaunt, S.J., Sharpe, P.T. and Duboule, D. (1988). Spatially restricted domains of homeo-gene transcripts in mouse embryos: relation to a segmented body plan. *Dev. Suppl.* **104**, 169–181.

Gaunt, S.J., Blum, M. and De Robertis, E. M. (1993). Expression of the mouse *goosecoid* gene during mid-embryogenesis may mark mesenchymal cell lineages in the developing head, limbs and body wall. *Development* **117**, 769–778.

Gawantka, V., Joos, T.O. and Hausen, P. (1994). A beta 1-integrin associated alpha-chain is differentially expressed during Xenopus embryogenesis. *Mech. Dev.* **47**, 199–211.

Geissler, E.N., McFarland, E.C. and Russell, E.S. (1981). Analysis of pleiotropism at the dominant white-spotting (*W*) locus of the house mouse: a description of ten new *W* alleles. *Genetics* **97**, 337–361.

Geissler, E.N., Ryan, M.A. and Housman, D.E. (1988). The dominant-white spotting (*W*) locus of the mouse encodes the *c-kit* proto-oncogene. *Cell* **55**, 185–192.

Gendron-Maguire, M., Mallo, M., Zhang, M. and Gridley, T. (1993). *Hoxa-2* mutant mice exhibit homeotic transformation of skeletal elements derived from cranial neural crest. *Cell* **75**, 1317–1331.

Gershon, M.D. (1995). Neural crest development: do developing enteric neurons need endothelins? *Curr. Biol.* **5**, 601–604.

Gershon, M.D. (1997). Genes and lineages in the formation of the enteric nervous system. *Curr. Opin. Neurobiol.* **7**, 101–109.

Gershon, M.D., Rothman, T.P., Joh, T.H. and Teitelman, G.N. (1984). Transient and differential expression of aspects of the catecholaminergic phenotype during development of the fetal bowel of rats and mice. *J. Neurosci.* **4**, 2269–2280.

Gershon, M.D., Chalazonitis, A. and Rothman, T.P. (1993). From neural crest to bowel: development of the enteric nervous system. *J. Neurobiol.* **24**, 199–214.

Geschwind, I.I. (1966). Change in hair color in mice induced by injection of alpha-MSH. *Endocrinology* **79**, 1165–1167.

Geschwind, I.I., Huseby, R.A. and Nishioka, R. (1972). The effect of melanocyte-stimulating hormone on coat color in the mouse. *Rec. Prog. Horm. Res.* **28**, 91–130.

Ghysen, A. and Dambly-Chaudière, C. (1988). From DNA to form: the *achaete–scute* complex. *Genes Dev.* **2**, 495–501.

Gibson, G. and Gehring, W.J. (1988). Head and thoracic transformations caused by ectopic expresion of *antennapedia* during *Drosophila* development. *Development* **102**, 657–675.

Giebel, L.B. and Spritz, R.A. (1992). The molecular basis of type-I (tyrosinase-deficient) human oculocutaneous albinism. *Pigment Cell Res.*, Suppl. **2**, 101–106.

Giguère, V., Ong, E.S., Segui, P. and Evans, R.M. (1987). Identification of a receptor for the morphogen retinoic acid. *Nature* **330**, 624–629.

Giguère, V., Lyn, S.., Yip, P., Siu, C.-H. and Amin, S. (1990). Molecular cloning of cDNA encoding a second cellular retinoic acid-binding protein. *Proc. Natl. Acad. Sci. USA* **87**, 6233–6237.

Gilardi-Hebenstreit, P., Nieto, M.A., Frain, M., Mattei, M.G., Chestier, A., Wilkinson, D.G. and Charnay, P. (1992). An Eph-related receptor protein tyrosine kinase gene segmentally expressed in the developing mouse hindbrain. *Oncogene* **7**, 2499–2506.

Gimlich, R.L. and Braun, J. (1985). Improved fluorescent compounds for tracing cell lineage. *Dev. Biol.* **109**, 509–514.

Girdlestone, J. and Weston, J.A. (1985). Identification of early neuronal subpopulations in avian neural crest cell cultures. *Dev. Biol.* **109**, 274–287.

Glass, D.J., Nye, S.H., Hantzopoulos, P., MacChi, M.J., Squinto, S.P., Goldfarb, M. and Yancopoulos, G.D. (1991). Trkb mediates BDNF/NT-3-dependent survival and proliferation in fibroblasts lacking the low affinity NGF receptor. *Cell* **66**, 405–413.

Glimelius, B. and Weston, J.A (1981). Analysis of developmentally homogenous neural crest cell populations *in vitro*. III. Role of culture environment in cluster formation and differentiation. *Cell Differ.* **10**, 57–67.

Gluksmann, A. (1951). Cell death in normal vertebrate development. *Biol. Rev.* **26**, 59–86.

Goldstein, R.S. (1993). Axial level-dependent differences in size of avian dorsal root ganglia are present from gangliogenesis. *J. Neurobiol.* **24**, 1121–1129.

Goldstein, R.S. and Kalcheim, C. (1991). Normal segmentation and size of the primary sympathetic ganglia depend upon the alternation of rostrocaudal properties of the somites. *Development* **112**, 327–334.

Goldstein, R.S., Teillet, M.-A. and Kalcheim, C. (1990). The microenvironment created by grafting rostral half-somites is mitogenic for neural crest cells. *Proc. Natl. Acad. Sci. USA* **87**, 4476–4480.

Goldstein, R.S., Avivi, C. and Geffen, R. (1995). Initial axial level-dependent differences in size of avian dorsal root ganglia are imposed by the sclerotome. *Dev. Biol.* **168**, 214–222.

Gonzalez-Reyes, A. and Morata, G. (1990). The developmental effect of overexpressing a *Ubx* product in Drosophila embryos is dependent on its interactions with other homeotic products. *Cell* **61**, 515–522.

Goodman, C.S. and Shatz, C.J. (1993). Developmental mechanisms that generate precise patterns of neuronal connectivity. *Cell* **72** (Suppl.) 77–98.

Goodrich, E.S. (1930). *Studies on Structure and Development of Vertebrates*. Macmillan, London.

Gordon, M. (1953). *Pigment Cell Growth*. Academic Press, New York.

Goronowitsch, N. (1892). Die axiale und die laterale Kopfmentamerie der Vögelembryonen. Die Rolle der sog. "Ganglienleisten" im Aufbaue der Nervenstämme. *Anat. Anz.* **7**, 454–464.

Goronowitsch, N. (1893). Weiters über die ektodermale Entstehung von Skeletanlagen im Kopfe der Wirbeltiere. *Morphol. Jahrb.* **20**, 425–428.

Gospodarowicz, D. (1991). Biological activities of fibroblast growth factors. *Ann. N. Y. Acad. Sci.* **638**, 1–8.

Gotz, R., Koster, R., Winkler, C., Raulf, F., Lottspeich, F., Schartl, M. and Thoenen, H. (1994). Neurotrophin-6 is a new member of the nerve growth factor family. *Nature* **372**, 266–269.

Goulding, M.D, Chalepakis, G., Deutch, U., Erselius, J.R. and Gruss, P. (1991). Pax-3, a novel murine DNA binding protein expressed during early neurogenesis. *EMBO J.* **10**, 1135–1147.

Gown, A.M., Vogel, A.M., Hoak, D., Gough, F. and McNutt, M.A. (1986). Monoclonal antibodies specific for melanocytic tumors distinguish subpopulations of melanocytes. *Am. J. Pathol.* **123**, 195–203.

Graham, A. and Lumsden, A. (1996). Interactions between rhombomeres modulate *Krox-20* and *follistatin* expression in the chick embryo hindbrain. *Development* **122**, 473–480.

Graham, A., Papalopulu, N. and Krumlauf, R. (1989). The murine and Drosophila homeobox gene complexes have common features of organization and expression. *Cell* **57**, 367–378.

Graham, A., Heyman, I. and Lumsden, A. (1993). Even-numbered rhombomeres control the apoptotic elimination of neural crest cells from odd-numbered rhombomeres in the chick hindbrain. *Development* **119**, 233–245.

Graham, A., Francis-West, P., Brickell, P. and Lumsden, A. (1994). The signalling molecule BMP4 mediates apoptosis in the rhombencephalic neural crest. *Nature* **372**, 684–686.

Grant, W.C., Jr. (1961). Special aspects of the metamorphic process: second metamorphosis. *Am. Zool.* **1**, 163–171.

Gräper, L. (1913). Die Rhombomeren und ihre Nervenbeziehungen. *Arch. Mikrosk. Anat.* **83**, 371–426.

Grapin-Botton, A., Bonnin, M.-A., McNaughton, L.A., Krumlauf, R. and Le Douarin, N.M. (1995). Plasticity of transposed rhombomeres: *Hox* gene induction is correlated with phenotypic modifications. *Development* **121**, 2707–2721.

Grapin-Botton, A., Bonnin, M.-A. and Le Douarin, N.M. (1997). *Hox* gene induction in the neural tube depends on three parameters: competence, signal supply and paralogue group. *Development* **124**, 849–859.

Grapin-Botton, A., Bonnin, M.-A., Sieweke, M. and Le Douarin, N.M. (1998). Defined concentrations of a posteriorizing signal are critical for *MafB/Kreisler* segmental expression in the hindbrain. *Development* **125**, 1173–1181.

Graus, F., Elkon, K.B., Cordon Cardo, C. and Posner, J.B. (1986). Pure sensory neuropathy. Restriction of the Hu antibody to the sensory neuropathy associated with small cell carcinoma of the lung. *Neurologia* **1**, 11–15.

Greene, L.A. and Tischler, A.S. (1976). Establishment of a noradrenergic clonal cell line of rat adrenal pheochromocytoma cells which respond to nerve growth factor. *Proc. Natl. Acad. Sci. USA* **73**, 2424–2428.

Greenspan, R.J. (1990). The *Notch* gene, adhesion, and developmental fate in the Drosophila embryo. *New Biol.* **2**, 595–600.

Greenstein-Baynash, A., Hosoda, K., Giaid, A., Richardson, J.A., Emoto, N., Hammer, R.E. and Yanagisawa, M. (1994). Interaction of endothelin-3 with endothelin-B receptor is essential for development of neural crest-derived melanocytes and enteric neurons: missense mutation of endothelin-3 gene in *lethal spotting* mice. *Cell* **79**, 1277–1285.

Gregory, R.A. and Tracy, H.J. (1964). The constitution and properties of two gastrins extracted from hog antral mucosa. *Gut* **5**, 103–117.

Grillo, M.A. (1966). Electron microscopy of sympathetic tissues. *Pharmacol. Rev.* **18**, 387–399.

Groves, A.K., George, K.M., Tissier-Seta, J.-P., Engel, J.D., Brunet, J.-F. and Anderson, D.J. (1995). Differential regulation of transcription factor gene expression and phenotypic markers in developing sympathetic neurons. *Development* **121**, 887–901.

Grueneberg, D.A., Natesan, S., Alexandre, C. and Gilman, M.Z. (1992). Human and Drosophila homeodomain proteins that enhance the DNA-binding activity of serum response factor. *Science* **257**, 1089–1095.

Gruss, P. and Walther, C. (1992). *Pax* in development. *Cell* **69**, 719–722.

Guillemin, R. (1978a). Biochemical and physiological correlates of hypothalamic peptides. The new endocrinology of the neuron. In *The Hypothalamus*, eds. S. Reichlin, R.J. Baldessarini and J.B. Martin, pp. 155–194. Raven Press, New York.

Guillemin, R. (1978b). Peptides in the brain: the new endocrinology of the neuron. *Science* **202**, 390–402.

Guillemot, F. and Joyner, A.L. (1993). Dynamic expression of the murine *Achaete–scute* homologue *Mash-1* in the developing nervous system. *Mech. Dev.* **42**, 171–185.

Guillemot, F., Lo, L.-C., Johnson, J.E., Auerbach, A. and Anderson, D.J. (1993). Mammalian *achaete–scute* homolog-1 is required for the early development of olfactory and autonomic neurons. *Cell* **75**, 463–476.

Guillory, G. and Bronner-Fraser, M. (1986). An *in vitro* assay for neural crest cell migration through the somites. *J. Embryol. Exp. Morphol.* **98**, 85–97.

Gustafson, A.-E., Dencker, L. and Eriksson, U. (1993). Non-overlapping expression of CRBP I and CRABP I during pattern formation of limbs and craniofacial structures in the early mouse embryo. *Development* **117**, 451–460.

Guthrie, S. and Lumsden, A. (1991). Formation and regeneration of rhombomere boundaries in the developing chick hindbrain. *Development* **112**, 221–229.

Gutmann, D.H. and Collins, F.S. (1992). Recent progress toward understanding the molecular biology of von Recklinghausen neurofibromatosis. *Ann. Neurol.* **31**, 555–561.

Guz, Y., Montminy, M.R., Stein, R., Leonard, J., Gamer, L.W., Wright, C.V. and Teitelman, G. (1995). Expression of murine STF-1, a putative insulin gene transcription factor, in beta cells of pancreas, duodenal epithelium and pancreatic exocrine and endocrine progenitors during ontogeny. *Development* **121**, 11–18.

Hadley, M.E. and Quevedo W.C., Jr. (1966). Vertebrate epidermal melanin unit. *Nature* **209**, 1334–1335.

Hafen, E., Levine, M. and Gehring, W.J. (1984). Regulation of *Antennapedia* transcript distribution by the *bithorax* complex in Drosophila. *Nature* **307**, 287–289.

Halaban, R., Moellmann, G., Tamura, A., Kwon, B.S., Kuklinska, E., Pomerantz, S.H. and Lerner, A.B. (1988). Tyrosinases of murine melanocytes with mutations at the albino locus. *Proc. Natl. Acad. Sci. USA* **85**, 7241–7245.

Halder, G., Callaerts, P. and Gehring, W.J. (1995). Induction of ectopic eyes by targeted expression of the eyeless gene in *Drosophila. Science* **267**, 1788–1792.

Halfter, W., Chiquet-Ehrismann, R. and Tucker, R.P. (1989). The effect of tenascin and embryonic basal lamina on the behavior and morphology of neural crest cells *in vitro. Dev. Biol.* **132**, 14–25.

Hall, A.K. and MacPhedran, S.E. (1995). Multiple mechanisms regulate sympathetic neuronal phenotype. *Development* **121**, 2361–2371.

Hall, B.K. (1983). Embryogenesis: cell–tissue interactions. In *Skeletal Research: An Experimental Approach*, eds. A.S. Kunin and D.J. Simmons, pp. 53–87. Academic Press, New York.

Hall, B.K. (1984). Developmental mechanisms underlying the formation of atavisms. *Biol. Rev.* **59**, 89–124.

Hall, B.K. (1987). Tissue interactions in the development and evolution of the vertebrate head. In *Developmental and Evolutionary Aspects of the Neural Crest*, ed. P.F.A. Maderson, pp. 215–259. John Wiley & Sons, New York.

Hall, B.K. (1990). Evolutionary issues in craniofacial biology. *Cleft Palate J.* **27**, 95–100.

Hall, B.K. (1991). Cellular interactions during cartilage and bone development. *J. Craniofac. Genet. Dev. Biol.* **11**, 238–250.

Hallböök, F., Ibanez, C.F. and Persson, H. (1991). Evolutionary studies of the nerve growth factor family reveal a novel member abundantly expressed in Xenopus ovary. *Neuron* **6**, 845–858.

Hallet, M.-M. and Ferrand, R. (1984). Quail melanoblast migration in two breeds of fowl and in their hybrids: evidence for a dominant genic control of the mesodermal pigment cell pattern through the tissue environment. *J. Exp. Zool.* **230**, 229–238.

Hallonet, M.E., Teillet, M.-A. and Le Douarin, N.M. (1990). A new approach to the development of the cerebellum provided by the quail–chick marker system. *Development* **108**, 19–31.

Haltmeier, H. and Rohrer, H. (1990). Distinct and different effects of the oncogenes *v-myc* and *v-src* on avian sympathetic neurons: retroviral transfer of *v-myc* stimulates neuronal proliferation whereas *v-src* transfer enhances neuronal differentiation. *J. Cell Biol.* **110**, 2087–2098.

Hamburger, V. (1961). Experimental analysis of the dual origin of the trigeminal ganglion in the chick embryo. *J. Exp. Zool.* **148**, 91–124.

Hamburger, V. (1992). History of the discovery of neuronal death in embryos. *J. Neurobiol.* **23**, 1116–1123.

Hamburger, V. and Hamilton, H.L. (1951). A series of normal stages in the development of chick embryo. *J. Morphol.* **88**, 49–92.

Hamburger, V. and Levi-Montalcini, R. (1949). Proliferation, differentiation and degeneration in the spinal ganglia of the chick embryo under normal and experimental conditions. *J. Exp. Zool.* **111**, 457–502.

Hamburger, V., Brunso-Bechtold, J.K. and Yip, J.W. (1981). Neuronal death in the spinal ganglia of the chick embryo and its reduction by nerve growth factor. *J. Neurosci.* **1**, 60–71.

Hammond, W.S. and Yntema, C.L. (1947). Depletions in the thoracolumbar sympathetic system following removal of neural crest in chick. *J. Comp. Neurol.* **86**, 237–265.

Hammond, W. and Yntema, C. (1958). Origin of ciliary ganglia in the chick. *J. Comp. Neurol.* **110**, 367–389.

Hammond, W.S. and Yntema, C.L. (1964). Depletions of pharyngeal arch cartilages following extirpation of cranial neural crest in chick embryos. *Acta Anat.* **56**, 21–34.

Hanaoka, Y. (1967). The effects of posterior hypophysectomy upon the growth and metamorphosis of the tadpole of *Rana pipiens. Gen. Comp. Endocrinol.* **8**, 417–431.

Harrison, R.G. (1935). The origin and development of the nervous system studied by the methods of experimental embryology. *Proc. R. Soc. Lond. Ser. B* **118**, 155–196.

Harrison, R.G. (1938). Die Neuralleiste. *Anat. Anz.* **85**, 3–30.

Harvey, S.C. and Burr, H.S. (1926). The development of the meninges. *Arch. Neurol. Psychiatr.* **15**, 545–565.

Harvey, S.C., Burr, H.S. and Van Campenhout, E. (1933). Development of the meninges. Further experiments. *Arch. Neurol. Psychiatr.* **29**, 683–690.

Haskell, B.E., Stach, R.W., Werrbach-Perez, K. and Perez-Polo, J.R. (1987). Effect of retinoic acid on nerve growth factor receptors. *Cell Tissue Res.* **247**, 67–73.

Hatada, Y. and Stern, C.D. (1994). A fate map of the epiblast of the early chick embryo. *Development* **120**, 2879–2889.

Hauptmann, G. and Gerster, T. (1995). *Pou-2*—a zebrafish gene active during cleavage stages and in the early hindbrain. *Mech. Dev.* **51**, 127–138.

Hay, E.D. (1980). Development of vertebrate cornea. *Int. Rev. Cytol.* **63**, 263–322.

Hay, E.D. and Revel, J.P. (1969). *Fine Structure of the Developing Avian Cornea.* Karger, Basel.

He, X., Treacy, M.N., Simmons, D.M., Ingraham, H.A., Swanson, L.W. and Rosenfeld, M.G. (1989). Expression of a large family of POU-domain regulatory genes in mammalian brain development. *Nature* **340**, 35–41.

Hearing, V.J. and Tsukamoto, K. (1991). Enzymatic control of pigmentation in mammals. *FASEB J.* **5**, 2903–2909.

Heath, L., Wild, A. and Thorogood, P. (1992). Monoclonal antibodies raised against pre-migratory neural crest reveal population heterogeneity during crest development. *Differentiation* **49**, 151–165.

Heller, S., Huber, J., Finn, T.P., Nishi, R. and Rohrer, H. (1993). GPA and CNTF produce similar effects in sympathetic neurons but differ in receptor binding. *Neuroreports* **5**, 357–360.

Heller, S., Finn, T.P., Huber, J., Nishi, R., Geissen, M., Puschel, A.W. and Rohrer, H. (1995). Analysis of function and expression of the chick GPA receptor (GPAR alpha) suggests multiple roles in neuronal development. *Development* **121**, 2681–2693.

Hellmich, H., Kos, L., Cho, E.., Mahou, K. and Zimmer, A. (1996). Embryonic expression of glial cell-derived neurotrophic factor (GDNF) suggests multiple developmental roles in neural differentiation and epithelial-mesenchymal interactions. *Mech. Dev.* **54**, 95–105.

Helms, J.A., Kim, C.H., Hu, D., Minkoff, R., Thaller, C. and Eichele, G. (1997). Sonic hedgehog participates in craniofacial morphogenesis and is down-regulated by teratogenic doses of retinoic acid. *Dev. Biol.* **187**, 25–35.

Hemesath, T.J., Steingrimsson, E., McGill, G., Hansen, M.J., Vaught, J., Hodgkinson, C.A., Arnheiter, H., Copeland, N.G., Jenkins, N.A. and Fisher, D.E. (1994). *microphthalmia*, a critical factor in melanocyte development, defines a discrete transcription factor family. *Genes Dev.* **8**, 2770–2780.

Hemmati-Brivanlou, A. and Melton, D.A. (1994). Inhibition of activin receptor signaling promotes neuralization in Xenopus. *Cell* **77**, 273–281.

Hemmati-Brivanlou, A., Kelly, O.G. and Melton, D.A. (1994). Follistatin, an antagonist of activin, is expressed in the Spemann organizer and displays direct neuralizing activity. *Cell* **77**, 283–295.

Hendry, I.A., Stockel, K., Thoenen, H. and Iversen, L.L. (1974). The retrograde axonal transport of nerve growth factor. *Brain Res.* **68**, 103–121.

Henion, P.D. and Weston, J.A. (1994). Retinoic acid selectively promotes the survival and proliferation of neurogenic precursors in cultured neural crest cell populations. *Dev. Biol.* **161**, 243–250.

Henion, P. and Weston, J.A. (1997). Timing and pattern of cell fate-restrictions in the neural crest lineage. *Development* **124**, 4351–4359.

Henion, P.D., Garner, A.S., Large, T.H. and Weston, J.A. (1995). *trk*C-mediated NT-3 signaling is required for the early development of a subpopulation of neurogenic neural crest cells. *Dev. Biol.* **172**, 603–613.

Henrique, D., Adam, J., Myat, A., Chitnis, A., Lewis, J. and Ish-Horowicz, D. (1995). Expression of a *delta* homologue in prospective neurons in the chick. *Nature* **375**, 787–790.

Herberg, L. and Coleman, D.L. (1977). Laboratory animals exhibiting obesity and diabetes syndromes. *Metabolism* **26**, 59–99.

Hertwig, P. (1944). Die Genese der Hirn- und Gehörorganmissbildungen bei rontgenmurtierten Kreisler-Maüsen. *Z. Konstlehre* **28**, 327–354.

Heuer, J.G., Von Bartheld, C.S., Kinoshita, Y., Evers, P.C. and Bothwell, M. (1990). Alternating phases of FGF receptor and NGF receptor expression in the developing chicken nervous system. *Neuron* **5**, 283–296.

Heumann, R., Korsching, S., Scott, J. and Thoenen, H. (1984). Relationship between levels of nerve growth factor (NGF) and its messenger RNA in sympathetic ganglia and peripheral target tissues. *EMBO J.* **3**, 3183–3189.

Hilber, H. (1943). Experimentelle Studien zum Schicksal des Rumpfganglienleistenmaterials. *Roux Arch. Entwicklungsmech. Org.* **142**, 100–120.

Hill, J., Clarke, J.D.W., Vargesson, N., Jowett, T. and Holder, N. (1995). Exogenous retinoic acid causes specific alterations in the development of the midbrain and hindbrain of the zebrafish embryo including positional respecification of the Mauthner neuron. *Mech. Dev.* **50**, 3–16.

Hill, R.E., Jones, P.F., Rees, A.R., Sime, C.M., Justice, M.J., Copeland, N.G., Jenkins, N.A., Graham, E. and Davidson, D.R. (1989). A new family of mouse homeobox-containing genes: molecular structure, chromosomal location, and developmental expression of *Hox-7.1*. *Genes Dev.* **3**, 26–37.

Hill, R.E., Favor, J., Hogan, B.L., Ton, C.C., Saunders, G.F., Hanson, I.M., Prosser, J., Jordan, T., Hastie, N.D. and Van Heyningen, V. (1991). Mouse small eye results from mutations in a paired-like homeobox-containing gene. *Nature* **354**, 523–525.

Hill, R.J. and Sternberg, P.W. (1993). Cell fate patterning during *C. elegans* vulval development. *Development* **suppl.** 9–18.

Hirai, H., Maru, Y., Hagiwara, K., Nishida, J. and Takaku, F. (1987). A novel putative tyrosine kinase receptor encoded by the *eph* gene. *Science* **238**, 1717–1720.

His, W. (1868). *Untersuchungen über die erste Anlage des Wirbeltierleibes. Die erste Entwicklung des Hühnchens im Ei.* F.C.W. Vogel, Leipzig.

His, W., Jr. (1903). Die Häute und Höhlen des Körpers. *Arch. Anat. Physiol., Anat. Abt.*, 368.

Hockenbery, D., Nunez, G., Milliman, C., Schreiber, R.D. and Korsmeyer, S.J. (1990). Bcl-2 is an inner mitochondrial membrane protein that blocks programmed cell death. *Nature* **348**, 334–336.

Hodgkinson, C.A., Moore, K.J., Nakayama, A., Steingrimsson, E., Copeland, N.G., Jenkins, N.A. and Arnheiter, H. (1993). Mutations at the mouse microphthalmia locus are associated with defects in a gene encoding a novel basic-helix–loop–helix-zipper protein. *Cell* **74**, 395–404.

Hofer, M.M. and Barde, Y.A. (1988). Brain-derived neurotrophic factor prevents neuronal death *in vivo*. *Nature* **331**, 261–262.

Hoffmann, B., Lehmann, J.M., Zhang, X.K., Hermann, T., Graupner, G. and Pfahl, M. (1990). A retinoic acid receptor specific element controls the retinoic acid receptor-beta promotor. *J. Mol. Endocrinol.* **4**, 1734–1743.

Hofmann, C. and Eichele, G. (1994). Retinoids in development. In *The Retinoids, Biology, Chemistry and Medicine*, eds. M.B. Sporn, A.B. Roberts and D.S. Goodman, pp. 387–441. Raven Press, New York.

Hofstra, R.M.W., Osinga, J., Tan-Sindhunata, G., Wu, Y., Kamsteeg, E.-J., Stulp, R.P., Van Ravenswaaij-Arts, C., Majoor-Krakauer, D., Angrist, M., Chakravarti, A., Meijers, C. and Buys, C.H.C.M. (1996). A homozygous mutation in the endothelin-3 gene associated with a combined Waardenburg type 2 and Hirschsprung phenotype (Shah-Waardenburg syndrome). *Nat. Genet.* **12**, 445–447.

Hogan, B.L.M. (1996). Bone morphogenetic proteins: multifunctional regulators of vertebrate development. *Genes Dev.* **10**, 1580–1594.

Hogan, B.L.M., Thaller, C. and Eichele, G. (1992). Evidence that Hensen's node is a site of retinoic acid synthesis. *Nature* **359**, 237–241.

Hohn, A., Leibrock, J., Bailey, K. and Barde, Y.-A. (1990). Identification and characterization of a novel member of the nerve growth factor/brain-derived neurotrophic factor family. *Nature* **344**, 339–341.

Hökfelt, T., Elde, R., Johansson, O., Luft, R. and Arimura, A. (1975a) Immunohistochemical evidence of the presence of somatostatin, a powerful inhibitory peptide, in some primary sensory neurons. *Neurosci. Lett.* **1**, 231–235.

Hökfelt, T., Kellerth, J.O., Nilsson, G. and Pernow, B. (1975b). Substance P: localization in the central nervous system and in some primary sensory neurons. *Science* **190**, 889–890.

Hökfelt, T., Kellerth, J.O., Nilsson, G. and Pernow, B. (1975c). Experimental immunohistochemical studies on the localization and distribution of Substance P in cat primary sensory neurons. *Brain Res.* **100**, 235–252.

Hökfelt, T., Elde, R., Johansson, O., Nilsson, G. and Arimura, A. (1976). Immunohistochemical evidence for separate populations of somatostatin-containing and Substance P-containing primary afferent neurons in the rat. *Neuroscience* **1**, 131–136.

Hökfelt, T., Johansson, O., Ljungdahl, A., Lundberg, J.M. and Schultzberg, M. (1980). Peptidergic neurons. *Nature* **284**, 515–521.

Holder, N. and Hill, J. (1991). Retinoic acid modifies development of the midbrain–hindbrain border and affects cranial ganglion formation in zebrafish embryos. *Development* **113**, 1159–1170.

Holland, P.W., Garcia-Fernandez, J., Williams, N.A. and Sidow, A. (1994). Gene duplications and the origins of vertebrate development. *Development*, Suppl., 125–133.

Hollyday, M., McMahon, J.A. and McMahon, A.P. (1995). *Wnt* expression patterns in chick embryo nervous system. *Mech. Dev.* **52**, 9–25.

Holtfreter, J. (1929). Uber die Aufzucht isolierter Teile des Amphibienkeimes. I. Methode einer Gewebezüchtung *in vivo*. *Roux Arch. Entwicklungsmech. Org.* **117**, 421–510.

Holtfreter, J. (1933). Der Einfluss von Wirtsalter und verschiedenen Organbezirken auf die Differenzierung von angelagerten Gastrulaektoderm. *Roux Arch. Entwicklungsmech. Org.* **127**, 619–775.

Holtfreter, J. (1935). Morphologische Beeinflussung von Urodelenektoderm bei xenoplastischer Transplantation. *Roux' Arch. Entwicklungsmech. Org.* **133**, 367–426.

Holton, B. and Weston, J.A. (1982a). Analysis of glial cell differentiation in peripheral nervous tissue I. S100 accumulation in quail embryo spinal ganglion cultures. *Dev. Biol.* **89**, 64–71.

Holton, B. and Weston, J.A. (1982b). Analysis of glial cell differentiation in peripheral nervous tissue II. Neurons promote S100 synthesis by purified glial precursor cell population. *Dev. Biol.* **89**, 73–81.

Hopwood, N.D., Pluck, A. and Gurdon, J.B. (1989). A Xenopus mRNA related to Drosophila *twist* is expressed in response to induction in the mesoderm and the neural crest. *Cell* **59**, 893–903.

Hornbruch, A. and Wolpert, L. (1986). Positional signalling by Hensen's node when grafted to the chick limb bud. *J. Embryol. Exp. Morphol.* **94**, 257–265.

Hörstadius, S. (1950). *The Neural Crest: Its Properties and Derivatives in the Light of Experimental Research*. Oxford University Press, London.

Hörstadius, S. and Sellman, S. (1941). Experimental studies on the determination of the chondrocranium in *Amblystoma mexicanum*. *Ark. Zool. Stockholm* **33A** (13).

Hörstadius, S. and Sellman, S. (1946). Experimentelle Untersuchungen über die Determination des Knorpeligen Kopfskelettes bei Urodelen. *Nova Acta Soc. Scie. Uppsaliensis, Ser. 4* **13**, 1–170.

Horton, C. and Maden, M. (1995). Endogenous distribution of retinoids during normal development and teratogenesis in the mouse embryo. *Dev. Dyn.* **202**, 313–323.

Hory-Lee, F., Russell, M., Lindsay, R.M. and Frank, E. (1993). Neurotrophin 3 supports the survival of developing muscle sensory neurons in culture. *Proc. Natl. Acad. Sci. USA* **90**, 2613–2617.

Hosoda, K., Hammer, R.E., Richardson, J.A., Greenstein-Baynash, A., Cheung, J.C., Giaid, A. and Yanagisawa, M. (1994). Targeted and natural (piebald-lethal) mutations of endothelin-B receptor gene produce megacolon associated with spotted coat color in mice. *Cell* **79**, 1267–1276.

Hosoya, T., Takizawa, K., Nitta, K. and Hotta, Y. (1995). Glial cells missing: a binary switch between neuronal and glial determination in Drosophila. *Cell* **82**, 1025–1036.

Hotary, K.B. and Tosney, K.W. (1996). Cellular interactions that guide sensory and motor neurites identified in an embryo slice preparation. *Dev. Biol.* **176**, 23–35.

Houillon, C. and Bagnara, J.T. (1996). Insights into pigmentary phenomena provided by grafting and chimera formation in the axolotl. *Pigment Cell Res.* **9**, 281–288.

Howard, M.J. and Bronner-Fraser, M. (1985). The influence of neural tube-derived factors on differentiation of neural crest cells *in vitro*. *J. Neurosci.* **5**, 3302–3309.

Howard, M.J. and Bronner-Fraser, M. (1986). Neural tube-derived factors influence differentiation of neural crest cells *in vitro*: effects on activity of neurotransmitter biosynthetic enzymes. *Dev. Biol.* **117**, 45–54.

Howard, M.J. and Gershon, M.D. (1993). Role of growth in catecholaminergic expression by neural crest cells: *in vitro* effects of transforming growth factor beta 1. *Dev. Dyn.* **196**, 1–10.

Howard, T.D., Paznekas, W.A., Green, E.D., Chiang, L.C., Ma, N., Ortiz de Luna, R.I., Garcia Delgado, C., Gonzalez-Ramos, M., Kline, A.D. and Jabs, E.W.

(1997). Mutations in *TWIST*, a basic helix–loop–helix transcription factor, in Saethre–Chotzen syndrome. *Nat. Genet.* **15**, 36–41.

Huang, E., Nocka, K., Beier, D.R., Chu, T.Y., Buck, J., Lahm, H.W., Wellner, D., Leder, P. and Besmer, P. (1990). The hematopoietic growth factor KL is encoded by the *Sl* locus and is the ligand of the c-*kit* receptor, the gene product of the *W* locus. *Cell* **63**, 225–233.

Huang, E.J., Nocka, K.H., Buck, J. and Besmer, P. (1992). Differential expression and processing of two cell associated forms of the kit-ligand—kl-1 and kl-2. *Mol. Biol. Cell* **3**, 349–362.

Hughes, A. (1957). The development of the primary sensory system in *Xenopus laevis* (Daudin). *J. Anat.* **91**, 323–338.

Hughes, M.J., Lingrel, J.B., Krakowsky, J.M. and Anderson, K.P. (1993). A helix–loop–helix transcription factor-like gene is located at the *mi* locus. *J. Biol. Chem.* **268**, 20687–20690.

Hui, C.C., Slusarski, D., Platt, K.A., Holmgren, R. and Joyner, A.L. (1994). Expression of three mouse homologs of the Drosophila segment polarity gene *Cubitus interruptus*, *Gli, Gli-2*, and *Gli-3*, in ectoderm-derived and mesoderm-derived tissues suggests multiple roles during postimplantation development. *Dev. Biol.* **162**, 403–413.

Huizinga, J.D., Thuneberg, L., Kluppel, M., Malysz, J., Mikkelsen, H.B. and Bernstein, A. (1995). *W/kit* gene required for intestitial cells of Cajal and for intestinal pacemaker activity. *Nature* **373**, 347–349.

Hulley, P.A., Stander, C.S. and Kidson, S.H. (1991). Terminal migration and early differentiation of melanocytes in embryonic chick skin. *Dev. Biol.* **145**, 183–194.

Hultén, M.A., Honeyman, M.M., Mayne, A.J. and Tarlow, M.J. (1987). Homozygosity in piebald trait. *J. Med. Genet.* **24**, 568–571.

Hume, C.R. and Dodd, J. (1993). *Cwnt-8C*: a novel *Wnt* gene with a potential role in primitive streak formation and hindbrain organization. *Development* **119**, 1147–1160.

Hunt, P., Gulisano, M., Cook, M., Sham, M.H., Faiella, A., Wilkinson, D., Boncinelli, E. and Krumlauf, R. (1991a). A distinct Hox code for the branchial region of the vertebrate head. *Nature* **353**, 861–864.

Hunt, P., Wilkinson, D. and Krumlauf, R. (1991b). Patterning the vertebrate head: murine *Hox 2* genes mark distinct subpopulations of premigratory and migrating cranial neural crest. *Development* **112**, 43–50.

Hunt, P., Ferretti, P., Krumlauf, R. and Thorogood, P. (1995). Restoration of normal Hox code and branchial arch morphogenesis after extensive deletion of hindbrain neural crest. *Dev. Biol.* **168**, 584–597.

Huszar, D., Sharpe, A. and Jaenisch, R. (1991). Migration and proliferation of cultured neural crest cells in W mutant neural crest chimeras. *Development* **112**, 131–141.

Huszar, D., Lynch, C.A., Fairchild-Huntress, V., Dunmore, J.H., Fang, Q., Berkemeier, L.R., Gu, W., Kesterson, R.A., Boston, B.A., Cone, R.D., Smith, F.J., Campfield, L.A., Burn, P. and Lee, F. (1997). Targeted disruption of the melanocortin-4 receptor results in obesity in mice. *Cell* **88**, 131–141.

Huynh, D.P., Nechiporuk, T. and Pulst, S.M. (1994). Differential expression and tissue distribution of type I and type II neurofibromins during mouse fetal development. *Dev. Biol.* **161**, 538–551.

Ibanez, C.F., Ernfors, P., Timmusk, T., Ip, N.Y., Arenas, E., Yancopoulos, G.D. and Persson, H. (1993). Neurotrophin-4 is a target-derived neurotrophic factor for neurons of the trigeminal ganglion. *Development* **117**, 1345–1353.

Imai, H., Osumi-Yamashita, N., Ninomiya, Y. and Eto, K. (1996). Contribution of early-emigrating midbrain crest cells to the dental mesenchyme of mandibular molar teeth in rat embryos. *Dev. Biol.* **176**, 151–165.

Inagaki, H., Bishop, A.E., Yura, J. and Polak, J.M. (1991). Localization of endothelin-1 and its binding sites to the nervous system of the human colon. *J. Cardiovasc. Pharmacol.* **17**, 455–457.

Ingham, P.W. (1988). The molecular genetics of embryonic pattern formation in *Drosophila*. *Nature* **335**, 25–34.

Inoue, T., Chisaka, O., Matsunami, H. and Takeichi, M. (1997). Cadherin-6 expression transiently delineates specific rhombomeres, other neural tube subdivisions, and neural crest subpopulations in mouse embryos. *Dev. Biol.* **183**, 183–194.

Ip, N.Y. and Yancopoulos, G.D. (1994). Neurotrophic factors and their receptors. *Ann. Neurol.* **35** (Suppl.), S13–S16.

Ip, N.Y. and Yancopoulos, G.D. (1996). The neurotrophins and CNTF: two families of collaborative neurotrophic factors. *Annu. Rev. Neurosci.* **19**, 491–515.

Ip, N.Y., Boulton, T.G., Li, Y., Verdi, J.M., Birren, S.J. and Anderson, D.J. (1994). CNTF, FGF, and NGF collaborate to drive the terminal differentiation of MAH cells into postmitotic neurons. *Neuron* **13**, 443–455.

Irving, C., Nieto, M.A., Das Gupta, R., Charnay, P. and Wilkinson, D.G. (1996). Progressive spatial restriction of *Sek-1* and *Krox-20* gene expression during hind-brain segmentation. *Dev. Biol.* **173**, 26–38.

Itasaki, N., Sharpe, J., Morrison, A. and Krumlauf, R. (1996). Reprogramming *Hox* expression in the vertebrate hindbrain: influence of paraxial mesoderm and rhom-bomere transposition. *Neuron* **16**, 487–500.

Ito, K. and Sieber-Blum, M. (1991). *In vitro* clonal analysis of quail cardiac neural crest development. *Dev. Biol.* **148**, 95–106.

Ito, K. and Sieber-Blum, M. (1993). Pluripotent and developmentally restricted neural-crest-derived cells in posterior visceral arches. *Dev. Biol.* **156**, 191–200.

Ito, K., Morita, T. and Sieber-Blum, M. (1993). *In vitro* clonal analysis of mouse neural crest development. *Dev. Biol.* **157**, 517–525.

Itoh, K. and Sokol, S.Y. (1997). Graded amounts of Xenopus dishevelled specify dis-crete anteroposterior cell fates in prospective ectoderm. *Mech. Dev.* **61**, 113–125.

Iversen, L.L. (1967). *The Uptake and Storage of Noradrenaline in Sympathetic Nerves.* Cambridge University Press, Cambridge.

Jacks, T., Shih, T.S., Schmitt, E.M., Bronson, R.T., Bernards, A. and Weinberg, R.A. (1994). Tumor predisposition in mice heterozygous for a targeted mutation of NF1. *Nat. Genet.* **7**, 353–361.

Jackson, I.J. (1988). A cDNA encoding tyrosinase-related protein maps to the *brown* locus in mouse. *Proc. Natl. Acad. Sci. USA* **85**, 4393–4396.

Jackson, I.J., Chambers, D.M., Tsukamoto, K., Copeland, N.G., Gilbert, D.J., Jenkins, N.A. and Hearing, V. (1992). A second tyrosinase-related protein, TRP-2, maps to and is mutated at the mouse *slaty* locus. *EMBO J.* **11**, 527–535.

Jacobs-Cohen, R.J., Payette, R.F., Gershon, M.D. and Rothman, T.P. (1987). Inability of neural crest cells to colonize the presumptive aganglionic bowel of ls/ls mutant mice: requirement for a permissive microenvironment. *J. Comp. Neurol.* **255**, 425–438.

Jacobson, A.G., Oster, G.F., Odell, G.M. and Cheng, L.Y. (1986). Neurulation and the cortical tractor model for epithelial folding. *J. Embryol. Exp. Morphol.* **96**, 19–49.

Jacobson, C.O. (1959). The localization of the presumptive cerebral regions in the neural plate of the *Axolotl* larva. *J. Embryol. Exp. Morphol.* **7**, 1–21.

Jacobson, M. (1984). Cell lineage analysis of neural induction: origins of cells forming the induced nervous system. *Dev. Biol.* **102**, 123–129.

Jaenisch, R. (1985). Mammalian neural crest cells participate in normal embryonic development on microinjection into post-implantation mouse embryos. *Nature* **318**, 181–183.

Jan, Y.N. and Jan, L.Y. (1993). Functional gene cassettes in development. *Proc. Natl. Acad. Sci. USA* **90**, 8305–8307.

Jasoni, C.L., Walker, M.B., Morris, M.D. and Reh, T.A. (1994). A chicken *achaete–scut* homolog (*CASH-1*) is expressed in a temporally and spatially discrete manner in the developing central nervous system. *Development* **120**, 769–783.

Jessell, T.M., Hynes, M.A. and Dodd, J. (1990). Carbohydrates and carbohydrate-binding proteins in the nervous system. *Annu. Rev. Neurosci.* **13**, 227–255.

Jessen, K.R., Morgan, L., Brammer, M. and Mirsky, R. (1985). Galactocerebroside is expressed by non-myelin-forming Schwann cells *in situ. J. Cell Biol.* **101**, 1135–1143.

Jessen, K.R., Morgan, L., Stewart, H.J.S. and Mirsky, R. (1990). Three markers of adult non-myelin-forming Schwann cells, 217c(Ran-1), A5E3 and GFAP: development and regulation by neuron–Schwann cell interactions. *Development* **109**, 91–103.

Jessen, K.R., Brennan, A., Morgan, L., Mirsky, R., Kent, A., Hashimoto, Y. and Gavrilovic, J. (1994). The Schwann cell precursor and its fate: a study of cell death and differentiation during gliogenesis in rat embryonic nerves. *Neuron* **12**, 509–527.

Jesuthasan, S. (1996). Contact inhibition collapse and pathfinding of neural crest cells in the zebrafish trunk. *Development* **122**, 381–389.

Jimbow, K., Szabo, G. and Fitzpatrick, T.B. (1974). Ulstructural investigation of auto-phagocytosis of melanosomes and programmed death of melanocytes in White Leghorn feathers: a study of morphogenetic events leading to hypomelanosis. *Dev. Biol.* **36**, 8–23.

Johnson, E.M., Jr., Andres, R.Y. and Bradshaw, R.A. (1978). Characterization of the retrograde transport of nerve growth factor using high specific activity [125]NGF. *Brain Res.* **150**, 319–331.

Johnson, J. and Deckwerth, T.L. (1993). Molecular mechanisms of developmental neuronal death. *Annu. Rev. Neurosci.* **16**, 31–46.

Johnson, J., Rich, K.M. and Yip, H.K. (1986). The role of NGF in sensory neurons *in vivo. Trends Neurosci.* **9**, 33–37.

Johnson, J.E., Birren, S.J. and Anderson, D.J. (1990). Two rat homologues of drosophila *achaete–scute* specifically expressed in neuronal precursors. *Nature* **346**, 858–861.

Johnson, R.L., Laufer, E., Riddle, R.D. and Tabin, C. (1994). Ectopic expression of *Sonic hedgehog* alters dorsal–ventral patterning of somites. *Cell* **79**, 1165–1173.

Johnson, S.L., Africa, D., Walker, C. and Weston, J.A. (1995). Genetic control of adult pigment stripe development in zebrafish. *Dev. Biol.* **167**, 27–33.

Johnson, W.C., Samaraweera, P., Zuasti, A., Law, J.H. and Bagnara, J.T. (1992). Preliminary biological characterization of a melanization stimulating factor (MSF) from the dorsal skin of the channel catfish, *Ictalurus punctatus. Life Sci.* **51**, 1229–1236.

Johnston, M.C. (1966). A radioautographic study of the migration and fate of cranial neural crest cells in the chick embryo. *Anat. Rec.* **156**, 143–156.

Johnston, M.C. (1974). Regional embryology: aspects relevant to the embryogenesis of craniofacial malformations. In *Proceedings of the International Conference on Craniofacial Malformations*, eds. J.M. Converse and S. Pruzansky. C.V. Mosby, St Louis.

Johnston, M.C. and Hazelton, R.D. (1972). Embryonic origins of facial structures related to oral sensory and motor function. In *Third Symposium on Oral Sensation and Perception: The Mouth of the Infant*, ed. J.F. Bosma, pp. 76–97. Charles C. Thomas, Springfield, IL.

Johnston, M.C., Bhakdinaronk, A. and Reid, Y.C. (1974). An expanded role of the neural crest in oral and pharyngeal development. In *Oral Sensation and Perception Development in the Fetus and Infant*, ed. J.F. Bosma, pp. 37–52. US Government Printing Office, Washington, DC.

Johnston, M.C., Noden, D.M., Hazelton, R.D., Coulombe, J.L. and Coulombre, A.J. (1979). Origin of avian ocular and periocular tissues. *Exp. Eye Res.* **29**, 27–44.

Jonakait, G.M., Markey, K.A., Goldstein, M. and Black, I.B. (1984). Transient expression of selected catecholaminergic traits in cranial sensory and dorsal root ganglia of the embryonic rat. *Dev. Biol.* **101**, 51–60.

Jones, B.W., Fetter, R.D., Tear, G. and Goodman, C.S. (1995). Glial cells missing: a genetic switch that controls glial versus neuronal fate. *Cell* **82**, 1013–1023.

Jones, K.R., Fariñas, I., Backus, C. and Reichardt, L.F. (1994). Targeted disruption of the BDNF gene perturbs brain and sensory neuron development but not motor neuron development. *Cell* **76**, 989–999.

Jonsson, J., Carlsson, L., Edlund, T. and Edlund, H. (1994). Insulin-promoter-factor 1 is required for pancreas development in mice. *Nature* **371**, 606–609.

Joyner, A.L. (1996). *Engrailed*, *Wnt* and *Pax* genes regulate midbrain hindbrain development. *Trends Genet.* **12**, 15–20.

Joyner, A.L. and Martin, G.R. (1987). *En-1* and *En-2*, two mouse genes with sequence homology to the Drosophila *engrailed* gene: expression during embryogenesis. *Genes Dev.* **1**, 29–38.

Joyner, A.L., Kornberg, T., Coleman, K.G., Cox, D.R. and Martin, G.R. (1985). Expression during embryogenesis of a mouse gene with sequence homology to the Drosophila *engrailed* gene. *Cell* **43**, 29–37.

Joyner, A.L., Herrup, K., Auerbach, B.A., Davis, C.A. and Rossant, J. (1991). Subtle cerebellar phenotype in mice homozygous for a targeted deletion of the *En-2* homeobox. *Science* **251**, 1239–1243.

Kaartinen, V., Voncken, J.W., Shuler, C., Warbuton, D., Bu, D., Heisterkamp, N. and Groffen, J. (1995). Abnormal lung development and cleft palate in mice lacking TGF-beta3 indicates defects of epithelial-mesenchymal interaction. *Nat. Genet.* **11**, 415–421.

Kahane, N. and Kalcheim, C. (1994). Expression of trkC receptor mRNA during development of the avian nervous system. *J. Neurobiol.* **25**, 571–584.

Kahane, N., Shelton, D.L. and Kalcheim, C. (1996). Expression and regulation of brain-derived neurotrophic factor and neurotrophin-3 mRNAs in distinct avian motoneuron subsets. *J. Neurobiol.* **29**, 277–292.

Kajiwara, Y., Inouye, M., Kuwana, T. and Fujimoto, T. (1988). Interspecific melanocyte chimaeras made by introducing rat cells into postimplantation mouse embryos *in utero*. *Dev. Biol.* **129**, 586–589.

Kalcheim, C. (1989). Basic fibroblast growth factor stimulates survival of nonneuronal cells developing from trunk neural crest. *Dev. Biol.* **134**, 1–10.

Kalcheim, C. (1996). The role of neurotrophins in development of neural crest cells that become sensory ganglia. *Philos. Trans. R. Soc. Lond. B* **351**, 375–383.

Kalcheim, C. (1998). Multiple roles of neurotrophins in early neural development. In *Neurotrophins and the Neural Crest*, ed. M. Sieber-Blum, pp. 217–242. CRC Press, Boca Raton, FL.

Kalcheim, C. and Gendreau, M. (1988). Brain-derived neurotrophic factor enhances survival and differentiation of neuronal precursors in cultured neural crest. *Dev. Brain Res.* **41**, 79–86.

Kalcheim, C. and Goldstein, R. (1991). Segmentation of sensory and sympathetic ganglia: Interaction between neural crest and somite cells. *J. Physiol.* **85**, 110–117.

Kalcheim, C. and Le Douarin, N.M. (1986). Requirement of a neural tube signal for the differentiation of neural crest cells into dorsal root ganglia. *Dev. Biol.* **116**, 451–456.

Kalcheim, C. and Leviel, V. (1988). Stimulation of collagen production *in vitro* by ascorbic acid released from explants of migrating avian neural crest. *Cell Differ.* **22**, 107–114.

Kalcheim, C. and Neufeld, G. (1990). Expression of basic fibroblast growth factor in the nervous system of early avian embryos. *Development* **109**, 203–215.

Kalcheim, C. and Teillet, M.-A. (1989). Consequences of somite manipulation on the pattern of dorsal root ganglion development. *Development* **106**, 85–93.

Kalcheim, C., Barde, Y.A., Thoenen, H. and Le Douarin, N.M. (1987). *In vivo* effect of brain-derived neurotrophic factor on the survival of developing dorsal root ganglia. *EMBO J.* **6**, 2871–2873.

Kalcheim, C., Carmeli, C. and Rosenthal, A. (1992). Neurotrophin 3 is a mitogen for cultured neural crest cells. *Proc. Natl. Acad. Sci. USA* **89**, 1661–1665.

Kanno, J., Matsubara, O. and Kasuga, T. (1987). Induction of melanogenesis in Schwann cell and perineural epithelium by 9,10–dimethyl-1,3–benzanthracene (DMBA) and 13–o–tetradecanoylphorbol-13–acetate (TPA) in BDF1 mice. *Acta Pathol. Jpn.* **37**, 1297–1304.

Kaplan, D.R., Matsumoto, K., Lucarelli, E. and Thiele, C.J. (1993). Induction of TrkB by retinoic acid mediates biologic responsiveness to BDNF and differentiation of human neuroblastoma cells. *Neuron* **11**, 321–331.

Kapur, R.P., Yost, C. and Palmiter, R.D. (1992). A transgenic model for studying development of the enteric nervous system in normal and aganglionic mice. *Development* **116**, 167–175.

Kapur, R.P., Yost, C. and Palmiter, R.D. (1993). Aggregation chimeras demonstrate that the primary defect responsible for aganglionic megacolon in *lethal-spotted* mice is not neuroblast autonomous. *Development* **117**, 993–999.

Kapur, R.P., Sweetser, D.A., Doggett, B., Siebert, J.R. and Palmiter, R.D. (1995). Intercellular signals downstream of endothelin receptor-B mediate colonization of the large intestine by enteric neuroblasts. *Development* **121**, 3787–3795.

Kastner, P., Grondona, J.M., Mark, M., Gansmuller, A., Lemeur, M., Decimo, D., Vonesch, J.L., Dollé, P. and Chambon, P. (1994). Genetic analysis of RXR alpha, developmental function: convergence of RXR and RAR signaling pathways in heart and eye morphogenesis. *Cell* **78**, 987–1003.

Kastschenko, N. (1888). Zur Entwicklungsgeschichte der Selachierembryos. *Anat. Anz.* **3**, 445–467.

Kato, A.C. and Lindsay, R.M. (1994). Overlapping and additive effects of neurotrophins and CNTF on cultured human spinal cord neurons. *Exp. Neurol.* **130**, 196–201.

Katz, D.M. (1991). A catecholaminergic sensory neuron phenotype in cranial derivatives of the neural crest: regulation by cell aggregation and nerve growth factor. *J. Neurosci.* **11**, 3991–4002.

Katz, D.M. and Erb, M.J. (1990). Developmental regulation of tyrosine hydroxylase expression in primary sensory neurons of the rat. *Dev. Biol.* **137**, 233–242.

Katz, D.M., Markey, K.A., Goldstein, M. and Black, I.B. (1983). Expression of catecholaminergic characteristics by primary sensory neurons in the normal adult rat in vivo. *Proc. Natl. Acad. Sci. USA* **80**, 3526–3530.

Katz, D.M., Erb, M., Lillis, R. and Neet, K. (1990). Trophic regulation of nodose ganglion cell development: evidence for an expanded role of nerve growth factor during embryogenesis in the rat. *Exp. Neurol.* **110**, 1–10.

Katz, D.M., White, M.E. and Hall, A.K. (1995). Lectin binding distinguishes between neuroendocrine and neuronal derivatives of the sympathoadrenal neural crest. *J. Neurobiol.* **26**, 241–252.

Keller, R.E., Löfberg, J. and Spieth, J. (1982). Neural crest cell behavior in white and dark embryos of *Ambystoma mexicanum*: epidermal inhibition of pigment cell migration in the white axolotl. *Dev. Biol.* **89**, 179–195.

Kelsh, R.N., Brand, M., Jiang, Y.J., Heisenberg, C.P., Lin, S., Haffter, P., Odenthal, J., Mullins, M.C., Van Eeden, F.J.M., Furutani-Seiki, M., Granato, M., Hammerschmidt, M., Kane, D.A., Warga, R.M., Beuchle, D., Vogelsang, L. and Nüsslein-Volhard, C. (1996). Zebrafish pigmentation mutations and the processes of neural crest development. *Development* **123**, 369–389.

Kenyon, C. and Wang, B. (1991). A cluster of Antennapedia-class homeobox genes in a nonsegmented animal. *Science* **253**, 516–517.

Kern, M.J., Witte, D.P., Valerius, M.T., Aronow, B.J. and Potter, S.S. (1992). A novel murine homeobox gene isolated by a tissue specific PCR cloning strategy. *Nucleic Acids Res.* **20**, 5189–5195.

Keshet, E., Lyman, S.D., Williams, D.E., Anderson, D.M., Jenkins, N.A., Copeland, N.G. and Parada, L.F. (1991). Embryonic RNA expression patterns of the c-kit receptor and its cognate ligand suggest multiple functional roles in mouse development. *EMBO J.* **10**, 2425–2435.

Kessel, M. and Gruss, P. (1990). Murine developmental control genes. *Science* **249**, 374–379.

Kessel, M. and Gruss, P. (1991). Homeotic transformations of murine vertebrae and concomitant alteration of Hox codes induced by retinoic acid. *Cell* **67**, 89–104.

Keynes, R. and Cook, G.M.W. (1995). Axon guidance molecules. *Cell* **83**, 161–169.

Keynes, R. and Lumsden, A. (1990). Segmentation and the origin of regional diversity in the vertebrate central nervous system. *Neuron* **2**, 1–9.

Keynes, R. and Stern, C.D. (1984). Segmentation in the vertebrate nervous system. *Nature* **310**, 786–789.

Kil, S.H., Lallier, T. and Bronner-Fraser, M. (1996). Inhibition of cranial neural crest adhesion *in vitro* and migration *in vivo* using integrin antisense oligonucleotides. *Dev. Biol.* **179**, 91–101.

Kimmel, C.B., Kane, D.A., Walker, C., Warge, R.M. and Rothman, M.B. (1989). A mutation that changes cell movement and cell fate in the zebrafish embryo. *Nature* **337**, 358–362.

King, R.A., Hearing, V.J., Creel, D.J. and Oetting, W.S. (1995). Albinism. In *The Metabolic Basis of Inherited Diseases*, eds. C.R. Scriver, A.L. Beaudet, W.S. Sly and D.V. Valle, pp. 4353–4392. McGraw-Hill, New York.

Kingsley, D.M. (1994). The TGFbeta superfamily: new members, new receptors, and new genetic tests of function in different organisms. *Genes Dev.* **8**, 133–146.

Kinutani, M., Coltey, M. and Le Douarin, N.M. (1986). Postnatal development of a demyelinating disease in avian spinal cord chimeras. *Cell* **45**, 307–314.

Kinutani, M., Tan, K., Desaki, J., Coltey, M., Kitaoka, K., Nagano, Y., Takashima, Y. and Le Douarin, N.M. (1989). Avian spinal cord chimeras. Further studies on the neurological syndrome affecting the chimeras after birth. *Cell Differ. Dev.* **26**, 145–162.

Kirby, M.L. (1989). Plasticity and predetermination of mesencephalic and trunk neural crest transplanted into the region of the cardiac neural crest. *Dev. Biol.* **134**, 403–412.

Kirby, M.L. and Gilmore, S.A. (1976). A correlative histofluorescence and light microscopic study of the formation of the sympathetic trunks in chick embryos. *Anat. Rec.* **186**, 437–450.

Kirby, M.L., Gale, T.F. and Stewart, D.E. (1983). Neural crest cells contribute to normal aorticopulmonary septation. *Science* **220**, 1059–1061.

Kirby, M.L., Turnage, K.L. and Hayes, B.M. (1985). Characterization of conotruncal malformations following ablation of "cardiac" neural crest. *Anat. Rec.* **213**, 87–93.

Kirby, M.L., Kumiski, D.H., Myers, T., Cerjan, C. and Mishima, N. (1993). Backtransplantation of chick cardiac neural crest cells cultured in LIF rescues heart development. *Dev. Dyn.* **198**, 296–311.

Kirchgessner, A.L., Adlersberg, M.A. and Gershon, M.D. (1992). Colonization of the developing pancreas by neural precursors from the bowel. *Dev. Dyn.* **194**, 142–154.

Kitamura, K., Takiguchi-Hayashi, K., Sezaki, M., Yamamoto, H. and Takeuchi, T. (1992). Avian neural crest cells express a melanogenic trait during early migration from the neural tube: observations with the new monoclonal antibody, "MEBL-1." *Development* **114**, 367–378.

Kjellen, L. and Lindahl, U. (1991). Proteoglycans: structures and interactions. *Annu. Rev. Biochem.* **60**, 443–475.

Klar, A., Baldassare, M. and Jessell, T.M. (1992). F-*spondin*: a gene expressed at high levels in the floor plate encodes a secreted protein that promotes neural cell adhesion and neurite extension. *Cell* **69**, 95–110.

Klein, R., Nanduri, V., Jing, S., Lamballe, F., Tapley, P., Bryant, S., Cordon-Cardo, C., Jones, K.R., Reichardt, L.F. and Barbacid, M. (1991). The *trk*B tyrosine protein kinase is a receptor for brain-derived neurotrophic factor and neurotrophin-3. *Cell* **66**, 395–403.

Klein, R., Smeyne, R.J., Wurst, W., Long, L.K., Auerbach, B.A., Joyner, A.L. and Barbacid, M. (1993). Targeted disruption of the trkB neurotrophin receptor gene results in nervous system lesions and neonatal death. *Cell* **75**, 113–122.

Klein, R., Silos-Santiago, I., Smeyne, R.J., Lira, S.A., Brambilla, R., Bryant, S., Zhang, L., Snider, W.D. and Barbacid, M. (1994). Disruption of the neurotrophin-3 receptor gene trkC eliminates la muscle afferents and results in abnormal movements. *Nature* **368**, 249–251.

Klein, R.D., Sherman, D., Ho, W.H., Stone, D., Bennett, G.L., Moffat, B., Vandlen, R., Simmons, L., Gu, Q.M., Hongo, J.A., Devaux, B., Poulsen, K., Armanini, M., Nozaki, C., Asai, N., Goddard, A., Phillips, H., Henderson, C.E., Takahashi, M. and Rosenthal, A. (1997). A GPI-linked protein that interacts with Ret to form a candidate neurturin receptor. *Nature* **387**, 717–721.

Klein, S.L. and Graziadei, P.P.C. (1983). The differentiation of the olfactory placode in *Xenopus laevis*: a light and electron microscope study. *J. Comp. Neurol.* **217**, 17–30.

Kliewer, S.A., Forman, B.M., Blumberg, B., Ong, E.S., Borgmeyer, U., Mangelsdorf, D.J., Umesono, K. and Evans, R.M. (1994). Differential expression and activation of a family of murine peroxisome proliferator-activated receptors. *Proc. Natl. Acad. Sci. USA* **91**, 7355–7359.

Knouff, R.A. (1927). The origin of the cranial ganglia of *Rana. J. Comp. Neurol.* **44**, 259–361.

Knouff, R.A. (1935). The developmental pattern of ectodermal placodes in *Rana pipiens. J. Comp. Neurol.* **62**, 17–71.

Kobayashi, M., Kurihara, K. and Matsuoka, I. (1994c). Retinoic acid induces BDNF responsiveness of sympathetic neurons by alteration of Trk neurotrophin receptor expression. *FEBS Lett.* **356**, 60–65.

Kobayashi, T., Urabe, K., Orlow, S.J., Higashi, K., Imokawa, G., Kwon, B.S., Potterf, B. and Hearing, V.J. (1994a). The Pmel 17/*silver* locus protein. Characterization and investigation of its melanogenic function. *J. Biol. Chem.* **269**, 29198–29205.

Kobayashi, T., Urabe, K., Winder, A., Jimenez-Cervantes, C., Imokawa, G., Brewington, T., Solano, F., Garcia-Borron, J.C. and Hearing, V.J. (1994b). Tyrosinase related protein 1 (TRP1) functions as a DHICA oxidase in melanin biosynthesis. *EMBO J.* **13**, 5818–5825.

Koerber, H.R. and Mendell, L.M. (1992). Functional heterogeneity of dorsal root ganglion cells. In *Sensory Neurons. Diversity, Development and Plasticity*, ed. S.A. Scott, pp. 77–97. Oxford University Press, Oxford.

Kölliker, H. (1880). *Grundriss der Entwicklungsgeschichte des Menschen.* Engelmann, Leipzig.

Köntges, G. and Lumsden, A. (1996). Rhombencephalic neural crest segmentation is preserved throughout craniofacial ontogeny. *Development* **122**, 3229–3242.

Korade, Z. and Frank, E. (1996). Restriction in cell fates of developing spinal cord cells transplanted to neural crest pathways. *J. Neurosci.* **16**, 7638–7648.

Korsching, S. and Thoenen, H. (1983a). Nerve growth factor in sympathetic ganglia and corresponding target organs of the rat: correlation with density of sympathetic innervation. *Proc. Natl. Acad. Sci. USA* **80**, 3513–3516.

Korsching, S. and Thoenen, H. (1983b). Quantitative demonstration of the retrograde axonal transport of endogenous Nerve Growth Factor. *Neurosci. Lett.* **39**, 1–4.

Kotzbauer, P.T., Lampe, P.A., Heuckeroth, R.O., Golden, J.P., Creedon, D.J., Johnson, E.M. and Milbrandt, J. (1996). Neurturin, a relative of glial-cell-line-derived neurotrophic factor. *Nature* **384**, 467–470.

Kraft, J.C., Schuh, T., Juchau, M. and Kimelman, D. (1994). The retinoid X receptor ligand, 9–cis-retinoic acid, is a potential regulator of early Xenopus development. *Proc. Natl. Acad. Sci. USA* **91**, 3067–3071.

Kramer, H., Cagan, R.L. and Zipursky, S.L. (1991). Interaction of bride of sevenless membrane-bound ligand and the sevenless tyrosine-kinase receptor. *Nature* **352**, 207–212.

Krotoski, D. and Bronner Fraser, M. (1990). Distribution of integrins and their ligands in the trunk of *Xenopus laevis* during neural crest cell migration. *J. Exp. Zool.* **253**, 139–150.

Krotoski, D.M., Domingo, C. and Bronner Fraser, M. (1986). Distribution of a putative cell surface receptor for fibronectin and laminin in the avian embryo. *J. Cell Biol.* **103**, 1061–1071.

Krotoski, D.M., Fraser, S.E. and Bronner Fraser, M. (1988). Mapping of neural crest pathways in *Xenopus laevis* using inter- and intra-specific cell markers. *Dev. Biol.* **127**, 119–132.

Krull, C.E., Collazo, A., Fraser, S.E. and Bronner-Fraser, M. (1995). Segmental migration of trunk neural crest: time-lapse analysis reveals a role for PNA-binding molecules. *Development* **121**, 3733–3743.

Krumlauf, R. (1993). Hox genes and pattern formation in the branchial region of the vertebrate head. *Trends Genet.* **9**, 106–112.

Kuklenski, J. (1915). Uber das Vorkommen und die Verteilung des Pigmentes in den Organen und Geweben bei japonischen Seidenhühnen. *Anat. Entwicklungsgesch.* **87**, 1–37.

Kuntz, A. (1953). *The Autonomic Nervous System*, pp. 117–134. Baillière, Tindall and Cox, London.

Kurihara, Y., Kurihara, H., Suzuki, H., Kodama, T., Maemura, K., Nagai, R., Oda, H., Kuwaki, T., Cao, W.-H., Kamada, N., Jishage, K., Ouchi, Y., Azuma, S., Toyoda, Y., Ishikawa, T., Kumada, M. and Yazaki, Y. (1994). Elevated blood pressure and craniofacial abnormalities in mice deficient in endothelin-1. *Nature* **368**, 703–710.

Kwon, B.S., Haq, A.K., Pomerantz, S.H. and Halaban, R. (1987). Isolation and sequence of a cDNA clone for human tyrosinase that maps at the mouse c-albino locus. *Proc. Natl. Acad. Sci. USA* **84**, 7473–7477.

Kwon, B.S., Chintamaneni, C., Kozak, C.A., Copeland, N.G., Gilbert, D.J., Jenkins, N., Barton, D., Francke, U., Kobayashi, Y. and Kim, K.K. (1991). A melanocyte-specific gene, pmel-17, maps near the silver coat color locus on mouse chromosome-10 and is in a syntenic region on human chromosome-12. *Proc. Natl. Acad. Sci. USA* **88**, 9228–9232.

Lahav, R., Lecoin, L., Ziller, C., Nataf, V., Carnahan, J.F., Martin, F.H. and Le Douarin, N.M. (1994). Effect of the *Steel* gene product on melanogenesis in avian neural crest cell cultures. *Differentiation* **58**, 133–139.

Lahav, R., Ziller, C., Dupin, E. and Le Douarin, N.M. (1996). Endothelin 3 promotes neural crest cell proliferation and mediates a vast increase in melanocyte number in culture. *Proc. Natl. Acad. Sci. USA* **93**, 3893–3897.

Lajtha, L.G. (1979). Haemopoietic stem cells: concept and definitions. *Blood Cells* **5**, 447–455.

Lallier, T.E. and Bronner-Fraser, M. (1988). A spatial and temporal analysis of dorsal root and sympathetic ganglion formation in the avian embryo. *Dev. Biol.* **127**, 99–112.

Lallier, T. and Bronner-Fraser, M. (1991). Avian neural crest cell attachment to laminin: involvement of divalent cation dependent and independent integrins. *Development* **113**, 1069–1084.

Lallier, T. and Bronner-Fraser, M. (1992). alpha1 beta1 Integrin on neural crest cells recognizes some laminin substrata in a $Ca^{2+}$-independent manner. *J. Cell Biol.* **119**, 1335–1345.

Lallier, T., Leblanc, G., Artinger, K.B. and Bronner-Fraser, M. (1992). Cranial and trunk neural crest cells use different mechanisms for attachment to extracellular matrices. *Development* **116**, 531–541.

Lallier, T., Deutzmann, R., Perris, R. and Bronner-Fraser, M. (1994). Neural crest cell interactions with laminin: structural requirements and localization of the binding site for alpha1 beta1 integrin. *Dev. Biol.* **162**, 451–464.

Lamballe, F., Klein, R. and Barbacid, M. (1991). *trk*C, a new member of the *trk* family of tyrosine protein kinases, is a receptor for neurotrophin-3. *Cell* **66**, 967–979.

Lamballe, F., Tapley, P. and Barbacid, M. (1993). *trkC* encodes multiple neurotrophin-3 receptors with distinct biological properties and substrate specificities. *EMBO J.* **12**, 3083–3094.

Lamballe, F., Smeyne, R.J. and Barbacid, M. (1994). Developmental expression of *trkC*, the neurotrophin-3 receptor, in the mammalian nervous system. *J. Neurosci.* **14**, 14–28.

Lamers, C.H., Rombout, J.W. and Timmermans, L.P. (1981). An experimental study on neural crest migration in *Barbus conchonius* (Cyprinidae, Teleostei), with special reference to the origin of the enteroendocrine cells. *J. Embryol. Exp. Morphol.* **62**, 309–323.

Lamoreux, M.L. and Mayer, T.C. (1975). Site of gene action in the development of hair pigment in recessive yellow (*e/e*) mice. *Dev. Biol.* **46**, 160–166.

Lance-Jones, C.C. and Lagenaur, C.F. (1987). A new marker for identifying quail cells in embryonic avian chimeras: a quail-specific antiserum. *J. Histochem. Cytochem.* **35**, 771–780.

Lanctot, C., Lamolet, B. and Drouin, J. (1997). The bicoid-related homeoprotein Ptx1 defines the most anterior domain of the embryo and differentiates posterior from anterior lateral mesoderm. *Development* **124**, 2807–2817.

Landacre, F.L. (1910). The origin of the cranial ganglia in *Ameiurus*. *J. Comp. Neurol.* **20**, 309–411.

Landacre, F.L. (1921). The fate of the neural crest in the head of the Urodeles. *J. Comp. Neurol.* **33**, 1–43.

Landacre, F.L. and McLellan, M. (1912). The cerebral ganglia of the embryo of *Rana pipiens*. *J. Comp. Neurol.* **22**, 461–486.

Landis, S.C. (1976). Rat sympathetic neurons and cardiac myocytes developing in microcultures: correlation of the fine structure of endings with neurotransmitter function in single neurons. *Proc. Natl. Acad. Sci. USA* **73**, 4220–4224.

Landis, S.C. and Keefe, D. (1983). Evidence for neurotransmitter plasticity *in vivo*: developmental changes in the properties of cholinergic sympathetic neurons. *Dev. Biol.* **98**, 349–372.

Landis, S.C., Siegal, R.E. and Schwab, M. (1988). Evidence for neurotransmitter plasticity *in vivo*: II. Immunocytochemical studies of rat sweat gland innervation during development. *Dev. Biol.* **126**, 129–138.

Landmesser, L. (1992). The relationship of intramuscular nerve branching and synaptogenesis to motoneuron survival. *J. Neurobiol.* **23**, 1131–1139.

Landmesser, L.T. and Pilar, G. (1972). The onset and development of transmission in the chick ciliary ganglion. *J. Physiol.* **222**, 691–713.

Landmesser, L.T. and Pilar, G. (1974). Synaptic transmission and cell death during normal ganglionic development. *J. Physiol.* **241**, 737–740.

Landolt, R.M., Vaughan, L., Winterhalter, K.H. and Zimmermann, D.R. (1995). Versican is selectively expressed in embryonic tissues that act as barriers to neural crest cell migration and axon outgrowth. *Development* **121**, 2303–2312.

Lane, P.W. (1966). Association of megacolon with two recessive spotting genes in the mouse. *J. Hered.* **57**, 28–31.

Lane, P.W. and Liu, H.M. (1984). Association of megacolon with a new dominant spotting gene (*Dom*) in the mouse. *J. Hered.* **75**, 435–439.

Langille, R.M. and Hall, B.K. (1988a). Role of the neural crest in development of the cartilaginous cranial and visceral skeleton of the medaka, *Oryzias latipes* (Teleostei). *Anat. Embryol. (Berl.)* **177**, 297–305.

Langille, R.M. and Hall, B.K. (1988b). Role of the neural crest in development of the trabeculae and branchial arches in embryonic sea lamprey, *Petromyzon marinus* (L). *Development* **102**, 301–310.

Langley, J.N. (1921). *The Autonomic Nervous System*. Part 1. W. Heffer, Cambridge.

Langstone, A.W. and Gudas, L.J. (1992). Identification of a retinoic acid responsive enhancer 3′ of the murine homeobox gene *Hox-1.6*. *Mech. Dev.* **38**, 217–228.

LaRosa, G.J. and Gudas, L.J. (1988). Early retinoic acid-induced F9 teratocarcinoma stem cell gene ERA-1: alternate splicing creates transcripts for a homeobox-containing protein and one lacking the homeobox. *Mol. Cell. Biol.* **8**, 3906–3917.

Lawrence, P.A. (1973). A clonal analysis of segment development in *Oncopeltus* (Hemiptera). *J. Embryol. Exp. Morphol.* **30**, 681–699.

Lawrence, P.A. (1981). The cellular basis of segmentation in insects. *Cell* **26**, 3–10.

Lawrence, P.A. (1989). Cell lineage and cell states in the Drosophila embryo. *Ciba Found. Symp.* **144**, 131–140.

Lawrence, P.A. and Morata, G. (1994). Homeobox genes: their function in Drosophila segmentation and pattern formation. *Cell* **78**, 181–189.

Lawson, S.N., Caddy, K.W.T. and Biscoe, T.J. (1974). Development of root ganglion neurones. Studies on cell birthdays and changes in mean cell diameter. *Cell Tissue Res.* **153**, 399–413.

Lawson, S.N., Perry, M.J., Prabhakar, E. and McCarthy, P.W. (1993). Primary sensory neurones: neurofilament, neuropeptides, and conduction velocity. *Brain Res. Bull.* **30**, 239–243.

Layer, P.G. and Willbold, E. (1995). Novel functions of cholinesterases in development, physiology and disease. *Prog. Histochem. Cytochem.* **29**, 1–94.

Le Douarin, N.M. (1969). Particularités du noyau interphasique chez la Caille japonaise (*Coturnix coturnix japonica*). Utilisation de ces particularités comme "marquage biologique" dans les recherches sur les interactions tissulaires et les migrations cellulaires au cours de l'ontogenèse. *Bull. Biol. Fr. Belg.* **103**, 435–452.

Le Douarin, N.M. (1970). Etude ultrastructurale du corps ultimobranchial de l'embryon de caille et du jeune cailleteau et comparaison avec le poulet *C. R. Soc. Biol.* **164**, 884–888.

Le Douarin, N.M. (1971a). Comparative ultrastructural study of the interphasic nucleus in the quail (*Coturnix coturnix japonica*) and the chicken (*Gallus gallus*) by the regressive EDTA staining method. *C. R. Acad. Sci., Ser. III* **272**, 2334–2337.

Le Douarin, N.M. (1971b). Caractéristiques ultrastructurales du noyau interphasique chez la caille et chez le poulet et utilisation de cellules de caille comme "marqueurs biologiques" en embryologie expérimentale. *Ann. Embryol. Morphog.* **4**, 125–135.

Le Douarin, N.M. (1973). A Feulgen-positive nucleolus. *Exp. Cell Res.* **77**, 459–468.

Le Douarin, N.M. (1974). Cell recognition based on natural morphological nuclear markers. *Med. Biol.* **52**, 281–319.

Le Douarin, N.M. (1978). The embryological origin of the endocrine cells associated with the digestive tract: experimental analysis based on the use of a stable cell marking technique. In *Gut Hormones*, ed. S.R. Bloom, pp. 49–56. Churchill Livingstone, Edinburgh.

Le Douarin, N. (1982). *The Neural Crest*. Cambridge University Press, Cambridge.

Le Douarin, N.M. (1986). Cell line segregation during peripheral nervous system ontogeny. *Science* **231**, 1515–1522.

Le Douarin, N.M. (1993). Embryonic neural chimaeras in the study of brain development. *Trends Neurosci.* **16**, 64–72.

Le Douarin, N.M. and Dupin, E. (1993). Cell lineage analysis in neural crest ontogeny. *J. Neurobiol.* **24**, 146–161.

Le Douarin, N.M. and Jotereau, F.V. (1975). Tracing of cells of the avian thymus through embryonic life in interspecific chimeras. *J. Exp. Med.* **142**, 17–40.

Le Douarin, N.M. and Le Lièvre, C. (1971). Sur l'origine des cellules à calcitonine du corps ultimobranchial de l'embryon d'oiseau. *C. R. Assoc. Anat.* **152**, 558–568.

Le Douarin, N.M. and Le Lièvre, C. (1976). Recherches expérimentales sur l'organogenèse du corps ultimobranchial chez l'embryon d'oiseau. *Actes 97e Congrès National Sociétés Savantes, section Sciences* **3**, 427–441.

Le Douarin, N.M. and Teillet, M.-A. (1971). Localisation, par la méthode des greffes interspécifiques, du territoire neural dont dérivent les cellules adrénales surrénaliennes chez l'embryon d'oiseau. *C. R. Acad. Sci., Ser. III* **272**, 481–484.

Le Douarin, N.M. and Teillet, M.-A. (1973). The migration of neural crest cells to the wall of the digestive tract in avian embryo. *J. Embryol. Exp. Morphol.* **30**, 31–48.

Le Douarin, N.M. and Teillet, M.-A. (1974). Experimental analysis of the migration and differentiation of neuroblasts of the autonomic nervous system and of neurectodermal mesenchymal derivatives, using a biological cell marking technique. *Dev. Biol.* **41**, 163–184.

Le Douarin, N., Le Lièvre, C. and Fontaine, J. (1972). Recherches expérimentales sur l'origine embryologique du corps carotidien chez les Oiseaux. *C. R. Acad. Sci., Ser. III* **275**, 583–586.

Le Douarin, N., Fontaine, J. and Le Lièvre, C. (1974). New studies on the neural crest origin of the avian ultimobranchial glandular cells. Interspecific combinations and cytochemical characterization of C cells based on the uptake of biogenic amine precursors. *Histochemistry* **38**, 297–305.

Le Douarin, N.M., Renaud, D., Teillet, M.-A. and Le Douarin, G.H. (1975). Cholinergic differentiation of presumptive adrenergic neuroblasts in interspecific chimeras after heterotopic transplantations. *Proc. Natl. Acad. Sci. USA* **72**, 728–732.

Le Douarin, N.M., Teillet, M.-A., Ziller, C. and Smith, J. (1978). Adrenergic differentiation of cells of the cholinergic ciliary and Remak ganglia in avian embryo after *in vivo* transplantation. *Proc. Natl. Acad. Sci. USA* **75**, 2030–2034.

Le Douarin, N.M., Le Lièvre, C.S., Schweizer, G. and Ziller, C. (1979). An analysis of cell line segregation in the neural crest. In *Cell Lineage, Stem Cells and Cell Determination*, ed. N. Le Douarin, pp. 353–365. Elsevier/North-Holland, Amsterdam.

Le Douarin, N.M., Guillemot, F., Oliver, P. and Péault, B. (1983). Distribution and origin of Ia-positive cells in the avian thymus analyzed by means of monoclonal antibodies in heterospecific chimeras. In *Progress in Immunology V*, eds. Y. Yamamura and T. Tada, pp. 613–631. Academic Press, New York.

Le Douarin, N.M., Cochard, P., Vincent, M., Duband, J.L., Tucker, G.C., Teillet, M.-A. and Thiery, J.P. (1984a). Nuclear, cytoplasmic and membrane markers to follow neural crest cell migration: a comparative study. In *The Role of Extracellular Matrix in Development*, ed. R.L. Treslstad, pp. 373–398. Alan R. Liss, New York.

Le Douarin, N.M., Jotereau, F.V., Houssaint, E. and Thiery, J.P. (1984b). Primary lymphoid organ ontogeny in birds. In *Chimeras in Developmental Biology*, eds. N.M. Le Douarin and A. McLaren, pp. 179–216. Academic Press, Orlando, FL.

Le Douarin, N.M., Dulac, C., Dupin, E. and Cameron-Curry, P. (1991). Glial cell lineages in the neural crest. *Glia* **4**, 175–184.

Le Douarin, N.M., Ziller, C. and Couly, G.F. (1993). Patterning of neural crest derivatives in the avian embryo: *in vivo* and *in vitro* studies. *Dev. Biol.* **159**, 24–49.

Le Douarin, N.M., Dupin, E. and Ziller, C. (1994). Genetic and epigenetic control in neural crest development. *Curr. Opin. Genet. Dev.* **4**, 685–695.

Le Douarin, N., Dieterlen-Lièvre, F. and Teillet, M.-A. (1996). Quail–chick transplantations. In *Methods in Cell Biology*, ed. M. Bronner-Fraser, pp. 23–61. Academic Press, San Diego, CA.

Le Douarin, N.M., Catala, M. and Batini, C. (1997). Embryonic neural chimeras in the study of vertebrate brain and head development. *Int. Rev. Cytol.* **175**, 241–309.

Le Lièvre, C. (1974). Rôle des cellules mésectodermiques issues des crêtes neurales céphaliques dans la formation des arcs branchiaux et du squelette viscéral. *J. Embryol. Exp. Morphol.* **31**, 453–477.

Le Lièvre, C. (1976). Contribution des crêtes neurales à la genèse des structures céphaliques et cervicales chez les oiseaux. Doctorat d'Etat, Nantes.

Le Lièvre, C.S. (1978). Participation of neural crest-derived cells in the genesis of the skull in birds. *J. Embryol. Exp. Morphol.* **47**, 17–37.

Le Lièvre, C. and Le Douarin, N. (1974). Ectodermic origin of the derma of the face and neck, demonstrated by interspecific combinations in the bird embryo. *C. R. Acad. Sci., Ser. III* **278**, 517–520.

Le Lièvre, C.S. and Le Douarin, N.M. (1975). Mesenchymal derivatives of the neural crest: analysis of chimaeric quail and chick embryos. *J. Embryol. Exp. Morphol.* **34**, 125–154.

Le Lièvre, C.S., Schweizer, G.G., Ziller, C.M. and Le Douarin, N.M. (1980). Restriction of developmental capabilities in neural crest cell derivatives as tested by *in vivo* transplantation experiments. *Dev. Biol.* **77**, 363–378.

Leblanc, G. and Landis, S.C. (1986). Development of choline acetyltransferase activity in the cholinergic sympathetic innervation of sweat glands. *J. Neurosci.* **6**, 260–265.

Lecoin, L., Lahav, R., Martin, F.H., Teillet, M.-A. and Le Douarin, N.M. (1995). *Steel* and *c-kit* in the development of avian melanocytes: a study of normally pigmented birds and of the hyperpigmented mutant silky fowl. *Dev. Dyn.* **203**, 106–118.

Lecoin, L., Gabella, G. and Le Douarin, N. (1996). Origin of the c-*kit*-positive interstitial cells in the avian bowel. *Development* **122**, 725–733.

Lecoin, L., Lahav, R., Dupin, E. and Le Douarin, N.M. (1998a). Development of melanocytes from neural crest progenitors. In *Molecular Basis of Epithelial Appendage Morphogenesis*, ed. C.M. Chuong, pp. 131–154. R.G. Landes Company, Georgetown, TX.

Lecoin, L., Sakurai, T., Ngo, M.-T., Abe, Y., Yanagisawa, M. and Le Douarin, N.M. (1998b). Cloning and characterization of a novel endothelin receptor subtype in the avian class. *Proc. Natl. Acad. Sci. USA* **95**, 3024–3029.

Lee, J.E., Hollenberg, S.M., Snider, L., Turner, D.L., Lipnick, N. and Weintraub, H. (1995). Conversion of Xenopus ectoderm into neurons by NeuroD, a basic helix–loop–helix protein. *Science* **268**, 836–844.

Lee, K.-F., Davies, A.M. and Jaenisch, R. (1994). p75-deficient embryonic dorsal root sensory and neonatal sympathetic neurons display a decreased sensitivity to NGF. *Development* **120**, 1027–1033.

Lee, Z.H., Hou, L., Moellmann, G., Kuklinska, E., Antol, K., Fraser, M., Halaban, R. and Kwon, B.S. (1996). Characterization and subcellular localization of human Pmel 17/silver, a 110-kDa (pre)melanosomal membrane protein associated with 5,6,-dihydroxyindole-3–carboxylic acid (DHICA) converting activity. *J. Invest. Dermatol.* **106**, 605–610.

Lefcort, F., Clary, D.O., Rusoff, A.C. and Reichardt, L.F. (1996). Inhibition of the NT-3 receptor trkC, early in chick embryogenesis, results in severe reductions in multiple neuronal subpopulations in the dorsal root ganglia. *J. Neurosci.* **16**, 3704–3713.

Lehman, F. (1927). Further studies on the morphogenetic role of the somites in the development of the nervous system of amphibians. The differentiation and arrangement of the spinal ganglia in *Pleurodeles waltii*. *J. Exp. Zool.* **49**, 93–131.

Leibrock, J., Lottspeich, F., Hohn, A., Hofer, M., Hengerer, B., Masiakowski, P., Thoenen, H. and Barde, Y. (1989). Molecular cloning and expression of brain-derived neurotrophic factor. *Nature* **341**, 149–152.

Lemke, G.E. and Brockes, J.P. (1984). Identification and purification of glial growth factor. *J. Neurosci.* **4**, 75–83.

Leonard, D.G.B., Ziff, E.B. and Greene, L.A. (1987). Identification and characterization of mRNAs regulated by nerve growth factor in PC12 cells. *Mol. Cell. Biol.* **4**, 3156–3157.

Leonard, L., Horton, C., Maden, M. and Pizzey, J.A. (1995). Anteriorization of CRABP-I expression by retinoic acid in the developing mouse central nervous system and its relationship to teratogenesis. *Dev. Biol.* **168**, 514–528.

Leptin, M. and Grunewald, B. (1990). Cell shape changes during gastrulation in Drosophila. *Development* **110**, 73–84.

Leung, D.W., Parent, A.S., Cachianes, G., Esch, F., Coulombe, J.N., Nikolics, K., Eckenstein, F.P. and Nishi, R. (1992). Cloning, expression during development, and evidence for release of a trophic factor for ciliary ganglion neurons. *Neuron* **8**, 1045–1053.

Levi-Montalcini, R. (1987). The Nerve Growth Factor 35 years later. *Science* **237**, 1154–1162.

Levinson, B., Vulpe, C., Elder, B., Martin, C., Verley, F., Packman, S. and Gitschier, J. (1994). The *mottled* gene is the mouse homologue of the Menkes disease gene. *Nat. Genet.* **6**, 369–373.

Lewis, E.B. (1978). A gene complex controlling segmentation in Drosophila. *Nature* **276**, 565–570.

Li, L., Krantz, I.D., Deng, Y., Genin, A., Banta, A.B., Collins, C.C., Qi, M., Trask, B.J., Kuo, W.L., Cochran, J., Costa, T., Pierpont, M.E., Rand, E.B., Piccoli, D.A., Hood, L. and Spinner, N.B. (1997). Alagille syndrome is caused by mutations in human Jagged1, which encodes a ligand for *Notch1*. *Nat. Genet.* **16**, 243–251.

Liem, K.F., Jr., Tremml, G., Roelink, H. and Jessell, T.M. (1995). Dorsal differentiation of neural plate cells induced by BMP-mediated signals from epidermal ectoderm. *Cell* **82**, 969–979.

Lillien, L.E. and Claude, P. (1985). Nerve growth factor is a mitogen for cultured chromaffin cells. *Nature* **317**, 633–634.

Lim, T.M., Lunn, E.R., Keynes, R.J. and Stern, C.D. (1987). The differing effects of occipital and trunk somites on neural development in the chick embryo. *Development* **100**, 525–533.

Lin, L.F., Armes, L.G., Sommer, A., Smith, D.J. and Collins, F. (1990). Isolation and characterization of ciliary neurotrophic factor from rabbit sciatic nerves. *J. Biol. Chem.* **265**, 8943–8947.

Lin, L.-F.H., Doherty, D.H., Lile, J.D., Bektesh, S. and Collins, F. (1993). GDNF: a glial-cell derived neurotrophic factor for midbrain dopaminergic neurons. *Science* **260**, 1130–1132.

Lin, L.-F.H., Zhang, T.J., Collins, F. and Armes, L.G. (1994). Purification and initial characterization of rat B49 glial cell line-derived neurotrophic factor. *J. Neurochem.* **63**, 758–768.

Lindsay, R.M. (1988). Nerve growth factors (NGF, BDNF) enhance axonal regeneration but are not required for survival of adult sensory neurons. *J. Neurosci.* **8**, 2394–2405.

Lindsay, R.M. and Rohrer, H. (1985). Placodal sensory neurons in culture: nodose ganglion neurons are unresponsive to NGF, lack NGF receptors but are supported by a liver-derived neurotrophic factor. *Dev. Biol.* **112**, 30–48.

Lindsay, R.M., Barde, Y.-A., Davies, A.M. and Rohrer, H. (1985). Differences and similarities in the neurotrophic growth factor requirements of sensory neurons derived from neural crest and neural placode. *J. Cell Sci. Suppl.* **3**, 115–129.

Little, C.C. (1915). The inheritance of *black-eyed white* spotting in mice. *Am. Nat.* **49**, 727–740.

Liu, Y.H., Kundu, R., Wu, L., Luo, W., Ignelzi Ma, J.R., Snead, M.L. and Maxson RE, Jr. (1995). Premature suture closure and ectopic cranial bone in mice expressing *Msx2* transgenes in the developing skull. *Proc. Natl. Acad. Sci. USA* **92**, 6137–6141.

Liu, Y.H., Ma, L., Kundu, R., Ignelzi, M., Sangiorgi, F., Wu, L., Luo, W., Snead, M.L. and Maxson, R. (1996). Function of the *Msx2* gene in the morphogenesis of the skull. *Ann. N. Y. Acad. Sci.* **785**, 48–58.

Livrea, M.A. and Packer, L. (1993). *Retinoids.* Marcel Dekker, New York.

Ljungberg, O., Cederquist, E. and Studnitz, W. (1967). Medullary thyroid carcinoma and phaeochrocytoma: a familial chromaffinomatosis. *Br. Med. J.* 279–281.

Lo, L.-C., Birren, S.J. and Anderson, D.J. (1991a). V-*myc* immortalization of early rat neural crest cells yields a clonal cell line which generates both glial and adrenergic progenitor cells. *Dev. Biol.* **145**, 139–153.

Lo, L.-C., Johnson, J.E., Wuenschell, C.W., Saito, T. and Anderson, D.J. (1991b). Mammalian *achaete-scute* homolog 1 is transiently expressed by spatially restricted subsets of early neuroepithelial and neural crest cells. *Genes Dev.* **5**, 1524–1537.

Löfberg, J., Ahlfors, K. and Fällström, C. (1980). Neural crest cell migration in relation to extracellular matrix organization in the embryonic axolotl trunk. *Dev. Biol.* **75**, 148–167.

Löfberg, J., Nynas McCoy, A., Olsson, C., Jonsson, L. and Perris, R. (1985). Stimulation of initial neural crest cell migration in the axolotl embryo by tissue grafts and extracellular matrix transplanted on microcarriers. *Dev. Biol.* **107**, 442–459.

Löfberg, J., Perris, R. and Epperlein, H.H. (1989). Timing in the regulation of neural crest cell migration: retarded "maturation" of regional extracellular matrix inhibits pigment cell migration in embryos of the white axolotl mutant. *Dev. Biol.* **131**, 168–181.

Lohnes, D., Mark, M., Mendelsohn, C., Dollé, P., Dierich, A., Gorry, P., Gansmuller, A. and Chambon, P. (1994). Function of the retinoic acid receptors (RARs) during development. 1. Craniofacial and skeletal abnormalities in RAR double mutants. *Development* **120**, 2723–2748.

Lohnes, D., Mark, M., Mendelsohn, C., Dollé, P., Decimo, D., Lemeur, M., Dierich, A., Gorry, P. and Chambon, P. (1995). Developmental roles of the retinoic acid receptors. *J. Ster. Biochem. Mol. Biol.* **53**, 475–486.

Lopashov, G.V. (1944). Origin of pigment cells and visceral cartilage in teleosts. *C.R. Acad. Sci. URSS* **44**, 169–172.

Lopez-Contreras, A.M., Martinez-Liarte, J.H., Solano, F., Samaraweera, P., Newton, J.M. and Bagnara, J.T. (1996). The amphibian melanization inhibiting factor (MIF) blocks the alpha-MSH effect on mouse malignant melanocytes. *Pigment Cell Res.* **9**, 311–316.

Loring, J.F. and Erickson, C.A. (1987). Neural crest cell migratory pathways in the trunk of the chick embryo. *Dev. Biol.* **121**, 220–236.

Lu, D., Willard, D., Patel, I.R., Kadwell, S., Overton, L., Kost, T., Luther, M., Chen, W., Woychik, R.P., Wilkinson, W.O. and Cone, R.D. (1994). Agouti protein is an antagonist of the melanocyte-stimulating-hormone receptor. *Nature* **371**, 799–802.

Lufkin, T., Dierich, A., Lemeur, M., Mark, M. and Chambon, P. (1991). Disruption of the *hox-1.6* homeobox gene results in defects in a region corresponding to its rostral domain of expression. *Cell* **66**, 1105–1119.

Lufkin, T., Mark, M., Hart, C.P., Dollé, P., Lemeur, M. and Chambon, P. (1992). Homeotic transformation of the occipital bones of the skull by ectopic expression of a homeobox gene. *Nature* **359**, 835–841.

Lumsden, A.G.S. (1984). Tooth morphogenesis: contributions of the cranial neural crest in mammals. In *Tooth Morphogenesis and Differentiation*, eds. A. Belcourt and J.V. Ruch, pp. 19–27. Colloque INSERM 125. INSERM, Paris.

Lumsden, A.G.S. (1987). The neural crest contribution to tooth development in the mammalian embryo. In *Developmental and Evolutionary Aspects of the Neural Crest*, ed. P.F.A. Maderson, pp. 261–300. John Wiley & Sons, New York.

Lumsden, A.G. (1988). Spatial organization of the epithelium and the role of neural crest cells in the initiation of the mammalian tooth germ. *Development* **103** (Suppl.), 155–169.

Lumsden, A. (1990). The cellular basis of segmentation in the developing hindbrain. *Trends Neurosci.* **13**, 329–335.

Lumsden, A.G.S. and Buchanan, J.A.G. (1986). An experimental study of timing and topography of early tooth development in the mouse embryo with an analysis of the role of innervation. *Arch. Oral Biol.* **31**, 301–311.

Lumsden, A. and Keynes, R. (1989). Segmental patterns of neuronal development in the chick hindbrain. *Nature* **337**, 424–428.

Lumsden, A., Sprawson, N. and Graham, A. (1991). Segmental origin and migration of neural crest cells in the hindbrain region of the chick embryo. *Development* **113**, 1281–1291.

Lundberg, J.M. and Hökfelt, T. (1986). *Multiple Coexistence of Peptides and Classical Neurotransmitters in Peripheral Autonomic and Sensory Neurons: Functional and Pharmacological Implications.* Elsevier, Amsterdam.

Lundberg, J.M., Angaard, A., Fahrenkrug, J., Hökfelt, T. and Mutt, V. (1980). Vasoactive intestinal polypeptide in cholinergic neurons of exocrine glands: functional significance of coexisting transmitters for vasodilation and secretion. *Proc. Natl. Acad. Sci. USA* **77**, 1651–1655.

Lundberg, J.M., Hökfelt, T., Angaard, A., Terenius, L., Elde, R., Markey, K., Goldstein, M. and Kimmel, J. (1982a). Organizational principles in the peripheral nervous system: subdivisions by coexisting peptides (somatostatin, avian pancreatic polypeptide and vasoactive intestinal polypeptide-like materials). *Proc. Natl. Acad. Sci. USA* **79**, 1303–1307.

Lundberg, J.M., Terenius, L., Hökfelt, T., Martling, C.R., Tatemoto, K., Mutt, V., Polak, J., Bloom, S. and Goldstein, M. (1982b). Neuropeptide Y (NPY)-like immunoreactivity in peripheral noradrenergic neurons and effects of NPY on sympathetic function. *Acta Physiol. Scand.* **116**, 477–480.

Lunn, E.R., Scourfield, J., Keynes, R.J. and Stern, C.D. (1987). The neural tube origin of ventral root sheath cells in the chick embryo. *Development* **101**, 247–254.

Luo, G., Hofmann, C., Bronckers, A.L.J.J., Sohocki, M., Bradley, A. and Karsenty, G. (1995). BMP-7 is an inducer of nephrogenesis, and is also required for eye development and skeletal patterning. *Genes Dev.* **9**, 2808–2820.

Lyon, M.F. and Searle, A.G. (1990). *Genetic Variants and Strains of the Laboratory Mouse.* Oxford University Press, Oxford.

Lyon, M.F., Fisher, G. and Glenister, P.H. (1985). A recessive allele of the mouse *agouti* locus showing lethality with yellow. *Adv. Genet. Res.* **46**, 95–99.

Ma, L., Golden, S., Wu, L. and Maxson, R. (1996a). The molecular basis of Boston-type craniosynostosis: the Pro148–His mutation in the N-terminal arm of the MSX2 homeodomain stabilizes DNA binding without altering nucleotide sequence preferences. *Hum. Mol. Genet.* **5**, 1915–1920.

Ma, Q.F., Kintner, C. and Anderson, D.J. (1996b). Identification of *neurogenin*, a vertebrate neuronal determination gene. *Cell* **87**, 43–52.

MacKenzie A., Leeming, G.L., Jowett, A.K., Ferguson, M.W.J. and Sharpe, P.T. (1991a). The homeobox gene *Hox 7.1* has specific regional and temporal expression patterns during early murine craniofacial embryogenesis, especially tooth development *in vivo* and *in vitro*. *Development* **111**, 269–285.

MacKenzie, A., Ferguson, M.W.J. and Sharpe, P.T. (1991b). *Hox-7* expression during murine craniofacial development. *Development* **113**, 601–611.

MacKenzie, A., Ferguson, M.W.J. and Sharpe, P.T. (1992). Expression patterns of the homeobox gene, *Hox-8*, in the mouse embryo suggest a role in specifying tooth initiation and shape. *Development* **115**, 403–420.

Mackie, E.J., Tucker, R.P., Halfter, W., Chiquet-Ehrismann, R. and Epperlein, H.H. (1988). The distribution of tenascin coincides with pathways of neural crest cell migration. *Development* **102**, 237–250.

Mackie, K., Sorkin, B.C., Nairn, A.C., Greengard, P., Edelman, G.M. and Cunningham, B.A. (1989). Identification of two protein kinases that phosphorylate the neural cell-adhesion molecule, N-CAM. *J. Neurosci.* **9**, 1883–1896.

Maden, M., Hunt, P., Eriksson, U., Kuroiwa, A., Krumlauf, R. and Summerbell, D (1991). Retinoic acid-binding protein, rhombomeres and the neural crest. *Development* **111**, 35–43.

Maden, M., Horton, C., Graham, A., Leonard, L., Pizzey, J., Siegenthaler, G., Lumsden, A. and Eriksson, U. (1992). Domains of cellular retinoic acid-binding protein I (CRABP I) expression in the hindbrain and neural crest of the mouse embryo. *Mech. Dev.* **37**, 13–23.

Maden, M., Gale, E., Kostetskii, I. and Zile, M. (1996). Vitamin A-deficient quail embryos have half a hindbrain and other neural defects. *Curr. Biol.* **6**, 417–426.

Maeda, H., Yamagata, A., Nishikawa, S., Yoshinaga, K., Kobayashi, S. and Nishi, K. (1992). Requirement of *c-kit* for development of intestinal pacemaker system. *Development* **116**, 369–375.

Mahmood, R., Mason, I.J. and Morriss-Kay, G.M. (1996). Expression of Fgf-3 in relation to hindbrain segmentation, otic pit position and pharyngeal arch morphology in normal and retinoic acid-exposed mouse embryos. *Anat. Embryol.* **194**, 13–22.

Maisonpierre, P.C., Belluscio, L., Friedman, B., Alderson, R.F., Wiegand, S.J., Furth, M.E., Lindsay, R.M. and Yancopoulos, G.D. (1990a). NT-3, BDNF, and NGF in the developing rat nervous system: parallel as well as reciprocal patterns of expression. *Neuron* **5**, 501–509.

Maisonpierre, P.C., Belluscio, L., Squinto, S., Ip, N.Y., Furth, M.E., Lindsay, R.M. and Yancopoulos, G.D. (1990b). Neurotrophin-3: a neurotrophic factor related to NGF and BDNF. *Science* **247**, 1446–1451.

Majumdar, M.K., Feng, L., Medlock, E., Toksoz, D. and Williams, D.A. (1994). Identification and mutation of primary and secondary proteolytic cleavage sites in murine stem cell factor cDNA yields biologically active, cell-associated protein. *J. Biol. Chem.* **269**, 1237–1242.

Malvadi, G., Mencacci, P. and Viola-Magni, M.P. (1968). Mitoses in the adrenal medullary cells. *Experientia* **24**, 475–477.

Mancilla, A. and Mayor, R. (1996). Neural crest formation in *Xenopus laevis*: mechanisms of *Xslug* induction. *Dev. Biol.* **177**, 580–589.

Mangano, F.T., Fukuzawa, T., Johnson, W.C. and Bagnara, J.T. (1992). Intrinsic pigment cell stimulating activity in the skin of the leopard frog, *Rana pipiens*. *J. Exp. Zool.* **263**, 112–118.

Mangold, O. (1929). Experimente zur Analyse der Determination und Induktion der Medullarplatte. *Roux Arch. Entwicklungsmech. Org.* **117**, 586–696.

Manley, N.R. and Capecchi, M.R. (1995). The role of *Hoxa-3* in mouse thymus and thyroid development. *Development* **121**, 1989–2003.

Manova, K. and Bachvarova, R.F. (1991). Expression of *c-kit* encoded at the *W* locus of mice in developing embryonic germ cells and presumptive melanoblasts. *Dev. Biol.* **146**, 313–324.

Mansouri, A., Hallonet, M. and Gruss, P. (1996a). *Pax* genes and their roles in cell differentiation and development. *Curr. Opin. Cell Biol.* **8**, 851–857.

Mansouri, A., Stoykova, A., Torres, M. and Gruss, P. (1996b). Dysgenesis of cephalic neural crest derivatives in *Pax7(–/–)* mutant mice. *Development* **122**, 831–838.

Manthorpe, M., Barbin, G. and Varon, S. (1982). Isoelectric focusing of the chick eye ciliary neuronotrophic factor. *J. Neurosci. Res.* **8**, 233–239.

Manzanares, M., Cordes, S., Kwan, C.T., Sham, M.H., Barsh, G.S. and Krumlauf, R. (1997). Segmental regulation of *Hoxb-3* by *Kreisler*. *Nature* **387**, 191–195.

Marangos, P.J., Zis, A.P., Clark, R.L. and Goodwin, F.K. (1978). Neuronal, non-neuronal and hybrid forms of enolase in brain: structural, immunological and functional comparisons. *Brain Res.* **150**, 117–133.

Marangos, P.J., Campbell, J.C., Schmechel, D.E., Murphy, D.L. and Goodwin, E.K. (1980). Blood platelets contain a neuron-specific enolase subunit. *J. Neurochem.* **34**, 1254–1258.

Margolis, F.L., Roffi, J. and Jost, A. (1966). Norepinephrine methylation in fetal rat adrenals. *Science* **154**, 275–276.

Mark, M., Lufkin, T., Vonesch, J.L., Ruberte, E., Olivo, J.C., Dollé, P., Gorry, P., Lumsden, A. and Chambon, P. (1993). Two rhombomeres are altered in *Hoxa-1* mutant mice. *Development* **119**, 319–338.

Mark, M., Lohnes, D., Mendelsohn, C., Dupe, V., Vonesch, J.L., Kastner, P., Rijli, F., Blochzupan, A. and Chambon, P. (1995). Roles of retinoic acid receptors and of *Hox* genes in the patterning of the teeth and of the jaw skeleton. *Int. J. Dev. Biol.* **39**, 111–121.

Market, C.L. and Silvers, W.K. (1956). The effects of genotype and cell environment on melanoblast differentiation in the house mouse. *Genetics* **41**, 429.

Marshall, H., Nonchev, S., Sham, M.H., Muchamore, I., Lumsden, A. and Krumlauf, R. (1992). Retinoic acid alters hindbrain *Hox* code and induces transformation of rhombomeres 2/3 into a 4/5 identity. *Nature* **360**, 737–741.

Marshall, H., Studer, M., Pöpperl, H., Aparicio, S., Kuroiwa, A., Brenner, S. and Krumlauf, R. (1994). A conserved retinoic acid response element required for early expression of the homeobox gene *Hoxb-1*. *Nature* **370**, 567–571.

Marti, E., Takada, R., Bumcrot, D.A., Sasaki, H. and McMahon, A.P. (1995). Distribution of Sonic hedgehog peptides in the developing chick and mouse embryo. *Development* **121**, 2537–2547.

Martin, D.P., Ito, A., Horigome, K., Lampe, P.A. and Johnson, E.M., Jr. (1992). Biochemical characterization of programmed cell death in NGF-deprived sympathetic neurons. *J. Neurobiol.* **23**, 1205–1220.

Martin, F.H., Suggs, S.V., Langley, K.E., Lu, H.S., Ting, J., Okino, K.H., Morris, C.F., McNiece, I.K., Jacobsen, F.W., Mendiaz, E.A., Birkett, N.C., Smith, K.A., Johnson, M.J., Parker, V.P., Flores, J.C., Patel, A.C., Fisher, E.F., Erjavec, H.O., Herrera, C.J., Wypych, J., Sachdev, R.K., Pope, J.A., Leslie, I., Wen, D., Lin, C.-H., Cupples, R.L. and Zsebo, K.M. (1990). Primary structure and functional expression of rat and human stem cell factor DNAs. *Cell* **63**, 203–211.

Martin, J.F., Bradley, A. and Olson, E.N. (1995). The *paired*-like homeo box gene *MHox* is required for early events of skeletogenesis in multiple lineages. *Genes Dev.* **9**, 1237–1249.

Martin-Zanca, D., Hughes, S.H. and Barbacid, M. (1986). A human oncogene formed by the fusion of truncated tropomyosin and protein tyrosine kinase sequences. *Nature* **319**, 743–748.

Martin-Zanca, D., Oskam, R., Mitra, G., Copeland, T. and Barbacid, M. (1989). Molecular and biochemical characterization of the human trk protooncogene. *Mol. Cell. Biol.* **9**, 24–33.

Martini, R. and Schachner, M. (1986). Immunoelectron microscopic localization of neural cell adhesion molecules (L1, N-CAM, and MAG) and their shared carbohydrate epitope and myelin basic protein in developing sciatic nerve. *J. Cell Biol.* **103**, 2439–2448.

Martins-Green, M. and Erickson, C.A. (1987). Basal lamina is not a barrier to neural crest cell emigration: documentation by TEM and by immunofluorescent and immunogold labelling. *Development* **101**, 517–533.

Marusich, M.F. and Weston, J.A. (1992). Identification of early neurogenic cells in the neural crest lineage. *Dev. Biol.* **149**, 295–306.

Marusich, M.F., Pourmehr, K. and Weston, J.A. (1986). A monoclonal antibody (SN1) identifies a subpopulation of avian sensory neurons whose distribution is correlated with axial level. *Dev. Biol.* **118**, 494–504.

Marusich, M.F., Furneaux, H.M., Henion, P.D. and Weston, J.A. (1994). Hu neuronal proteins are expressed in proliferating neurogenic cells. *J. Neurobiol.* **25**, 143–155.

Marwitt, R., Pilar, G. and Weakley, J.N. (1971). Characterization of two cell populations in avian ciliary ganglia. *Brain Res.* **25**, 317–334.

Masu, Y., Wolf, E., Holtmann, B., Sendtner, M., Brem, G. and Thoenen, H. (1993). Disruption of the *CNTF* gene results in motor neuron degeneration. *Nature* **365**, 27–32.

Matsuo, I., Kuratani, S., Kimura, C., Takeda, N. and Aizawa, S. (1995). Mouse *Otx2* functions in the formation and patterning of rostral head. *Genes Dev.* **9**, 2646–2658.

Matsuo, T., Osumi-Yamashita, N., Noji, S., Ohuchi, H., Koyama, E., Myokai, F., Matsuo, N., Taniguchi, S., Doi, H., Iseki, S., Ninomiya, Y., Fujiwara, M., Watanabe, T. and Eto, K. (1993). A mutation in the *Pax-6* gene in rat *small eye* is associated with impaired migration of midbrain crest cells. *Nat. Genet.* **3**, 299–304.

Matzuk, M.M., Kumar, T.R., Vassalli, A., Bickenbach, J.R., Roop, D.R., Jaenisch, R. and Bradley, A. (1995a). Functional analysis of activins during mammalian development. *Nature* **374**, 354–356.

Matzuk, M.M., Kumar, T.R. and Bradley, A. (1995b). Different phenotypes for mice deficient in either activins or activin receptor type II. *Nature* **374**, 356–360.

Maubert, E., Ciofi, P., Tramu, G., Mazzuca, M. and Dupouy, J.-P. (1992). Early transient expression of somatostatin (SRIF) immunoreactivity in dorsal root ganglia during ontogenesis in the rat. *Brain Res.* **573**, 153–156.

Mavilio, F. (1993). Regulation of vertebrate homeobox-containing genes by morphogens. *Eur. J. Biochem.* **212**, 273–288.

Maxwell, G.D. and Forbes, M.E. (1988). Adrenergic development of neural crest cells grown in a defined medium under a reconstituted basement-membrane-like matrix. *Neurosci. Lett.* **95**, 64–68.

Maxwell, G.D. and Forbes, M.E. (1990). The phenotypic response of cultured quail trunk neural crest cells to a reconstituted basement membrane-like matrix is specific. *Dev. Biol.* **141**, 233–237.

Maxwell, G.D. and Sietz, P.D. (1985). Development of cells containing catecholamines and somatostatin-like immunoreactivity in neural crest cultures: relationship of DNA synthesis to phenotypic expression. *Dev. Biol.* **108**, 203–209.

Maxwell, G.D., Forbes, M.E. and Christie, D.S. (1988). Analysis of the development of cellular subsets present in the neural crest using cell sorting and cell culture. *Neuron* **1**, 557–568.

Maxwell, G.D., Reid, K., Elefanty, A., Bartlett, P.F. and Murphy, M. (1996). Glial cell line-derived neurotrophic factor promotes the development of adrenergic neurons in mouse neural crest cultures. *Proc. Natl. Acad. Sci. USA* **93**, 13274–13279.

Mayer, T.C. (1965). The development of *piebald* spotting in mice. *Dev. Biol.* **11**, 319–334.

Mayer, T.C. (1970). A comparison of pigment cell development in albino, steel, and dominant spotting mutant mouse embryos. *Dev. Biol.* **23**, 297–309.

Mayer, T.C. (1973). The migratory pathway of neural crest cells into the skin of mouse embryos. *Dev. Biol.* **34**, 39–46.

Mayer, T.C. and Maltby, E.L. (1964). An experimental investigation of pattern development in lethal spotting and belted mouse embryos. *Dev. Biol.* **9**, 269–286.

Mayor, R., Morgan, R. and Sargent, M.G. (1995). Induction of the prospective neural crest of *Xenopus*. *Development* **121**, 767–777.

McClearn, D. and Noden, D.M. (1988). Ontogeny of architectural complexity in embryonic quail visceral arch muscles. *Am. J. Anat.* **183**, 277–293.

McEvilly, R.J., Erkman, L., Luo, L., Sawchenko, P.E., Ryan, A.F. and Rosenfeld, M.G. (1996). Requirement for Brn-3.0 in differentiation and survival of sensory and motor neurons. *Nature* **384**, 574–577.

McGinnis, W. and Krumlauf, R. (1992). Homeobox genes and axial patterning. *Cell* **68**, 283–302.

McGinnis, W., Levine, M.S., Hafen, E., Kuroiwa, A. and Gehring, W.J. (1984). A conserved DNA sequence in homoeotic genes of the *Drosophila* Antennapedia and bithorax complexes. *Nature* **308**, 428–433.

McKay, I.J., Muchamore, L., Krumlauf, R., Maden, M., Lumsden, A. and Lewis, J. (1994). The *kreisler* mouse: a hindbrain segmentation mutant that lacks two rhombomeres. *Development* **120**, 2199–2211.

McLaren, A. and Bowman, P. (1969). Mouse chimaeras derived from fusion of embryos differing by nine genetic factors. *Nature* **224**, 238–240.

McMahon, S.B., Armanini, M.P., Ling, L.H. and Phillips, H.S. (1994). Expression and coexpression of Trk receptors in subpopulations of adult primary sensory neurons projecting to identified peripheral targets. *Neuron* **12**, 1161–1171.

McMillan-Carr, V. and Simpson, J. (1978). Proliferative and degenerative events in the early development of chick dorsal root ganglia. I. Normal development. *J. Comp. Neurol.* **182**, 727–740.

Meek, A. (1910). The cranial segments and nerves of the rabbit with some remarks on the phylogeny of the nervous system. *Anat. Anz.* **36**, 573–605.

Meier, S. (1979). Development of the chick embryo mesoblast. Formation of the embryonic axis and establishment of the metameric pattern. *Dev. Biol.* **73**, 25–45.

Meier, S. (1981). Development of the chick embryo mesoblast: morphogenesis of the prechordal plate and cranial segments. *Dev. Biol.* **83**, 49–61.

Meier-Ruge, W. (1974). Hirschprung's disease: its aetiology, pathogenesis and differential diagnosis. *Curr. Top. Pathol.* **59**, 131–179.

Memberg, S.P. and Hall, A.K. (1995). Proliferation, differentiation, and survival of rat sensory neuron precursors *in vitro* require specific trophic factors. *Mol. Cell. Neurosci.* **6**, 323–335.

Mendelsohn, C., Ruberte, E., Lemeur, M., Morriss-Kay, G. and Chambon, P. (1991). Developmental analysis of the retinoic acid-inducible RAR-beta 2 promoter in transgenic animals. *Development* **113**, 723–734.

Mendelsohn, C., Larkin, S., Mark, M., Lemeur, M., Clifford, J., Zelent, A. and Chambon, P. (1994a). RAR beta isoforms: distinct transcriptional control by retinoic acid and specific spatial patterns of promoter activity during mouse embryonic development. *Mech. Dev.* **45**, 227–241.

Mendelsohn, C., Lohnes, D., Decimo, D., Lufkin, T., Lemeur, M., Chambon, P. and Mark, M. (1994b). Function of the retinoic acid receptors (RARs) during development. II. Multiple abnormalities at various stages of organogenesis in RAR double mutants. *Development* **120**, 2749–2771.

Mendelsohn, C., Mark, M., Dollé, P., Dierich, A., Gaub, M.P., Krust, A., Lampron, C. and Chambon, P. (1994c). Retinoic acid receptor beta 2 (RAR beta 2) null mutant mice appear normal. *Dev. Biol.* **166**, 246–258.

Meyer, V., Wenk, H. and Grosse, G. (1973). Zur Histogenese und Chemodifferenzierung des Ganglion trigeminale beim Hühnerembryo. *Z. Mikrosk-Anat. Forsch.* **87**, 147–169.

Michaud, E.J., Bultman, S.J., Klebig, M.L., Van Vugt, M.J., Stubbs, L.J., Russell, L.B. and Woychik, R.P. (1994a). A molecular model for the genetic and phenotypic characteristics of the mouse lethal yellow ($A^y$) mutation. *Proc. Natl. Acad. Sci. USA* **91**, 2563–2566.

Michaud, E.J., Van Vugt, M.J., Bultman, S.J., Sweet, H.O., Davisson, M.T. and Woychik, R.P. (1994b). Differential expression of a new dominant agouti allele ($A^{iapy}$) is correlated with methylation state and is influenced by parental lineage. *Genes Dev.* **8**, 1463–1472.

Milaire, J. (1959). Prédifférenciation cytochimique des diverses ébauches céphaliques chez l'embryon de souris. *Arch. Biol.* **70**, 587–730.

Miller, C.P., McGehee R.E, Jr. and Habener, J.F. (1995). IDX-1: a new homeodomain transcription factor expressed in rat pancreatic islets and duodenum that transactivates the somatostatin gene. *EMBO J.* **13**, 1145–1156.

Miller, M.W., Duhl, D.M.J., Vrieling, H., Cordes, S.P., Ollmann, M.M., Winles, B.M. and Barsh, G.S. (1993). Cloning of the mouse *agouti* gene predicts a secreted protein ubiquitously expressed in mice carrying the lethal *yellow* mutation. *Genes Dev.* **7**, 454–467.

Mina, M. and Kollar, M. (1987). The induction of odontogenesis in non-dental mesenchyme combined with early murine mandibular arch epithelium. *Arch. Oral Biol.* **32**, 123–127.

Minichiello, L., Piehl, F., Vazquez, E., Schimmang, T., Hokfelt, T., Represa, J. and Klein, R. (1995). Differential effects of combined trk receptor mutations on dorsal root ganglion and inner ear sensory neurons. *Development* **121**, 4067–4075.

Mintz, B. (1962a). Formation of genotypically mosaic mouse embryos. *Am. Zool.* **2**, 432.

Mintz, B. (1962b). Experimental recombination of cells in the developing mouse egg: normal and lethal mutant genotypes. *Am. Zool.* **2**, 541–542.

Mintz, B. (1964). Formation of genetically mosaic mouse embryos, and early development of "*lethal $t^{12}/t^{12}$-normal*" mosaics. *J. Exp. Zool.* **157**, 273–292.

Mintz, B. (1967). Gene control of mammalian pigmentary differentiation. I. Clonal origin of melanocytes. *Proc. Natl. Acad. Sci. USA* **58**, 344–351.

Mintz, B. (1970). Gene expression in allophenic mice. In *Control Mechanisms in the Expression of Cellular Phenotypes*, ed. H.A. Padykula, pp. 15–42. Academic Press, New York.

Mintz, B. (1971). The clonal basis of mammalian differentiation. In *Control of Mechanisms of Growth and Differentiation*, eds. D.D. Davies and M. Balls, pp. 345–370. Cambridge University Press, London.

Mintz, B. and Silvers, W.K. (1970). Histocompatibility antigens on melanoblasts and hair follicle cells. *Transplantation* **9**, 497–505.

Mirsky, R., Winter, J., Abney, E.R., Pruss, R.M., Gavrilovic, J. and Raff, M.C. (1980). Myelin-specific proteins and glycolipids in rat Schwann cells and oligodendrocytes in culture. *J. Cell Biol.* **84**, 483–494.

Misuraca, G., Prota, G., Bagnara, J.T. and Frost, S.K. (1977). Identification of the leaf-frog melanophore pigment, rhodomelanochrome, as pterorhodin. *Comp. Biochem. Physiol.* **57B**, 41–43.

Mitchell, P.J., Timmons, P.M., Hebert, J.M., Rigby, P.W.J. and Tjian, R. (1991). Transcription factor ap-2 is expressed in neural crest cell lineages during mouse embryogenesis. *Genes Dev.* **5**, 105–119.

Mitsiadis, T.A., Henrique, D., Thesleff, I. and Lendahl, U. (1997). Mouse *Serrate-1 (Jagged-1)*: expression in the developing tooth is regulated by epithelial-mesenchymal interactions and fibroblast growth factor-4. *Development* **124**, 1473–1483.

Miyagawa-Tomita, S., Waldo, K., Tomita, H. and Kirby, M.L. (1991). Temporospatial study of the migration and distribution of cardiac neural crest in quail–chick chimeras. *Am. J. Anat.* **192**, 79–88.

Miyanaga, K. and Shimasaki, S. (1993). Structural and functional characterization of the rat follistatin (activin-binding protein) gene promoter. *Mol. Cell. Endocrinol.* **92**, 99–109.

Mo, R., Freer, A.M., Zinyk, D.L., Crackower, M.A., Michaud, J., Heng, H.H.Q., Chik, K.W., Shi, X.M., Tsui, L.C., Cheng, S.H., Joyner, A.L. and Hui, C.C. (1997). Specific and redundant functions of *Gli2* and *Gli3* zinc finger genes in skeletal patterning and development. *Development* **124**, 113–123.

Moase, C.E. and Trasler, D.G. (1990). Delayed neural crest cell emigration from *Sp* and *Spd* mouse neural tube explants. *Teratology* **42**, 171–182.

Monaghan, A.P., Davidson, D.R., Sime, C., Graham, E., Baldock, R., Bhattacharya, S.S. and Hill, R.E. (1991). The *Msh*-like homeobox genes define domains in the developing vertebrate eye. *Development* **112**, 1053–1061.

Monsoro-Burq, A.-H., Bontoux, M., Teillet, M.-A. and Le Douarin, N.M. (1994). Heterogeneity in the development of the vertebra. *Proc. Natl. Acad. Sci. USA* **91**, 10435–10439.

Moore, M.W., Klein, R.D., Fariñas, I., Sauer, H., Armanini, M., Phillips, H., Reichardt, L.F., Ryan, A.M., Carver-Moore, K. and Rosenthal, A. (1996). Renal and neuronal abnormalities in mice lacking GDNF. *Nature* **382**, 76–79.

Morin-Kensicki, E.M. and Eisen, J.S. (1997). Sclerotome development and peripheral nervous system segmentation in embryonic zebrafish. *Development* **124**, 159–167.

Morrison-Graham, K. and Weston, J.A. (1993). Transient steel factor dependence by neural crest-derived melanocyte precursors. *Dev. Biol.* **159**, 346–352.

Morrison-Graham, K., Schatteman, G.C., Bork, T., Bowen-Pope, D.F. and Weston, J.A. (1992). A PDGF receptor mutation in the mouse (*Patch*) perturbs the development of a non-neuronal subset of neural crest-derived cells. *Development* **115**, 133–142.

Morriss, G.M. (1972). Morphogenesis of the malformations induced in rat embryos by maternal hypervitaminosis A. *J. Anat.* **113**, 241–250.

Morriss, G.M. and Thorogood, P.V. (1978). An approach to cranial neural crest cell migration and differentiation in mammalian embryos. In *Development in Mammals*, Vol. 3, ed. M.H. Johnson, pp. 363–412. Elsevier/North-Holland, Amsterdam.

Morriss-Kay, G. (1992). *Retinoids in Normal Development and Teratogenesis*. Oxford University Press, Oxford.

Morriss-Kay, G.M., Hill, R.E. and Davidson, D.R. (1991). Effects of retinoic acid excess on expression of *Hox-2.9* and *Krox-20* and on morphological segmentation in the hindbrain of mouse embryos. *EMBO J.* **10**, 2985–2995.

Mountjoy, K.G., Robbins, L.S., Mortrud, M.T. and Cone, R.D. (1992). The cloning of a family of genes that encode the melanocortin receptors. *Science* **257**, 1248–1251.

Moury, J.D. and Jacobson, A.G. (1989). Neural fold formation at newly created boundaries between neural plate and epidermis in the axolotl. *Dev. Biol.* **133**, 44–57.

Moury, J.D. and Jacobson, A.G. (1990). The origins of neural crest cells in the axolotl. *Dev. Biol.* **141**, 243–253.

Mu, X., Silos-Santiago, I., Carroll, S.L. and Snider, W.D. (1993). Neurotrophin receptor genes are expressed in distinct patterns in developing dorsal root ganglia. *J. Neurosci.* **13**, 4029–4041.

Mucchielli, M.L., Mitsiadis, T.A., Raffo, S., Brunet, J.F., Proust, J.P. and Goridis, C. (1997). Mouse *Otlx2/RIEG* expression in the odontogenic epithelium precedes tooth initiation and requires mesenchyme-derived signals for its maintenance. *Dev. Biol.* **189**, 275–284.

Mudge, A.W. (1981). Effect of chemical environment on levels of Substance P and somatostatin in cultured sensory neurones. *Nature* **292**, 764–767.

Murillo-Ferrol, N.L. (1967). The development of the carotid body in *Gallus domesticus*. *Acta Anat.* **68**, 103–126.

Murphy, M., Reid, K., Hilton, D.J. and Bartlett, P.F. (1991). Generation of sensory neurons is stimulated by leukemia inhibitory factor. *Proc. Natl. Acad. Sci. USA* **88**, 3498–3501.

Murphy, M., Reid, K., Williams, D.E., Lyman, S.D. and Bartlett, P.F. (1992). Steel factor is required for maintenance, but not differentiation, of melanocyte precursors in the neural crest. *Dev. Biol.* **153**, 396–401.

Murphy, M., Reid, K., Brown, M.A. and Bartlett, P.F. (1993). Involvement of leukemia inhibitory factor and nerve growth factor in the development of dorsal root ganglion neurons. *Development* **117**, 1173–1182.

Murphy, M., Reid, K., Ford, M., Furness, J.B. and Bartlett, P.F. (1994). FGF2 regulates proliferation of neural crest cells, with subsequent neuronal differentiation regulated by LIF or related factors. *Development* **120**, 3519–3528.

Murphy, P., Davidson, D.R. and Hill, R.E. (1989). Segment specific expression of a homeobox-containing gene in the mouse hindbrain. *Nature* **341**, 156–159.

Murray, S.S., Glackin, C.A., Winters, K.A., Gazit, D., Kahn, A.J. and Murray, E.J. (1992). Expression of helix–loop–helix regulatory genes during differentiation of mouse osteoblastic cells. *J. Bone Miner. Res.* **10**, 1133–1138.

Murre, C., McCaw, P.S., Vaessin, H., Caudy, M., Jan, L.Y., Jan, Y.N., Cabrera, C.V., Buskin, J.N., Hauschka, S.D., Lassar, A.B., Weintraub, H. and Baltimore, D. (1989). Interactions between heterologous helix–loop–helix proteins generate complexes that bind specifically to a common DNA sequence. *Cell* **58**, 537–544.

Mutt, V., Jorpes, J.E. and Magnusson, S. (1970). Structure of porcine secretin. The aminoacid sequence. *Eur. J. Biochem.* **15**, 513–519.

Nagle, D.L., Martin-Deleon, P., Hough, R.B. and Bucan, M. (1994). Structural analysis of chromosomal rearrangements associated with the developmental mutations *Ph*, $W^{19H}$, and *Rw* on mouse chromosome 5. *Proc. Natl. Acad. Sci. USA* **91**, 7237–7241.

Nagy, J.I. and Hunt, S.P. (1982). Fluoride-resistent acid phosphatase-containing neurons in dorsal root ganglia are separate from those containing Substance P or somatostatin. *Neuroscience* **7**, 89–97.

Nakagawa, S. and Takeichi, M. (1995). Neural crest cell–cell adhesion controlled by sequential and subpopulation-specific expression of novel cadherins. *Development* **121**, 1321–1332.

Nakamoto, M., Cheng, H.J., Friedman, G.C., McLaughlin, T., Hansen, M.J., Yoon, C.H., O'Leary, D.D.M. and Flanagan, J.G. (1996). Topographically specific effects of ELF-1 on retinal axon guidance *in vitro* and retinal axon mapping *in vivo*. *Cell* **86**, 755–766.

Nakamura, H. and Ayer-Le Lièvre, C.S. (1982). Mesectodermal capabilities of the trunk neural crest of birds. *J. Embryol. Exp. Morphol.* **70**, 1–18.

Nakamura, S., Stock, D.W., Wydner, K., Zhao, Z., Minowada, J., Lawrence, J.B., Weiss, K.M. and Ruddle, F.H. (1996). Genomic analysis of a new mammalian distal-less gene: *Dlx7*. *Genomics* **38**, 314–324.

Nakamura, T., Takio, K., Eto, Y., Shibai, H., Titani, K. and Sugino, H. (1990). Activin-binding protein from rat ovary is follistatin. *Science* **247**, 836–838.

Narayanan, C.H. and Narayanan, Y. (1978a). Determination of the embryonic origin of the mesencephalic nucleus of the trigeminal nerve in birds. *J. Embryol. Exp. Morphol.* **43**, 85–105.

Narayanan, C.H. and Narayanan, Y. (1978b). On the origin of the ciliary ganglion in birds studied by the method of interspecific transplantation of embryonic brain region between quail and chick. *J. Embryol. Exp. Morphol.* **47**, 137–148.

Narayanan, C.H. and Narayanan, Y. (1980). Neural crest and placodal contributions in the development of the glossopharyngeal-vagal complex in the chick. *Anat. Rec.* **196**, 71–82.

Nataf, V. and Monier, S. (1992). Effect of insulin and insulin-like growth factor I on the expression of the catecholaminergic phenotype by neural crest cells. *Dev. Brain Res.* **69**, 59–66.

Nataf, V., Mercier, P., Ziller, C. and Le Douarin, N.M. (1993). Novel markers of melanocyte differentiation in the avian embryo. *Exp. Cell Res.* **207**, 171–182.

Nataf, V., Mercier, P., De Néchaud, B., Guillemot, J.C., Capdevielle, J., Lapointe, F. and Le Douarin, N.M. (1995). Melanoblast/melanocyte early marker (MelEM) is a glutathione S-transferase subunit. *Exp. Cell Res.* **218**, 394–400.

Nataf, V., Lecoin, L., Eichmann, A. and Le Douarin, N.M. (1996). Endothelin-B receptor is expressed by neural crest cells in the avian embryo. *Proc. Natl. Acad. Sci. USA* **93**, 9645–9650.

Nataf, V., Amemiya, A., Yanagisawa, M. and Le Douarin, N.M. (1998a). The expression pattern of endothelin 3 in the avian embryo. *Mech. Dev.* **73**, 217–220.

Nataf, V., Grapin-Botton, A., Champeval, D., Amemiya, A., Yanagisawa, M. and Le Douarin, N.M. (1998b). The expression patterns of endothelin-A receptor and endothelin 1 in the avian embryo. *Mech. Dev.* **75**, 145–149.

Neal, H.V. (1918). The history of the eye muscles. *J. Morphol.* **30**, 433–453.

Nelson, L.B., Spaeth, G.L., Nowinski, T.S., Margo, C.E. and Jackson, L. (1984). Aniridia. A review. *Surv. Ophthalmol.* **28**, 621–642.

New, H.V. and Mudge, A.W. (1986). Distribution and ontogeny of SP, CGRP, SOM, and VIP in chick sensory and sympathetic ganglia. *Dev. Biol.* **116**, 337–346.

Newgreen, D. (1984). Spreading of explants of embryonic chick mesenchymes and epithelia on fibronectin and laminin. *Cell Tissue Res.* **236**, 265–277.

Newgreen, D.F. (1989). Physical influences on neural crest cell migration in avian embryos: contact guidance and spatial restriction. *Dev. Biol.* **131**, 136–148.

Newgreen, D. and Gibbins, I. (1982). Factors controlling the time of onset of the migration of neural crest cells in the fowl embryo. *Cell Tissue Res.* **224**, 145–160.

Newgreen, D. and Gooday, D. (1985). Control of the onset of migration of neural crest cells in avian embryos. Role of CA[++]-dependent cell adhesion. *Cell Tissue Res.* **239**, 329–336.

Newgreen, D.F. and Minichiello, J. (1995). Control of epitheliomesenchymal transformation. I. Events in the onset of neural crest cell migration are separable and inducible by protein kinase inhibitors. *Dev. Biol.* **170**, 91–101.

Newgreen, D. and Thiery, J.P. (1980). Fibronectin in early avian embryo: synthesis and distribution along the migration pathways of neural crest cells. *Cell Tissue Res.* **211**, 269–291.

Newgreen, D.F., Scheel, M. and Kastner, V. (1986). Morphogenesis of sclerotome and neural crest in avian embryos *in vivo* and *in vitro* studies on the role of notochordal extracellular material. *Cell Tissue Res.* **244**, 299–313.

Newth, D.R. (1951). A remarkable embryonic tissue. *Br. Med. J.* **ii**, 96–106.

Newth, D.R. (1956). On the neural crest of the lamprey embryo. *J. Embryol. Exp. Morphol.* **4**, 358–375.

Ng, N.F.L. and Shooter, E.M. (1993). Activation of p21$^{ras}$ by nerve growth factor in embryonic sensory neurons and PC12 cells. *J. Biol. Chem.* **268**, 25329–25333.

Nichols, D.H. (1981). Neural crest formation in the head of the mouse embryo as observed using a new histological technique. *J. Embryol. Exp. Morphol.* **64**, 105–120.

Nichols, D.H. (1986). Formation and distribution of neural crest mesenchyme to the first pharyngeal arch region of the mouse embryo. *Am. J. Anat.* **176**, 221–231.

Nichols, D.H. and Weston, J.A. (1977). Melanogenesis in cultures of peripheral nervous tissue. I. The origin and prospective fate of cells giving rise to melanocytes. *Dev. Biol.* **60**, 217–225.

Nichols, D.H., Kaplan, R.A. and Weston, J.A. (1977). Melanogenesis in cultures of peripheral nervous tissues. II. Environmental factors determining the fate of pigment-forming cells. *Dev. Biol.* **60**, 226–237.

Nieto, M.A., Bradley, L.C. and Wilkinson, D.G. (1991). Conserved segmental expression of *Krox-20* in the vertebrate hindbrain and its relationship to lineage restriction. *Development* Suppl. 2, 59–62.

Nieto, M.A., Gilardi-Hebenstreit, P., Charnay, P. and Wilkinson, D.G. (1992). A receptor protein tyrosine kinase implicated in the segmental patterning of the hindbrain and mesoderm. *Development* **116**, 1137–1150.

Nieto, M.A., Sargent, M.G., Wilkinson, D.G. and Cooke, J. (1994). Control of cell behavior during vertebrate development by *Slug,* a zinc finger gene. *Science* **264**, 835–839.

Nieto, M.A., Sechrist, J., Wilkinson, D.G. and Bronner-Fraser, M. (1995). Relationship between spatially restricted *Krox-20* gene expression in branchial neural crest and segmentation in the chick embryo hindbrain. *EMBO J.* **14**, 1697–1710.

Nishi, R. and Berg, D.K. (1977). Dissociated ciliary ganglion neurons *in vitro*: survival and synapse formation. *Proc. Natl. Acad. Sci. USA* **74**, 5171–5175.

Nishi, R. and Berg, D.K. (1981). Two components from eye tissue that differentially stimulate the growth and development of ciliary ganglion neurons in cell culture. *J. Neurosci.* **1**, 505–513.

Nishikawa, S., Kusakabe, M., Yoshinaga, K., Ogawa, M., Hayashi, S., Kunisada, T., Era, T., Sakakura, T. and Nishikawa, S. (1991). *In utero* manipulation of coat color formation by a monoclonal anti-*c-kit* antibody: two distinct waves of *c-kit*-dependency during melanocyte development. *EMBO J.* **10**, 2111–2118.

Niu, M.C. (1947). The axial organization of the neural crest studied with particular reference to its pigmentary component. *J. Exp. Zool.* **105**, 79–114.

Nocka, K., Buck, J., Levi, E. and Besmer, P. (1990a). Candidate ligand for the c-kit transmembrane kinase receptor: KL, a fibroblast derived growth factor stimulates mast cells and erythroid progenitors. *EMBO J.* **9**, 3287–3294.

Nocka, K., Tan, J.C., Chiu, E., Chu, T.Y., Ray, P., Traktman, P. and Besmer, P. (1990b). Molecular bases of dominant negative and loss of function mutations at the murine *c-kit*/white spotting locus: $W^{37}$, $W^v$, $W^{41}$ and $W$. *EMBO J.* **9**, 1805–1813.

Noden, D.M. (1975). An analysis of migratory behavior of avian cephalic neural crest cells. *Dev. Biol.* **42**, 106–130.

Noden, D.M. (1976). Cytodifferentiation in heterotopically transplanted neural crest cells. *J. Gen. Physiol.* **68**, 13a.

Noden, D.M. (1978a). The control of avian cephalic neural crest cytodifferentiation. I. Skeletal and connective tissues. *Dev. Biol.* **67**, 296–312.

Noden, D.M. (1978b). The control of avian cephalic neural crest cytodifferentiation. II. Neural tissues. *Dev. Biol.* **67**, 313–329.

Noden, D.M. (1982). Patterns and organization of craniofacial skeletogenic and myogenic mesenchyme: a perspective. *Prog. Clin. Biol. Res.* **101**, 167–203.

Noden, D.M. (1983a). The embryonic origins of avian cephalic and cervical muscles and associated connective tissues. *Am. J. Anat.* **168**, 257–276.

Noden, D.M. (1983b). The role of the neural crest in patterning of avian cranial skeletal, connective, and muscle tissues. *Dev. Biol.* **96**, 144–165.

Noden, D.M. (1986a). Origins and patterning of craniofacial mesenchymal tissues. *J. Craniofac. Genet. Dev. Biol. (Suppl.)* **2**, 15–31.

Noden, D.M. (1986b). Patterning of avian craniofacial muscles. *Dev. Biol.* **116**, 347–356.

Noden, D.M. (1988). Interactions and fates of avian craniofacial mesenchyme. *Development* **103** (Suppl.), 121–140.

Nunez, G., London, L., Hockenbery, D., Alexander, M., McKearn, J.P. and Korsmeyer, S.J. (1990). Deregulated *bcl-2* gene expression selectively prolongs survival of growth factor-deprived hemopoietic cell lines. *J. Immunol.* **144**, 3603–3610.

Nüsslein-Volhard, C., Wieschaus, E. and Kluding, H. (1984). Mutations affecting the pattern of the larval cuticle in *Drosophila melanogaster*. Zygotic loci on the second chromosome. *Roux Arch. Dev. Biol.* **193**, 267–282.

Oakley, R.A. and Tosney, K.W. (1991). Peanut agglutinin and chondroitin-6–sulfate are molecular markers for tissues that act as barriers to axon advance in the avian embryo. *Dev. Biol.* **147**, 187–206.

Oakley, R.A., Lasky, C.J., Erickson, C.A. and Tosney, K.W. (1994). Glycoconjugates mark a transient barrier to neural crest migration in the chicken embryo. *Development* **120**, 103–114.

Oakley, R.A., Garner, A.S., Large, T.H. and Frank, E. (1995). Muscle sensory neurons require neurotrophin-3 from peripheral tissues during the period of normal cell death. *Development* **121**, 1341–1350.

Oberling, C. (1922). Les tumeurs des méninges. *Bull. Assoc. Franç. Etude Canc.* **2**, 365.

Obika, M. (1963). Association of pteridines with amphibian larval pigmentation and their biosynthesis in developing chromatophores. *Dev. Biol.* **6**, 99–112.

Obika, M. and Bagnara, J.T. (1964). Pteridines as pigments in amphibians. *Science* **143**, 485–487.

Ockel, M., Lewin, G.R. and Barde, Y.A. (1996). *In vivo* effects of neurotrophin-3 during sensory neurogenesis. *Development* **122**, 301–307.

Oda, T., Elkahloun, A.G., Pike, B.L., Okajima, K., Krantz, I.D., Genin, A., Piccoli, D.A., Meltzer, P.S., Spinner, N.B., Collins, F.S. and Chandrasekharappa, S.C. (1997). Mutations in the human *Jagged1* gene are responsible for Alagille syndrome. *Nat. Genet.* **16**, 235–242.

Odenthal, J., Rossnagel, K., Haffter, P., Kelsh, R.N., Vogelsang, E., Brand, M., Van Eeden, F.J., Furutani-Seiki, M., Granato, M., Hammerschmidt, M., Heisenberg, C.P., Jiang, Y.J., Kane, D.A., Mullins, M.C. and Nüsslein-Volhard, C. (1996). Mutations affecting xanthophore pigmentation in the zebrafish, *Danio rerio*. *Development* **123**, 391–398.

O'Farrell, P.H. (1994). Unanimity waits in the wings. *Nature* **368**, 188–189.

Ogawa, M., Nishikawa, S., Yoshinaga, K., Hayashi, S.-I., Kunisada, T., Nakao, J., Kina, T., Sudo, T., Kodama, H. and Nishikawa, S.-I. (1993). Expression and function of c-Kit in fetal hemopoietic progenitor cells: transition from the early c-Kit-independent to the late c-Kit-dependent wave of hemopoiesis in the murine embryo. *Development* **117**, 1089–1098.

Ohki, H., Martin, C., Corbel, C., Coltey, M. and Le Douarin, N.M. (1987). Tolerance induced by thymic epithelial grafts in birds. *Science* **237**, 1033–1035.

Ohki, H., Martin, C., Corbel, C., Coltey, M. and Le Douarin, N.M. (1989). Effect of early embryonic grafting of foreign tissues on the immune response of the host. In *Recent Advances in Avian Immunology Research*, ed. B.S. Boghal, pp. 3–17. Alan R. Liss Inc., New York.

Ohlsson, H., Karlsson, K. and Edlund, T. (1993). IPF1, a homeodomain-containing transactivator of the insulin gene. *EMBO J.* **12**, 4251–4259.

Ohsugi, K. and Ide, H. (1983). Melanophore differentiation in *Xenopus laevis*, with special reference to dorsoventral pigment pattern formation. *J. Embryol. Exp. Morphol.* **75**, 141–150.

Okabe, H., Chijiiwa, Y., Nakamura, K., Yoshinaga, M., Akiho, H., Harada, N. and Nawata, H. (1995). Two endothelin receptors (ETA and ETB) expressed on circular smooth muscle cells of guinea pig cecum. *Gastroenterology* **108**, 51–57.

O'League, P.H., Obata, K., Claude, P., Furshpan, E.J. and Potter, D.D. (1974). Evidence for cholinergic synapses between dissociated rat sympathetic neurons in cell culture. *Proc. Natl. Acad. Sci. USA* **71**, 3602–3606.

Ollmann, M.M. and Barsh, G.S. (1996). Pharmacologic studies of the agouti protein using Xenopus melanophores. *Pigment Cell Res.* Suppl. 5, 58–59.

Ollmann, M.M., Wilson, B.D., Yang, Y.K., Kerns, J.A., Chen, Y.R., Gantz, 1. and Barsh, G.S. (1997). Antagonism of central melanocortin receptors *in vitro* and *in vivo* by Agouti-related protein. *Science* **278**, 135–138.

Olson, E.C. (1959). The evolution of mammalian characters. *Evolution* **13**, 344–353.

Olson, L. (1970). Fluorescence histochemical evidence for axonal growth and secretion from transplanted adrenal medullary tissue. *Histochemie* **22**, 1–7.

Olsson, L., Stigson, M., Perris, R., Sorrell, J.M. and Lofberg, J. (1996a). Distribution of keratan sulphate and chondroitin sulphate in wild type and white mutant axolotl embryos during neural crest cell migration. *Pigment Cell Res.* **9**, 5–17.

Olsson, L., Svensson, K. and Perris, R. (1996b). Effects of extracellular matrix molecules on subepidermal neural crest cell migration in wild type and white mutant (dd) axolotl embryos. *Pigment Cell Res.* **9**, 18–27.

Opdecamp, K., Nakayama, A., Nguyen, M.T., Hodgkinson, C.A., Pavan, W.J. and Arnheiter, H. (1997). Melanocyte development *in vivo* and in neural crest cell cultures: crucial dependence on the Mitf basic-helix–loop–helix–zipper transcription factor. *Development* **124**, 2377–2386.

Oppenheim, R.W. (1989). The neurotrophic theory and naturally occurring motoneuron death. *Trends Neurosci.* **12**, 252–255.

Oppenheim, R.W., Prevette, D., Qin-Wei, Y., Collins, F. and MacDonald, J. (1991). Control of embryonic motoneuron survival *in vivo* by ciliary neurotrophic factor. *Science* **251**, 1616–1618.

Oppenheim, R.W., Prevette, D. and Fuller, F. (1992a). The lack of effect of basic and acidic fibroblast growth factors on the naturally occurring death of neurons in the chick embryo. *J. Neurosci.* **12**, 2726–2734.

Oppenheim, R.W., Schwartz, L.M. and Shatz, C.J. (1992b). Neuronal death, a tradition of dying. *J. Neurobiol.* **23**, 1111–1115.

Orr, H. (1887). Contribution to the embryology of the lizard. *J. Morphol.* **1**, 311–372.

Ortmann, R. (1948). Über Placoden und Neuralleiste beim Entenembryo: ein Beitrag zum Kopfproblem. *Z. Anat. Entwicklungsgesch.* **112**, 537–587.

Osborn, J.W. (1978). Morphogenetic gradients: fields versus clones. In *Development, Function and Evolution of Teeth*, eds. P.M. Butler and K.A. Joysey, pp. 171–201. Academic Press, London.

Osumi-Yamashita, N., Ninomiya, Y., Doi, H. and Eto, K. (1994). The contribution of both forebrain and midbrain crest cells to the mesenchyme in the frontonasal mass of mouse embryos. *Dev. Biol.* **164**, 409–419.

Osumi-Yamashita, N., Ninomiya, Y., Doi, H. and Eto, K. (1996). Rhombomere formation and hind-brain crest cell migration from prorhombomeric origins in mouse embryos. *Dev. Growth Differ.* **38**, 107–118.

Osumi-Yamashita, N., Ninomiya, Y. and Eto, K. (1997). Mammalian craniofacial embryology *in vitro. Int. J. Dev. Biol.* **41**, 187–194.

Owens, G.C. and Bunge, R.P. (1989). Evidence for an early role for myelin-associated glycoprotein in the process of myelination. *Glia* **2**, 119–128.

Oxtoby, E. and Jowett, T. (1993). Cloning of the zebrafish *krox-20* gene (*krx-20*) and its expression during hindbrain development. *Nucleic Acids Res.* **21**, 1087–1095.

Pachnis, V., Mankoo, B. and Constantini, F. (1993). Expression of the *c-ret* proto-oncogene during mouse embryogenesis. *Development* **119**, 1005–1017.

Palmiter, R.D., Behringer, R.R., Quaife, C.J., Maxwell, F., Maxwell, I.H. and Brinster, R.L. (1987). Cell lineage ablation in transgenic mice by cell-specific expression of a toxin gene. *Cell* **50**, 435–443.

Pannese, E. (1974). The histogenesis of the spinal ganglia. *Adv. Anat. Cell Biol.* **47**, 1–97.

Papaioannou, V.E. and Mardon, H. (1983). Lethal nonagouti (*a^x*): description of a second embryonic lethal at the *agouti* locus. *Dev. Genet.* **4**, 21–29.

Papalopulu, N., Clarke, J.D.W., Bradley, L., Wilkinson, D., Krumlauf, R. and Holder, N. (1991). Retinoic acid causes abnormal development and segmental patterning of the anterior hindbrain in Xenopus embryos. *Development* **113**, 1145–1158.

Pardanaud, L., Altmann, C., Kitos, P., Dieterlen-Lièvre, F. and Buck, C.A. (1987). Vasculogenesis in the early quail blastodisc as studied with a monoclonal antibody recognizing endothelial cells. *Development* **100**, 339–349.

Parikh, D.H., Tam, P.K., Lloyd, D.A., Van-Velzen, D. and Edgar, D.H. (1992). Quantitative and qualitative analysis of the extracellular matrix protein, laminin, in Hirschprung's disease. *J. Pediatr. Surg.* **27**, 991–995.

Parker, L.N. and Noble, E.P. (1967). Prenatal glucocorticoid administration and the development of the epinephrine-forming enzyme. *Proc. Soc. Exp. Biol.* **126**, 734–737.

Passarge, E. (1973). Genetics of Hirschprung's disease. *Clin. Gastroenterol* **2**, 507–513.

Patterson, P.H. (1990). Control of cell fate in a vertebrate neurogenic lineage. *Cell* **62**, 1035–1038.

Patterson, P.H. and Chun, L.L.Y. (1977). The induction of acetylcholine synthesis in primary cultures of dissociated rat sympathetic neurons. I. Effects of conditioned medium. *Dev. Biol.* **56**, 263–280.

Patterson, P.H., Reichardt, L.F. and Chun, L.L.Y. (1975). Biochemical studies on the development of primary sympathetic neurons in cell culture. *Cold Spring Harbor Symp. Quant. Biol.* **40**, 389–397.

Paul, G. and Davies, A.M. (1995). Trigeminal sensory neurons require extrinsic signals to switch neurotrophin dependence during the early stages of target field innervation. *Dev. Biol.* **171**, 590–605.

Pavan, W.J. and Tilghman, S.M. (1994). Piebald lethal (*s^l*) acts early to disrupt the development of neural crest-derived melanocytes. *Proc. Natl. Acad. Sci. USA* **91**, 7159–7163.

Payette, R.F., Tennyson, V.M., Pomeranz, H.D., Pharm, T.D., Rothman, T.P. and Gershon, M.D. (1988). Accumulation of components of basal laminae: association with the failure of neural crest cells to colonize the presumptive aganglionic bowel of ls/ls mutant mice. *Dev. Biol.* **125**, 341–360.

Pearse, A.G.E. (1966a). Common cytochemical properties of cells producing polypeptide hormones with particular reference to calcitonin and the thyroid C cells. *Vet. Res.* **79**, 587–590.

Pearse, A.G.E. (1966b). 5-Hydroxytryptophan uptake by dog thyroid "C" cells, and its possible significance in polypeptide hormone production. *Nature* **211**, 598–600.

Pearse, A.G.E. (1968). Common cytochemical and ultrastructural characteristics of cells producing polypeptide hormones (the APUD series) and their relevance to thyroid and ultimobranchial C cells and calcitonin. *Proc. R. Soc. Lond., Ser. B* **170**, 71–80.

Pearse, A.G.E. (1969). The cytochemical and ultrastructure of polypeptide hormone-producing cells of the APUD series and the embryologic, physiologic and pathologic implications of the concept. *J. Histochem. Cytochem.* **17**, 303–313.

Pearse, A.G.E. (1976). Peptides in brain and intestine. *Nature* **262**, 93–94.

Pearse, A.G. (1980). APUD: concept, tumours, molecular markers and amyloid. *Mikroskopie* **36**, 257–269.

Pearse, A.G.E. and Carvalheira, A. (1967). Cytochemical evidence for an ultimobranchial origin of rodent thyroid C cells. *Nature* **214**, 929–930.

Pearse, A.G. and Polak, J.M. (1971a). Neural crest origin of the endocrine polypeptide (APUD) cells of the gastrointestinal tract and pancreas. *Gut* **12**, 783–788.

Pearse, A.G. and Polak, J.M. (1971b). Cytochemical evidence for the neural crest origin of mammalian ultimobranchial C cells. *Histochemie* **27**, 96–102.

Pearse, A.G., Polak, J.M., Rost, F.W., Fontaine, J., Le Lièvre, C. and Le Douarin, N. (1973). Demonstration of the neural crest origin of type I (APUD) cells in the avian carotid body, using a cytochemical marker system. *Histochemie* **34**, 191–203.

Péault, B.M., Thiery, J.P. and Le Douarin, N.M. (1983). Surface marker for hemopoietic and endothelial cell lineages in quail that is defined by a monoclonal antibody. *Proc. Natl. Acad. Sci. USA* **80**, 2976–2980.

Pehlemann, F.W. (1967). Der morphologische Farbwechsel von *Xenopus laevis*-larven. *Z. Zellforsch.* **78**, 484–510.

Pendleton, J.W., Nagai, B.K., Murtha, M.T. and Ruddle, F.H. (1993). Expansion of the *Hox* gene family and the evolution of chordates. *Proc. Natl. Acad. Sci. USA* **90**, 6300–6304.

Perez-Castro, A.V., Toth-Rogler, L.E., Wei, L.-N. and Nguyen-Huu, M.C. (1989). Spatial and temporal pattern of expression of the cellular retinoic acid-binding protein and the cellular retinol-binding protein during mouse embryogenesis. *Proc. Natl. Acad. Sci. USA* **86**, 8813–8817.

Perrimon, N. (1995). *Hedgehog* and beyond. *Cell* **80**, 517–529.

Perris, R. (1997). The extracellular matrix in neural crest-cell migration. *Trends Neurosci.* **20**, 23–31.

Perris, R. and Löfberg, J. (1986). Promotion of chromatophore differentiation in isolated premigratory neural crest cells by extracellular material explanted on microcarriers. *Dev. Biol.* **113**, 223–238.

Perris, R., Paulsson, M. and Bronner-Fraser, M. (1989). Molecular mechanisms of avian neural crest cell migration on fibronectin and laminin. *Dev. Biol.* **136**, 223–238.

Perris, R., Löfberg, J., Fallström, C., Von Boxberg, Y., Olsson, L. and Newgreen, D.F. (1990). Structural and compositional divergencies in the extracellular matrix encountered by neural crest cells in the white mutant axolotl embryo. *Development* **109**, 533–551.

Perris, R., Krotoski, D., Lallier, T., Domingo, C., Sorrell, J.M. and Bronner-Fraser, M. (1991a). Spatial and temporal changes in the distribution of proteoglycans during avian neural crest development. *Development* **111**, 583–599.

Perris, R., Krotoski, D. and Bronner-Fraser, M. (1991b). Collagens in avian neural crest development: distribution *in vivo* and migration-promoting ability *in vitro*. *Development* **113**, 969–984.

Perris, R., Kuo, H.-J., Glanville, R.W., Leibold, S. and Bronner-Fraser, M. (1993a). Neural crest cell interaction with type VI collagen is mediated by multiple co-operative binding sites within triple-helix and globular domains. *Exp. Cell Res.* **209**, 103–117.

Perris, R., Syfrig, J., Paulsson, M. and Bronner-Fraser, M. (1993b). Molecular mechanisms of neural crest cell attachment and migration on types I and IV collagen. *J. Cell Sci.* **106**, 1357–1368.

Perris, R., Perissinotto, D., Pettway, Z., Bronner Fraser, M., Morgelin, M. and Kimata, K. (1996). Inhibitory effects of PG-H/aggrecan and PG-M/versican on avian neural crest cell migration. *FASEB J.* **10**, 293–301.

Petkovich, M. (1992). Regulation of gene expression by vitamin A: the role of nuclear retinoic acid receptors. *Annu. Rev. Nutr.* **12**, 443–471.

Petkovich, M., Brand, N.J., Krust A. and Chambon, P. (1987). A human retinoic acid receptor which belongs to the family of nuclear receptors. *Nature* **330**, 444–450.

Pettway, Z., Guillory, G. and Bronner Fraser, M. (1990). Absence of neural crest cells from the region surrounding implanted notochords *in situ*. *Dev. Biol.* **142**, 335–345.

Pettway, Z., Domowicz, M., Schwartz, N.B. and Bronner-Fraser, M. (1996). Age-dependent inhibition of neural crest migration by the notochord correlates with alterations in the S103L chondroitin sulfate proteoglycan. *Exp. Cell Res.* **225**, 195–206.

Phillips, M.T., Kirby, M.L. and Forbes, G. (1987). Analysis of cranial neural crest distribution in the developing heart using quail–chick chimeras. *Circ. Res.* **60**, 27–30.

Pichel, J.G., Shen, L.Y., Sheng, H 7, Granholm, A.C., Drago, J., Grinberg, A., Lee, E.J., Huang, S.P., Saarma, M., Hoffer, B.J., Sariola, H. and Westphal, H. (1996). Defects in enteric innervation and kidney development in mice lacking GDNF. *Nature* **382**, 73–76.

Pictet, R.L., Rall, L.B., Phelps, P. and Rutter, W.J. (1976). The neural crest and the origin of the insulin-producing and other gastrointestinal hormone-producing cells. *Science* **191**, 191–192.

Pilar, G., Landmesser, L. and Burstein, L. (1980). Competition for survival among developing ciliary ganglion cells. *J. Neurophysiol.* **43**, 233–254.

Pinco, O., Carmeli, C., Rosenthal, A. and Kalcheim, C. (1993). Neurotrophin-3 affects proliferation and differentiation of distinct neural crest cells and is present in the early neural tube of avian embryos. *J. Neurobiol.* **24**, 1626–1641.

Pirvola, U., Ylikoski, J., Palgi, J., Lehtonen, E., Arumäe, U. and Saarma, M. (1992). Brain-derived neurotrophic factor and neurotrophin 3 mRNAs in the peripheral target fields of developing inner ear ganglia. *Proc. Natl. Acad. Sci. USA* **89**, 9915–9919.

Platt, J.B. (1893). Ectodermic origin of the cartilage of the head. *Anat. Anz.* **8**, 506–509.

Platt, J.B. (1896). Ontogenetic differentiation of the ectoderm in the *Necturus*. II. On the development of the peripheral nervous system. *Q. J. Microsc. Sci.* **38**, 485–547.

Platt, J.B. (1897). The development of the cartilaginous skull and of the branchial and hypoglossal musculature in *Necturus*. *Morphol. Jahrb.* **25**, 377–464.

Polak, J.M. and Marangos, P.J. (1984). Neuron-specific enolase: a marker for neuroendocrine cells. In *Evolution and Tumor Pathology of the Neuroendocrine System*, eds. S. Falkmer, R. Hakanson and F. Sundler, pp. 433–452. Elsevier Science, Amsterdam.

Polak, J.M., Rost, F.W. and Pearse, A.G. (1971). Fluorogenic amine tracing of neural crest derivatives forming the adrenal medulla. *Gen. Comp. Endocrinol.* **16**, 132–136.

Polak, J.M., Pearse, A.G.E., Le Lièvre, C., Fontaine, J. and Le Douarin, N.M. (1974). Immunocytochemical confirmation of the neural crest origin of avian calcitonin producing cells. *Histochemistry* **40**, 209–214.

Pomeranz, H.D. and Gershon, M.D. (1990). Colonization of the avian hindgut by cells derived from the sacral neural crest. *Dev. Biol.* **137**, 378–394.

Pomeranz, H.D., Rothman, T.P. and Gershon, M.D. (1991a). Colonization of the post-umbilical bowel by cells derived from the sacral neural crest: direct tracing of cell migration using an intercalating probe and a replication-deficient retrovirus. *Development* **111**, 647–655.

Pomeranz, H.D., Sherman, D.L., Smalheiser, N.R., Tennyson, V.M. and Gershon, M.D. (1991b). Expression of a neurally related laminin binding protein by neural crest-derived cells that colonize the gut: relationship to the formation of enteric ganglia. *J. Comp. Neurol.* **313**, 625–642.

Pomeranz, H.D., Rothman, T.P., Chalazonitis, A., Tennyson, V.M. and Gershon, M.D. (1993). Neural crest-derived cells isolated from the gut by immunoselection develop neuronal and glial phenotypes when cultured on laminin. *Dev. Biol.* **156**, 341–361.

Poole, T.J. and Thiery, J.P. (1986). Antibodies and a synthetic peptide that block cell-fibronectin adhesion arrest neural crest cell migration *in vivo*. *Prog. Clin. Biol. Res.* **217B**, 235–238.

Popperl, H. and Featherstone, M.S. (1993). Identification of a retinoic acid response element upstream of the murine *Hox-4.2* gene. *Mol. Cell. Biol.* **13**, 257–265.

Porteus, M.H., Bulfone, A., Ciaranello, R.D. and Rubenstein, J.L.R. (1991). Isolation and characterization of a novel cDNA clone encoding a homeodomain that is developmentally regulated in the ventral forebrain. *Neuron* **7**, 221–229.

Potter, D.D., Andis, S.C., Matsumoto, S.G. and Furshpan, E.J. (1986). Synaptic functions in rat sympathetic neurons in microcultures. II. Adrenergic/cholinergic dual status and plasticity. *J. Neurosci.* **6**, 1080–1098.

Pourquié, O., Corbel, C., Le Caer, J.P., Rossier, J. and Le Douarin, N.M. (1992). BEN, a surface glycoprotein of the immunoglobulin superfamily, is expressed in a variety of developing systems. *Proc. Natl. Acad. Sci. USA* **89**, 5261–5265.

Pourquié, O., Coltey, M., Teillet, M.-A., Ordahl, C. and Le Douarin, N.M. (1993). Control of dorsoventral patterning of somitic derivatives by notochord and floor plate. *Proc. Natl. Acad. Sci. USA* **90**, 5243–5246.

Price, J. (1985). An immunohistochemical and quantitative examination of dorsal root ganglion neuronal subpopulations. *J. Neurosci.* **5**, 2051–2059.

Price, M., Lemaistre, M., Pischetola, M., Di Lauro, R. and Duboule, D. (1991). A mouse gene related to *Distal-less* shows a restricted expression in the developing forebrain. *Nature* **351**, 748–751.

Prince, V. and Lumsden, A. (1994). *Hoxa-2* expression in normal and transposed rhombomeres: independent regulation in the neural tube and neural crest. *Development* **120**, 911–923.

Prota, G. (1992). *Melanins and Melanogenesis*. Academic Press Inc., San Diego, CA.

Provencio, I., Jiang, G., de Grip, W.J., Hayes, W.P. and Rollag, M.D. (1998). Melanopsin: an opsin in melanophores, brain, and eye. *Proc. Natl. Acad. Sci. USA* **95**, 340–345.

Pruginin-Bluger, M., Shelton, D.L. and Kalcheim, C. (1997). A paracrine effect for neuron-derived BDNF in development of dorsal root ganglia: stimulation of Schwann cell Myelin Protein expression by glial cells. *Mech. Dev.* **61**, 99–111.

Pueblitz, S., Weinberg, A.G. and Albores-Saavedra, J. (1993). Thyroid C cells in the Di George anomaly: a quantitative study. *Pediatr. Pathol.* **13**, 463–473.

Puelles, L. and Rubenstein, J.L.R. (1993). Expression patterns of homeobox and other putative regulatory genes in the embryonic mouse forebrain suggest a neuromeric organization. *Trends Neurosci.* **16**, 473–479.

Puffenberger, E.G., Hosoda, K., Washington, S.S., Nakao, K., Dewit, D., Yanagisawa, M. and Chakravarti, A. (1994). A missense mutation of the endothelin-B receptor gene in multigenic Hirschsprung's disease. *Cell* **79**, 1257–1266.

Qiu, F.H., Ray, P., Brown, K., Barker, P.E., Jhanwar, S., Ruddle, F.H. and Besmer, P. (1988). Primary structure of c-*kit:* relationship with the CSF-1/PDGF receptor kinase family-oncogenic activation of v-*kit* involves deletion of extracellular domain and C terminus. *EMBO J.* **7**, 1003–1011.

Qiu, M., Bulfone, A., Martinez, S., Meneses, J.J., Shimamura, K., Pedersen, R.A. and Rubenstein, J.L. (1995). Null mutation of *Dlx-2* results in abnormal morphogenesis of proximal first and second branchial arch derivatives and abnormal differentiation in the forebrain. *Genes Dev.* **9**, 2523–2538.

Qiu, M.S., Bulfone, A., Ghattas, I., Meneses, J.J., Christensen, L., Sharpe, P.T., Presley, R., Pedersen, R.A. and Rubenstein, J.L.R. (1997). Role of the Dlx homeobox genes in proximodistal patterning of the branchial arches: mutations of *Dlx-1*, *Dlx-2*, and *Dlx-1* and *-2* alter morphogenesis of proximal skeletal and soft tissue structures derived from the first and second arches. *Dev. Biol.* **185**, 165–184.

Rabl, W. (1922). Entwicklung der Carotisdrüse bei Meerschweinchen. *Arch. Mikrobiol. Anat.* **96**, 315–339.

Raff, M. (1992). Social controls on cell survival and cell death. *Nature* **356**, 397–400.

Raible, D.W. and Eisen, J.S. (1992). Spatiotemporal restriction of trunk neural crest cell lineage in the embryonic zebrafish. *Soc. Neurosci. Abstr.* 18.

Raible, D.W. and Eisen, J.S. (1994). Restriction of neural crest cell fate in the trunk of the embryonic Zebrafish. *Development* **120**, 495–503.

Raible, D.W. and Eisen, J.S. (1996). Regulative interactions in zebrafish neural crest. *Development* **122**, 501–507.

Raible, D.W., Wood, A., Hodsdon, W., Henion, P.D., Weston, J.A. and Eisen, J.S. (1992). Segregation and early dispersal of neural crest cells in the embryonic Zebrafish. *Dev. Dyn.* **195**, 29–42.

Rama-Sastry, B.V. and Sadavongvivad, G. (1979). Cholinergic systems in non-nervous tissues. *Pharmacol. Rev.* **30**, 65–132.

Rambourg, A., Clermont, Y. and Beaudet, A. (1983). Ultrastructural features of six types of neurons in dorsal root ganglia. *J. Neurocytol.* **12**, 47–66.

Ranscht, B. and Bronner-Fraser, M. (1991). T-cadherin expression alternates with migrating neural crest cells in the trunk of the avian embryo. *Development* **111**, 15–22.

Ranscht, B. and Dours-Zimmermann, M.T. (1991). T-cadherin, a novel cadherin cell adhesion molecule in the nervous system lacks the conserved cytoplasmic region. *Neuron* **7**, 391–402.

Rao, M. and Landis, S.C. (1990). Characterization of a target-derived neuronal cholinergic differentiation factor. *Neuron* **5**, 899–910.

Rao, M.S. and Landis, S.C. (1993). Cell interactions that determine sympathetic neuron transmitter phenotype and the neurokines that mediate them. *J. Neurobiol.* **24**, 215–233.

Rao, M.S., Landis, S.C. and Patterson, P.H. (1990). The cholinergic neuronal differentiation factor from heart cell conditioned medium is different from the cholinergic factors in sciatic nerve and spinal cord. *Dev. Biol.* **139**, 65–74.

Rao, M.S., Patterson, P.H. and Landis, S.C. (1992a). Multiple cholinergic differentiation factors are present in footpad extracts: comparison with known cholinergic factors. *Development* **116**, 731–744.

Rao, M.S., Tyrrell, S., Landis, S.C. and Patterson, P.H. (1992b). Effects of ciliary neurotrophic factor (CNTF) and depolarization on neuropeptide expression in cultured sympathetic neurons. *Dev. Biol.* **150**, 281–293.

Rao, M., Patterson, P.H. and Landis, S.C. (1992c). Membrane-associated neurotransmitter stimulating factor is very similar to ciliary neurotrophic factor. *Dev. Biol.* **153**, 411–416

Rapraeger, A.C, Guimond, S., Krufka, A. and Olwin, B.B. (1994). Regulation by heparan sulfate in fibroblast growth factor signalling. *Methods Enzymol.* **245**, 219–240.

Rau, A.S. and Johnson, P.H. (1923). Observations on the development of the sympathetic system and suprarenal bodies in the sparrow. *Proc. Zool. Soc. Lond.* **3**, 741–768.

Raven, C.P. (1931). Zur Entwicklung der Ganglienleiste. I. Die Kinematik der Ganglienleistenentwicklung bei den Urodelen. *Roux Arch. Entwicklungsmech. Org.* **125**, 210–292.

Raven, C.P. (1933). Zur Entwicklung der Ganglienleiste. II. Uber das Differenzierungsvermögen des Kopfganglienleistenmaterials von Urodelen. *Roux Arch. Entwicklungsmech. Org.* **129**, 179–198.

Raven, C.P. (1935). Zur Entwicklung der Ganglienleiste. IV. Untersuchungen über Zeitpunkt und Verlauf der "materiellen Determination" des präsumptiven Kopfganglienleistenmaterials der Urodelen. *Roux Arch. Entwicklungsmech. Org.* **132**, 509–575.

Raven, C.P. (1936). Zur Entwicklung der Ganglienleiste. V. Uber die Differenzierung des Rumpfganglienleistenmaterials. *Roux Arch. Entwicklungsmech. Org.* **134**, 123–145.

Raven, C.P. (1937). Experiments on the origin of the sheath cells and sympathetic neuroblasts in Amphibia. *J. Comp. Neurol.* **67**, 220–240.

Raven, C.R. and Kloos, J. (1945). Induction by medial and lateral pieces of the archenteronroof, with special reference to the determination of the neural crest. *Acta Neerl. Morphol.* **5**, 348–362.

Rawles, M.E. (1940a). The pigment forming potency of early chick blastoderm. *Proc. Natl. Acad. Sci. USA* **26**, 86–94.

Rawles, M. (1940b). The development of melanophores from embryonic mouse tissue grown in the coelom of chick embryos. *Proc. Natl. Acad. Sci. USA* **26**, 673–680.

Rawles, M.E. (1945). Behavior of melanoblasts derived from the coelomic lining in interbreed grafts of wing skin. *Physiol. Zool.* **18**, 1–16.

Rawles, M.E. (1947). Some observations of the developmental properties of the presumptive hind-limb area of the chick. *Anat. Rec.* **99**, 648–649.

Rawles, M.E. (1948). Origin of melanophores and their role in development of color patterns in vertebrates. *Physiol. Rev.* **28**, 383–408.

Regan, L.J., Dodd, J., Barondes, S.H. and Jessell, T.M. (1986). Selective expression of endogenous lactose-binding lectins and lactoseries glycoconjugates in subsets of rat sensory neurons. *Proc. Natl. Acad. Sci. USA* **83**, 2248–2252.

Reichardt, L.F. and Patterson, P.H. (1977). Neurotransmitter synthesis and uptake by isolated sympathetic neurones in microcultures. *Nature* **270**, 147–151.

Reid, K., Nishikawa, S., Bartlett, P.F. and Murphy, M. (1995). Steel factor directs melanocyte development *in vitro* through selective regulation of the number of c-kit+ progenitors. *Dev. Biol.* **169**, 568–579.

Reid, K., Turnley, A.M., Maxwell, G.D., Kurihara, Y., Kurihara, H., Bartlett, P.F. and Murphy, M. (1996). Multiple roles for endothelin in melanocyte development: regulation of progenitor number and stimulation of differentiation. *Development* **122**, 3911–3919.

Reissmann, E., Ernsberger, U., Francis-West, P.H., Rueger, D., Brickell, P.M. and Rohrer, H. (1996). Involvement of bone morphogenetic protein-4 and bone mor-phogenetic protein-7 in the differentiation of the adrenergic phenotype in develop-ing sympathetic neurons. *Development* **122**, 2079–2088.

Reith, A.D., Rottapel, R., Giddens, E., Brady, C., Forester, L. and Bernstein, A. (1990). *W* mutant mice with mild or severe developmental defects contain distinct point mutations in the kinase domain of the c-*kit* transmembrane receptor. *Genes Dev.* **4**, 390–400.

Reynolds, K., Mezey, E. and Zimmer, A. (1991). Activity of the beta-retinoic acid receptor promoter in transgenic mice. *Mech. Dev.* **36**, 15–29.

Riccardi, V.M. (1991). Neurofibromatosis: past, present and future. *N. Engl. J. Med.* **324**, 1283–1285.

Richardson, M.K. and Sieber-Blum, M. (1993). Pluripotent neural crest cells in the developing skin of the quail embryo. *Dev. Biol.* **157**, 348–358.

Richardson, M.K., Hornbruch, A. and Wolpert, L. (1989). Pigment pattern expression in the plumage of the quail embryo and the quail–chick chimaera. *Development* **107**, 805–818.

Richardson, M.K., Hornbruch, A. and Wolpert, L. (1990). Mechanisms of pigment pattern formation in the quail embryo. *Development* **109**, 81–89.

Richardson, P.M. and Ebendal, T. (1982). Nerve growth activities in rat peripheral nerve. *Brain Res.* **246**, 57–64.

Richman, J.M., Herbert, M., Matovinovic, E. and Walin, J. (1997). Effect of fibroblast growth factors on outgrowth of facial mesenchyme. *Dev. Biol.* **189**, 135–147.

Rickmann, M., Fawcett, J.W. and Keynes, R.J. (1985). The migration of neural crest cells and the growth of motor axons through the rostral half of the chick somite. *J. Embryol. Exp. Morphol.* **90**, 437–455.

Rijli, F.M., Mark, M., Lakkaraju, S., Dierich, A., Dollé, P. and Chambon, P. (1993). A homeotic transformation is generated in the rostral branchial region of the head by disruption of *Hoxa-2*, which acts as a selector gene. *Cell* **75**, 1333–1349.

Rinchik, E.M., Bultman, S.J., Horsthemke, B., Lee, S.T., Strunk, K.M., Spritz, R.A., Avidano, K.M., Jong, M.T. and Nicholls, R.D. (1993). A gene for the mouse pink-eyed dilution locus and for human type II oculocutaneous albinism. *Nature* **361**, 73–76.

Ring, C., Hassell, J. and Halfter, W. (1996). Expression pattern of collagen IX and potential role in the segmentation of the peripheral nervous system. *Dev. Biol.* **180**, 41–53.

Ris, H. (1941). An experimental study of the origin of melanophores in birds. *Physiol. Zool.* **14**, 48–66.

Rivera-Pérez, J.A., Mallo, M., Gendron-Maguire, M., Gridley, T. and Behringer, R.R. (1995). *goosecoid* is not an essential component of the mouse gastrula organizer but is required for craniofacial and rib development. *Development* **121**, 3005–3012.

Robbins, L.S., Nadeau, J.H., Johnson, K.R., Kelly, M.A., Roselli-Rehfuss, L., Baack, E., Mountjoy, K.G. and Cone, R.D. (1993). Pigmentation phenotypes of variant extension locus alleles result from point mutations that alter MSH receptor function. *Cell* **72**, 827–834.

Robert, B., Sassoon, D., Jacq, B., Gehring, W. and Buckingham, M. (1989). *Hox-7*, a mouse homeobox gene with a novel pattern of expression during embryogenesis. *EMBO J.* **8**, 91–100.

Roberts, V.J. and Barth, S.L. (1994). Expression of messenger ribonucleic acids encoding the inhibin/activin system during mid- and late-gestation rat embryogenesis. *Endocrinology* **134**, 914–923.

Robertson, K. and Mason, I. (1995). Expression of *ret* in the chicken embryo suggests roles in regionalisation of the vagal neural tube and somites and in development of multiple neural crest and placodal lineages. *Mech. Dev.* **53**, 329–344.

Robinson, G.W. and Mahon, K.A. (1994). Differential and overlapping expression domains of *Dlx-2* and *Dlx-3* suggest distinct roles for *Distal-less* homeobox genes in craniofacial development. *Mech. Dev.* **48**, 199–215.

Robinson, G.W., Wray, S. and Mahon, K.A. (1991). Spatially restricted expression of a member of a new family of murine distal-less homeobox genes in the developing forebrain. *New Biol.* **3**, 1183–1194.

Rockwood, J.M. and Maxwell, G.D. (1996). An analysis of the effects of retinoic acid and other retinoids on the development of adrenergic cells from the avian neural crest. *Exp. Cell Res.* **223**, 250–258.

Rodriguez-Tebar, A. and Rohrer, H. (1991). Retinoic acid induces NGF-dependent survival response and high affinity NGF receptors in immature chick sympathetic neurons. *Development* **112**, 813–820.

Roelink, H., Augsburger, A., Heemskerk, J., Korzh, V., Norlin, S., Ruiz i Altaba, A., Tanabe, Y., Placzek, M., Edlund, T., Jessell, T.M. and Dodd, J. (1994). Floor plate and motor neuron induction by *vhh-1*, a vertebrate homolog of *hedgehog* expressed by the notochord. *Cell* **76**, 761–775.

Rogers, D.C. (1965). The development of the rat carotid body. *J. Anat.* **99**, 89–101.

Rogers, S.L., Bernard, L. and Weston, J.A. (1990). Substratum effects on cell dispersal, morphology, and differentiation in cultures of avian neural crest cells. *Dev. Biol.* **141**, 173–182.

Rogers, S.L., Edson, K.J., Letourneau, P.C. and McLoon, S. (1986). Distribution of laminin in the developing peripheral nervous system of the chick. *Dev. Biol.* **113**, 429–435.

Rohlich, K. (1929). Experimentelle Untersuchungen uber der Zeitpunkt der determination der gehorblase bei *Amblystoma* embryonen. *Arch. Entwicklungsmech.* **118**, 164–199.

Rohrer, H. (1992). Cholinergic neuronal differentiation factors: evidence for the presence of both CNTF-like and non-CNTF-like factors in developing rat footpad. *Development* **114**, 689–698.

Rohrer, H. and Thoenen, H. (1987). Relationship between differentiation and terminal mitosis: chick sensory and ciliary neurons differentiate after terminal mitosis of precursor cells, whereas sympathetic neurons continue to divide after differentiation. *J. Neurosci.* **7**, 3739–3748.

Rohrer, H., Heumann, R. and Thoenen, H. (1988a). The synthesis of nerve growth factor (NGF) in developing skin is independent of innervation. *Dev. Biol.* **128**, 240–244.

Rohrer, H., Hofer, M., Hellweg, R., Korsching, S., Stehle, A.D., Saadat, S. and Thoenen, H. (1988b). Antibodies against mouse nerve growth factor interfere *in vivo* with the development of avian sensory and sympathetic neurones. *Development* **103**, 545–552.

Rollhauser ter Horst, J. (1979). Artificial neural crest formation in amphibia. *Anat. Embryol. Berl.* **157**, 113–120.

Rong, P.M., Teillet, M.-A., Ziller, C. and Le Douarin, N.M. (1992). The neural tube/notochord complex is necessary for vertebral but not limb and body wall striated muscle differentiation. *Development* **115**, 657–672.

Ros, M.A., Sefton, M. and Nieto, M.A. (1997). *Slug*, a zinc finger gene previously implicated in the early patterning of the mesoderm and the neural crest, is also involved in chick limb development. *Development* **124**, 1821–1829.

Rosen, O., Geffen, R., Avivi, C. and Goldstein, R.S. (1996). Growth, proliferation, and cell death in the ontogeny of transient DRG (Froriep's ganglia) of chick embryos. *J. Neurobiol.* **30**, 219–230.

Rosenthal, A., Goeddel, D.V., Nguyen, T., Lewis, M., Shih, A., Laramee, G.R., Nikolics, K. and Winslow, J. W. (1990). Primary structure and biological activity of a novel human neurotrophic factor. *Neuron* **4**, 767–773.

Rossant, J., Zirngibl, R., Cado, D., Shago, M. and Giguère, V. (1991). Expression of a retinoic acid response element-*hsplacZ* transgene defines specific domains of transcriptional activity during mouse embryogenesis. *Genes Dev.* **5**, 1333–1344.

Rothman, T. and Gershon, M.D. (1982). Phenotypic expression in the developing murine enteric nervous system. *J. Neurosci.* **2**, 381–393.

Rothman, T.P. and Gershon, M.D. (1984). Regionally defective colonization of the terminal bowel by the precursors of enteric neurons in lethal spotted mutant mice. *Neuroscience* **12**, 1293–1311.

Rothman, T., Gershon, M.D. and Holtzer, H. (1978). The relationship of cell division to the acquisition of adrenergic characteristics by developing sympathetic ganglion cell precursors. *Dev. Biol.* **65**, 323–341.

Rothman, T.P., Specht, L.A., Gershon, M.D., Joh, T.H., Teitelman, G., Pickel, V.M. and Reis, D.J. (1980). Catecholamine biosynthetic enzymes are expressed in replicating cells of the peripheral but not the central nervous system. *Proc. Natl. Acad. Sci. USA* **77**, 6221–6225.

Rothman, T.P., Sherman, D., Cochard, P. and Gershon, M.D. (1986). Development of the monoaminergic innervation of the avian gut: transient and permanent expression of phenotypic markers. *Dev. Biol.* **116**, 357–380.

Rothman, T.P., Le Douarin, N.M., Fontaine-Pérus, J.C. and Gershon, M.D. (1990). Developmental potential of neural crest-derived cells migrating from segments of developing quail bowel back-grafted into younger chick host embryos. *Development* **109**, 411–423.

Rothman, T.P., Goldowitz, D. and Gershon, M.D. (1993a). Inhibition of migration of neural crest-derived cells by the abnormal mesenchyme of the presumptive aganglionic bowel of *Ls/Ls* mice: analysis with aggregation and interspecies chimeras. *Dev. Biol.* **159**, 559–573.

Rothman, T.P., Le Douarin, N.M., Fontaine-Pérus, J.C. and Gershon, M.D. (1993b). Colonization of the bowel by neural crest-derived cells re-migrating from foregut backtransplanted to vagal or sacral regions of host embryos. *Dev. Dyn.* **196**, 217–233.

Rothman, T.P., Chen, J., Howard, M.J., Costantini, F., Schuchardt, A., Pachnis, V. and Gershon, M.D. (1996). Increased expression of laminin-1 and collagen (IV) subunits in the aganglionic bowel of ls/ls, but not c-ret –/– mice. *Dev. Biol.* **178**, 498–513.

Rovasio, R.A., Delouvée, A., Yamada, K.M., Timpl, R. and Thiery, J.P. (1983). Neural crest cell migration: requirements for exogenous fibronectin and high cell density. *J. Cell Biol.* **96**, 462–473.

Rubenstein, J.L.R. and Puelles, L. (1994). Homeobox gene expression during development of the vertebrate brain. In *Current Topics in Developmental Biology*, ed. R.A. Pedersen, pp. 1–63. Academic Press, San Diego.

Rubenstein, J.L.R., Martinez, S., Shimamura, K. and Puelles, L. (1994). The embryonic vertebrate forebrain: the prosomeric model. *Science* **266**, 578–580.

Ruberte, E., Dollé, P., Krust, A., Zelent, A., Morriss-Kay, G. and Chambon, P. (1990). Specific spatial and temporal distribution of retinoic acid receptor gamma transcripts during mouse embryogenesis. *Development* **108**, 213–222.

Ruberte, E., Dollé, P., Chambon, P. and Morriss-Kay, G. (1991). Retinoic acid receptors and cellular retinoid binding proteins. II. Their differential pattern of transcription during early morphogenesis in mouse embryos. *Development* **111**, 45–60.

Ruberte, E., Friederich, V., Morriss-Kay, G. and Chambon, P. (1992). Differential distribution patterns of CRABP-I and CRABP-II transcripts during mouse embryogenesis. *Development* **115**, 973–987.

Ruberte, E., Friederich, V., Chambon, P. and Morriss-Kay, G. (1993). Retinoic acid receptors and cellular retinoid-binding proteins. III. Their differential transcript distribution during mouse nervous system development. *Development* **118**, 267–282.

Rubin, E. (1985). Development of the rat superior cervical ganglion: ganglion cell maturation. *J. Neurosci.* **5**, 673–684.

Ruch, J.V. (1984). Tooth morphogenesis and differentiation. In *Dentin and Dentinogenesis*, ed. A. Linde, pp. 47–79. CRC Press, Boca Raton, FL.

Ruch, J.V. (1995). *Odontogenesis. Embryonic dentition as a tool for developmental biology*. Ruch ed., special issue of *Int. J. Dev. Biol.* **39**, 1–297.

Rudel, C. and Rohrer, H. (1992). Analysis of glia cell differentiation in the developing chick peripheral nervous system: sensory and sympathetic satellite cells express different cell surface antigens. *Development* **115**, 519–526.

Ruit, K.G., Osborne, P.A., Schmidt, R.E., Johnson, E.M., Jr. and Snider, W.D. (1990). Nerve growth factor regulates sympathetic ganglion cell morphology and survival in the adult mouse. *J. Neurosci.* **10**, 2412–2419.

Ruit, K.G., Elliott, J.L., Osborne, P.A., Yan, Q. and Snider, W.D. (1992). Selective dependence of mammalian dorsal root ganglion neurons on nerve growth factor during embryonic development. *Neuron* **8**, 573–587.

Ruiz i Altaba, A. and Jessell, T.M. (1991). Retinoic acid modifies the pattern of cell differentiation in the central nervous system of neurula stage *Xenopus* embryos. *Development* **112**, 945–958.

Ruiz, J.C. and Robertson, E.J. (1994). The expression of the receptor-protein tyrosine kinase gene, *eck*, is highly restricted during early mouse development. *Mech. Dev.* **46**, 87–100.

Russell, E.S. (1979). Hereditary anemias of the mouse: a review for geneticists. *Adv. Genet.* **20**, 357–459.

Russell, W.L. (1947). *Splotch*, a new mutation in the house mouse *Mus musculus*. *Genetics* **32**, 107–111.

Saadat, S., Sendtner, M. and Rohrer, H. (1989). Ciliary neurotrophic factor induces cholinergic differentiation of rat sympathetic neurons in culture. *J. Cell Biol.* **108**, 1807–1816.

Sadaghiani, B. and Thiebaud, C.H. (1987). Neural crest development in the *Xenopus laevis* embryo, studied by interspecific transplantation and scanning electron microscopy. *Dev. Biol.* **124**, 91–110.

Sadaghiani, B. and Vielkind, J.R. (1989). Neural crest development in *Xiphophorus* fishes: scanning electron and light microscopic studies. *Development* **105**, 487–504.

Sadaghiani, B. and Vielkind, J.R. (1990). Distribution and migration pathways of HNK-1-immunoreactive neural crest cells in teleost fish embryos. *Development* **110**, 197–209.

Sadler, T.W., Liu, E.T. and Augustine, K.A. (1995). Antisense targeting of *engrailed-1* causes abnormal axis formation in mouse embryos. *Teratology* **51**, 292–299.

Saga, Y., Yagi, T., Ikawa, Y., Sakakura, T. and Aizawa, S. (1992). Mice develop normally without tenascin. *Genes Dev.* **6**, 1821–1831.

Saksela, O. and Rifkin, D.B. (1990). Release of basic fibroblast growth factor-heparan sulfate complexes from endothelial cells by plasminogen activator-mediated proteolytic activity. *J. Cell Biol.* **110**, 767–775.

Sakurai, T., Yanagisawa, M., Takuwa, Y., Miyazaki, H., Kimura, S., Goto, K. and Masaki, T. (1990). Cloning of a cDNA encoding a non-isopeptide-selective subtype of the endothelin receptor. *Nature* **348**, 733–735.

Saldivar, J.R., Sechrist, J.W., Krull, C.E., Ruffins, S. and Bronner-Fraser, M. (1997). Dorsal hindbrain ablation results in rerouting of neural crest migration and changes in gene expression, but normal hyoid development. *Development* **124**, 2729–2739.

Samaraweera, P., Fukuzawa, T., Law, J.T. and Bagnara, J.T. (1991). A monoclonal antibody to frog MIF. *Pigment Cell Res.* **4**, 133.

Sanchez, M.P., Silos-Santiago, I., Frisen, J., He, B., Lira, S.A. and Barbacid, M. (1996). Renal agenesis and the absence of enteric neurons in mice lacking GDNF. *Nature* **382**, 70–73.

Sanford, L.P., Ormsby, I., Gittenberger De Groot, A.C., Sariola, H., Friedman, R., Boivin, G.P., Cardell, E.L. and Doetschman, T. (1997). TGF beta 2 knockout mice have multiple developmental defects that are nonoverlapping with other TGF beta knockout phenotypes. *Development* **124**, 2659–2670.

Santoro, M., Rosati, R., Grieco, M., Berlingieri, M.T., D'Amato, G.L., de Franciscis, V. and Fusco, A. (1990). The ret proto-oncogene is consistently expressed in human pheochromocytomas and thyroid medullary carcinomas. *Oncogene* **5**, 1595–1598.

Sarvella, P.A. and Russell, L.B. (1956). *Steel*, a new dominant gene in the house mouse with effects on coat pigment and blood. *J. Heredity* **47**, 123–128.

Sasaki, E., Okamura, H., Chikamune, T., Kanai, Y., Watanabe, M., Naito, M. and Sakurai, M. (1993). Cloning and expression of the chicken *c-kit* proto-oncogene. *Gene* **128**, 257–261.

Satokata, I. and Maas, R. (1994). *Msx1* deficient mice exhibit cleft palate and abnormalities of craniofacial and tooth development. *Nat. Genet.* **6**, 348–356.

Schaible, R.H. (1969). Clonal distribution of melanocytes in piebald-spotted and variegated mice. *J. Exp. Zool.* **172**, 181–200.

Schecterson, L.C. and Bothwell, M. (1992). Novel roles for neurotrophins are suggested by BDNF and NT-3 mRNA expression in developing neurons. *Neuron* **9**, 449–463.

Scherson, T., Serbedzija, G., Fraser, S. and Bronner-Fraser, M. (1993). Regulative capacity of the cranial neural tube to form neural crest. *Development* **118**, 1049–1061.

Schilling, T.F. and Kimmel, C.B. (1994). Segment and cell type lineage restrictions during pharyngeal arch development in the zebrafish embryo. *Development* **120**, 483–494.

Schimmang, T., Minichiello., L., Vazquez, E., San-Jose, I., Giraldez, F., Klein, R. and Represa, J. (1995). Developing inner ear sensory neurons require TrkB and trkC receptors for innervation of their peripheral targets. *Development* **121**, 3381–3391.

Schmechel, D., Marangos, P.J. and Brightman, M. (1978). Neurone-specific enolase is a molecular marker for peripheral and central neuroendocrine cells. *Nature* **276**, 834–836.

Schmidt, C., Bladt, F., Goedecke, S., Brinkmann, V., Zschiesche, W., Sharpe, M., Gherardi, E. and Birchmeier, C. (1995). Scatter factor/hepatocyte growth factor is essential for liver development. *Nature* **373**, 699–702.

Schneider-Maunoury, S., Topilko, P., Seitanidou, T., Levi, G., Cohen-Tannoudji, M., Pournin, S., Babinet, C. and Charnay, P. (1993). Disruption of *Krox-20* results in alteration of rhombomere 3 and rhombomere 5 in the developing hindbrain. *Cell* **75**, 1199–1214.

Schoenwolf, G.C. (1991). Cell movements driving neurulation in avian embryos. *Development* Suppl. 2, 157–168.

Schoenwolf, G.C. and Smith, J.L. (1990). Mechanisms of neurulation: traditional viewpoint and recent advances. *Development* **109**, 243–270.

Schorle, H., Meier, P., Buchert, M., Jaenisch, R. and Mitchell, P.J. (1996). Transcription factor AP-2 essential for cranial closure and craniofacial development. *Nature* **381**, 235–238.

Schotzinger, R.J. and Landis, S.C. (1988). Cholinergic phenotype developed by noradrenergic sympathetic neurons after innervation of a novel cholinergic target *in vivo*. *Nature* **335**, 637–639.

Schotzinger, R. and Landis, S.C. (1990). Acquisition of cholinergic and peptidergic properties by the sympathetic innervation of rat sweat glands requires interaction with normal target. *Neuron* **5**, 91–100.

Schotzinger, R., Yin, X. and Landis, S. (1994). Target determination of neurotransmitter phenotype in sympathetic neurons. *J. Neurobiol.* **25**, 620–639.

Schroeder, T.E. (1970). Neurulation in *Xenopus laevis*. An analysis and model based upon light and electron microscopy. *J. Embryol. Exp. Morphol.* **23**, 427–462.

Schrott, A. and Spoendlin, H. (1987). Pigment anomaly-associated inner ear deafness. *Acta Oto-Laryngol.* **103**, 451–457.

Schubert, F.R., Fainsod, A., Gruenbaum, Y. and Gruss, P. (1995). Expression of the novel murine homeobox gene *Sax-1* in the developing nervous system. *Mech. Dev.* **51**, 99–114.

Schuchardt, A., D'Agati, V., Larsson-Blomberg, L., Constantini, F. and Pachnis, V. (1994). Defects in the kidney and enteric nervous system of mice lacking the tyrosine kinase receptor Ret. *Nature* **367**, 380–383.

Schuchardt, A., D'Agati, V., Larsson-Blomberg, L., Constantini, F. and Pachnis, V. (1995a). RET-deficient mice: an animal model for Hirschprung's disease and renal agenesis. *J. Int. Med.* **238**, 327–332.

Schuchardt, A., Srinivas, S., Pachnis, V. and Constantini, F. (1995b) Isolation and characterization of a chicken homolog of the c-ret proto-oncogene. *Oncogene* **10**, 641–649.

Schwarting, G.A., Story, C.M. and Deutch, G. (1992). A monoclonal anti-glycoconjugate antibody defines a stage and position-dependent gradient in the developing sympathoadrenal system. *Histochem. J.* **24**, 842–851.

Schweizer, G., Ayer-Le Lièvre, C. and Le Douarin, N.M. (1983). Restrictions of developmental capacities in the dorsal root ganglia during the course of development. *Cell Differ.* **13**, 191–200.

Scott, M.P. and Weiner, A.J. (1984). Structural relationships among genes that control development: sequence homology between the Antennapedia, Ultrabithorax, and fushi tarazu loci of *Drosophila*. *Proc. Natl. Acad. Sci. USA* **81**, 4115–4119.

Scott, M.P., Weiner, A.J., Hazelrigg, T.I., Polisky, B.A., Pirrotta, V., Scalenghe, F. and Kaufman, T.C. (1983). The molecular organization of the *Antennapedia* locus of Drosophila. *Cell* **35**, 763–776.

Scott, S.A. (1993). Ontogeny, characterization and trophic dependence of AC4/anti-SSEA-1-positive sensory neurons in the chick. *Dev. Brain Res.* **75**, 175–184.

Scott, S.A., Patel, N. and Levine, J.M. (1990). Lectin binding identifies a subpopulation of neurons in chick dorsal root ganglia. *J. Neurosci.* **10**, 336–345.

Searle, A.G. (1968a). *Comparative Genetics of Coat Colour in Mammals*. Logos Press, London.

Searle, A.G. (1968b). An extension series in the mouse. *J. Heredity* **59**, 341–342.

Sechrist, J., Serbedzija, G.N., Scherson, T., Fraser, S.E. and Bronner-Fraser, M. (1993). Segmental migration of the hindbrain neural crest does not arise from its segmental generation. *Development* **118**, 691–703.

Sechrist, J., Nieto, M.A., Zamanian, R.T. and Bronner-Fraser, M. (1995). Regulative response of the cranial neural tube after neural fold ablation: spatiotemporal nature of neural crest regeneration and up-regulation of *Slug*. *Development* **121**, 4103–4115.

Seidl, K. and Unsicker, K.T. (1989). The determination of the adrenal medullary cell fate during embryogenesis. *Dev. Biol.* **136**, 481–490.

Seilheimer, B. and Schachner, M. (1987). Regulation of neural cell adhesion molecule expression on cultured mouse schwann cells by nerve growth factor. *EMBO J.* **6**, 1611–1616.

Sejvar, J.J., Landis, S.C. and Hall, A.K. (1993). SA-1 antigen expression in small intensely fluorescent cells is associated with proliferation. *Dev. Biol.* **157**, 547–552.

Sekido, Y., Takahashi, T., Ueda, R., Takahashi, M., Suzuki, H., Nishida, K., Tsukamoto, T., Hida, T., Shimokata, K., Zsebo, K.M. and Takahashi, T. (1993). Recombinant human stem cell factor mediates chemotaxis of small-cell lung cancer cell lines aberrantly expressing the *c-kit* protooncogene. *Cancer Res.* **53**, 1709–1714.

Selleck, M.A.J. and Bronner-Fraser, M. (1995). Origins of the avian neural crest: the role of neural plate-epidermal interactions. *Development* **121**, 525–538.

Selleck, M.A. and Stern, C.D. (1991). Fate mapping and cell lineage analysis of Hensen's node in the chick embryo. *Development* **112**, 615–626.

Selleck, M.A.J. and Stern, C.D. (1992). Commitment of mesoderm cells in Hensen's node of the chick embryo to notochord and somite. *Development* **114**, 403–415.

Semina, E.V., Reiter, R., Leysens, N.J., Alward, W.L.M., Small, K.W., Datson, N.A., Siegelbartelt, J., Bierkenelson, D., Bitoun, P., Zabel, B.U., Carey, J.C. and Murray, J.C. (1996). Cloning and characterization of a novel bicoid-related homeobox transcription factor gene, *RIEG*, involved in Rieger syndrome. *Nature Genet.* **14**, 392–399.

Sendtner, M., Kreutzberg, G.W. and Thoenen, H. (1990). Ciliary neurotrophic factor prevents the degeneration of motor neurons after axotomy. *Nature* **345**, 440–441.

Sendtner, M., Schmalbruch, H., Stöckli, K.A., Carroll, P., Kreutzberg, G.W. and Thoenen, H. (1992). Ciliary neurotrophic factor prevents degeneration of motor neurons in mouse mutant progressive motor neuronopathy. *Nature* **358**, 503–504.

Sendtner, M., Carroll, P., Holtmann, B., Hughes, R.A. and Thoenen, H. (1994). Ciliary neurotrophic factor. *J. Neurobiol.* **25**, 1436–1453.

Sengel, P. (1976). *Morphogenesis of the Skin.* Cambridge University Press, Cambridge.

Serbedzija, G.N. and McMahon, A.P. (1997). Analysis of neural crest cell migration in *Splotch* mice using a neural crest-specific LacZ reporter. *Dev. Biol.* **185**, 139–148.

Serbedzija, G.N., Bronner Fraser, M. and Fraser, S.E. (1989). A vital dye analysis of the timing and pathways of avian trunk neural crest cell migration. *Development* **106**, 809–816.

Serbedzija, G.N., Fraser, S.E. and Bronner-Fraser, M. (1990). Pathways of trunk neural crest cell migration in the mouse embryo as revealed by vital dye labelling. *Development* **108**, 605–612.

Serbedzija, G.N., Burgan, S., Fraser, S.E. and Bronner-Fraser, M. (1991). Vital dye labelling demonstrates a sacral neural crest contribution to the enteric nervous system of chick and mouse embryos. *Development* **111**, 857–866.

Serbedzija, G.N., Bronner-Fraser, M. and Fraser, S.E. (1992). Vital dye analysis of cranial neural crest cell migration in the mouse embryo. *Development* **116**, 297–307.

Servetnick, M. and Grainger, R.M. (1991). Changes in neural and lens competence in *Xenopus* ectoderm: evidence for an autonomous developmental timer. *Development* **112**, 177–188.

Sextier-Sainte-Claire Deville, F., Ziller, C. and Le Douarin, N.M. (1992). Developmental potentialities of cells derived from the truncal neural crest in clonal cultures. *Dev. Brain Res.* **66**, 1–10.

Sextier-Sainte-Claire Deville, F., Ziller, C. and Le Douarin, N.M. (1994). Developmental potentials of enteric neural crest-derived cells in clonal and mass cultures. *Dev. Biol.,* **163**, 141–151.

Shah, N.M., Marchionni, M.A., Isaacs, I., Stroobant, P. and Anderson, D.J. (1994). Glial growth factor restricts mammalian neural crest stem cells to a glial fate. *Cell* **77**, 349–360.

Shah, N.M., Groves, A.K. and Anderson, D.J. (1996). Alternative neural crest cell fates are instructively promoted by TGFb superfamily members. *Cell* **85**, 331–343.

Sham, M.H., Vesque, C., Nonchev, S., Marshall, H., Frain, M., Das Gupta, R., Whiting, J., Wilkinson, D., Charnay, P. and Krumlauf, R. (1993). The zinc finger gene *Krox20* regulates *HoxB2* (*Hox2.8*) during hindbrain segmentation. *Cell* **72**, 183–196.

Sharma, K., Korade, Z. and Frank, E. (1995). Late-migrating neuroepithelial cells from the spinal cord differentiate into sensory ganglion cells and melanocytes. *Neuron* **14**, 143–152.

Sharpe, P.T. (1995). Homeobox genes and orofacial development. *Connect. Tissue Res.* **32**, 17–25.

Shelton, D.L. and Reichardt, L.F. (1984). Expression of the *Beta-nerve growth factor* gene correlates with the density of sympathetic innervation in effector organs. *Proc. Natl. Acad. Sci. USA* **81**, 7951–7955.

Sherman, L., Stocker, K.M., Morrison, R. and Ciment, G. (1993). Basic fibroblast growth factor (bFGF) acts intracellularly to cause the transdifferentiation of avian neural crest-derived Schwann cell precursors into melanocytes. *Development* **118**, 1313–1326.

Shibahara, S., Tomita, Y., Sakakura, T., Nager, C., Chaudhuri, B. and Müller, R. (1986). Cloning and expression of cDNA encoding mouse tyrosinase. *Nucleic Acids Res.* **14**, 2413–2427.

Shimamura, K., Martinez, S., Puelles, L. and Rubenstein, J.L.R. (1997). Patterns of gene expression in the neural plate and neural tube subdivide the embryonic forebrain into transverse and longitudinal domains. *Dev. Neurosci.* **19**, 88–96.

Shimeld, S.M., McKay, I.J. and Sharpe, P.T. (1996). The murine homeobox gene *Msx-3* shows highly restricted expression in the developing neural tube. *Mech. Dev.* **55**, 201–210.

Sieber-Blum, M. (1989a). SSEA-1 is a specific marker for the spinal sensory neuron lineage in the quail embryo and in neural crest cell cultures. *Dev. Biol.* **134**, 363–375.

Sieber-Blum, M. (1989b). Commitment of neural crest cells to the sensory neuron lineage. *Science* **243**, 1608–1611.

Sieber-Blum, M. (1989c). Inhibition of the adrenergic phenotype in cultured neural crest cells by norepinephrine uptake inhibitors. *Dev. Biol.* **136**, 373–380.

Sieber-Blum, M. (1990). Mechanisms of neural crest diversification. *Comments Dev. Neurobiol.* **1**, 225–249.

Sieber-Blum, M. (1991). Role of the neurotrophic factors BDNF and NGF in the commitment of pluripotent neural crest cells. *Neuron* **6**, 949–955.

Sieber-Blum, M. and Cohen, A.M. (1980). Clonal analysis of quail neural crest cells: they are pluripotent and differentiate *in vitro* in the absence of noncrest cells. *Dev. Biol.* **80**, 96–106.

Sieber-Blum, M., Ito, K., Richardson, M.K., Langtimm, C.J. and Duff, R.S. (1993). Distribution of pluripotent neural crest cells in the embryo and the role of brain-derived neurotrophic factor in the commitment to the primary sensory neuron lineage. *J. Neurobiol.* **24**, 173–184.

Silver, L.M. (1995). *Mouse Genetics. Concepts and Applications*. Oxford University Press, New York.

Silvers, W.K. (1958a). An experimental approach to action of gene at the *agouti* locus in the mouse. II. Transplants of newborn *aa* ventral skin to $a^t a$, $A^w a$, $A–$, and *aa* hosts. *J. Exp. Zool.* **137**, 181–188.

Silvers, W.K. (1958b). An experimental approach to action of genes at the *agouti* locus in the mouse. III. Transplants of newborns $A^w–$, $A–$, and $a'–$ skin to $A^y–$, $A^w–$, $A–$, and *aa* hosts. *J. Exp. Zool.* **137**, 189–196.

Silvers, W.K. (1979). White spotting: piebald, lethal spotting, and belted. In *The Coat Color of Mice: A Model for Gene Action and Interaction*, ed. W.K. Silvers, pp. 185–205. Springer Verlag, New York.

Silvers, W.K. and Russel, E.S. (1955). An experimental approach to action of genes at the *agouti* locus in the mouse. *J. Exp. Zool.* **130**, 199–220.

Simeone, A., Gulisano, M., Acampora, D., Stornaiuolo, A., Rambaldi, M. and Boncinelli, E. (1992). Two vertebrate homeobox genes related to the Drosophila empty spiracles gene are expressed in the embryonic cerebral cortex. *EMBO J.* **11**, 2541–2550.

Simeone, A., Acampora, D., Mallamaci, A., Stornaiuolo, A., Dapice, M.R., Nigro, V. and Boncinelli, E. (1993). A vertebrate gene related to orthodenticle contains a homeodomain of the bicoid class and demarcates anterior neuroectoderm in the gastrulating mouse embryo. *EMBO J.* **12**, 2735–2747.

Simeone, A., Acampora, D., Pannese, M., Desposito, M., Stornaiuolo, A., Gulisano, M., Mallamaci, A., Kastury, K., Druck, T., Huebner, K. and Boncinelli, E. (1994). Cloning and characterization of two members of the vertebrate *Dlx* gene family. *Proc. Natl. Acad. Sci. USA* **91**, 2250–2254.

Simpson, P. (1983). Maternal-zygotic gene interactions during formation of the dorso-ventral pattern in *Drosophila* embryos. *Genetics* **105**, 615–632.

Siracusa, L.D. (1991). Genomic organization and molecular genetics of the *agouti* locus in the mouse. *Ann. N. Y. Acad. Sci.* **642**, 419–430.

Siracusa, L.D. (1994). The *agouti* gene: turned on to yellow. *Trends Genet.* **10**, 423–428.

Sive, H.L., Draper, B.W., Harland, R.M. and Weintraub, H. (1990). Identification of a retinoic acid-sensitive period during primary axis formation in *Xenopus laevis*. *Genes Dev.* **4**, 932–942.

Sjokvist, F. (1963a). Pharmacological analysis of acetylcholinesterase-rich ganglion cells in the lumbosacral sympathetic system of the cat. *Acta Physiol. Scand.* **57**, 352–362.

Sjokvist, F. (1963b). The correlation between the occurrence and localization of acetyl-cholinesterase-rich cell bodies in the stellate ganglion and the outflow of cholinergic sweat secretory fibers to the fore paw of the cat. *Acta Physiol. Scand.* **57**, 339–351.

Skorstengaard, K., Jensen, M.S., Petersen, T.E. and Magnusson, S. (1986). Purification and complete primary structures of the heparin-, cell-, and DNA-binding domains of bovine plasma fibronectin. *Eur. J. Biochem.* **154**, 15–29.

Slominski, A., Costantino, R., Wortsman, J., Paus, R. and Ling, N. (1992). Melanotropic activity of gamma MSH peptides in melanoma cells. *Life Sci.* **50**, 1103–1108.

Slotkin, T.A., Smith, P.G., Lau, C. and Barsis, D.L. (1980) Functional aspects of development of catecholamine biosynthesis and release in the sympathetic nervous system. In *Biogenic Amines in Development*, eds. H. Parvez and S. Parvez, pp. 29–48. Elsevier, Amsterdam.

Smeyne, R.J., Klein, R., Schnapp, A., Long, L.K., Bryant, S., Lewin, A., Lira, S.A. and Barbacid, M. (1994). Severe sensory and sympathetic neuropathies in mice carrying a disrupted Trk/NGF receptor gene. *Nature* **368**, 246–249.

Smith, A., Robinson, V., Patel, K. and Wilkinson, D.G. (1997). The EphA4 and EphB1 receptor tyrosine kinases and ephrin-B2 ligand regulate targeted migration of bran-chial neural crest cells. *Curr. Biol.* **7**, 561–570.

Smith, J. and Fauquet, M. (1984). Glucocorticoids stimulate adrenergic differentiation in cultures of migrating and premigratory neural crest. *J. Neurosci.* **4**, 2160–2172.

Smith, J., Cochard, P. and Le Douarin, N.M. (1977). Development of choline acetyl-transferase and cholinesterase activities in enteric ganglia derived from presumptive adrenergic and cholinergic levels of the neural crest. *Cell Differ.* **6**, 199–216.

Smith, J., Fauquet, M., Ziller, C. and Le Douarin, N.M. (1979). Acetylcholine synthesis by mesencephalic neural crest cells in the process of migration *in vivo*. *Nature* **282**, 853–855.

Smith, M., Hickman, A., Amanze, D., Lumsden, A. and Thorogood, P. (1994). Trunk neural crest origin of caudal fin mesenchyme in the zebrafish *Brachydanio rerio*. *Proc. R. Soc. Lond. Ser. B* **256**, 137–145.

Smith, S.M. and Eichele, G. (1991). Temporal and regional differences in the expression pattern of distinct retinoic acid receptor-beta transcripts in the chick embryo. *Development* **111**, 245–252.

Snider, W.D. and Wright, D.E. (1996). Neurotrophins cause a new sensation. *Neuron* **16**, 229–232.

Snipes, G.J., Suter, U., Welcher, A.A. and Shooter, E.M. (1992). Characterization of a novel peripheral nervous system myelin protein (PMP-22/SR13). *J. Cell Biol.* **117**, 225–238.

Solter, D. and Knowles, B.B. (1978). Monoclonal antibody defining a stage-specific mouse embryonic antigen (SSEA-1). *Proc. Natl. Acad. Sci. USA* **75**, 5565–5569.

Sommer, E.W., Kazimierczak, J. and Droz, B. (1985). Neuronal subpopulations in the dorsal root ganglion of the mouse as characterized by combination of ultrastructural and cytochemical features. *Brain Res.* **346**, 310–326.

Sommer, L., Shah, N., Rao, M. and Anderson, D.J. (1995). The cellular function of MASH1 in autonomic neurogenesis. *Neuron* **15**, 1245–1258.

Soppet, D., Escandon, E., Maragos, J., Middlemas, D.S., Reid, S.W., Blair, J., Burton, L.E., Stanton, B.R., Kaplan, D.R., Hunter, T., Nikolics, K. and Parada, L.F. (1991). The neurotrophic factors brain-derived neurotrophic factor and neurotrophin-3 are ligands for the trkb tyrosine kinase receptor. *Cell* **65**, 895–903.

Sosa-Pineda, B., Chowdhury, K., Torres, M., Oliver, G. and Gruss, P. (1997). The *Pax4* gene is essential for differentiation of insulin-producing beta cells in the mammalian pancreas. *Nature* **386**, 399–402.

Speight, J.L., Yao, L., Rozenberg, I. and Bernd, P. (1993). Early embryonic quail dorsal root ganglia exhibit high affinity nerve growth factor binding and NGF responsiveness: absence of NGF receptors on migrating neural crest cells. *Dev. Brain Res.* **75**, 55–64.

Sporn, M.B., Roberts, A.B. and Goodman, D.S. (1984). *The Retinoids*, Vols 1 and 2. Academic Press, Orlando, FL.

Sprenger, F., Stevens, L.M. and Nüsslein-Vollard, C. (1989). The *Drosophila* gene *torso* encodes a putative receptor tyrosine kinase. *Nature* **338**, 478–483.

Squinto, S.P., Stitt, T.N., Aldrich, T.H., Davis, S., Bianco, S.M., Radziejewski, C., Glass, D.J., Masiakowski, P., Furth, M.E., Valenzuela, D.M., DiStefano, P.S. and Yancopoulos, G.D. (1991). *trk*B encodes a functional receptor for brain-derived neurotrophic factor and neurotrophin-3 but not nerve growth factor. *Cell* **65**, 885–893.

Steel, K.P. (1991). Similarities between mice and human with hereditary deafness. *Ann. N. Y. Acad. Sci.* **630**, 68–79.

Steel, K.P. and Barkway, C. (1989). Another role for melanocytes: their importance for normal *stria vascularis* development in the mammalian inner ear. *Development* **107**, 453–463.

Steel, K.P., Barkway, C. and Bock, G.R. (1987). Strial dysfunction in mice with cochleo-saccular abnormalities. *Hear Res.* **27**, 11–26.

Steel, K.P., Davidson, D.R. and Jackson, I.J. (1992). TRP-2/DT, a new early melanoblast marker, shows that steel growth factor (c-kit ligand) is a survival factor. *Development* **115**, 1111–1119.

Stein, R., Orit, S. and Anderson, D.J. (1988). The induction of a neural-specific gene, *SCG10*, by nerve growth factor in PC12 cells is transcriptional, protein synthesis dependent, and glucocorticoid inhibitable. *Dev. Biol.* **127**, 316–325.

Steingrimsson, E., Moore, K.J., Lamoreux, M.L., Ferre-D'Amare, A.R., Burley, S.K., Zimring, D.C., Skow, L.C., Hodgkinson, C.A., Arnheiter, H., Copeland, N.G. and Jenkins, N.A. (1994). Molecular basis of mouse microphthalmia (*mi*) mutations helps explain their developmental and phenotypic consequences. *Nat. Genet.* **8**, 256–263.

Stemple, D.L. and Anderson, D.J. (1992). Isolation of a stem cell for neurons and glia from the mammalian neural crest. *Cell* **71**, 973–985.

Stemple, D.L., Mahanthappa, N.K. and Anderson, D.J. (1988). Basic FGF induces neuronal differentiation, cell division, and NGF dependence in chromaffin cells: a sequence of events in sympathetic development. *Neuron* **1**, 517–525.

Stephenson, D.A., Mercola, M., Anderson, E., Wang, C.Y., Stiles, C.D., Bowen-Pope, D.F. and Chapman, V.M. (1991). Platelet-derived growth factor receptor alpha-subunit gene (*Pdgfra*) is deleted in the mouse patch (*Ph*) mutation. *Proc. Natl. Acad. Sci. USA* **88**, 6–10.

Stern, C.D. (1990). The marginal zone and its contribution to the hypoblast and primitive streak of the chick embryo. *Development* **109**, 667–682.

Stern, C.D., Norris, W.E., Bronner-Fraser, M., Carlson, G.J., Faissner, A., Keynes, R.J. and Schachner, M. (1989). J1/tenascin-related molecules are not responsible for the segmented pattern of neural crest cells or motor axons in the chick embryo. *Development* **107**, 309–319.

Stern, C.D., Artinger, K.B. and Bronner-Fraser, M. (1991). Tissue interactions affecting the migration and differentiation of neural crest cells in the chick embryo. *Development* **113**, 207–216.

Sternberg, J. and Kimber, S.J. (1986a). Distribution of fibronectin, laminin and entactin in the environment of migrating neural crest cells in early mouse embryos. *J. Embryol. Exp. Morphol.* **91**, 267–282.

Sternberg, J. and Kimber, S.J. (1986b). The relationship between emerging neural crest cells and basement membranes in the trunk of the mouse embryo: a TEM and immunocytochemical study. *J. Embryol. Exp. Morphol.* **98**, 251–268.

Sternfeld, M., Ming, M., Song, H., Sela, K., Tomberg, R., Poo, M. and Soreq, H. (1998). Acetylcholinesterase enhances neurite growth and synapse development through alternative contributions of its hydrolytic capacity, core protein and variable C termini. *J. Neurosci.* **15**, 1240–1249.

Stevens, L.M. and Landis, S.C. (1987). Development and properties of the secretory response in rat sweat glands: relationship to the induction of cholinergic function in sweat gland innervation. *Dev. Biol.* **123**, 179–190.

Stevens, L.M. and Landis, S.C. (1990). Target influences on transmitter choice by sympathetic neurons developing in the anterior chamber of the eye. *Dev. Biol.* **137**, 109–124.

Stocker, K.M., Sherman, L., Rees, S. and Ciment, G. (1991). Basic FGF and TGF-beta1 influence commitment to melanogenesis in neural crest-derived cells of avian embryos. *Development* **111**, 635–645.

Stocker, K.M., Baizer, L., Coston, T., Sherman, L. and Ciment, G. (1995). Regulated expression of neurofibromin in migrating neural crest cells of avian embryos. *J. Neurobiol.* **27**, 535–552.

Stoeckel, M.E. and Porte, A. (1969). Etude ultrastructurale des corps ultimobranchiaux du poulet. I. Aspect normal et développement embryonnaire. *Z. Zellforsch.* **94**, 495–512.

Stoeckel, M.E. and Porte, A. (1970). Origine embryonnaire et différenciation sécrétoire des cellulkes à calcitonine (cellules C) dans la thyroïde foetale du rat. Etude au microscope électronique. *Z. Zellforsch.* **106**, 251–268.

Stohr, P. (1955). Zusammenfassende Ergebnisse über die Endingunsweise des vegetativen Nervesnsystems. *Acta Neuroveg.* **10**, 21–109.

Stone, L.S. (1922). Experiments on the development of the cranial ganglia and the lateral line sense organs in *Amblystoma punctatum*. *J. Exp. Zool.* **35**, 421–496.

Stone, L.S. (1926). Further experiments on the extirpation and transplantation of mesectoderm in *Amblystoma punctatum*. *J. Exp. Zool.* **44**, 95–131.

Stone, L.S. (1929). Experiments showing the role of migrating neural crest (mesectoderm) in the formation of head skeleton and loose connective tissue in *Rana palustris*. *Wilhelm Roux' Arch. Entwicklungsmech. Org.* **118**, 40–77.

Stone, L.S. (1932). Selective staining of the neural crest and its preservation for microscopic study. *Anat. Rec.* **51**, 267–273.

St-Onge, L., Sosa-Pineda, B., Chowdhury, K., Mansouri, A. and Gruss, P. (1997). *Pax6* is required for differentiation of glucagon-producing alpha-cells in mouse pancreas. *Nature* **387**, 406–409.

Streeter, G.L. (1908). The nuclei of origin of the cranial nerves in the 10mm human embryos. *Anat. Rec.* **2**, 111–115.

Strudel, G. (1953). Conséquences de l'excision de tronçons du tube nerveux sur la morphogenèse de l'embryon de poulet et sur la différenciation de ses organes: contribution à la genèse de l'orthosympatique. *Ann. Sci. Nat. Zool.* **15**, 251–329.

Struhl, G. (1983). Role of the *esc* + gene product in ensuring the selective expression of segment-specific homeotic genes in *Drosophila*. *J. Embryol. Exp. Morphol.* **76**, 297–331.

Struhl, G. and White, R.A.H. (1985). Regulation of the *Ultrabithorax* gene of Drosophila by other bithorax complex genes. *Cell* **43**, 507–519.

Studer, M., Popperl, H., Marshall, H., Kuroiwa, A. and Krumlauf, R. (1994). Role of a conserved retinoic acid response element in rhombomere restriction of *Hoxb-1*. *Science* **265**, 1728–1732.

Sucov, H.M., Murakami, K.K. and Evans, R.M. (1990). Characterization of an auto-regulated response element in the mouse retinoic acid receptor type beta gene. *Proc. Natl. Acad. Sci. USA* **87**, 5392–5396.

Suda, T., Suda, J. and Ogawa, M. (1983). Single-cell origin of mouse hemopoietic colonies expressing multiple lineages in variable combinations. *Proc. Natl. Acad. Sci. USA* **80**, 6689–6693.

Summerbell, D. (1983). The effect of local application of retinoic acid to the anterior margin of the developing chick limb. *J. Embryol. Exp. Morphol.* **78**, 269–289.

Sundin, O.H. and Eichele, G. (1990). A homeo domain protein reveals the metameric nature of the developing chick hindbrain. *Genes Dev.* **4**, 1267–1276.

Sundin, O. and Eichele, G. (1992). An early marker of axial pattern in the chick embryo and its respecification by retinoic acid. *Development* **114**, 841–852.

Sutherland, A.E., Calarco, P.G. and Damsky, C.H. (1993). Developmental regulation of integrin expression at the time of implantation in the mouse embryo. *Development* **119**, 1175–1186.

Suvanto, P., Hiltunen, J.O., Arumae, U., Moshnyakov, M., Sariola, H., Sainio, K. and Saarma, M. (1996). Localization of glial cell line-derived neurotrophic factor (GDNF) mRNA in embryonic rat by *in situ* hybridization. *Eur. J. Neurosci.* **8**, 816–822.

Swiatek, P.J. and Gridley, T. (1993). Perinatal lethality and defects in hindbrain development in mice homozygous for a targeted mutation of the zinc finger gene *Krox20*. *Genes Dev.* **7**, 2071–2084.

Tabin, C.J. (1991). Retinoids, homeoboxes, and growth factors: toward molecular models for limb development. *Cell* **66**, 199–217.

Takagi, S., Tsuji, T., Kinutani, M. and Fujisawa, H. (1989). Monoclonal antibodies against species-specific antigens in the chick central nervous system: putative application as transplantation markers in the chick-quail chimeras. *J. Histochem. Cytochem.* **37**, 177–184.

Takahashi, C., Akiyama, N., Matsuzaki, T., Takai, S., Kitayama, H. and Noda, M. (1996). Characterization of a human MSX-2 cDNA and its fragment isolated as a transformation suppressor gene against v-Ki-ras oncogene. *Oncogene* **12**, 2137–2146.

Takahashi, M., Ritz, J. and Cooper, G.M. (1985). Activation of a novel human transforming gene, *ret*, by DNA rearrangement. *Cell* **42**, 581–588.

Takahashi, Y. and Le Douarin, N.M. (1990). cDNA cloning of a quail homeobox gene and its expression in neural crest-derived mesenchyme and lateral plate mesoderm. *Proc. Natl. Acad. Sci. USA* **87**, 7482–7486.

Takahashi, Y., Bontoux, M. and Le Douarin, N.M. (1991). Epithelio-mesenchymal interactions are critical for *Quox 7* expression and membrane bone differentiation in the neural crest-derived mandibular mesenchyme. *EMBO J.* **10**, 2387–2393.

Takahashi, Y., Monsoro-Burq, A.-H., Bontoux, M. and Le Douarin, N.M. (1992). A role for *QUOX-8* in the establishment of the dorsoventral pattern during vertebrate development. *Proc. Natl. Acad. Sci. USA* **89**, 10237–10241.

Takayama, H., La Rochelle, W.J., Anver, M., Bockman, D.E. and Merlino, G. (1996). Scatter factor/hepatocyte growth factor as a regulator of skeletal muscle and neural crest development. *Proc. Natl. Acad. Sci. USA* **93**, 5866–5871.

Takeichi, M. (1988). The cadherins: cell–cell adhesion molecules controlling animal morphogenesis. *Development* **102**, 639–655.

Takeichi, M., Matsunami, H., Inoue, T., Kimura, Y., Suzuki, S. and Tanaka, T. (1997). Roles for cadherins in patterning of the developing brain. *Dev. Neurosci.* **19**, 86–87.

Tan, J.C., Nocka, K., Ray, P., Traktman, P. and Besmer, P. (1990). The dominant *W42 spotting* phenotype results from a missense mutation in the c-*kit* receptor kinase. *Science* **247**, 209–212.

Tan, S.A., Deglon, N., Zurn, A.D., Baetge, E.E., Bamber, B., Kato, A.C. and Aebischer, P. (1996). Rescue of motoneurons from axotomy-induced cell death by polymer encapsulated cells genetically engineered to release CNTF. *Cell Transplant.* **5**, 577–587.

Tan, S.S. and Morriss-Kay, G.M. (1986). Analysis of cranial neural crest cell migration and early fates in postimplantation rat chimaeras. *J. Embryol. Exp. Morphol.* **98**, 21–58.

Tan, S.S., Crossin, K.L., Hoffman, S. and Edelman, G.M. (1987). Asymmetric expression in somites of cytotactin its proteoglycan ligand is correlated with neural crest cell distribution. *Proc. Natl. Acad. Sci. USA* **84**, 7977–7981.

Tan, S.S., Prieto, A.L., Newgreen, D.F., Crossin, K.L. and Edelman, G.M. (1991). Cytotactin expression in somites after dorsal neural tube and neural crest ablation in chicken embryos. *Proc. Natl. Acad. Sci. USA* **88**, 6398–6402.

Tanaka, H., Kinutani, M., Agata, A., Takashima, Y. and Obata, K. (1990). Pathfinding during spinal tract formation in Quail–Chick chimera analysed by species specific monoclonal antibodies. *Development* **110**, 565–571.

Taneja, R., Thisse, B., Rijli, F.M., Thisse, C., Bouillet, P., Dollé, P. and Chambon, P. (1996). The expression pattern of the mouse receptor tyrosine kinase gene MDK1 is conserved through evolution and requires *Hoxa-2* for rhombomere-specific expression in mouse embryos. *Dev. Biol.* **177**, 397–412.

Tapscott, S.J., Davies, R.L., Thayer, M.J., Cheng, P., Weintraub, H. and Lassar, A.B. (1988). MyoD1: a nuclear phosphoprotein requiring a Myc homology to convert fibroblasts to myoblasts. *Science* **242**, 405–411.

Tarkowski, A.K. (1961). Mouse chimaeras developed from fused eggs. *Nature* **190**, 857–860.

Tarkowski, A.K. (1963). Studies on mouse chimaeras developed from eggs fused *in vitro*. *Natl. Cancer Inst. Monogr.* **11**, 51–71.

Tarkowski, A.K. (1964). Patterns of pigmentation in experimentally produced mouse chimaeras. *J. Embryol. Exp. Morphol.* **12**, 575–585.

Tassabehji, M., Read, A.P., Newton, V.E., Harris, R., Balling, R., Gruss, P. and Strachan, T. (1992). Waardenburg's syndrome patients have mutations in the human homologue of the *Pax-3* paired box gene. *Nature* **355**, 635–636.

Tassabehji, M., Newton, V.E. and Read, A.P. (1994). Waardenburg syndrome type 2 caused by mutations in the human microphthalmia (*MITF*) gene. *Nat. Genet.* **8**, 251–255.

Teillet, M.-A. (1971a). Recherches sur le mode de migration et la différenciation des mélanoblastes cutanés chez l'embryon d'oiseau: étude expérimentale par la méthode des greffes hétérospécifiques entre embryons de Caille et de Poulet. *Ann. Embryol. Morphog.* **4**, 95–109.

Teillet, M.-A. (1971b). Niveau d'origine de plusieurs dérivés des crêtes neurales cervicales postérieures, thoraciques et lombaires chez l'embryon d'oiseau. *C. R. Assoc. Anat.* **152**, 734–743.

Teillet, M.-A. (1978). Evolution of the lumbo-sacral neural crest in the avian embryo: origin and differentiation of the ganglionated nerve of Remak studied in interspecific quail–chick chimerae. *Roux Arch. Dev. Biol.* **184**, 251–268.

Teillet, M.-A. and Le Douarin, N.M. (1970). La migration des cellules pigmentaires étudiée par la méthode des greffes hétérospécifiques de tube nerveux chez l'embryon d'oiseau. *C.R. Acad. Sci., Ser B* **270**, 3095–3098.

Teillet, M.A. and Le Douarin, N. (1974). Determination of the level of the origin of the adrenal medulla cells in the neural axis using heterospecific grafts of quail neural rudiments on chick embryos. *Arch. Anat. Microsc. Morphol. Exp.* **63**, 51–62.

Teillet, M.A. and Le Douarin, N.M (1983). Consequences of neural tube and notochord excision on the development of the peripheral nervous system in the chick embryo. *Dev. Biol.* **98**, 192–211.

Teillet, M.-A., Cochard, P. and Le Douarin, N.M. (1978). Relatives roles of the mesenchymal tissues and of the complex neural tube–notochord on the expression of adrenergic metabolism in neural crest cells. *Zoon* **6**, 115–122.

Teillet, M.-A., Kalcheim, C. and Le Douarin, N.M. (1987). Formation of the dorsal root ganglia in the avian embryo: segmental origin and migratory behavior of neural crest progenitor cells. *Dev. Biol.* **120**, 329–347.

Teillet, M.-A., Lapointe, F. and Le Douarin, N.M. (1998a). The relationships between notochord and floor plate in vertebrate development revisited. *Proc. Natl. Acad. Sci. USA*, 11733–11738.

Teillet, M.-A., Watanabe, Y., Jeffs, P., Duprez, D., Lapointe, F. and Le Douarin, N.M. (1998b). *Sonic hedgehog* is required for survival of both myogenic and chondrogenic somitic lineages. *Development* **125**, 2019–2030.

Teitelman, G. (1990). Insulin cells of pancreas extend neurites but do not arise from the neuroectoderm. *Dev. Biol.* **142**, 368–379.

Teitelman, G. and Lee, J.K. (1987). Cell lineage analysis of pancreatic islet cell development: glucagon and insulin cells arise from catecholaminergic precursors present in the pancreatic duct. *Dev. Biol.* **121**, 454–466.

Teitelman, G., Gershon, M.D., Rothman, T.P., Joh, T. and Reis, D.J. (1981). Proliferation and distribution of cells that transiently express a catecholaminergic phenotype during development in mice and rats. *Dev. Biol.* **86**, 348–355.

Tello, J.F. (1925). Sur la formation des chaînes primaires et secondaires du grand sympathique dans l'embryon de poulet. *Trab. Lab. Invest. Biol. Univ. Madrid* **23**, 1–28.

Tello, J.F. (1946). Sobre la formacion de los ganglios nerviosos craneales y el mesectodermo cefalico en los embriones de pollo. *Trab. Inst. Cajal Invest. Biol.* **38**, 1–40.

Tennyson, V.M., Pham, T.D., Rothman, T.P. and Gershon, M.D. (1986). Abnormalities of smooth muscle, basal laminae, and nerves in the aganglionic segments of the bowel of lethal spotted mutant mice. *Anat. Rec.* **215**, 267–281.

Tennyson, V.M., Howard, M., Pomeranz, H.D., Rothman, T.P. and Gershon, M.D. (1991). Acquisition of the immunoreactivity of a 110kDa cellular laminin binding protein by the neural crest-derived precursors of enteric neurons. *Anat. Rec.* **229**, 89.

Tessarollo, L., Tsoulfas, P., Martinzanca, D., Gilbert, D.J., Jenkins, N.A., Copeland, N.G. and Parada, L.F. (1993). trkC, a receptor for neurotrophin-3, is widely expressed in the developing nervous system and in non-neuronal tissues. *Development* **118**, 463–475.

Tessier-Lavigne, M. (1995). Eph receptor tyrosine kinases, axon repulsion, and the development of topographic maps. *Cell* **82**, 345–348.

Thaler, C.D., Suhr, L., Ip, N. and Katz, D.M. (1994). Leukemia inhibitory factor and neurotrophins support overlapping populations of rat nodose sensory neurons in culture. *Dev. Biol.* **161**, 338–344.

Thaller, C. and Eichele, G. (1987). Identification and spatial distribution of retinoids in the developing chick limb bud. *Nature* **327**, 625–628.

Thesleff, I. and Nieminen, P. (1996). Tooth morphogenesis and cell differentiation. *Curr. Opin. Cell Biol.* **8**, 844–850.

Thesleff, I. and Sharpe, P. (1997). Signalling networks regulating dental development. *Mech. Dev.* **67**, 111–123.

Thesleff, I., Vaahtokari, A. and Partanen, A.M. (1995). Regulation of organogenesis. Common molecular mechanisms regulating the development of teeth and other organs. *Int. J. Dev. Biol.* **39**, 35–50.

Thesleff, I., Vaahtokari, A., Vainio, S. and Jowett, A. (1996). Molecular mechanisms of cell and tissue interactions during early tooth development. *Anat. Rec.* **245**, 151–161.

Thiebaud, C.H. (1983). A reliable new cell marker in *Xenopus*. *Dev. Biol.* **98**, 245–249.

Thiery, J.P., Duband, J.L. and Delouvée, A. (1982). Pathways and mechanism of avian trunk neural crest cell migration and localization. *Dev. Biol.* **93**, 324–343.

Thisse, B., Stoetzel, C., Gorostiza-Thisse, C. and Perrin-Schmitt, F. (1988). Sequence of the *twist* gene and nuclear localization of its protein in endomesodermal cells of early Drosophila embryos. *EMBO J.* **7**, 2175–2183.

Thoenen, H. (1991). The changing scene of neurotrophic factors. *Trends Neurosci* **14**, 165–170.

Thomas, B.L., Porteus, M.H., Rubenstein, J.H.R. and Sharp, P.T. (1995). The spatial localization of *Dlx-2* during tooth development. *Conn. Tissue Res.* **32**, 27–34.

Thompson, M., Fleming, K.A., Evans, D.J., Fundele, R., Surani, M.A. and Wright, N.A. (1990). Gastric endocrine cells share a clonal origin with other gut cell lineages. *Development* **110**, 477–481.

Thorogood, P. (1998). The head and the face. In *Embryos, Genes and Birth Defects*, ed. P. Thorogood, pp. 197–229. Wiley, New York.

Thuneberg, L. (1989) Interstitial cells of Cajal. In *Handbook of Physiology, the Gastrointestinal System*, eds. G.S. Schultz, J.D. Wood and B.B. Rauner, pp. 349–386. American Physiological Society, Bethesda, MD.

Tickle, C., Alberts, B., Wolpert, L. and Lee, J. (1982). Local application of retinoic acid to the limb bud mimics the action of the polarizing region. *Nature* **296**, 564–566.

Tischler, A.S., Ruzicka, L.A., Donahue, S.R. and DeLallis, R.A. (1989). Chromaffin cell proliferation in the adult rat adrenal medulla. *Int. J. Dev. Neurosci.* **7**, 439–448.

Tischler, A.S., Riseberg, J.C. and Cherington, V. (1994). Multiple mitogenic signalling pathways in chromaffin cells: a model for cell cycle regulation in the nervous system. *Neurosci. Lett.* **168**, 181–184.

Tojo, H., Kaisho, Y., Nakata, M., Matsuoka, K., Kitagawa, M., Abe, T., Takami, K., Yamamoto, M., Shino, A., Igarashi, K. *et al.* (1995). Targeted disruption of the neurotrophin-3 gene with lacZ induces loss of trkC-positive neurons in sensory ganglia but not in spinal cords. *Brain Res.* **16**, 163–175.

Tojo, H., Takami, K., Kaisho, Y., Nakata, M., Abe, T., Shiho, O. and Igarashi, K. (1996). Analysis of neurotrophin-3 expression using the *lacZ* reporter gene suggests its local mode of neurotrophic activity. *Neuroscience* **71**, 221–230.

Ton, C.C.T., Hirvonen, H., Miwa, H., Weil, M.M., Monaghan, P., Jordan, T., Vanheyningen, V., Hastie, N.D., Meijersheijboer, H., Drechsler, M., Royerpokora, B., Collins, F., Swaroop, A., Strong, L.C. and Saunders, G.F. (1991). Positional cloning and characterization of a paired box-containing and homeobox-containing gene from the aniridia region. *Cell* **67**, 1059–1074.

Tongiorgi, E., Bernhardt, R.R., Zinn, K. and Schachner, M. (1995). Tenascin-C mRNA is expressed in cranial neural crest cells, in some placodal derivatives, and in discrete domains of the embryonic zebrafish brain. *J. Neurobiol.* **28**, 391–407.

Torihashi, S., Ward, S.M., Nishikawa, S., Nishi, K., Kobayashi, S. and Sanders, K.M. (1995). c-kit-dependent development of interstitial cells and electrical activity in the murine gastrointestinal tract. *Cell Tissue Res,* **280**, 97–111.

Tosney, K.W. (1978). The early migration of neural crest cells in the trunk region of the avian embryo: an electron microscopic study. *Dev. Biol.* **62**, 317–333.

Tosney, K.W. (1982). The segregation and early migration of cranial neural crest cells in the avian embryo. *Dev. Biol.* **89**, 13–24.

Tosney, K.W. (1987). Proximal tissues and patterned neurite outgrowth at the lumbosacral level of the chick embryo: deletion of the dermamyotome. *Dev. Biol.* **122**, 540–558.

Tosney, K.W. (1988). Proximal tissues and patterned neurite outgrowth at the lumbosacral level of the chick embryo: partial and complete deletion of the somite. *Dev. Biol.* **127**, 266–286.

Tosney, K.W. (1992). A long-distance cue from emerging dermis stimulates neural crest migration. *Soc. Neurosci. Abstr.* **18**, 1284.

Tosney, K.W. and Landmesser, L.T. (1984). Pattern and specificity of axonal outgrowth following varying degrees of chick limb bud ablation. *J. Neurosci.* **4**, 2518–2527.

Tosney, K.W. and Oakley, R.A. (1990). The perinotochordal mesenchyme acts as a barrier to axon advance in the chick embryo: implications for a general mechanism of axonal guidance. *Exp. Neurol.* **109**, 75–89.

Tosney, K.W., Dehnbostel, D.B. and Erickson, C.A. (1994). Neural crest cells prefer the myotome's basal lamina over the sclerotome as a substratum. *Dev. Biol.* **163**, 389–406.

Trapp, B.D., Dubois-Dalcq, M. and Quarles, R.H. (1984). Ultrastructural localization of P2 protein in actively myelinating rat Schwann cells. *J. Neurochem.* **43**, 944–948.

Treanor, J.J., Goodman, L., de Sauvage, F., Stone, D.M., Poulsen, K.T., Beck, C.D., Gray, C., Armanini, M.P., Pollock, R.A., Hefti, F., Phillips, H.S., Goddard, A., Moore, M.W., Buj-Bello, A., Davies, A.M., Asai, N., Takahashi, M., Vandlen, R., Henderson, C.E. and Rosenthal, A. (1996). Characterization of a multicomponent receptor for GDNF. *Nature* **382**, 80–83.

Triplett, E.L. (1958). The development of the sympathetic ganglia, sheath cells, and meninges in amphibians. *J. Exp. Zool.* **138**, 283–312.

Trupp, M., Arenas, E., Fainzilber, M., Nilsson, A.S., Sieber, B.A., Grigoriou, M., Kilkenny, C., Salazar-Grueso, E., Pachnis, V., Arumae, U., Sariola, H., Saarma, M. and Ibanez, C.F. (1996). Functional receptor for GDNF encoded by the c-ret proto-oncogene. *Nature* **381**, 785–788.

Tsoulfas, P., Soppet, D., Escandon, E., Tessarollo, L., Mendoza-Ramirez, J.L., Rosenthal, A., Nikolics, K. and Parada, L.F. (1993). The rat *trk*C locus encodes multiple neurogenic receptors that exhibit differential response to neurotrophin-3 in PC12 cells. *Neuron* **10**, 975–990.

Tsujimura, T., Hirota, S., Nomura, S., Niwa, Y., Yamazaki, M., Tono, T., Morii, E., Kim, H.M., Kondo, K., Nishimune, Y. and Kitamura, Y. (1991). Characterization of *Ws* mutant allele of rats: a 13–base deletion in tyrosine kinase domain of c-*kit* gene. *Blood* **78**, 1942–1946.

Tsukamoto, K., Jackson, I.J., Urabe, K., Montague, P.M. and Hearing, V.J. (1992). A second tyrosinase-related protein, TRP-2, is a melanogenic enzyme termed DOPAchrome tautomerase. *EMBO J.* **11**, 519–526.

Tucker, G.C. (1984). Distribution tissulaire et caractérisation biochimique d'un antigène associé à des mouvements morphogénétiques chez l'embryon de vertébré. Thèse de docteur Ingénieur. Paris VI: Pierre et Marie Curie.

Tucker, G.C., Aoyama, H., Lipinski, M., Tursz, T. and Thiery, J.P. (1984). Identical reactivity of monoclonal antibodies HNK-1 and NC-1: conservation in vertebrates on cells derived from the neural primordium and on some leukocytes. *Cell Differ.* **14**, 223–230.

Tucker, G.C., Ciment, G. and Thiery, J.P. (1986). Pathways of avian neural crest cell migration in the developing gut. *Dev. Biol.* **116**, 439–450.

Tucker, R.P. and Erickson, C.A. (1984). Morphology and behavior of quail neural crest cells in artificial three-dimensional extracellular matrices. *Dev. Biol.* **104**, 390–405.

Tucker, R.P. and Erickson, C.A. (1986a). The control of pigment cell pattern formation in the California newt, *Taricha torosa*. *J. Embryol. Exp. Morphol.* **97**, 141–168.

Tucker, R.P. and Erickson, C.A. (1986b). Pigment cell pattern formation in *Taricha torosa*: the role of the extracellular matrix in controlling pigment cell migration and differentiation. *Dev. Biol.* **118**, 268–285.

Turner, W.A., Jr., Chen, S.-T., Wahn, H., Lightbody, L.T., Bagnara, J.T., Taylor, J.D. and Chen, T.T. (1977). Trophic effect of MSH on melanophores. *Front Hormone Res.* **4**, 105–116.

Tuzi, N.L. and Gullick, W.J. (1994). *eph*, the largest known family of putative growth factor receptors. *Br. J. Cancer* **69**, 417–421.

Twitty, V.C. (1936). Correlated genetic and embryological experiments on *Triturus*: I and II. *J. Exp. Zool.* **74**, 239–302.

Twitty, V.C. and Bodenstein, D. (1939). Correlated genetic and embryological experiments on *Triturus*. III: Further transplantation experiments on pigment development. IV: The study of pigment cell behavior *in vitro*. *J. Exp. Zool.* **81**, 357–398.

Twitty, V.C. and Bodenstein, D. (1941). Experiments on the determination problem. I: The roles of ectoderm and neural crest in the development of the dorsal fin in Amphibia. II: Changes in ciliary polarity associated with the induction of fin epidermis. *J. Exp. Zool.* **86**, 343–380.

Tyrrell, S. and Landis, S.C. (1994). The appearance of NPY and VIP in sympathetic neuroblasts and subsequent alterations in their expression. *J. Neurosci.* **14**, 4529–4547.

Uchida, S. (1927). Uber die Entwicklung des sympatischen Nervensystems bei den Vögeln. *Acta Sch. Med. Univ. Kyoto* **10**, 63–136.

Uehara, Y., Minowa, O., Mori, C., Shiota, K., Kuno, J., Noda, T. and Kitamura, N. (1995). Placental defect and embryonic lethality in mice lacking hepatocyte growth factor/scatter factor. *Nature* **373**, 702–705.

Ullrich, A. and Schlessinger, J. (1990). Signal transduction by receptors with tyrosine kinase activity. *Cell* **61**, 203–212.

Unsicker, K. (1996). GDNF: a cytokine at the interface of TGF-bs and neurotrophins. *Cell Tissue Res.* **286**, 175–178.

Unsicker, K., Reichert-Preibsch, H., Schmidt, R., Pettmann, B., Labourdette, G. and Sensebrenner, M. (1987). Astroglial and fibroblast growth factors have neurotrophic functions for cultured peripheral and central nervous system neurons. *Proc. Natl. Acad. Sci. USA* **84**, 5459–5463.

Upchurch, B.H., Aponte, G.W. and Leiter, A.B. (1994). Expression of peptide YY in all four islet cell types in the developing mouse pancreas suggests a common peptide YY-producing progenitor. *Development* **120**, 245–252.

Vaage, S. (1969). The segmentation of the primitive neural tube in chick embryos (*Gallus domesticus*). A morphological, histochemical and autoradiographical investigation. In *Advances in Anatomy, Embryology, and Cell Biology*, eds. A. Brodal, K.T.H. Schiebler and W.G. Töndury, pp. 1–88. Springer Verlag, Berlin.

Valenzuela, D.M., Maisonpierre, P.C., Glass, D.J., Rojas, E., Nunez, L., Kong, Y., Gies, D R., Stitt, T.N., Ip, N.Y. and Yancopoulos, G.D. (1993). Alternative forms of rat trkC with different functional capabilities. *Neuron* **10**, 963–974.

Valinsky, J.E. and Le Douarin, N.M. (1985). Production of plasminogen activator by migrating cephalic neural crest cells. *EMBO J.* **4**, 1403–1406.

Van Campenhout, E. (1931). Le développement du système nerveux sympathique chez le poulet. *Arch. Biol.* **42**, 479–507.

Van Campenhout, E. (1932). Further experiments on the origin of the enteric nervous system in the chick. *Physiol. Zool.* **5**, 333–353.

Van Campenhout, E. (1937). Le développement du système nerveux crânien chez le poulet. *Arch. Biol.* **48**, 611–666.

Van Oostrom, C.G. and Verwoerd, C.D.A. (1972). The origin of the olfactory placode. *Acta Morphol. Neerl. Scand.* **9**, 160.

Varley, J.E. and Maxwell, G.D. (1996). BMP-2 and BMP-4, but not BMP-6, increase the number of adrenergic cells which develop in quail trunk neural crest cultures. *Exp. Neurol.* **140**, 84–94.

Varley, J.E., Wehby, R.G., Rueger, D.C. and Maxwell, G.D. (1995). Number of adrenergic and islet-1 immunoreactive cells is increased in avian trunk neural crest cultures in the presence of human recombinant osteogenic protein-1. *Dev. Dyn.* **203**, 434–437.

Vastardis, H., Karimbux, N., Guthua, S.W., Seidman, J.G. and Seidman, C.E. (1996). A human *MSX1* homeodomain missense mutation causes selective tooth agenesis. *Nat. Genet.* **13**, 417–421.

Verdi, J.M. and Anderson, D.J. (1994). Neurotrophins regulate sequential changes in neurotrophin receptor expression by sympathetic neuroblasts. *Neuron* **13**, 1359–1372.

Verdi, J.M., Ip, N., Yancopoulos, G.D. and Anderson, D.J. (1994). Expression of *trk* in MAH cells lacking the p75 low-affinity nerve growth factor receptor is sufficient to permit nerve growth factor-induced differentiation to postmitotic neurons. *Proc. Natl. Acad. Sci. USA* **91**, 3949–3953.

Verdi, J.M., Groves, A.K., Fariñas, I., Jones, K., Marchionni, M.A., Reichardt, L.F. and Anderson, D.J. (1996). A reciprocal cell–cell interaction mediated by NT-3 and neuregulins controls the early survival and development of sympathetic neuroblasts. *Neuron* **16**, 515–527.

Verhofstad, A.A., Coupland, R.E. and Colenbrander, B. (1989). Immunohistochemical and biochemical analysis of the development of the noradrenaline- and adrenaline-storing cells in the adrenal medulla of the rat and pig. *Arch. Histol. Cytol.* **52** (Suppl.), 351–360.

Villares, R. and Cabrera, C.V. (1987). The *achaete-scute* gene complex of *D. melanogaster*: conserved domains in a subset of genes required for neurogenesis and their homology to myc. *Cell* **50**, 415–424.

Vincent, M. and Thiery, J.P. (1984). A cell surface marker for neural crest and placodal cells: further evolution in peripheral and central nervous system. *Dev. Biol.* **103**, 468–481.

Vogel, A., Rodriguez, C. and Izpisua-Belmonte, J.C. (1996). Involvement of FGF-8 in initiation, outgrowth and patterning of the vertebrate limb. *Development* **122**, 1737–1750.

Vogel, K.S. and Weston, J.A. (1988). A subpopulation of cultured avian neural crest cells has transient neurogenic potential. *Neuron* **1**, 569–577.

Vogel, K.S. and Weston, J.A. (1990). The sympathoadrenal lineage in avian embryos. I. Adrenal chromaffin cells lose neuronal traits during embryogenesis. *Dev. Biol.* **139**, 1–12.

Vogel, K.S., Brannan, C.I., Jenkins, N.A., Copeland, N.G. and Parada, L.F. (1995). Loss of neurofibromin results in neurotrophin-independent survival of embryonic sensory and sympathetic neurons. *Cell* **82**, 733–742.

Volpe, E.P. (1964). Fate of neural crest homotransplants in pattern mutants of the leopard frog. *J. Exp. Zool.* **157**, 179–196.

Von Baer, K.E. (1828). *Ueber Entwicklungsgeschichte der Tiere*. Bornträger, Königsberg.

Vonesch, J.-L., Nakshatri, H., Philippe, M., Chambon, P. and Dollé, P. (1994). Stage and tissue-specific expression of the alcohol deshydrogenase 1 (*Adh-1*) gene during mouse development. *Dev. Dyn.* **199**, 199–213.

Vortkamp, A., Gessler, M. and Grzeschik, K.H. (1991). *GLI3* zinc-finger gene interrupted by translocations in Greig syndrome families. *Nature* **352**, 539–540.

Vrieling, H., Duhl, D.M., Millar, S.E., Miller, K.A. and Barsh, G.S. (1994). Differences in dorsal and ventral pigmentation result from regional expression of the mouse *agouti* gene. *Proc. Natl. Acad. Sci. USA* **91**, 5667–5671.

Wagner, G. (1949). Die Bedeutung der Neuralleiste für die Kopfgestaltung der Amphibienlarven. *Rev. Suisse Zool.* **56**, 519–620.

Wagner, G. (1955). Chimaerische Zahnlagen aus *Triton-Schmelzorgan* und *Bombinator-Papille*. *J. Embryol. Exp. Morphol.* **3**, 160–188.

Wagner, M., Thaller, C., Jessell, T.M. and Eichele, G. (1990). Polarizing activity of retinoid synthesis in the floor plate of the neural tube. *Nature* **345**, 819–822.

Wagner, M., Han, B. and Jessell, T.M. (1992). Regional differences in retinoid release from embryonic neural tissue detected by an *in vitro* reporter assay. *Development* **116**, 55–66.

Wahn, H.L., Lightbody, L.T., Tchen, T.T. and Taylor, J.D. (1976). Adenosine 3'5'-monophosphate, morphogenetic movements and embryonic neural differentiation in *Pleurodeles waltlii*. *J. Exp. Zool.* **196**, 125–130.

Wakamatsu, Y., Watanabe, Y., Nakamura, H. and Kondoh, H. (1997). Regulation of the neural crest cell fate by *N-myc*: promotion of ventral migration and neuronal differentiation. *Development* **124**, 1953–1962.

Wald, G. (1968). Molecular basis of visual excitation. *Science* **162**, 230–239.

Waldo, K.L., Kumiski, D.H. and Kirby, M.L. (1994). Association of the cardiac neural crest with development of the coronary arteries in the chick embryo. *Anat. Rec.* **239**, 315–331.

Waldo, K.L., Kumiski, D. and Kirby, M.L. (1996). Cardiac neural crest is essential for the persistence rather than the formation of an arch artery. *Dev. Dyn.* **205**, 281–292.

Walker, W.F., Jr. (1987). *Functional Anatomy of the Vertebrates. An Evolutionary Perspective.* Saunders College Publishing, Philadelphia.

Wall, N.A. and Hogan, B.L.M. (1995). Expression of bone morphogenetic protein-4 (BMP-4), bone morphogenetic protein-7 (BMP-7), fibroblast growth factor-8 (FGF-8) and sonic hedgehog (SHH) during branchial arch development in the chick. *Mech. Dev.* **53**, 383–392.

Walterhouse, D., Ahmed, M., Slusarski, D., Kalamaras, J., Boucher, D., Holmgren, R. and Iannaccone, P. (1993). *gli*, a zinc finger transcription factor and oncogene, is expressed during normal mouse development. *Dev. Dyn.* **196**, 91–102.

Wang, H.U. and Anderson, D.J. (1997). Eph family transmembrane ligands can mediate repulsive guidance of trunk neural crest migration and motor axon outgrowth. *Neuron* **18**, 383–396.

Ward, S.M., Xue, C., Shuttleworth C.W.R., Bredt, D.S., Snyder, S.A. and Senders, K.M. (1992). NADPH-diaphorase and nitric oxide synthase co-localization in enteric neurons of the canine proximal colon. *Am. J. Physiol.* **263**, G277–G284.

Ward, S.M, Burns, A.J, Torihasi, S. and Sanders, K.M. (1994). Mutation of the proto-oncogene c-kit blocks development of interstitial cells and electrical rhythmicity in murine intestine. *J. Physiol.* **480**, 91–97.

Warga, R.M. and Kimmel, C.B. (1990). Cell movements during epiboly and gastrulation in zebrafish. *Development* **108**, 569–580.

Watterson, R.L. (1938). On the production of feather color pattern by mesodermal grafts between Barred Plymouth Rock and White Leghorn chick embryos. *Anat. Rec. Suppl.* **72**, 100–101.

Watterson, R.L. (1942). The morphogenesis of down feathers with special reference to the developmental history of melanophores. *Physiol. Zoöl.* **15**, 234–259.

Webster, B., Johnston, M., Lammer, E. and Sulik, K. (1986). Isotretinoin embryopathy and the cranial neural crest: an *in vivo* and *in vitro* study. *J. Craniofac. Genet. Dev. Biol.* **6**, 211–222.

Wehrle-Haller, B. and Weston, J.A. (1995). Soluble and cell-bound forms of steel factor activity play distinct roles in melanocyte precursor dispersal and survival on the lateral neural crest migration pathway. *Development* **121**, 731–742.

Wehrle-Haller, B. and Weston, J.A. (1997). Receptor tyrosine kinase-dependent neural crest migration in response to differentially localized growth factors. *BioEssays* **19**, 337–345.

Wehrle-Haller, B., Morrison-Graham, K. and Weston, J.A. (1996). Ectopic c-kit expression affects the fate of melanocyte precursors in *Patch* mutant embryos. *Dev. Biol.* **177**, 463–474.

Weinmaster, G., Roberts, V.J. and Lemke, G. (1991). A homolog of Drosophila *Notch* expressed during mammalian development. *Development* **113**, 199–205.

Weinstein, D.C., Rahman, S.M., Ruiz, J.C. and Hemmati-Brivanlou, A. (1996). Embryonic expression of *eph* signalling factors in Xenopus. *Mech. Dev.* **57**, 133–144.

Weisblat, D.A. and Shankland, M. (1985). Cell lineage and segmentation in the leech. *Phil. Trans. Roy. Soc. London - Ser. B - Biol. Sci.* **312**, 39–56.

Weiss, K.M., Bollekens, J., Ruddle, F.H. and Takashita, K. (1994). Distal-less and other homeobox genes in the development of the dentition. *J. Exp. Zool.* **270**, 273–284.

Weiss, K.M., Ruddle, F.H. and Bollekens, J. (1995). *Dlx* and other homeobox genes in the morphological development of the dentition. *Connect. Tissue Res.* **32**, 35–40.

Weiss, P. and Andres, G. (1952). Experiments on the fate of embryonic cells (chick) disseminated by the vascular route. *J. Exp. Zool.* **121**, 449–487.

Weston, J.A. (1963). A radioautographic analysis of the migration and localization of trunk neural crest cells in the chick. *Dev. Biol.* **6**, 279–310.

Weston, J.A. (1967). Cell marking. In *Methods in Developmental Biology*, eds F.H. Wessels and N.K. Wilt, pp. 723–736. Thomas Y. Crowell, New York.

Weston, J.A. (1970). The migration and differentiation of neural crest cells. *Adv. Morphog.* **8**, 41–114.

Weston, J.A. (1991). Sequential segregation and fate of developmentally restricted intermediate cell populations in the neural crest lineage. *Curr. Top. Dev. Biol.* **149**, 133–153.

Weston, J.A. and Butler, S.L. (1966). Temporal factors affecting localization of neural crest cells in tbe chicken embryo. *Dev. Biol.* **14**, 246–266.

Wheeler, E.F. and Bothwell, M. (1992). Spatiotemporal patterns of expression of NGF and the low-affinity NGF receptor in rat embryos suggest functional roles in tissue morphogenesis and myogenesis. *J. Neurosci.* **12**, 930–945.

Wiedenmann, B., Franke, W.W., Kuhn, C., Moll, R. and Gould, V.E. (1986). Synaptophysin: a marker protein for neuroendocrine cells and neoplasms. *Proc. Natl. Acad. Sci. USA* **83**, 3500–3504.

Wijnholds, J., Chowdhury, K., Wehr, R. and Gruss, P. (1995). Segment-specific expression of the neuronatin gene during early hindbrain development. *Dev. Biol.* **171**, 73–84.

Wilkie, A.O. (1997). Craniosynostosis: genes and mechanisms. *Hum. Mol. Genet.* **6**, 1647–1656.

Wilkinson, D.G., Peters, G., Dickson, C. and McMahon, A.P. (1988). Expression of the FGF-related proto-oncogene *int-2* during gastrulation and neurulation in the mouse. *EMBO J.* **7**, 691–695.

Wilkinson, D.G., Bhatt, S., Chavrier, P., Bravo, R. and Charnay, P. (1989a). Segment-specific expression of a zinc-finger gene in the developing nervous system of the mouse. *Nature* **337**, 461–464.

Wilkinson, D.G., Bhatt, S., Cook, M., Boncinelli, E. and Krumlauf, R. (1989b). Segmental expression of *Hox-2* homoeobox-containing genes in the developing mouse hindbrain. *Nature* **341**, 405–409.

Willard, D.H., Bodnar, W., Harris, C., Kiefer, L., Nichols, J.S., Blanchard, S., Hoffman, C., Moyer, M., Burkhart, W., Weiel, J., Luther, M.A., Wilkison,

W.O. and Rocque, W.J. (1995). Agouti structure and function: characterization of a potent alpha-melanocyte stimulating hormone receptor antagonist. *Biochemistry* **34**, 12341–12346.

Williams, D.E., Eisenman, J., Baird, A., Rauch, C., Van Ness, K., March, C.J., Park, L.S., Martin, U., Mochizuki, D.Y., Boswell, H.S., Burgess, G.S., Cosman, D. and Lyman, S.D. (1990). Identification of a ligand for the c-*kit* proto-oncogene. *Cell* **63**, 167–174.

Williams, D.E., de Vries, P., Namen, A.E., Widmer, M.B. and Lyman, S.D. (1992). The Steel factor. *Dev. Biol.* **151**, 368–376.

Williams, R., Bäckström, A., Ebendal, T. and Hallböök, F. (1993). Molecular cloning and cellular localization of *trk*C in the chicken embryo. *Dev. Brain Res.* **75**, 235–252.

Willier, B.H. (1930). A study of the origin and differentiation of the suprarenal gland in the chick embryo by chorio-allantoic grafting. *Physiol. Zool.* **3**, 201–225.

Willier, B.H. and Rawles, M.E. (1940). The control of feather color pattern by melanophores grafted from one embryo to another of a different breed of fowl. *Physiol. Zool.* **13**, 177–202.

Wilson, J.G., Roth, C.B. and Warkany, J. (1953). An analysis of the syndrome of malformations induced by maternal vitamin A deficiency. Effects of restoration of vitamin A at various times during gestation. *Am. J. Anat.* **92**, 189–217.

Winklbauer, R., Selchow, A., Nagel, M. and Angres, B. (1992). Cell interaction and its role in mesoderm cell migration during *Xenopus* gastrulation. *Dev. Dyn.* **195**, 290–302.

Winning, R.S. and Sargent, T.D. (1994). Pagliaccio, a member of the Eph family of receptor tyrosine kinase genes, has localized expression in a subset of neural crest and neural tissues in *Xenopus laevis* embryos. *Mech. Dev.* **46**, 219–229.

Wintzen, M. and Gilchrest, B.A. (1996). Proopiomelanocortin, its derived peptides, and the skin. *J. Invest. Dermatol.* **106**, 3–10.

Wolbach, S.B. and Howe, P.R. (1925). Tissue changes following deprivation of fat-soluble A vitamin. *J. Exp. Med.* **42**, 753–777.

Wolf, E., Black, I.B. and DiCicco-Bloom, E. (1996). Mitotic neuroblasts determine neuritic patterning of progeny. *J. Comp. Neurol.* **367**, 623–635.

Wolgemuth, D.J., Behringer, R.R., Mostoller, M.P., Brinster, R.L. and Palmiter, R.D. (1989). Transgenic mice overexpressing the mouse homeobox-containing gene *Hox-1.4* exhibit abnormal gut development. *Nature* **337**, 464–467.

Wolpert, L. (1981). Positional information and pattern formation. *Philos. Trans. R. Soc. Lond. B. Biol. Sci.* **295**, 441–450.

Wong, V. and Kessler, J.A. (1987). Solubilization of a membrane factor that stimulates levels of Substance P and choline acetyltransferase in sympathetic neurons. *Proc. Natl. Acad. Sci. USA* **84**, 8726–8729.

Wright, E.M., Vogel, K.S. and Davies, A.M. (1992). Neurotrophic factors promote the maturation of developing sensory neurons before they become dependent on these factors for survival. *Neuron* **9**, 139–150.

Wurst, W., Auerbach, A.B. and Joyner, A.L. (1994). Multiple developmental defects in Engrailed-1 mutant mice: an early mid-hindbrain deletion and patterning defects in forelimbs and sternum. *Development* **120**, 2065–2075.

Wurtman, R.J. and Axelrod, J. (1965). Adrenaline synthesis: control by the pituitary gland and adrenal glucocorticoids. *Science* **150**, 1464–1465.

Wurtman, R.J. and Axelrod, J. (1966). Control of enzymatic synthesis of adrenaline in the adrenal medulla by adrenal cortical steroids. *J. Biol. Chem.* **241**, 2301–2305.

Xiang, M., Gans, L., Zhou, L., Klein, W.H. and Nathans, J. (1996). Targeted deletion of the mouse POU domain gene *Brn-3a* causes a selective loss of neurons in the brainstem and trigeminal ganglion, uncoordinated limb movement, and impaired suckling. *Proc. Natl. Acad. Sci. USA* **93**, 11950–11955.

Xu, D., Emoto, N., Giaid, A., Slaughter, C., Kaw, S., Dewit, D. and Yanagisawa, M. (1994). ECE-1: a membrane-bound metalloprotease that catalyzes the proteolytic activation of big endothelin-1. *Cell* **78**, 473–485.

Xue, Z.-G., Smith, J. and Le Douarin, N.M. (1985). Differentiation of catecholaminergic cells in cultures of embryonic avian sensory ganglia. *Proc. Natl. Acad. Sci. USA* **82**, 8800–8804.

Xue, Z.-G., Smith, J. and Le Douarin, N.M. (1987). Developmental capacities of avian embryonic dorsal root ganglion cells: neuropeptides and tyrosine hydroxylase in dissociated cell cultures. *Dev. Brain Res.* **34**, 99–109.

Xue, Z.-G., Xue, X.J., Fauquet, M., Smith, J. and Le Douarin, N. (1992). Expression of the gene encoding tyrosine hydroxylase in a subpopulation of quail dorsal root ganglion cells cultured in the presence of insulin or chick embryo extract. *Dev. Brain Res.* **69**, 23–30.

Yamada, G., Mansouri, A., Torres, M., Stuart, E.T., Blum, M., Schultz, M., de Robertis, E.M. and Gruss, P. (1995). Targeted mutation of the murine *goosecoid* gene results in craniofacial defects and neonatal death. *Development* **121**, 2917–2922.

Yamamori, T., Fukada, K., Aebersold, R., Korsching, S., Fann, M.J. and Patterson, P.H. (1989). The cholinergic neuronal differentiation factor from heart cells is identical to leukemia inhibitory factor. *Science* **246**, 1412–1416.

Yanagisawa, H., Yanagisawa, M., Kapur, R.P., Richardson, J.A., Williams, S.C., Clouthier, D.E., de Wit, D., Emoto, N. and Hammer, R.E. (1998). Dual genetic pathways of endothelin-mediated intercellular signaling revealed by targeted disruption of endothelin converting enzyme-1 gene. *Development* **125**, 825–836.

Yanagisawa, M. (1994). The endothelin system. A new target for therapeutic intervention. *Circulation* **89**, 1320–1322.

Yanagisawa, M., Kurihara, H., Kimura, S., Tomobe, Y., Kobayashi, M., Mitsui, Y., Yazaki, Y., Goto, K. and Masaki, T. (1988). A novel potent vasoconstrictor peptide produced by vascular endothelial cells. *Nature* **332**, 411–415.

Yao, K.M., Samson, M.L., Reeves, R. and White, K. (1993). Gene *elav* of *Drosophila melanogaster*: a prototype for neuronal-specific RNA binding protein gene family that is conserved in flies and humans. *J. Neurobiol.* **24**, 723–739.

Yao, L., Zhang, D. and Bernd, P. (1994). The onset of neurotrophin and *trk* mRNA expression in early embryonic tissues of the quail. *Dev. Biol.* **165**, 727–730.

Yarden, Y. and Ullrich, A. (1988). Growth factor receptor tyrosine kinases. *Annu. Rev. Biochem.* **57**, 443–478.

Yarden, Y., Kuang, W.J., Yang-Feng, T., Coussens, L., Munemitsu, S., Dull, T.J., Chen, E., Schlessinger, J., Francke, U. and Ullrich, A. (1987). Human proto-oncogene c-*kit*: a new cell surface receptor tyrosine kinase for an unidentified ligand. *EMBO J.* **6**, 3341–3351.

Yasumoto, K., Yokoyama, K., Takahashi, K., Tomita, Y. and Shibahara, S. (1997). Functional analysis of microphthalmia-associated transcription factor in pigment cell-specific transcription of the human tyrosinase family genes. *J. Biol. Chem.* **272**, 503–509.

Yavuzer, U., Keenan, E., Lowings, P., Vachtenheim, J., Currie, G. and Goding, C.R. (1995). The *Microphthalmia* gene product interacts with the retinoblastoma protein *in vitro* and is a target for deregulation of melanocyte-specific transcription. *Oncogene* **10**, 123–134.

Yip, H.K. and Johnson, E.M., Jr. (1984). Developing dorsal root ganglion neurons require trophic support from their central processes: evidence for a role of retrogradely transported nerve growth factor from the central nervous system to the periphery. *Proc. Natl. Acad. Sci. USA* **81**, 6245–6249.

Yip, J.P. (1986). Migratory pathways of sympathetic ganglioblasts and other neural crest derivatives in chick embryos. *J. Neurosci.* **6**, 3465–3473.

Yip, J.W. (1996). Specificity of sympathetic preganglionic projections in the chick is influenced by the somitic mesoderm. *J. Neurosci.* **16**, 613–620.

Yntema, C.L. (1937). An experimental study of the origin of the cells which constitute the VIIth and VIIIth ganglia and nerves in the embryo of *Amblystoma punctatum*. *J. Exp. Zool.* **75**, 75–101.

Yntema, C.L. (1943). An experimental study of the origin of the sensory neurons and sheath cells of the IXth and Xth cranial nerves in the embryo of *Amblystoma punctatum*. *J. Exp. Zool.* **92**, 93–120.

Yntema, C.L. (1944). Experiments on the origin of the sensory ganglia of the facial nerve of the chick. *J. Comp. Neurol.* **81**, 147–167.

Yntema, C.L. and Hammond, W.S. (1945). Depletions and abnormalities in the cervical sympathetic system of the chick following extirpation of the neural crest. *J. Exp. Zool.* **100**, 237–263.

Yntema, C.L. and Hammond, W.S. (1947). The development of the autonomic nervous system. *Biol. Rev.* **22**, 344–357.

Yntema, C.L. and Hammond, W.S. (1954). The origin of intrinsic ganglia of trunk viscera from vagal neural crest in the chick embryo. *J. Comp. Neurol.* **101**, 515–541.

Yntema, C.L. and Hammond, W.S. (1955). Experiments on the origin and development of the sacral autonomic nerves in the chick embryo. *J. Exp. Zool.* **129**, 375–414.

Yoshida, H., Hayashi, S., Shultz, L.D., Yamamura, K., Nishikawa, S., Nishikawa, S. and Kunisada, T. (1996a). Neural and skin cell-specific expression pattern conferred by steel factor regulatory sequence in transgenic mice. *Dev. Dyn.* **207**, 222–232.

Yoshida, H., Kunisada, T., Kusakabe, M., Nishikawa, S. and Nishikawa, S.I. (1996b). Distinct stages of melanocyte differentiation revealed by analysis of nonuniform pigmentation patterns. *Development* **122**, 1207–1214.

Young, H.M., Furness, J.B., Shuttleworth, C.W., Bredt, D.S. and Snyder, S.H. (1992). Co-localization of nitric oxide synthase immunoreactivity and NADPH diaphorase staining in neurons of the guinea-pig intestine. *Histochemistry* **97**, 375–378.

Young, H.M., Ciampoli, D., Southwell, B.R. and Newgreen, D.F. (1996). Origin of interstitial cells of Cajal in the mouse intestine. *Dev. Biol.* **180**, 97–107.

Zacchei, A.M. (1961). Lo sviluppo embrionale della quaglia giapponese (*Coturnix coturnix japonica*, T. e S.). *Arch. Ital. Anat. Embriol.* **66**, 36–62.

Zhang, D., Yao, L. and Bernd, P. (1994a). Expression of *trk* and neurotrophin mRNA in dorsal root and sympathetic ganglia of the quail during development. *J. Neurobiol.* **25**, 1517–1532.

Zhang, J.A., Hagopian-Donaldson, S., Serbedzija, G., Elsemore, J., Plehn-Dujowich, D., MacMahon, A.P., Flavell, R.A. and Williams, T. (1996). Neural tube, skeletal and body wall defects in mice lacking transcription factor AP-2. *Nature* **381**, 238–241.

Zhang, M.B., Kim, H.J., Marshall, H., Gendron-Maguire, M., Lucas, D.A., Baron, A., Gudas, L.J., Gridley, T., Krumlauf, R. and Grippo, J.F. (1994b). Ectopic *Hoxa-1* induces rhombomere transformation in mouse hindbrain. *Development* **120**, 2431–2442.

Zhao, G.-Q., Eberspaecher, H., Seldin, M.F. and de Crombrugghe, B. (1994). The gene for the homeodomain-containing protein Cart-1 is expressed in the cells that have a chondrogenic potential during embryonic development. *Mech. Dev.* **48**, 245–254.

Zhou, B.K., Kobayashi, T., Donatien, P.D., Bennett, D.C., Hearing, V.J. and Orlow, S.J. (1994). Identification of a melanosomal matrix protein encoded by the murine *si* (silver) locus using "organelle scanning". *Proc. Natl. Acad. Sci. USA* **91**, 7076–7080.

Ziller, C., Dupin, E., Brazeau, P., Paulin, D. and Le Douarin, N.M. (1983). Early segregation of a neuronal precursor cell line in the neural crest as revealed by culture in a chemically defined medium. *Cell* **32**, 627–638.

Ziller, C., Fauquet, M., Kalcheim, C., Smith, J. and Le Douarin, N.M. (1987). Cell lineages in peripheral nervous system ontogeny: medium-induced modulation of neuronal phenotypic expression in neural crest cell cultures. *Dev. Biol.* **120**, 101–111.

Zimmer, A. and Zimmer, A. (1992). Induction of a RAR-*beta3–lacZ* transgene by retinoic acid reflects the neuromeric organization of the central nervous system. *Development* **116**, 977–983.

Zollinger, R.M. and Ellison, E.H. (1955). Primary peptic ulceration of the jejunum associated with islet cell tumors of the pancreas. *Ann. Surg.* **142**, 709–728.

Zsebo, K.M., Wypych, J., McNiece, I.K., Lu, H.S., Smith, K.A., Karkare, S.B., Sachdev, R.K., Yuschenkoff, V.N., Birkett, N.C., Williams, L.R., Satyagal, V.N., Tung, W., Bosselman, R.A., Mendiaz, E.A. and Langley, K.E. (1990a). Identification, purification, and biological characterization of hematopoietic stem cell factor from buffalo rat liver-conditioned medium. *Cell* **63**, 195–201.

Zsebo, K.M., Williams, D.A., Geissler, E.N., Broudy, V.C., Martin, F.H., Atkins, H.L., Hsu, R.Y., Birkett, N.C., Okino, K.H., Murdock, D.C., Jacobsen, F.W., Langley, K.E., Smith, K.A., Takeishi, T., Cattanach, B.M., Galli, S.J. and Suggs, S.V. (1990b). Stem cell factor is encoded at the *Sl* locus of the mouse and is the ligand for the c-*kit* tyrosine kinase receptor. *Cell* **63**, 213–224.

Zuasti, A., Johnson, W.C., Samaraweera, P. and Bagnara, J.T. (1992). Intrinsic pigment cell stimulating activity in the catfish integument. *Pigment Cell Res.* **5**, 25–262.

Zuasti, A., Martinez-Liarte, J.H., Ferrer, C., Canizares, M., Newton, J. and Bagnara, J.T. (1993). Melanization stimulating activity in the skin of the gilthead porgy, *Sparus auratus*. *Pigment Cell Res.* **6**, 359–364.

Zurn, A. (1992). Fibroblast growth factor differentially modulates the neurotransmitter phenotype of cultured sympathetic neurons. *J. Neurosci.* **12**, 4195–4201.

# Author Index

# Subject Index

acetylcholine (ACh), 4, 211–213
acetylcholinesterase, 3–5, 16, 25, 205, 243
*Achaete-scute*, 225, 328
adrenal medulla, 203, 223–227
adrenaline, 203, 222–224, 227
adrenergic cell, 208, 213–216, 309, 316, 326
adrenergic differentiation, 213–215
allophenic (*see also* tetraparental), 250, 269, 271
*Amblystoma*, 3, 40, 46, 61, 63, 77, 119, 160
amine precursor uptake and decarboxylation
  (APUD), 241–245, 248–251
amphibian(s), 3, 9–10, 24–25, 40–41, 60–66,
  82–83, 147, 164, 253–254, 256
aneural hindgut, 20, 231
angular, 85–86, 111–112, 114, 127–128, 142
*Antennapedia*, 95
aortic arches, 34, 60–65, 70, 74, 110, 134, 337
aortic plexus, 35, 201, 203
apoptosis, 108, 146, 164, 182–183, 214, 219
articular, 110–113, 126–128

basal lamina, 28, 37–38
bone morphogenetic protein (BMP), 27, 81, 151
*Brachydanio rerio* (*see also* zebrafish), 1, 10, 28,
  40–43, 63, 65, 96, 103–104, 109, 125, 131,
  135–136, 143, 262, 266–267, 311, 320
brain-derived neurotrophic factor (BDNF),
  154, 166–168, 172–177, 181, 184–187,
  190–192, 194–196, 229, 231–232, 327–328,
  333, 335
branchial arches, 107, 110–115, 120–125

cadherin, 25, 29, 57
calcitonin, 21, 130, 165, 209, 244, 247, 251

calcitonin gene-related peptide (CGRP), 165,
  209, 212, 226, 244, 247
calcitonin-producing cells (C-producing cells),
  251
carotenoids, 252, 259
carotid body, 74, 110, 241, 245, 247–248, 251
catecholamine(s) (CA), 208–214, 224, 241–242,
  273
cellular retinoic acid binding protein (CRABP),
  130–140
cellular retinol binding protein (CRBP),
  130–140
cephalic mesoderm, 19, 85–90, 142
cephalic neural crest, 7, 61, 65–68, 71, 73, 79,
  107–108, 112, 115–116, 119, 142, 145–146,
  150, 274–275
cephalic neural fold, 7–8, 18, 67, 82
cephalic paraxial mesoderm, 66, 71, 85, 87,
  89–90, 108–109
choline acetyltransferase (CAT), 4, 211–212
cholinergic, 20, 68, 197, 208, 211–213, 309
chondrocranium, 68–69, 86, 179
chondroitin-6-sulfate, 38, 50
chromaffin, 35, 212, 197, 203, 208, 210, 216,
  218–227, 325
ciliary neurotrophic factor (CNTF), 167, 183,
  188–190, 192, 211–212, 217, 220–221, 226,
  228–230
clonal analysis, 223, 305, 313–319, 326, 332
cochlear neuron, 187
collagen, 28–29, 49, 51
columella, 86, 111–112, 128
cranial ganglia, 9, 34, 78, 121, 149, 160–161,
  163, 165, 203
cytotactin-binding protein, 50

442